An introduction to

Human Geography

Visit the *An introduction to Human Geography,* second edition
Companion Website at **www.pearsoned.co.uk/daniels** to access
a rich resource of valuable learning materials for **students**, including:

- Nine in-depth online **Tutorials**, offering valuable, practical advice on how
  to improve your human geography study skills
- Annotated **weblinks**, to relevant, specific Internet resources to facilitate
  in-depth independent research
- Searchable online **Glossary** of key terms
- Chapter-specific **Further Reading**, recommended by the authors
- Full version of each chapter's **References**, for you to print out and take to
  the library
- **Learning Outcomes** from each chapter

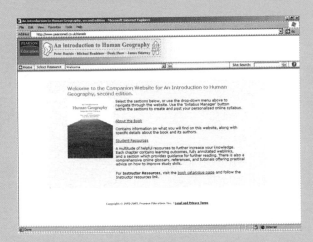

Second edition

An introduction to
Human Geography
Issues for the 21st Century

Edited by
Peter Daniels

School of Geography, Earth and Environmental Sciences, The University of Birmingham

Michael Bradshaw

Department of Geography, University of Leicester

Denis Shaw

School of Geography, Earth and Environmental Sciences, The University of Birmingham

James Sidaway

Department of Geography, National University of Singapore

Harlow, England • London • New York • Boston • San Francisco • Toronto • Sydney • Singapore • Hong Kong
Tokyo • Seoul • Taipei • New Delhi • Cape Town • Madrid • Mexico City • Amsterdam • Munich • Paris • Milan

Pearson Education Limited
Edinburgh Gate
Harlow
Essex CM20 2JE
England

and Associated Companies around the world

Visit us on the World Wide Web at:
www.pearsoned.co.uk

First published 2001
Second edition published 2005

© Pearson Education Limited 2001, 2005

ISBN 0 131 21766 6

British Library Cataloguing-in-Publication Data
A catalogue record for this book can be obtained from the British Library.

Library of Congress Cataloging-in-Publication Data
An introduction to human geography : issues for the 21st century / edited by Peter Daniels
   . . . [et al.].-- 2nd ed.
          p. cm.
     Rev. of: Human geography, c2001.
     Includes bibliographical references and index.
     ISBN 0-13-121766-6 (alk. paper)
        1. Human geography. I. Daniels, Peter (Peter W.) II. Human geography.

   GF41.I574 2004
   304.2--dc22

                                                                          2004054695

10 9 8 7 6 5 4 3 2 1
08 07 06 05

Typeset in 9.75pt Minion by 3
Printed and bound by Mateu-Cromo Artes Graficas, Spain

The publishers' policy is to use paper manufactured from sustainable forests.

BRIEF CONTENTS

Companion Website resources

Visit the *An introduction to Human Geography* Companion Website at www.pearsoned.co.uk/daniels to access a rich resource of valuable teaching and learning material, including the following content:

For students

- Nine in-depth online **Tutorials**, offering valuable, practical advice on how to improve your human geography study skills
- Annotated **weblinks**, to relevant, specific Internet resources to facilitate in-depth independent research
- Searchable online **Glossary** of key terms
- Chapter-specific **Further Reading**, recommended by the authors
- Full version of each chapter's **References**, for you to print out and take to the library
- **Learning Outcomes** from each chapter

For lecturers

- A secure, **password-protected** site offering downloadable teaching support
- **PowerPoint slides**, including useful figures and maps from the main text
- **Essay Questions** and **Exam Questions** with suggested answers for every chapter
- Suggested **Seminar Activities** based on the human geography study skills tutorials
- **Learning Outcomes** from each chapter
- Searchable online **Glossary**

Also: This site has a syllabus manager, search functions and email results functions.

CONTENTS

Contents

Contents

x

Thematic Case Studies	Page no.	Historical perspectives	Models	Social & cultural themes	Globalization	Population & migration	Resource management	Geopolitics & the nation-state	Spatial inequalities	Economy & development	Cities & urban spaces	Agriculture & rural spaces	Money & space	Environmental issues
11.1 Geographies of institutional disinvestment	251			●					●			●		●
11.2 Community opposition to asylum seekers	255			●								●		
12.1 The globalization of culture: some examples	269			●	●					●				
12.2 Berlusconi breaks ranks over Islam	274			●				●						
12.5 Englishness: bounded or hybrid culture?	277			●	●			●						
12.6 Hybridity/Diaspora – some examples	278	●		●		●								
13.1 Models of economic location	290	●	●						●	●				
13.2 An oligopoly: Nestlé	299				●					●				
14.4 Global Toyotaism	329	●		●										
14.5 Nike	330			●						●				
16.5 Skateboarding and the consumption of urban space	375		●								●			
17.1 The continuing Cold War?	399	●						●						
18.1 Territoriality on the streets	410			●							●			
18.3 The gendered spaces of football	415			●										
20.1 The uneven introduction of women's suffrage	446	●		●				●						
20.2 A simple model of the development of modern citizenship	449			●				●						
20.3 The spatial organization of electoral democracy: the power of boundaries	458							●						

CONTRIBUTORS

Ian Bowler Professor in Human Geography, Department of Geography, University of Leicester. Agricultural geography; rural development; food policy.

Michael Bradshaw Professor of Human Geography, Department of Geography, University of Leicester. Political and economic geography; economic transformation in the post-socialist states and regional change in the Russian Federation.

John R. Bryson Reader in Economic Geography, School of Geography, Earth and Environmental Sciences, The University of Birmingham. Economic geography; service worlds; alternative finance.

Tony Champion Emeritus Professor of Population Geography, School of Geography, Politics and Sociology, The University of Newcastle upon Tyne. Population distribution trends; migration and residential mobility; housing and land use planning.

Allan Cochrane Professor of Public Policy, Faculty of Social Sciences, The Open University. Urban studies, particularly urban policy; the local state and restructuring.

Philip Crang Reader in Human Geography, Department of Geography, Royal Holloway, The University of London. Cultural and economic geography, especially commodity culture.

P. W. Daniels Professor of Geography, School of Geography, Earth and Environmental Sciences, The University of Birmingham. Geography of service industries; services and globalization; trade, foreign direct investment and internationalization of services.

Carl Grundy-Warr Senior Lecturer, Department of Geography, National University of Singapore. Political geography; cross-border identities and movements in Southeast Asia; political geography of natural resource management.

Nick Henry Reader in Urban and Regional Studies, Centre for Urban and Regional Development Studies, University of Newcastle upon Tyne. Economic geography; regional development.

Phil Hubbard Reader in Urban Social Geography, Department of Geography, Loughborough University. Relations of sexuality and space; the social spaces of the post-industrial city; geographies of urban consumption.

Brian Ilbery Professor of Human Geography and Associate Dean (Research), School of Science and the Environment, Coventry University. Agricultural change and policy; alternative food economy.

Murray Low Lecturer in Human Geography, Department of Geography and Environment, The London School of Economics. Political geography; urban politics; social and political theory.

Cheryl McEwan Lecturer in Geography, School of Geography, University of Durham. Feminist and cultural geographies; postcolonial theory and multiculturalism; citizenship and difference in South Africa.

Andrew Millington Professor of Geography, Department of Geography, University of Leicester. Remote sensing; land use and land cover change; biogeography; natural resource management and sustainable development.

Jenny Pickerill Lecturer in Geography, Department of Geography, University of Leicester. Environmental politics, radical protest, and alternative lifestyles; internet activism; use of GIS in environmental debates; forms of political participation and models of democracy.

Jane Pollard Senior Lecturer in Urban and Regional Development Studies, Centre for Urban and Regional Development Studies, University of Newcastle upon Tyne. Geographies of money and finance; regional economic development.

Marcus Power Lecturer in Geography, School of Geographical Sciences, University of Bristol. Geographies of disability and development; politics of ethnicity and identity in Africa; globalization and critical geopolitics.

Denis J. B. Shaw Reader in Russian Geography, School of Geography, Earth and Environmental Sciences, The University of Birmingham. Historical geography of Russia; history of geographical thought.

James D. Sidaway Associate Professor in Geography, Department of Geography, National University of Singapore. Geopolitics; political economy and geographies of development; history of geographical thought.

T. R. Slater Reader in Historical Geography, School of Geography, Earth and Environmental Sciences, The University of Birmingham. Historical geography; urban morphology; medieval towns.

David Storey Senior Lecturer in Geography, Department of Applied Sciences, Geography and Archaeology, University College Worcester. Political geography; nationalism; rural social change; rural development.

GUIDED TOUR

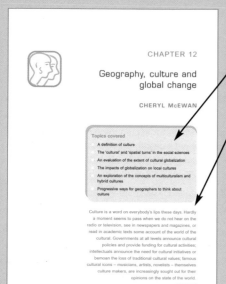

Topics covered introduce, at a glance, the key ideas and perspectives of the chapter

Memorable **Quotes** from important commentators past and present help contextualise and bring the subject to life

Figures and Tables graphically illustrate the main points, concepts and processes in the text, to help your learning

Regional Case Studies provide you with an insight into a real-life issue specific to a region

A wide range of **Maps**, both ancient and modern, create a clear, visual representation of worldwide geographical data

Snappy **Spotlight** boxes improve your understanding by expanding on the central concepts of the chapter

Thematic Case Studies, often accompanied by thought-provoking images, investigate crucial geographical themes and illuminate how they can be applied to real-life

Key Terms are highlighted throughout the text and are then collated to form a full **Glossary** at the back of the book. This helps you with unfamiliar concepts or jargon and is particularly helpful for revision purposes

The **Further Reading** section helps direct your independent study to the most appropriate texts, journal articles and websites

Learning Outcomes provide a key checklist of learning milestones, to test your understanding of the chapter

EDITORS' ACKNOWLEDGEMENTS

First Edition

Editing a book is a learning experience. If nothing else, we have discovered that even amongst just four human geographers there is considerable diversity in our academic interests and motivations, our approaches and priorities for understanding the world around us, and our philosophical and methodological approaches towards the study of human geography. The importance each of us attaches to the significance of key concepts and processes such as capitalism, globalization, technological change, resource management, culture and identity, or geopolitics is also equally varied. For those familiar with human geographers this perhaps comes as no surprise! Yet from such diversity emerges unity. It is built around a shared belief that at the beginning of the twenty-first century many of the issues outlined in this book remain dependent on the ways in which human activities are shaped by space, place and time. Other disciplines such as economics, sociology or political studies are, in our opinion rather belatedly, acknowledging this. Therefore the challenge for human geographers at the beginning of the new millennium is not just how to use their skills and insights to address some of the issues identified in this book, but to take the lead in providing explanations or at least posing different questions.

An academic rationale for a book (see Chapter 1) counts for little unless it is matched by a will to transform some initial and sketchy ideas into something more tangible. A very early source of encouragement was Matthew Smith, Social Sciences Editor at Pearson Education. He used some market analysis undertaken in 1996 to try to convince us that we should consider compiling a proposal for an introductory human geography textbook suitable for adoption as a course text in the UK. Ever since some exploratory discussions at the Institute of British Geographers' Conference in Exeter (1997) about the potential for such a project and how it might be delivered, Matthew has been unswerving in his commitment to making it happen. We hope that the faith he has consistently shown in all those involved will have proved worthwhile. This external influence coincided with some internal discussions amongst the human geographers in the School of Geography and Environmental Sciences at the University of Birmingham about the structure and contents of its first-year human geography modules. The approach that we were contemplating, elements of which we were already teaching, coincided with the views that came through via colleagues in the market analysis undertaken in the UK and in North America. We were already faced with large numbers of first-year students and the prospect of future increases. Taking account of the legal restrictions on producing course packs and the intense competition amongst our students for the learning resources in the library, there seemed to be a case for an introductory course text.

Having identified the opportunity, there was still a requirement to assemble the team of editors and authors who would subscribe to the book proposal to be considered by Pearson Education. It was hoped that it could be an exclusively 'Birmingham' project but it proved difficult to match the proposed themes in the book with those colleagues available and willing to be involved. We are extremely grateful to all our colleagues in the School who enthusiastically embraced the idea from the start as well as the one or two 'reluctants' who needed to be persuaded to come on board. We hope that they feel that their involvement has proved worthwhile. It did not prove difficult to identify the authors to cover the topics that could not be covered internally and we were delighted by their positive and enthusiastic responses to our invitations to join the team. The response to the book proposal from external reviewers was helpful and many of their comments and suggestions have been incorporated in the completed work.

The discharge of our responsibilities as editors has undoubtedly been helped by the willingness of all the authors in the team to keep to deadlines or to respond promptly to the inevitable queries and clarifications that arise as a book makes the transition from first draft chapters to page proofs. This has been especially important during the later stages of conversion from the final manuscript to finished product when the time schedule has been particularly tight. How did editors and publishers manage before the days of e-mail and the facsimile machine? A sense of common purpose has been fostered by a number of half-day meetings in Birmingham hosted by Matthew Smith and arranged with the help of Claire Clark at the School of Geography and Environmental Sciences. These meetings have facilitated the sharing of ideas, information about progress, opportunities to comment on the design and format of the book, and discussions about the design and structure of the companion web-site. Such is the pace of development in learning support and the expectations of course instructors that the latter was not an anticipated requirement when we accepted the original contract. This is probably the first introductory human geography textbook put together by UK-based human geographers to incorporate dedicated web support and we are grateful to all the authors for agreeing to undertake some of the additional tasks involved, both now and in the future. A glossary and a bibliography are more 'standard' features and we would like to thank Jon Oldfield for the prompt and efficient manner with which he liaised with the editors to compile the former. He also checked the entries in the consolidated bibliography at the end of the book.

We would like to acknowledge the invaluable contribution of several individuals whose specialist knowledge and skills have been vital, especially following delivery of the final manuscript to Pearson Education. Carol Gardiner, who channelled a steady stream of queries through the editors to authors, undertook copy editing with tact and patience; this ensured that there was adequate time to research and return the information requested. One of the more demanding tasks was undertaken by the picture researcher, Faith Perkins, who had to use suggestions/ideas at various levels of detail from the authors to prepare a portfolio of plates and slides from which those included in the book could be selected. She not only produced a wide selection to an extremely demanding schedule but also found time to provide an excellent lunch during 'selection day' at her home in Newbury. While some of the roughs of the figures in the book were capably prepared by Ann Anckorn and Kevin Burkhill in the Drawing Office in the School of Geography and Environmental Sciences, all the final artwork was undertaken by David Hoxley who often liaised patiently and constructively with individual authors about their detailed requirements. Suki Cheyne came to a half-day meeting in Birmingham to present her plans for the dedicated web page for the book and provoked a wide-ranging discussion about aspects of the content that has hopefully benefited the finished product.

Claire Brewer, Senior Editor in the Production Department at Pearson Education, deserves our particular thanks for the cheerful and ever helpful way in which she has acted as the 'go between' dealing with the numerous queries and problems that inevitably arise when working on a book of this kind. She always kept us fully informed about the production schedule and has given us deadlines for the completion of various tasks that suggested some empathy for the numerous demands on the editors' time, not least during the May/June examination period! There are no doubt numerous other 'unseen' individuals at Pearson Education to whom we owe our thanks, not least the book design team and those involved with the promotion and marketing of the book.

Of course, it is on the basis of the decisions made by the editors (with guidance and advice from Pearson Education) that the book will ultimately be judged. Throughout its gestation we have held regular meetings to pool ideas, to debate first drafts, to sort out the division of labour between us, to address and argue about problems or to agree priorities and deadlines for specific tasks. During all of this we have certainly learned something about each other; our strengths and weaknesses and how to work as an editorial team. It is pleasing to report that we are still smiling and talking to each other!

Peter Daniels
Michael Bradshaw
Denis Shaw
James Sidaway
Birmingham, July 2000

Second Edition

By mid-2002 there were encouraging indications that the First Edition had been widely adopted and Morten Fuglevand initiated the process of exploring the potential for a Second Edition. Although his remit spanned a wide range of disciplines, he had prepared much of the groundwork before relinquishing responsibility for the project in late 2002 to Andrew Taylor who was the Development Editor from September 2002. Andrew was, from the start, an enthusiastic champion for the book; he kept the Editors fully informed about the feedback received from colleagues who had adopted the book for their First Year modules and dispelled any reservations that we might have had about embarking on an extension of the original project. We are indebted to all our colleagues in Europe, North America, the UK, and elsewhere who gave their time to providing thought-provoking and constructive comments on the strengths, weaknesses and possible ways of enhancing the First Edition. Wherever practicable we have tried to take their suggestions on board, especially the advice on the need for more comprehensive coverage of some key themes that were not fully covered in the First Edition.

Andrew Taylor's sojourn as Development Editor for the book was relatively brief; he became Acquisitions Editor in April 2004 with David Cox taking over as Development Editor. This was certainly not accompanied by any change in approach or enthusiasm for the project at a time when the finishing touches were being made to the revised manuscript, including decisions about the format of the text and the illustrations. The task of tracking down suitable images was one of several duties undertaken with impressive patience and good humour by Aylene Rogers who has been required to liaise closely with the Editors during the later stages of converting the manuscript to the finished product. Her powers of persuasion have been considerable; getting four busy people to work to deadlines, not to change their minds too often, or to make a firm decision on a pressing editorial matter is no mean feat! Aylene also came up with some useful suggestions relating to the effort to make the book as user-friendly as possible; the fact that she is a Geography graduate helped. She worked closely on our behalf with the book design and marketing teams at Pearson and the success or otherwise of the Second Edition will also owe much to their vision, ideas, and suggestions. Our thanks also go to Emma Travis, Editorial Assistant, for all her excellent work in helping us research some of the new photographs in this edition.

There are a number of others to whom we owe a debt of gratitude, not least Allan Strachan who following his retirement from the Department of Geography, University of Leicester had the energy to accept the challenge of producing the annotated web pages that offer practical advice on how to improve study skills and provide interactive tutorials that support the material covered in the text. The collaboration with the web-site team at Pearson has been productive and has produced a web-based support facility that is a significant advance on its predecessor. Tracking down useful web-sites to accompany individual chapters has relied on a mix of internal and external knowledge and we are grateful to John Bourne, School of Historical Studies, University of Birmingham for some particularly useful guidance. Geoff Dowling, Media Services Manager, School of Geography, Earth and Environmental Sciences, University of Birmingham provided invaluable help on more than one occasion with our efforts to find some suitable replacement photographs suitable for the book. We also want to thank Kevin Burkhill, Drawing Office Technician, School of Geography, Earth and Environmental Sciences, University of Birmingham for his prompt and skilful assistance with the drawing and redrawing of several of the maps and charts that have been revised or added to this edition.

Finally, collaboration amongst the Editors and between them and the contributors has been much more at arms length for this edition. This reflects the nature of the editorial task, as well as the opportunities offered by e-mail and electronic file transfer. The experience has been no less interesting, although there have been times when a face-to-face meeting may actually have saved several hours of intensive e-mail activity as information about the latest query/problem was exchanged, debated, and resolved. Electronic communication undoubtedly mitigates against some of the disadvantages of significant geographical separation but does not necessarily enhance the efficiency of information transfer and decision-making. Nevertheless, we hope that even though the operational context for 'producing' the Second Edition has changed, the outcome is a step in the right direction.

Peter Daniels
Michael Bradshaw
Denis Shaw
James Sidaway
Philadelphia, March 2004

PUBLISHER'S ACKNOWLEDGEMENTS

We are grateful to the following for permission to reproduce copyright material:

Guardian News Services Limited for the articles 'Berlusconi breaks rank over Islam' by John Hooper and Kate Connolly published in *The Guardian* 27th September 2001 and 'The chill East wind at your doorstep' by Rory Carroll published in *The Guardian* 28th October 1998 © Guardian; Nike for an extract published on *www.nike.com*; and Motor Sport Valley for an extract published on *www.motorsportvalley.com*, Motorsport Valley™ is a trade mark of the Motorsport Industry Association.

Figure 1.1 from *Changing the Face of the Earth: Culture, Environment, History*, 2nd edition, Blackwell Publishing, (Simmons, I.G. 1996); Figure 1.2 from *The Cambridge Encyclopaedia of Archaeology*, Cambridge University Press, (Sherratt, A. (ed) 1980); Figure 1.3 after Major Agricultural Regions of the Earth in *Annuals of the Association of American Geographers*, Blackwell Publishing, (Whittlesey, D. 1936); Figure 1.5 from *Roman Britain (Oxford History of England Volume 1A)* © Oxford University Press, (Salway, P. (ed) 1981), by permission of Oxford University Press; Figures 1.6, 1.7 and 3.2 from *Times Atlas of World History*, © Collins Bartholomew Ltd 1989, reproduced by permission of HarperCollins Publishers; Figure 2.1 reprinted with the permission of the Free Press, a Division of Simon & Schuster Adult Publishing Group, from *Urban Development in Central Europe*, Volume 1 in the International History of City Development Series, The Free Press, (Gutkind, E.A. 1964), copyright renewed © 1992 by Peter C. W. Gutkind. All rights reserved; Figure 2.5 from *Black's General Atlas*, A&C Black Publishers; Figure 2.6 from *The Hamlyn Historical Atlas*, Hamlyn, Octopus Publishing Group, (Moore, R.I. (ed) 1981); Figure 2.7 from *Yorkshire Textile Mills 1770–1930*, HMSO, (Giles and Goodall, 1992) © Crown Copyright, NMR, reproduced by permission of National Monuments Record; Figure 2.8 courtesy of The Ironbridge Gorge Museum Trust; Figure 2.9 reprinted by permission of Popperfoto; Figure 2.10 from The morphological evolution of a nineteenth-century city centre: Lodz, Poland, 1825–1973 (Koter, M.) in *The Built Form of Western Cities*, Leicester University Press, copyright Cassell, (Slater, T.R. (ed) 1990), reproduced by permission of the Continuum International Publishing Group; Figure 2.12 from *An Historical Geography of Europe*, Cambridge University Press, (Pounds, N.J.G. 1990); Figure 2.13 (A) from *Poverty, Ethnicity and the American City, 1840–1925*, Cambridge University Press, (Ward, D. 1989); Figure 3.1 from *Great Britain: Geographical Essay*, Cambridge University Press, (Mitchell, J. B. 1962); Figure 4.5 from Space-time diffusion of the demographic transition model: the twentieth century patterns (Chung, R) in *Population Geography: A Reader*, © 1970 The McGraw-Hill Companies, Inc. Reprinted by permission of The McGraw-Hill Companies Inc. (Demko, G.J., Rose, H.M. and Schnell, G.A. 1970); Figures 5.1 and 5.3 from *Natural Resources: Allocation, Economics and Policy*, 2nd edition, Routledge, (Rees, J. 1985), by permission of Thomson Publishing Services; Figure 5.2 from Resources and environment: scarcity and sustainability (Rees, J.) in *Global Change and Challenge: Geography in the 1990s*, Routledge, (Bennett, R. and Estall, R. (eds) 1991) by permission of Thomson Publishing Services; Figure 5.4 reprinted by permission of Sage Publications Ltd from *Resources, Society and Environment Management*, Sage Publications, (Jones, G. and Hollier, G. (eds) Giles and Goodall 1992) © Sage Publications 1997); Figures 5.5, 5.6, 5.7, 5.8 and 5.9 reprinted by permission of BP plc. © Copyright BP plc 2003; Figure 5.11 from *Environmental Resources*, Pearson Education Limited, (Mather, A.S. and Chapman, K. 1995); Figure 6.1 from *The Root Causes of Biodiversity Loss*, Earthscan Publications, p. 14, (Wood, A. *et al.* 2000); Figure 6.2 from African Soil Erosion: Nature undone and the Limitations of Technology, *Land Degradation and Rehabilitation*, **1**, p. 285, Wiley & Sons Ltd., (Millington, A.C. et al, 1989); Figure 7.1 from Globalisation and the changing networks of food supply: the importation of fresh horticultural produce from Kenya into the UK in *Transactions of the Institute of British Geographers* Volume 24, Blackwell Publishing,

(Barrett et al 1999); Figure 7.2a Reprinted from *Journal of Rural Studies,* **16**, Ilbery, B. and Kneafsey, M., Producer constructions of quality in regional speciality food production: a case study from south west England, pp 217–30, Copyright 2000, with permission from Elsevier; Figure 8.1 from *The Bank, the President and the Pearl of Africa* International Broadcasting Trust (IBT)/Oxfam (1998); Figure 8.2 from *World Bank Atlas,* World Bank (1998). © The International Bank for Reconstruction and Development/The World Bank; Figure 8.3 from *Rethinking Development Geographies,* Routledge, (Power, M. 2003), by permission of Thomson Publishing Services; Figure 8.4 from 'The World Bank and the State: a Recipe for Change?', brettonwoodsproject.org, (Hildyard, N.); Figure 14.2 from *The United States: A Contemporary Human Geography,* Longman, (Knox P.L. et al 1998), reproduced by permission of Pearson Education Limited; Figure 14.3 from *Industrial Location: Principles and Policies,* Blackwell Publishing, (Chapman, K. and Walker, D. 1987); Figure 14.4 from *Japanese Technology and Innovation Management,* Edward Elgar Publishing Ltd, © Dr S. Harryson, (Harryson, S. 1998); Figure 15.1 from *Financial Geography,* School of Economics and Commercial Law, Göteborg University, (Laulajainen, R. 1998), a more recent figure is available from the 2003 edition, publisher Routledge; Figure 17.1 from the 1912 Minneapolis Journal, Copyright 2004 Star Tribune. Republished with permission of Star Tribune, Minneapolis-St. Paul. No further republication or redistribution is permitted without written consent of Star Tribune; Figure 17.2 from *The Fate of the Forest: Developers, Destroyers and Defenders of the Amazon,* Verso Ltd, (Hecht, S. and Cockburn, A. 1989); Figure 17.3 from *Political Geography,* John Wiley, (Glassner, M.I. 1993), Copyright © 1993 John Wiley & Sons Inc. This material is used by permission of John Wiley & Sons, Inc.; Figure 17.4 from The geographical pivot of history in *Geographical Journal* **23**, Blackwell Publishing, (Mackinder, H. 1904); Figure 17.5 from Political geography and panregions in *Geographical Review* **80**, American Geographical Society, (O'Loughlin, J. and van der Wusten, H. 1990); Figure 18.1 from *Apartheid in South Africa* 3rd edition, Cambridge University Press, (Smith, D.M. 1990); Figures 18.2 and 18.3 Adapted from *Tents: Architecture of the Nomads* by Torvald Faegre, copyright © 1979 by Torvald Faegre. Used by permission of Doubleday, a division of Random House, Inc.; Figure 19.1 from Introduction in *A People without a Country: the Kurds and Kurdestan,* 2nd edition, Zed Books, (Chailand, G. 1993); Figures 20.1 and 20.2 reprinted by permission of Mary Evans Picture Library Ltd and Figure 20.4 maps on pp. 666 and 823 from *The American People,* 3rd edition by Gary B. Nash, et al. Copyright © 1994 by HarperCollins College Publishers. Reprinted by permission of Addison-Wesley Educational Publishers, Inc.

Table 1.1 from Introduction: themes and concepts in the study of early agriculture in *The Origins and Spread of Agriculture and Pastoralism in Eurasia* UCL Press, reproduced by permission of Thomson Publishing Services, (Harris, D.R. (ed) 1996); Table 2.2 from *The Hamlyn Historical Atlas,* Hamlyn, Octopus Publishing Group, (Moore, R.I. (ed) 1981); Table 5.1 from Resources and environment: scarcity and sustainability (Rees, J.) in *Global Change and Challenge: Geography in the 1990s,* Routledge (Bennett, R. and Estell, R. (eds) 1991) by permission of Thomson Publishing Services; Tables 5.2, 5.3, 5.5, 5.6 and 5.7 reprinted by permission of BP plc. © copyright BP plc 2003; Table 5.4 from *World Development Indicators 1998,* World Bank. © The International Bank for Reconstruction and Development/The World Bank; Tables 5.8 and 5.9 from *www.iea.org,* © OECD/IEA, 2003, Key World Energy Statistics, reproduced by permission of the Organisation for Economic Co-operation and Development and the International Energy Agency; Table 6.4 from *Developing the Environment,* Longmans (Barrow, C.J. 1995), reproduced by permission of Pearson Education Limited; Table 6.6 from www.un.org/milleniumgoals, reprinted by permission of the United Nations; Table 7.2 reprinted by permission of Sage Publications Ltd from Product and place: promoting quality products and services in the lagging rural regions of the European Union in *European Urban and Regional Studies,* Vol. 5, Sage Publications, (Ilbery, B. and Kneafsey, M. © Sage Publications Ltd 1998); Table 9.1 from *Philips Atlas of the World,* 8th edition, Philips, Octopus Publishing Group, (George Philip in association with the RGS and the IBG), the most recent statistics will be in Atlas of the World 13th edition, published September 2003; Table 11.2 from *Dangerous disorder: Riots and violent disturbances in thirteen areas of Britain, 1991–92* by Anne Power and Rebecca Tunstall, published in 1997 by the Joseph Rowntree Foundation. Reproduced by permission of the Joseph Rowntree Foundation; Table 13.3 and table on p. 306 reprinted by permission of the United Nations from *United Nations Conference on Trade and Development*

World Investment Report 2003: Transnational Corporations and Export Competitiveness, UNCTAD, Geneva; Table 14.1 reprinted by permission of Sage Publications Ltd from *Global Shift: Transforming the World Economy*, Paul Chapman, (Dicken, P. © Sage Publications Ltd 1998) and Table 15.1 from www.world-exchanges.org reproduced by permission of the World Federation of Exchanges;

Plates 1.1, 9.6 and 9.8 reprinted by permission of Maggie Murray/Photofusion; Plate 1.2 reprinted by permission of Michael Holford and C.M. Dixon; Plate 1.3 reprinted by permission of Simmons Aerofilms; Plates 1.4 and 3.2 reprinted by permission of Mary Evans Picture Library Ltd; Plate 1.5 reprinted by permission of The Dean and Chapter of Hereford and the Hereford Mappa Mundi Trust; Plates 3.1, 3.3, 6.1, 6.2, 6.4, 8.4, 9.1, 9.2, 9.3, 9.4, 9.5, 9.10, 10.1, 12.1, 12.3, 13.2, 13.5, 15.1, 16.1, 16.3, 17.2, 17.4, 17.6, 19.2 and 19.5 reprinted by permission of Corbis UK; Plates 3.5, 3.7, 4.3, 4.5, 5.2, 5.6, 5.8, 14.2, 16.5, 17.1 and 20.4 reprinted by permission of Popperfoto; Plate 3.8 copyright Raissa Page; Plates 4.1, 4.2, 5.1, 10.3, and 16.4 reprinted by permission of Panos Pictures; Plate 4.4 ©Karen Robinson/Panos Pictures; Plate 4.6 reprinted by permission of Camera Press Ltd; Plates 5.3, 6.3 and 11.4 reprinted by permission of Getty Images; Plates 5.4, 11.2, 11.5, 13.6, 15.3, 20.1, 20.2 and 20.3 reprinted by permission of The Associated Press Ltd; Plate 5.5 reprinted by permission of Novosti/Science Photo Library; Plates 5.7 and 15.2 reprinted by permission of Sheila Gray; Plate 7.2 reproduced by permission of R.J. Smedley and Aardman Animations; Plate 8.1 is reproduced by permission of Still Pictures; Plate 8.2, The IMF and its hard to swallow 'bitter economic medicine, from Oxfam is reproduced with the permission of Oxfam, GB, 274 Banbury Road, Oxford, OX2 7DZ www.oxfam.org.uk; Plate 8.3 reprinted by permission of Ulrike Preuss/Photofusion; Plate 9.7 reproduced with the permission of Geophotos; Plate 9.9 reproduced by permission of LinkIndia, Link Picture Library; Plate 10.2 reproduced by kind permission from Pixzone; Plate 11.1 reprinted by permission of Network Photographers; Plate 11.3 reprinted by permission of Nottingham Evening Post; Plate 12.2 reprinted by permission of Joanne O'Brien/Photofusion; Plate 13.4 reprinted by permission of T. Paul Daniels; Plate 14.1 reprinted by permission of the Archives at Birmingham City Council; Plates 14.3 and 17.5 reprinted by permission of Topham Picturepoint; Plate 16.6 reprinted by permission of G.P. Dowling; Plate 16.7 reprinted by permission of Bullring; Plate 19.1 reprinted by permission of Art Directors & TRIP Photo Library and Plate 19.4 P. Vanriel/Robert Harding, reprinted by permission of Robert Harding Picture Library Ltd;

In some instances we have been unable to trace the owners of the copyright material, and we would appreciate any information that would enable us to do so.

INTRODUCTION

Geography: a discipline for the twenty-first century

PETER DANIELS
MICHAEL BRADSHAW
DENIS SHAW
JAMES SIDAWAY

Geography is indispensable to survival. All animals, including American students who consistently fail their geography tests, must be competent applied geographers. How else do they get around, find food and mate, avoid dangerous places?

(Yi-Fu Tuan 2002: 123)

. . . geographical knowledge is not – and should not attempt to be – static and detached from what is going on in the world, but is rather dynamic and profoundly influenced by events, struggles and politics beyond university life.

(Blunt and Wills 2000: x-xi)

With a minor modification to its title (the addition of *An Introduction...*), this book is the second edition of *Human Geography: Issues for the 21st Century*. The first edition was published in 2001, although it actually went on sale in the UK in late 2000. The task of compiling the first edition therefore dates back to 1997–98 so that it is now some six years since we wrote our first introduction. While the second edition incorporates substantial revisions, and we hope improvements – all the chapters have been revised and updated and a number of new chapters added – we have tried not to lose sight of our original goal, which was to provide an introduction to human geography through a survey of the field, focusing upon contemporary issues and adopting contemporary approaches.

One reviewer of the first edition criticized us for not providing more on the nature of human geography and its history. It is true that most 'North American style' textbooks spend a good deal of time explaining what geography is and what constitutes a geographical approach. For example, Knox and Marston (2004: 2) define human geography as: 'the study of the spatial organization of human activity and peoples' relationship with their environments'. Apart from adopting a very anthropocentric notion of environment, such a definition is relatively unproblematic. The reason why introductory textbooks in the United States need to define human geography is that many of the students taking a module in Introductory Human Geography are not geography majors, that is they won't go on to specialize in geography, and they may have had limited exposure to it as a discipline at high school. The institutional setting within which the present textbook has been constructed is very different: most students have chosen to specialize in geography at school and have made a deliberate choice to read for a degree in geography. Consequently, they have an idea of what the essential subject matter of the discipline is and, allowing for this context, the purpose of an introductory module in human geography is to explore how university-level geography differs from students' experiences of the discipline so far. In such modules students are challenged not to take the world for granted, to understand that human geography is not a direct reflection of objective reality, but a social construction. Although we may share certain experiences, beliefs and value systems vary and interpretations of the world differ. The very notion that there is a *single*, definable, notion of what human

geography is about is itself problematic. Whose human geography, where and when?

As we shall see, geography as a way of seeing the world is constantly evolving. In the five years since we embarked on this book a great deal has changed in the world; the dramatic events of 11 September 2001, for example, have left indelible marks. New divisions are being created, old conflicts revived and the world appears to be a more fragmented and contested place than it was five years ago. In such a context, we believe that human geography has much to offer in the process of critical understanding and enhancing the prospects of alternative interpretations and resolutions.

Human geography, as practised in the Anglophone realm, underwent numerous changes during the second half of the twentieth century (see Johnston and Sidaway 2004 for a history of postwar 'Anglo-American' human geography and also Hubbard *et al.* 2002: 22–56 for a brief history of geographic thought). Initially, in its desire to move from description to explanation of the forces that shape human society, human geography tended to overstate and then to overlook the importance of diversity and difference. Model building and the application of quantitative analysis arguably reinforced the latter tendency, at least until the last quarter of the century. Since then, human geographers have more consciously embraced diversity and difference amongst economies, societies and cultures. However, along the way, the modelling, data processing and visualizing capabilities of geographic information science continued to be refined (Fairbairn and Dorling 1997) and the Internet and mobile telecommunications produced new capacities for communication and altered relations and perceptions of proximity and distance (see http://www.zooknic.com/ for work on the geographies of the Internet).

In the 1970s the so-called 'crisis of relevance' in human geography kindled an interest in the underlying economic causes of inequality in capitalist societies. Reflecting the times, human geographers became more concerned with inequality, economic and political crises and contradictions. The associated development of so-called 'structuralist' approaches in geographic analysis was later criticized by many for being too deterministic and for placing too much emphasis on the economic as the explanation for inequality and its geographical manifestations. Political, social and cultural considerations were therefore increasingly incorporated into both contemporary and historical geographical studies. While human geographers remained within their sub-disciplines, such as urban, political, historical, or cultural geography, the boundaries between them started to become fuzzy. This was encouraged by 'a recent movement in philosophy, arts and social sciences characterized by scepticism towards the grand claims of the modern era, and their privileged vantage point, stressing in its place an openness to a range of voices in social enquiry' (Johnston *et al.*, *Dictionary of Human Geography*, 2000: 4). We refer to this movement as postmodernism; this fostered the breaking down of boundaries between sub-disciplines and the bringing together of urban, historical, economic and other geographies. Regional geography, out of favour since the mid-twentieth century, began to attract renewed interest in response to a desire to develop a more integrated approach to geographical explanation. For some, outside the discipline, the decline of the Cold War in the late 1980s brought the 'end of history'. For others, the development of information and telecommunications technology and the globalization of the economy were creating a 'borderless world'; some even proclaimed the 'end of geography'.

Such proclamations were premature: geography (still) matters. In fact we would argue that, given the challenges of the contemporary world, it matters even more than before. As we shall see, globalization is a geographically uneven and contested set of processes. It is a symptom of an increasingly competitive and interconnected world that has the potential to impose social, economic and cultural uniformity. Yet, a sense of place remains as critical as ever to all of us as a way of maintaining our identity and a meaningful role in the world. While the enhanced mobility of people and of capital is generating increased inequality between the rich and the poor, the diversity of, and differences between, places are now used to attract investment and to stake a claim to scarce and sometimes non-renewable resources.

For the most part, the evolution of human geography has been driven by the adoption of ideas, concepts and theories from other disciplines. The challenge for human geographers has been to apply these to a geographical context. Economic geography, for example, initially took ideas from economics and business studies and applied them to help explain how economic activity is distributed across space. The geography of economic activities was seen as the outcome of this process with the job of the geographer being to map and describe that pattern of economic activity. This often seemed to relegate the economic

geographer (and geographers more generally) to the role of suppliers of information for others to decipher.

More recently, however, the boot has shifted to the other foot. The rise of globalization and postmodernism has encouraged other social scientists to embrace geography. Of course, geographers have always claimed that *geography matters* for all the social sciences. But for some reason this is often acknowledged only when internationally renowned economists, political scientists and sociologists actually say that geography must be included in their analyses and interpretations. But as geographers we should beware; practitioners from other disciplines tend to adopt their own 'take' on geography, one that is often antiquated and simplistic – reinventing industrial location models in the name of 'geographical economics', for example. Nevertheless, geographers can take some comfort from the fact that other disciplines are at last recognizing the *spatiality of human society*.

Put simply, the fact that social and economic processes take place across space matters. Human geography is not just about *describing* the spatial manifestations of economy and society; it is about *explaining* how space is transformed and shapes economies, societies and social processes. Thus, geography is not a passive outcome; it is a critical component of social and economic processes. More than that, geography in its broader definition provides a unique interface between the human and the natural worlds. Without being too pompous, we would argue that geography is *the* subject for the twenty-first century, in part because many of the challenges that face humanity are at the interface between human societies and natural environments.

Partly because of the emergence of feminist geographies and postmodernism, human geographers have recently become more aware of the ways that knowledge is socially constructed. The way you see the world is a function of who you are. Prevailing views of the world, be they academic, scientific, artistic, political, economic or spiritual, can often be seen to be masculine constructions. Equally, human geographers have come to realize that much of the knowledge and understanding that they claim to be universal is in various ways 'Eurocentric'. Thus, 1 January 2000 was not the beginning of a new century or a new millennium for many of the world's populations; yet, an instant measured on the time scale of a particular religion that, until recently, dominated Western society, was widely taken to be a global event (something that we shall return to in the Conclusion). Similarly,

3

Eurocentrism has given a very 'white' view of a multi-ethnic world. This means that geographers must now confront the ways in which assumptions and value judgments shape the way they view the world. They must accept that *all* descriptions of the world are culturally determined, often politically motivated, and can always be contested.

As social scientists, few human geographers now actually believe that there is a straightforward objective reality around us that we are all seeking to describe and explain. Moreover, the fact that this book has been written by geographers whose lives and careers have largely been spent in the United Kingdom will have influenced its style and contents. The fact that one of the editors has been involved in producing the second edition while teaching at the National University of Singapore and that much of the research of two of the other editors is conducted in Australasia and Russia has undoubtedly increased our sensitivity towards the limits and assumptions of most geography written from an offshore island of Europe. A straw poll would probably quickly reveal that there remain differences between the authors, and among the editors, over the contents of the book, or how it should have been written. What you see is the result of compromise. The same type of book written by a group of geographers in Australia (see Wait *et al.* 2000), New Zealand (see Le Heron *et al.* 1999), India, Russia or South Africa would undoubtedly be different. At the same time, each individual reader of this book will interpret its contents in different ways. The position that you occupy in your society and where you are on the globe will tend to shape the way you look at the world. Given the anticipated market, the majority of the readers of this book will probably reside in the so-called 'western' world. The first edition was reprinted in India in 2003 as a local edition (authorized for sale in India, Bangladesh, Pakistan, Nepal, Sri Lanka and the Maldives). No doubt that readership will sometimes have different worldviews from those of students in the UK. For all readers the most important geographical fact shaping their life-chances is where they were born. You may think that being at a university or college places you in a privileged position in society; in fact being born in the 'developed world' puts you, from birth, in a relatively privileged position amongst humankind. So, think about your position and privilege; realize that this book presents the views of human geographers working at a particular place and at a particular time. We have set out to challenge the reader to think about the ways that human geography interprets the major social, cultural, economic, environmental, and related issues that face the world in the early years of the twenty-first century. It is for you to use this book to inform your own insights and opinions as geographers; for some it could be part of a life-long engagement with human geography.

Approach of the book

Writing an introductory textbook is a challenging task. No matter what its contents, or how they are organized, the finished product will not satisfy everyone. The book not only has to engage the attention and interest of the students who will learn with it, it also has to satisfy the instructors who will teach with it. Thus, as practising human geographers, we are caught on the horns of a dilemma: how to make the text as accessible and as user-friendly as possible to students, while not appearing to be too naïve to the instructor assigning it. We also have to acknowledge that each geography programme is different, at least in detail if not in overall objective, and that a textbook fulfilling every need would be very large indeed! Broadly speaking, however, most first-year geography programmes will have a course that offers an introduction to human geography. Such a course is nearly always compulsory. It is often supported with a 'world regional geography' course or a course devoted to topical 'issues in human geography'. Increasingly (though this was always the case in North America) such courses are required to meet the needs of non-geography students who are taking geography as an option or elective. This poses a major challenge for the instructor confronted by a class comprising 'wanna-be' human geographers, physical geographers who might rather be elsewhere and non-geographers who are there out of curiosity or requirement. Such a diverse audience also poses a challenge to the textbook writer.

If you browse the shelves in your university or college library, you will come across a variety of human geography textbooks. Broadly speaking these are of three types: systematic texts organized by major sub-discipline in human geography, topical texts organized by key topics or issues, and world regional texts using major regions to introduce human geography. It will also be evident that these books are clearly aimed at different markets or audiences. Assigning a course text is almost a matter of fact in the US (students are aghast if there is no 'course text' and now also expect a student guide and dedicated website), but elsewhere, certainly in the UK, assigned texts are a relatively recent phenomenon. Increasing student numbers, reduced library budgets, and the resultant pressure on library resources mean that an assigned text is seen as a necessary *starting point* for an introductory course in human geography. With students' budgets also becoming stretched, there is also a good case for purchasing a textbook that will cover more than one course and that will be of lasting use in later years. The benefit of regular updates in the form of new editions is that they will provide access to the most recent literature, although given the time it takes to a publish a book, textbook bibliographies should always be supplemented by library research.

There is no doubt that introductory textbooks do not always travel well. This is because they are often designed for a particular educational system; but more than that they also reflect a specific version of what is important in the world and what constitutes human geography. In addressing this problem, we have sought to produce a textbook that offers an *issue-oriented* introduction to human geography. The major sections focus on important issues facing the world at the beginning of the twenty-first century, but they also relate to major sub-disciplines in human geography. No introductory text can hope to be comprehensive. In revising the first edition we have added substantive chapters on social geography, the environment and environmentalism, and territoriality. The latter is part of a restructured section that deals with political geography. We believe that the issues we have selected are important and will continue to be so for some time. They also reflect the interests and expertise of the authors and our sense of what captures the enthusiasm and imagination of the students we have taught over the years. The individual chapters bring together ideas from a variety of fields in human geography. They also incorporate coverage of the globe, drawing ideas from a variety of places to reflect diversity and difference as well as the increasingly interdependent nature of the world in which we live. While this integrated approach may not sit well with the traditional ways of producing introductory human geography textbooks, we believe that it acknowledges the complexity of human society as well as the practice of contemporary human geography.

Structure of the book

This book is divided into five sections, supported by an Introduction and a brief Conclusion. Section 1 provides an essential historical context for the four issue-oriented sections that follow. While it is natural for us to proclaim that geography matters, it is also the case that historical geography matters. A cursory examination of an atlas printed at the turn of the last century will show you that over the past century the world has changed a great deal. Many of the contemporary issues discussed in this book, such as the geographies of development and of inequality, owe much to the way the world looked in 1900, or even earlier. This is not historical determinism, rather an acknowledgment that history is critical to an understanding of contemporary human geography. However, if we cast our minds even further back into the records of antiquity we are reminded that societies are transitory; nothing lasts forever. Great civilizations have come and gone and this should teach us that there is nothing permanent about the current world order. At the same time, what is past, and no longer visible, has a profound influence on the world we live in now. Section 1 highlights some of these issues and then focuses on the emergence of capitalism and its relationship with the making of the twentieth-century world that immediately preceded our own.

The remaining four sections comprise groups of chapters that focus on a related set of issues. It makes sense to work with Section 1 first; after that the sections can be utilized in whatever order you see fit. Each section starts with a brief summary of the important issues covered by each of the chapters and there is logic to the order in which the chapters are presented within each section. Section 2 examines the interrelationship between population change, resource production and consumption, environment and global economic development. Section 3 focuses on social and cultural issues within an urban and rural context and at a global scale. Section 4 examines the globalization of economic activity, the putative transitions of production from Fordism to post-Fordism, the emergence of a global financial system and means of exchange, and the importance of consumption to an understanding of the geography of economy. Section 5 considers a variety of political geographies at differing scales from the global, to the local, to the individual. In the Conclusion we reflect on the challenges facing human geography at the turn of the century (for most) and the new millennium (for many).

How to use the book

Each chapter is designed to introduce a set of specific issues and particular fields in human geography. Each begins with a list of topics covered. The text is broken into manageable sections with supporting materials; important concepts and specific examples are presented in boxes. Each chapter concludes with a list of learning outcomes. These provide a means of assessing whether you have understood the content of the chapter. To assist you in reading beyond the text, each chapter has a set of recommended readings that you should be able to find in your library. Where relevant, we have also included a list of web-sites that contain information you may find useful in your own research and writing. In addition, at the back of the book there is a glossary of important terms and an extensive bibliography. It is not normal to include such a lengthy bibliography in an introductory text; however, we want this book to remain an essential reference tool for the duration of your geography programme. Each chapter contains references to the bibliography giving you access to the specialist reading that is essential when you wish to research or otherwise follow through on specific topics in detail. One way of using this book is to read the chapters to support your lectures and seminars, the recommended readings to prepare for examinations, and the references, web-sites and bibliography to research and write assignments or term papers. In other words, reading this textbook is the beginning of your introduction to human geography, not the end!

Learning resources

As we noted above, each chapter provides a set of recommended readings and, where relevant, useful web-sites. This chapter is no exception. Here we consider other resources that provide an introduction to human geography. The readings focus on other introductory texts and the study skills necessary to conduct your own research. The web-site guide lists places to visit to find out more about geography as a discipline and as a profession.

Further reading

This first set of readings provides a sample of other text-books that you can use to supplement your reading. Individual chapters from them are cited elsewhere as references in this book, but they are also worth browsing to get a sense of the different ways in which human geography is presented.

Cloke, P., Crang, P. and **Goodwin, M.** (eds) (1999) *Introducing Human Geographies*, Arnold, London. An innovative text of 34 short chapters that reflect the diversity of approaches in contemporary human geography. Designed more for the UK market, but well worth delving into wherever you may be. A second edition is currently in preparation.

Knox, P. and **Marston, S.A.** (2004) *Human Geography: Places and Regions in Global Context*, 3rd edition, Prentice Hall, Upper Saddle River, NJ. A comprehensive text written for the US market with lots of colour pictures and diagrams. Probably the best US-style textbook on the market at present.

Le Heron, R., Murphy, L., Forer, P. and **Goldstone, M.** (1999) *Explorations in Human Geography: Encountering Place*, Oxford University Press, Auckland. This is a stimulating textbook written by a group of New Zealand geographers specifically for New Zealand geography students. It may be difficult to find in the northern hemisphere, but reading it could turn upside down your own taken-for-granted view of the world.

Wait, G., McGuirk, P., Dunn, K. and **Hartig, K.** (2000) *Introducing Human Geography: Globalization, Difference and Inequality*, Longman, French's Forest, NSW. Like the above, another rewarding Antipodean text; this one is published in New South Wales, Australia.

Study aids and further insights into human geography

This section is dedicated to publications that will assist you in conducting geographical research and/or expand your knowledge of debates and approaches in the discipline. The list is comprised of dictionaries, and books on methodology, approaches and theory in human geography as well as a few (such as autobiographies of influential geographers) that are hard to categorize. You should also seek out the various readers that have now been produced. Some bring together the most influential readings in a particular area of human geography, whilst others are collections of specially commissioned review essays that assess the status of particular areas of research.

Agnew, J., Livingstone, D. and **Rogers, A.** (eds) (1996) *Human Geography: An Essential Anthology*, Blackwell, Oxford. A collection of classic geographical writings, and a sober reminder that there is very little that is wholly new.

Benko, G. and **Strohmayer, U.** (2004) *Human Geography: A History for the Twenty First Century*, Arnold, London. Not yet published as we went to press, this book promises to address both histories of continental European and Anglophone geography and the relations between them.

Blunt, A. and **Wills, J.** (2000) *Dissident Geographies: An Introduction to Radical Ideas and Practice*, Prentice Hall, Harlow. The geographies explored in the text all share political commitments to critique and challenge prevailing relations of power. From gay and lesbian geographies to geographies of anarchism and anti-racism, this book brings out some of the ways in which the production of geographical knowledge is tied to politics and struggles outside, as well as within, universities.

Cloke, P., Crang, P. and **Goodwin, M.** (eds) (2004) *Envisioning Human Geographies*, Arnold, London. A collection that sets out future agendas and directions for human geography.

Cloke, P., Crang, P., Goodwin, M., Painter and **Philo** (2002) *Practising Human Geography*, Sage, London. A more advanced guide to research methods and writing strategies in human geography.

Daniels, S. and **Lee, R.** (eds) (1996) *Exploring Human Geography: A Reader*, Arnold, London. The individual articles were not intended for a student audience, but the collection provides a survey of recent critical human geography.

Dorling, D. and **Fairbairn, D.** (1997) *Mapping: Ways of Representing the World*, Prentice Hall, Harlow. Maps have always been associated with geography and geographers. This text will help you to understand how maps express the will to describe, understand and control.

Flowerdew, R. and **Martin, D.** (1997) *Methods in Human Geography: A Guide for Students Doing a Research Project*, Longman, Harlow. An indispensable guide to a wide variety of the research techniques employed by contemporary human geography.

Gould, P. and **Pitts, F. R.** (eds) (2002) *Geographical Voices: Fourteen Autobiographical Essays*, Syracuse University Press, Syracuse, NY. A collection of autobiographies reflecting the experiences and insights of some leading geographers of the past fifty years.

Hoggart, K., Lees, L. and **Davies, A.** (2002) *Researching Human Geography*, Arnold, London. One of a number of recent books that provide an introduction to research methods in human geography, a good place to look when starting your own dissertation or research project.

Hubbard, P., Kitchin, R., Bartley, B. and **Fuller, D.** (2002) *Thinking Geographically: Space, Theory and Contemporary Human Geography*, Continuum, London. This book represents an attempt to demystify the nature and role of theory in contemporary human geography.

Johnston, R.J., Gregory, D., Pratt, G. and **Watts, M.** (eds) (2000) *The Dictionary of Human Geography*, 4th edition, Blackwell, Oxford. An indispensable reference, but it is already getting dated so don't rely solely on the references. Do your own literature searches!

Johnston, R.J. and **Sidaway J.D.** (2004) *Geography and Geographers: Anglo-American Human Geography since 1945*, 6th edition, Arnold, London. Recently updated, this book provides an accessible survey of the major trends in human geography since 1945 in the English-speaking world. Includes an interesting discussion of geography as a discipline that might help you understand what academic geographers do.

Livingstone, D. (1992) *The Geographical Tradition: Issues in the History of a Contested Enterprise*, Blackwell, Oxford. A scholarly study of the history of human geography.

Rogers, A. and **Viles, H.** (eds) (2003) *The Student's Companion to Geography*, 2nd edition, Blackwell, Oxford. A new edition of a successful volume; the various sections contain information on the fields of geography, methods and data sources.

Women and Geography Study Group (1997) *Feminist Geographies: Explorations in Diversity and Difference*, Longman, Harlow. A good way to begin to understand how feminist geographies have reshaped the study of human geography.

Yi-Fu Tuan (1999) *Who am I? An Autobiography of Emotion, Mind and Spirit*, University of Wisconsin Press, Madison, WI. The autobiography of a Chinese-American who arrived in the United States as a 20-year-old graduate student in the 1950s and went on to become one of America's most original and respected geographers.

Useful web-sites

The Internet contains a wealth of information that is of relevance to an introductory course on human geography. It is impossible to cover all of the potential sources here. This section provides access to the sites of some of the major geographical societies in the UK and North America. All of these sites provide links to societies and organizations across the globe, as well as links to geography-related sites. In addition, many of these sites now carry links to electronic versions of their journals and other publications as well as information and links to resources such as maps, statistics, atlases or photographic collections. Your own department will have a web-site; visit it and find out what the academics who teach you are also up to. Visit the sites of other departments across the world and discover how geography is taught in different countries. Increasingly, academic publishers are also posting electronic versions of the journals that they produce. Thus, the Internet provides you with your own virtual guide to human geography. With the partial exception of the Canadian Association of Geographers, the links below focus on the Anglophone world. Use the International Geographical Union web-site to seek out geographers across the globe.

www.aag.org The web-site of the Association of American Geographers. Founded in 1904, the AAG is the primary academic geography organization in the United States. This site provides a wealth of information on the Association's activities, as well as access to its two scholarly journals: the *Annals of the Association of American Geographers* and the *Professional Geographer*, (not available electronically).

www.amergeog.org The web-site of the American Geographical Society. The Society was established in 1851 and is the oldest professional geographical organization in the United States. The society publishes the *Geographical Review* and *Focus*.

www.cag-acg.ca The web-site of the Canadian Association of Geographers. Founded in 1951, the CAG-

ACG is the primary academic organization in Canada. The association publishes *The Canadian Geographer* and organizes national and regional meetings.

www.geography.org.uk The web-site of the Geographical Association, which was founded in 1893 and is the national subject teaching organization for all geographers in the UK. The Association publishes three journals: *Primary Geographer*, *Teaching Geography*, and *Geography*.

www.iag.org.au The web-site of the Institute of Australian Geographers. The IAG was founded in Adelaide in 1958 and is the principal body representing geographers and promoting geography in Australia. The institute publishes the journal *Australian Geographical Studies*.

www.igu.net.org The web-site of the International Geographical Union, the international umbrella organization for academic geography. This web-site contains a comprehensive list of geography departments throughout the world.

www.nationalgeographic.org A public face of geography in the English-speaking world. Home to the *National Geographic* magazine, the site also provides access to the maps and photographs that appear in its magazine as well as a searchable index.

www.rgs.org The home of the Royal Geographical Society (with the Institute of British Geographers). The RGS was originally founded in 1830 and the RGS-IBG is the primary academic organization in UK geography. The web-site provides information on the societies as well as electronic access to its three learned journals: *Area*, the *Geographical Journal* and *Transactions*.

In addition to the web-sites of the various learned societies, there are now dedicated 'web portals' that provide access to a wealth of information to support your geographical studies. The two examples below are UK-based, but their outreach is international and their links may take you to portals or subject gateways based elsewhere.

Resource Guide for Geography and the Environment (www.jisc.ac.uk/resourceguides) This web-site professes to be the essential guide for exam revision and dissertations and theses. It provides access to online catalogues, publications, and subject gateways and data services. It requires an Athens username and password and therefore may be limited to UK users.

Geosource: Geography and Environment Information on the Internet (www.geosource.ac.uk) This is a free service led by the University of Manchester. It currently contains over 5,000 individually selected information-rich Internet resources covering five subject areas: the environment, general geography and environmental science, geographical techniques and approaches, human geography and physical geography.

 For annotated, clickable weblinks and useful tutorials full of practical advice on how to improve your study skills, visit this book's website at www.pearsoned.co.uk/daniels

SECTION 1

Worlds in the past: changing scales of experience

EDITED BY DENIS SHAW

If present-day globalization means anything, it refers to a world marked by increasing intercommunication and interaction at a global scale. The corollary is that people's lives in the past were not as immediately affected by what happens elsewhere in the world, they were more ignorant of such happenings, and they had fewer possibilities to travel and to communicate across great distances. Changing scales of experience imply the growing significance of larger scales of activity and communication to human life as time has gone on. However, it would be a mistake to imagine that linkages over large distances are an entirely recent phenomenon, even at the global scale. This section shows how centuries and even millennia ago people, ideas and phenomena travelled vast distances, though usually taking much longer to do so than would be the case today. Geographical diffusion has been an important engine in social change, and change has been the norm rather than the exception throughout human history.

The present-day world has been moulded and is being remoulded by the forces of capitalism (defined in Chapter 2, pp. 37–8). As modern capitalism spread across the globe it inevitably changed the societies which preceded it. But the pre-capitalist societies, while adapting to capitalism, did not necessarily lose all their distinctive features, and capitalism itself was changed in the process of interacting with such societies. The fact that human societies in the modern world differ among themselves in all kinds of ways is only partly the result of the way global capitalism works and of the ways in which societies are now adapting to the various forces affecting them. Societies also differ because of their past histories, histories that stretch back into pre-capitalist times. The world into which modern capitalism spread was itself already enormously varied and had been changing in innumerable ways over thousands of years.

Chapter 1 of this section considers what the world was like before capitalism and what if any generalizations can be made about this long period of human history. Chapter 2 analyzes the rise and spread of modern capitalism down to the end of the nineteenth century. Its history during this period is intimately linked to the rise of the European powers and to the worldwide spread of their influence. This, of course, is not to deny that capitalism probably sprouted, and then died, in other societies in other periods, or that it might have gained worldwide significance on the basis of a non-European core had circumstances been different. During the twentieth century, which is the subject of Chapter 3, capitalism became a truly global phenomenon and radically changed in the process. This chapter thus provides a more immediate background for the sections that follow.

In summary, this part of the book emphasizes the variety and complexity of the world before globalization. And it demonstrates that only by knowing something about that world can globalization itself be understood.

CHAPTER 1

Pre-capitalist worlds

DENIS SHAW

Topics covered

- Why geographers need to know about the past
- Ways of classifying past and present societies
- Bands, tribes, chiefdoms, regulated states, market-based states
- Hunting and gathering
- The invention and spread of agriculture
- The rise of cities, states and civilizations
- Medieval feudalism
- Problems of studying past societies

Much of this book is about how capitalism has spread across the globe and even now is remoulding human societies in new and unexpected ways. This chapter is about what the world was like before capitalism. In other words it is about how human beings have lived during most of their existence on this planet.

Capitalism has changed the world profoundly and often very quickly. In the first half of the twentieth century it was still possible for geographers and others to travel to regions where its influence was minimal and where people still lived in pre-capitalist societies. Today this is much less true. Even so the geographer can still find numerous areas where the way of life has many pre-capitalist traits, though perhaps increasingly influenced by the modern world. Let us consider some examples.

About 2 per cent of the population of Australia is composed of Aborigines whose ancestors first settled the continent up to 50,000 years before white Europeans (Broome 1994). Today their way of life varies between that of suburban professionals, and remote outstations and homelands where people live partly off the land in traditional style. When the British established their first penal colony on Australian soil in 1788, more Aborigines lived on the continent than today, and they were divided into some 500 groups displaying a wide variety of languages, cultures, economies and technologies. All were hunters and gatherers leading a semi-nomadic existence and using some similar tools, like stone core hammers, knives, scrapers and axe heads as well as wooden implements like spears, digging sticks and vessels. However, there was also a variety in technology and economy that reflected the diverse environments in which Aboriginal peoples lived. Whether living on coasts, riverbanks, in woodland or desert regions, they were efficient and skilled in exploiting the flora and fauna of their surroundings. Aboriginal peoples who hunt and fish today are often using skills honed by their ancestors over thousands of years. Their languages and cultures similarly reflect the experiences of many hundreds of generations.

The Inuit peoples living in the Arctic, mainly coastal, territories of Greenland, northern Canada, Alaska and the extreme eastern tip of Russia, are another group who in the past lived almost entirely by hunting and gathering (Sugden 1982). Hunting and gathering still play an important role in many of the remoter communities. Like the Australian Aborigines, the Inuit showed a remarkable ability to exploit the

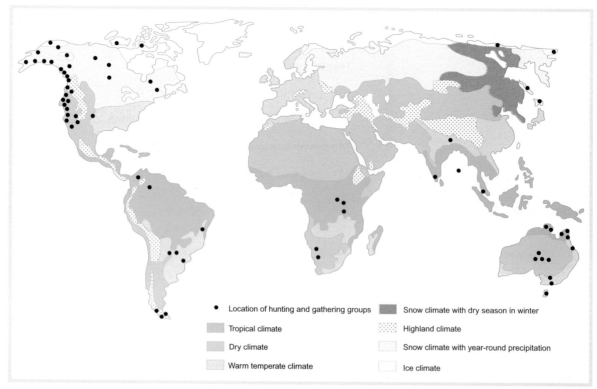

Figure 1.1 World distribution of hunter gatherers today and in the recent past, with indication of their habitats (based on a classification of climate).
Source: Adapted from Simmons (1996: 48)

resources of what in their case is an extremely harsh environment. Because of their environment, Inuit communities demonstrated less economic variation than the Aborigines but they were far from uniform. There were marked seasonal movements depending on the character of the local resources as groups moved between exploiting the fish and mammals of the open ocean, hunting seals on the ice, and seeking caribou, musk ox and similar fauna inland. The rich Inuit culture is also far from uniform but by no means as varied as the Aboriginal (there are only two interrelated languages, for example). Considering the vast geographical spaces that the Inuit occupy, the latter fact is remarkable and may reflect a relative ease of communication by sledge and boat as well as the frequency of migrations. Like the Aborigines, the Inuit have had to adapt in the recent period to an intrusive, mainly white culture, but are still sufficiently conscious of their distinctive histories and life-styles to have a strong sense of identity.

The Australian Aborigines and the Inuit are but two of many human groups existing either now or in the recent past who have lived by hunting and gathering (see Figure 1.1). The present-day world also contains other groups whose way of life to a greater or lesser extent reflects pre-capitalist characteristics. Pastoral nomads, for example, live mainly by raising and herding domestic animals (cattle, sheep, goats, camels, yak, reindeer and others), that provide them with food, clothing and other necessities. Pastoralists are particularly found in marginal (semi-arid, sub-arctic, sub-alpine) lands in parts of Eurasia and Africa (see Plate 1.1). Cases in point are some of the peoples (for example, Nentsy, Evenki, Chukchi) who live in northern Russia, especially in the northern coniferous forests and tundra (Vitebsky 1996). Many of these peoples have traditionally lived by herding reindeer (though hunting and freshwater fishing are also important to them). Because reindeer are very demanding on the environment, the herders are traditionally nomadic, following set routes in accordance with seasonal changes. In the twentieth century the northern peoples were subjected to considerable intrusions on to their territories, especially in the wake of the Soviet government's determination to exploit the region for its minerals and other resources. The environmental consequences were very serious. There were also ideologically inspired

Plate 1.1   Pastoral farmers with their cattle, West Africa. (Photofusion/Maggie Murray)

attempts to force the northern peoples to abandon their nomadic ways. Like the other cases discussed already, therefore, the northern minorities have had to make adjustments in their traditional ways, but reindeer herding is still vital to many of them.

As a final example of peoples whose way of life has yet to be completely changed by capitalism, one might mention the many peasant farmers still to be found in parts of sub-Saharan Africa, southern Asia and elsewhere. Some still grow much of their own food and market relationships have yet to become of central importance to their lives (Wolf 1966). To the extent that such people eat and otherwise depend on what they grow and raise rather than on what they can sell or earn outside the farm, their way of life is similar to that of peasants over the centuries. Of course nowadays the number of peasants is declining and they are increasingly exposed to the influences of the outside world. But again they remind us of a world before capitalism.

The point, then, is that today's world still contains many societies whose ways of life differ from those lived by most of this book's readers and recall earlier periods in human history. In order to understand those ways of life and the world as it is now, the geographer must know something about the world before capitalism. Learning about the world before capitalism helps us to see how capitalism has changed the world and the plusses and minuses of that process.

A number of arguments can be advanced to suggest why it is important to know something about the world before capitalism:

- Understanding the past helps geographers and others to understand themselves and their societies. Societies and individuals are products of the past, not just of the present. The present cannot be understood in ignorance of the past. Studying the past provides answers not merely to questions about how things were but also about how things are now (and, more tentatively, about how things will be).

- An example of the latter is the character of the physical environment in which people now live. Human beings have lived on this planet for so many thousands of years that over vast areas the physical environment has been profoundly modified. The world in which we now live is the product of thousands of generations of human activity.

- A further important point is that Aborigines, Inuit and other indigenous peoples (those peoples native to a particular territory that was later colonized,

particularly by Europeans) have recently been asserting their rights, for example, by claiming rights to local resources long ago usurped by outsiders and by trying to protect their distinctive cultures. They feel they are the victims of the past. By studying that past and trying to understand the variety of cultures and ways of life that exist in the world and how they came to be, geographers are more likely to understand and respect such feelings.

Thus, while human geography is primarily about the present, it cannot afford to ignore the past. Some of the above points will be illustrated in the following pages.

1.1 Making sense of the past

Despite the spread of capitalism and globalization, the world of today remains immensely varied and complex. The same is true of the world in the past. The question for the geographer is how to make sense of this complex past; how to make it amenable to geographical analysis and understanding.

Different scholars have tried different ways of answering this question, but all suffer from shortcomings. For example, the economist W.W. Rostow, in his famous theory of economic growth typical of capitalist development, simply described the pre-industrial period as 'traditional society' (Rostow 1971: 4–6). Yet the fact is that pre-industrial human societies ranged from the smallest communities of hunters and gatherers to societies as sophisticated and geographically extended as the Roman Empire, ancient China and feudal Europe. In as much as such societies existed in the past, and not today, it is not possible to observe them directly. But the more scholars learn about them, the more they realize how unlike one another these early societies were.

One reason why Rostow oversimplified the pre-industrial past was that he was not much interested in it. He was really interested in modernity, as exemplified by the United States and its Western allies. Furthermore his theory can be described as 'unilinear evolutionary' in as much as it suggested that all societies should pass through a series of set stages of economic development before finally arriving at the age of 'high mass consumption'. The United States (where Rostow lived) and other 'developed' countries have already reached this age. Evolutionary models have certainly long been a popular way whereby Western societies have tried to make sense of the past.

One of the attractions of such models has no doubt been their tendency to suggest that the West is the most 'developed' and thus most 'progressive' part of the world and, by implication, the best. But quite apart from the questionable assumption that all societies are seeking to imitate the West, this raises important issues about the meaning of 'progress' and 'development', two terms that certainly carry very positive connotations in the West. For example, while those parts of the world usually deemed most developed and wealthiest certainly use most energy and enjoy access to more material goods, they also make huge demands on the environment, perhaps ultimately to everyone's undoing. Thus what is progress in this context? Similarly, it has often been noted that the price people tend to pay for more 'development' and wealth is that they have to work harder and for more hours. Unless hard work is regarded as a virtue in itself, this again raises questions about the nature of progress. The point here is not to disparage all forms of evolutionary theory or idea (for example, to deny that human societies have, by and large, become more complex and spatially extended through time; see Dodgshon 1987). Rather, it is to suggest that what is most recent or new or complex is not necessarily best. Thus, in what follows, words like 'modern' do not imply 'better', nor do 'ancient' or 'primitive' imply 'worse'.

An alternative method of trying to make sense of the past, and one that does not have the unilinear evolutionary structure of, say, Rostow's theory, is to classify human societies into a series of types. Classification can be described as a way of simplifying a complex world by grouping together phenomena that are regarded as having some common feature or property, particularly where the latter is deemed especially significant. Thus Marxists have commonly grouped societies in accordance with their prevailing mode of production, or in other words with the way in which material production is related to social structure (Hindess and Hirst 1975). As a geographer, primarily interested in spatial structure, Robert Sack classified societies in accordance with their use of territoriality (Sack 1986). The point is that there is no right or wrong mode of classification. It all depends on what the purpose of the classification is. Most social theorists have wanted to claim that their mode of classifying societies is particularly significant for understanding how and why societies differ. The problem is that there is a large measure of disagreement about what the best mode is, and all have their shortcomings.

1.2 A classification of human societies

As an example of the kinds of difficulties that face any attempt to classify past and present human societies, this chapter will focus on one mode of classification that has had widespread appeal not only for human geographers but also for anthropologists and other scholars. This is the mode of classifying human societies into bands, tribes, chiefdoms and states (regulated and market-based) (see Bobek 1962, Service 1971, Dodgshon 1987). Spotlight boxes 1.1 to 1.5 describe, in simplified form, the major characteristics of each type of human society according to this classification. It will quickly be seen that the classification describes a range of societies from the simplest and most primitive to the capitalist societies of today. What it does not suggest is the circumstances under which one kind of society may change into, or be succeeded by another.

One reason why this mode of classification has had such appeal is because it suggests that over the course of human history larger and more complex societies have tended to appear. It also relates size and complexity to the way societies occupy space and their relationships to their physical environment. However, it is worth reiterating that this is only one way of classifying human societies. Indeed some scholars reject the whole idea of classifying human societies because they feel it leads to misapprehensions about how societies are formed and survive, and even to dangerous notions of unilinear evolution. It thus needs to be remembered that this scheme is suggestive only, and does not necessarily describe the way any given society has actually changed through time. In particular, it is not stipulating that societies must develop through a series of set stages, or that one type of society necessarily leads to the next.

Subsequent sections will spell out in detail the social processes that might have helped to produce, change and perhaps destroy the different kinds of human society described in Spotlight boxes 1.1 to 1.5. In so doing some of the limitations of the classification system used will become apparent. Before proceeding, however, it is worth making one important point. Our knowledge of prehistoric societies (societies which have left no written records) derives mainly from archaeological data, or material left by such societies and the traces of their environmental impacts. Methods for analysing such data have become ever more sophisticated, but the data themselves obviously become scarcer the older they are. Moreover, there are many kinds of questions, for example about the way early people viewed the world around them, which cannot be answered directly from the archaeological evidence. New discoveries are always liable to change our understanding of past societies, and especially of the earliest ones. Our knowledge of the latter societies, and of the dates attaching to them, must therefore be regarded as especially tentative.

1.3 Hunting and gathering

The time when the first human-type species (hominids) first appeared on earth is very uncertain, but a date of at least two and a half million years ago is often given. Homo erectus, forerunner of the modern human species, is believed to have appeared around 1.9 million years ago, and the modern species (Homo sapiens sapiens) at least 40,000 years ago (but see Foley 1995, for a contrary view). Since agriculture appears to have arisen less than 10,000 years ago, hunting and gathering in bands have been the basic occupations for much of humanity's existence (Spotlight box 1.1).

As noted already, only a limited amount is known about early human societies (see Plate 1.2). Much has to be inferred, for example from hunter gatherers who are still in existence or who existed until quite recently. Needless to say, such inference can be dangerous because modern hunter gatherers have probably been influenced by other, more 'advanced', societies that now share their world. Their way of life is thus unlikely to be identical with those who existed in the remote past.

The distribution of hunter gatherers in the recent

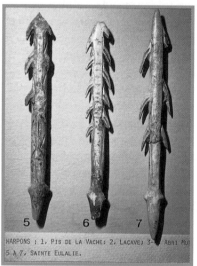

Plate 1.2 Artefacts from early human sites: flints (Michael Holford) and bone harpoons (C.M. Dixon).

past shows that they have existed in every major climatic zone and have practised enormously varied economies, depending on the nature and potentials of the local environment (Figure 1.1). This fact alone suggests that a simplistic evolutionary model of human societies will fail to do justice to their complex histories. Some generalizations do seem possible,

Spotlight Box 1.1

Bands

- Bands are societies of hunter gatherers. Most of human history has been lived in this way.
- Bands live by hunting wild animals and gathering food from the surrounding flora.
- Because of the pressures they place on the surrounding environment, bands are almost always nomadic.

- Bands are the smallest and simplest of human groups.
- Bands rarely exceed 500 people.
- For most of the time band members generally live in smaller groups of 25–100 people, which allows for efficient exploitation of the environment.
- Bands usually have a minimum of social differentiation and functional specialization.

however – for example, that in recent times groups living in the highest latitudes (above 60 degrees) have relied largely on hunting, and those below 39 degrees on plant collecting. Fishing seems to have played an important role for many groups living in intermediate latitudes (Lee 1968).

Hunter gatherers have lived in this world for such a long period that today's environment is the way it is partly as a result of their activities. For example, any notion that early human beings lived in complete harmony with their environments or always had 'sustainable' economies is simply untrue. Thus hunter gatherers are credited with the extinction of many animals. An outstanding case is North America where perhaps two-thirds of the large mammal fauna living there just before the arrival of the ancestors of modern native Americans (via the Bering land bridge, possibly about 12,000 years ago) subsequently disappeared. The most likely explanation is the effects of hunting. The discovery of fire, its use by hominids long predating the appearance of Homo sapiens sapiens, was also an important instrument whereby human beings changed their environments. Some scholars have argued that the present-day appearance of such major biomes as the African savannas and the prairies of North America is the result of the human use of fire over thousands of years (Simmons 1996). More recently, contacts between hunter gatherers and agricultural and industrial societies frequently had even more far-reaching environmental effects.

Scholars have wondered about processes whereby bands evolved into tribes, or hunter gathering economies into agricultural ones (Harris 1996a, b). It has been pointed out, for example, that where the

natural environment was especially favourable, the packing of bands often became denser, leading to less mobility and perhaps to semi-permanent settlement and closer interaction between bands. There may also have been a much greater degree of manipulation of the environment and its resources than might be expected in a pure hunter gatherer economy. Altogether the life of bands was much more variable and a good deal less stable than the above classification scheme suggests. The boundary between band and tribe is thus blurred.

1.4 Human settlement and agriculture

Just as the boundary line between band and tribe may not always be easy to define in practice, the same can be said of the boundary between tribe and chiefdom. Some scholars, for example, doubt whether the category tribe (Spotlight box 1.2) is particularly helpful, implying as it does the lack of a hierarchical social structure (Friedman and Rowlands 1977). Such scholars tend to believe that the appearance of agriculture will have sparked off competition for access to the best land, or at least rules whereby such land was allocated and inherited, and that some people and groups will inevitably have lost out. They therefore see a social hierarchy (characteristic of chiefdoms and states – see Spotlight boxes 1.3 and 1.4) beginning to emerge even as agriculture and the process of permanent settlement began.

Be that as it may, it is now widely accepted that the traditional picture of agriculture being invented in a

Tribes

- Tribes appeared with the invention and spread of agriculture.
- Agriculture usually demands considerable investment of human effort into a relatively small area. This reduces the propensity to migrate (but pastoral nomads are an exception).
- The appearance of agriculture usually allows more people to live in a smaller space. Agricultural soci-

eties thus tend to be bigger than bands, with higher population densities.
- Settlements tend to become more fixed than in bands.
- The greater spatial fixity of society encourages intermarriage and the development of extended family networks. Tribes thus develop a sense of kinship and of common descent.
- Socially, tribes are relatively egalitarian, at least as between kinship groups.

Spotlight Box 1.3

Chiefdoms

- Chiefdoms arose with the emergence of 'ranked and stratified' societies.
- 'Ranked' societies are societies where groups and individuals have, on a relatively permanent basis, different degrees of status and power.
- 'Stratified' societies are societies where groups and individuals have, on a relatively permanent basis, different degrees of material wealth.
- A chiefdom implies the presence of a permanent ruling group and/or individual, though kinship remains the chief bond between ruler and subject.
- Chiefdoms imply a greater degree of centralization and control within society than in a tribe (shown, for example, in the imposition of taxes and tribute on the ordinary subjects). Spatially this might be reflected in the greater importance accorded to a central settlement (proto-city) where the ruler resides.
- The first chiefdoms began to appear about 3000 BC in Europe, but earlier elsewhere.

Spotlight Box 1.4

Regulated (pre-modern) states

- Whereas the chiefdom is organized around the principle of kinship, the state is organized on the basis of territory (viz. one is subject to the state if one resides in the territory of that state).
- States, whilst also based on social inequality, tend to be larger and administratively more complex than chiefdoms. This implies greater functional specialization and greater probability of the rise of urban forms.
- Pre-modern states existed in a world before modern capitalism. The market was not central to their functioning and was often controlled ('regulated') in various ways. Thus economic relationships were generally subordinated to political, social and religious considerations.
- Other forms of social relationship, for instance landholding, also often tended to be controlled (regulated) rather than being determined by the market.

Spotlight Box 1.5

Market-based states

- Market-based states are modern states whose development is closely linked to that of capitalism and thus of the world economy.
- The market is of key importance to their functioning.

small number of 'hearths' and then spreading across the globe is far too simple. Bands of hunters and gatherers are known to develop an often intimate knowledge of their local environments, and no doubt human beings understood much about the factors influencing plant and animal development long before they began to practise full-blown agriculture. David Harris, for example, talks of a 'continuum of people–plant interaction', starting at the simplest level with wild plant food procurement or foraging and ending with cultivation and full-blown agriculture (Table 1.1). By the latter stage, the domestication of plants and animals had led to the modification of species which could then no longer survive in the wild (Harris 1989, 1996a, b). The important point is that Harris is not propounding an inevitable evolutionary path that is applicable to all societies. What he is saying is that the development of agriculture was a long drawn-out affair and that there must have been much trial and error before it finally emerged in some places in a fully recognizable form. Some prehistoric northern Australian Aborigines, for example, knew about agriculture, but never adopted it.

Something can, however, be said about when and perhaps why agriculture appeared. The archaeological evidence provides most information about where and when the latter stages of Harris's 'continuum' emerged. For example, it is possible to detect and date the remains of domesticated varieties of plants and animals. According to Simmons (1996: 93) there were probably three foci for the initial surges of domestication: around 7000 BC (to use the Christian chronology) in south-west Asia, around 6000 BC in south-east Asia, and around 5000 BC in Meso-America. Different species of plants were associated with each, for example: wheat, barley and oats in south-west Asia; rice in south-east Asia; and maize, squash and gourds in Meso-America. Domesticated animals also began to appear about the same time (though the dog, domesticated from the wolf, appeared before agriculture). Sheep, goats, pigs and cattle are all associated with south-west Asia (Simmons 1996: 87–134), though pigs may also have been domesticated independently in eastern Asia.

As to why agriculture was adopted, scholars differ. The simple answer, that it was more efficient than hunting and gathering, simply will not do in naturally productive environments where it clearly involved much greater effort relative to the return achieved. Some scholars favour population pressure. Others,

Table 1.1 An evolutionary classification of systems of (above) plant and (below) animal exploitation

Wild plant-food procurement	Plant-food production		
	Wild plant-food production dominant		**Crop-production dominant**
Gathering and collecting	Cultivation with small-scale clearance of vegetation and minimal tillage	Cultivation with larger-scale land clearance and systematic tillage	Agriculture based largely or exclusively on the cultivation of domesticated plants

Decreasing dependence on wild plants ⟶

Increasing dependence on domesticated plants ⟶

Time ─────────────⟶

Predation	Protection		Domestication
Generalized hunting (and scavenging) Specialized hunting (and scavenging) Fishing	Taming Protective herding Free-range management		Livestock raising by settled agriculturalists Transhumance Nomadic pastoralism

Decreasing dependence on wild animals ⟶

Increasing dependence on domesticated animals ⟶

Time ─────────────⟶

Source: After Harris (1996a: 4)

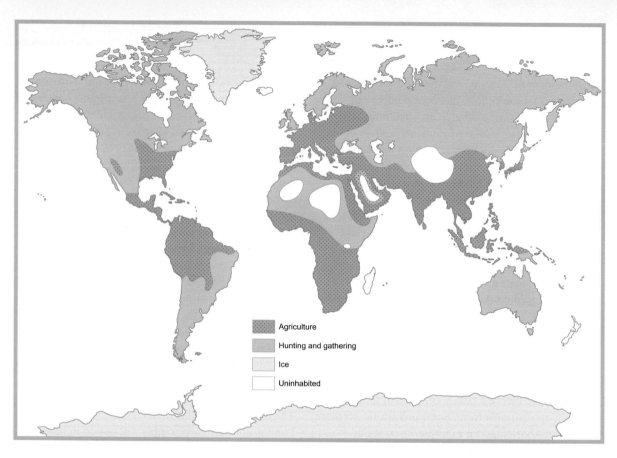

Figure 1.2 World distribution of agriculture and hunter gathering about AD 1(a) and AD 1500(b).
Source: Sherratt (1980: 97, 117)

*Flat polar quartic
equal-area projection*

1 Nomadic herding
2 Livestock
3 Primitive subsistence agriculture
4a Intensive subsistence, wet rice dominant
4b Intensive subsistence wet rice *not* dominant
5 Plantations and small farms
6 Mediterranean agriculture
7 Commercial grain farming
8 Crop and livestock farming
9 Commercial dairy farming
10 Commercial gardening and fruit
11 Little or no agriculture

0   2500 miles
0   4000 km

Figure 1.3   Agricultural regions of the world.
Source: After Whittlesey (1936)

however, argue that this view is over-deterministic and point out that even hunter gatherers had the means of controlling population growth. Such scholars tend to favour more complex explanations that embrace cultural preferences and choices as well as environmental and population pressures (Maisels 1993: 25–31).

Just as the factors that led up to the appearance and spread of agriculture are by no means straightforward, the same is also true of its further development. No simple evolutionary model, such as the one proposed by Esther Boserup involving an increasing intensification of agricultural land use through time, can account for how agriculture has changed and developed throughout the world (Boserup 1965). For example, it is no longer generally accepted, as it once was, that extensive forms of agriculture like shifting agriculture usually preceded farming in permanent fields. Britain may be unusual in Europe in having undergone a slow transition to permanent forms of agriculture (Dodgshon 1987: 66–70). Similarly pastoral nomadism is no longer regarded as more primitive than cultivation. Indeed, many scholars have argued that pastoralists cannot exist in isolation from cultivators and therefore cannot have preceded them. Throughout the world, traditional agriculturalists showed a remarkable ability to adapt their practices to local environmental conditions. Eventually they largely displaced the hunter gatherers (Figure 1.2) and gave rise to an enormous variety of agricultural systems (Figure 1.3).

Agriculture's effects on the landscape were much greater than the effects of hunting and gathering. It changed the vegetation cover of wide areas as forests and grasslands disappeared under fields, practices like terracing and irrigation were introduced, and the effects on soil composition abetted erosion in some places. The grazing of animals and other forms of resource use frequently changed the species composition of forests, grasslands and other areas. Agricultural systems, crops and domesticated animals spread way beyond their initial locations. An outstanding example is the cultivation of rice, which may have started in east Asia around 5000 BC, if not before, and spread as far as Egypt and Sicily by Roman times. Such developments long preceded the European overseas expansion beginning in the fifteenth century AD, which was to have even more far-reaching consequences both for the geography of agriculture and for the environment (see Chapter 2, pp. 41–5).

There is no doubt that the spread of agriculture

(whether as a result of migration or through cultural diffusion) had far-reaching effects on human society, encouraging permanent settlement and the emergence of tribal systems (though Crone, 1986, questions the nature of the link between agriculture and tribes). Tribal members, living permanently side by side, almost inevitably developed similar cultural traits such as common languages. As suggested in Spotlight box 1.2, tribes also typically develop myths of common ancestry (indeed, this seems to be part of their definition). Scholars have long pondered whether such myths might sometimes have a basis in biological fact, and whether early tribes might have formed the basis for some of the ethnic divisions apparent today. For example, some scholars now believe that the present-day geography of language groups (for example, that of the Indo-European languages, which are spread across much of Europe and parts of south Asia) can be partly explained by the prehistoric migrations of interrelated peoples speaking similar languages. They also spread agricultural practices into these regions. Such scholars see a correspondence between the present-day geography of languages and the geography of human genetic patterns that science is now beginning to reveal for Europe and other areas (Renfrew 1996). Other scholars, however, are more cautious. It has been argued, for example, that different human groups, settling in the same region, may gradually have adopted similar languages and cultural patterns, thus slowly fusing into one people or ethnic group even though they were not originally closely interrelated. Thus, according to this argument, the Slavs of eastern Europe might not be descendants of one original tribe, but of different groups who gradually adopted the same culture (Dolukhanov 1996). The question, then, is how far tribal (and nowadays ethnic) differences are purely cultural and how far biological. It is a question to which there is no sure answer. What can be said is that the migration of peoples, which certainly played a big role in history even after the adoption of permanent settlement and agriculture, must often have led to the intermixing of peoples and to frequent redefinitions of tribal and ethnic divisions.

1.5 Cities and civilization

The invention of agriculture was only one of the events that moved human societies on a road that ultimately led to the kinds of societies which predominate in the world today. Others included the appearance of cities,

Plate 1.3 A medieval walled town, Villeneuve-sur-Lot, showing evidence of planning. (Simmons Aerofilms)

states and civilizations. Cities and states usually go together. Maisels, for example, argues that the functional specialization associated with cities allowed rulers to rule more effectively and also put a greater social distance between them and their subjects (Maisels 1993: 302). The gathering of specialists around the person of a state's ruler fostered urban life, commerce and the many other things (such as writing, technology, religion, and speculative thought) with which is associated the word 'civilization'. Indeed, the very words 'city' and 'civilization' have a common Latin root (civitas, meaning 'state').

As is the case with the appearance of agriculture, the processes that eventually gave rise to cities and civilizations are far from straightforward. It is clear that the development of a social hierarchy was a necessary precursor, but what produced that? Some scholars have argued that competition for land in the face of population pressure, inequalities fostered by inheritance rules, and the emergence of powerful

leaders who organized inter-tribal exchanges, acted as 'tribal bankers' in times of food shortage, or proved their military prowess, are all possible explanations. But while these and other factors may have produced **chiefdoms** (see Spotlight box 1.3), something more seems to be required to explain the rise of cities, states and civilizations. As Maisels writes, chiefs enjoy only 'hegemony' over their peoples, and are bound by rules of kinship which imply mutual obligations. Chiefdoms were notoriously unstable. Kings and other state rulers, by contrast, exercise 'sovereignty' (ultimate power) over a **territory** and its peoples (Maisels 1993: 199).

Scholars are generally agreed that cities and states required, in the first place, a minimum threshold population density, which only settled agriculture would normally permit. Only settled agriculture can produce the reliable food surpluses needed to feed the specialists and non-agriculturalists who lived in the first cities. But equally something else was required to persuade the agriculturalists to yield up their surpluses to the cities. Trade was one possibility, but probably not the main one in what, after all, were still relatively simple societies. A much more likely candidate was the political power and influence exercised by incipient rulers. There is no agreed theory that explains how such power arose (Dodgshon 1987: 139–45). Some have argued that the origins of the state lie in the ability of an individual or group to monopolize control over some function that is of great value to the community as a whole. Examples might include trade and exchange (including the storage of needed goods), administration, religion or the engineering of irrigation works. Others prefer to see the explanation in the social divisions produced by such factors as conquest by outsiders, or struggles between groups over limited resources. The problem with all such monocausal explanations of the rise of the state, however, is to show how the kind of social inequality that might be produced by such processes then transforms itself into the stable and legitimate power that a ruling group exercises in the state. Other scholars, therefore, have proposed that states came into being as a result of different processes working together in a contingent way rather than being caused deterministically by a single overriding process (Maisels 1993: 199–220).

Whatever may be the explanation for the initial development of states and associated cities, they gave rise to major changes in human societies (see Spotlight box 1.4). Rather than being organized around the principle of kinship like the tribe and the chiefdom, for example, states were organized on the territorial principle. This means that the rulers of states exercised their rule over defined territories and their inhabitants, no matter who the latter happened to be. This proved an extremely powerful way of organizing human societies. States thus appeared in different parts of the globe. The first organized states are believed to have developed in Sumer in present-day Iraq around 3000 BC. According to Maisels, these took the form of city-states. In China the first state, associated with the Shang dynasty, arose about 1500 BC, and by about 1000 BC the Maya civilization had appeared in Meso-America. In Europe the first states are commonly taken to be the Greek city-states that began to develop in eastern Greece and the neighbouring islands about the eighth century BC. Whether the state arose independently in different places across the globe, or whether the idea of the state was somehow diffused from one or two initial foci, is uncertain.

Early states took many different forms. All faced difficulty in enforcing and maintaining their territoriality in the context of poor communications, and different states approached this problem in different ways. The Greek city-states, for example, typically occupied only small areas and restricted their population sizes. When population in any one state exceeded a certain threshold, that state would found a self-governing colony elsewhere on the sea coast. Travel by sea was generally easier than overland travel and thus settlements could keep in touch (for an analogous example, see Plate 1.4). In this way the Greeks eventually colonized much of the Mediterranean coast (Figure 1.4). The huge Roman Empire, by contrast, could not rely solely on sea travel. This was a much more centralized polity and only by having superb military organization, an elaborate system of roads and fortifications, good administration, and a state-controlled religion centred on the person of the emperor, could it survive for several centuries (Figure 1.5 and Case study 1.1). Another huge empire was the Mongol one that controlled much of Asia and a good slice of central and eastern Europe in the thirteenth and fourteenth centuries AD. The Mongols were pastoral nomads whose formidable warrior skills were based on their horsemanship and their ability to outmanoeuvre their enemies. But nomadic empires had a tendency to contract as quickly as they arose. It was difficult for the emperor to maintain control over his swift-moving armies of mounted warriors. Finally, medieval Europe's answer to the problem of territoriality was to decentralize political power through kings to territorial lords and minor nobles (see Case study 1.2). In return for their rights to their land and other privileges, lesser nobles owed their lords a duty of military support in time of need, which meant having to appear with an army of knights, retainers and other soldiers when ordered to do so. The lords in turn owed loyalty (or fealty, hence the term 'feudalism') to their overlords, and so on in a hierarchy that ended with kings or other rulers. Medieval European states were thus quite decentralized and unstable, as regional lords and their underlings were often tempted to rebel (Figure 1.6). Only after centuries of struggle between rulers, lords and ordinary subjects did the European states of today finally appear.

A modern geographer, looking at maps of past states, might be surprised at how large some of them were, given their many problems. They are a lesson in how human ingenuity can cope with geographical difficulty. But it is important to remember that before

Plate 1.4 A nineteenth century Polynesian canoe. Travel by water was the most efficient form in pre-industrial times, allowing vast distances to be covered in remarkably short periods of time. (Mary Evans Picture Library)

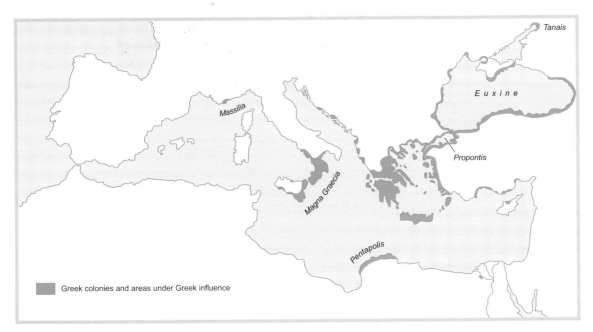

Greek colonies and areas under Greek influence

Figure 1.4 The Greek colonization of the ancient Mediterranean.
Source: After Pounds (1947: 49)

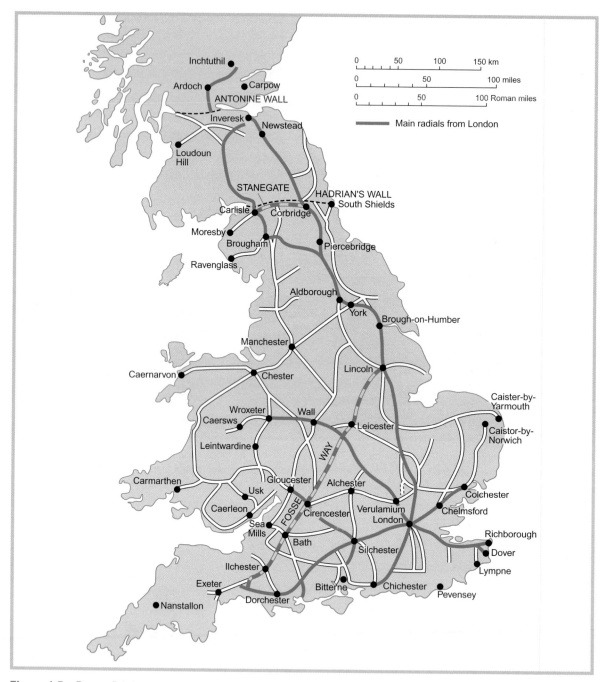

Figure 1.5 Roman Britain.

Source: © Oxford University Press 1981. Redrawn from map VI from *Roman Britain* (Oxford History of England) edited by Peter Salway (1981) by permission of Oxford University Press

modern communications the geographical constraints to state power were formidable. Pre-modern states and empires were inherently unstable, and the bigger they were, the more unstable they tended to be. Only modern communications have allowed states to overcome this problem to some extent, but even then imperfectly, as future chapters will show.

1.6 Pre-capitalist societies

It should be clear by now that pre-capitalist societies were by no means simple. They differed among themselves in all kinds of ways. At the same time, when compared with present-day capitalist societies, they appear to have a number of distinguishing features. The final section will consider some of the more obvious of these features.

Geographers and others have sometimes been tempted to think that past societies, lacking modern communications, were essentially small in scale and localized, with life based around the community. It is hoped that enough has been said already to suggest how over-simplified this view is. Huge empires had to be administered and defended, great cities like classical Rome had to be fed. Classical Rome imported some 17 million bushels of wheat each year from Egypt, North Africa and Sicily (Simmons 1996: 109). In the Roman Empire and other societies, high-value goods often travelled much greater distances. Even the Inuit, living the lives of hunters and gatherers, sometimes travelled

Thematic Case Study 1.1

The Roman Empire at its zenith (first to fourth centuries AD)

The Roman Empire was one of the greatest achievements of state-building in the pre-modern period and is testimony to the way in which huge distances could be crossed and enormous territories controlled in spite of the absence of modern industrial technologies. At its greatest extent the empire's east–west axis stretched from the Caucasus to Cape Finisterre in north-west Spain, a distance of about 2,800 miles. North–south it reached from Hadrian's Wall, near the present-day Anglo-Scottish border, to the fringes of the Sahara in North Africa, some 1,600 miles. The empire was characterized by two major languages (Latin and Greek), low tariff barriers, a common currency, a common code of laws, and the basis of a common system of education and culture for the Romanized elite. Rome was a truly 'universal state', which lasted in its mature form for half a millennium, and in its eastern (Byzantine) manifestation until the capture of Constantinople by the Turks in 1453.

Like all early states, Rome had problems in enforcing its territoriality – the sheer scale of the empire meant that there were always difficulties in maintaining control over its far-flung provinces. That the empire was able to endure for so long (albeit with many vicissitudes) was a tribute to its highly developed capacity for organization and an astuteness in gaining and retaining local loyalties. The acquisition of new territories proceeded by a mixture of conquest, colonization by Roman and Latin colonists, the establishment of vassal kingdoms and powers, alliances (often forced), and other means. The Roman army, reformed by the Emperor Augustus after 13 BC, was a masterpiece of military organization superior to any of Rome's enemies in this period. The seas and waterways were patrolled by the navy, protecting merchant shipping. Towns and military bases were interconnected by a network of fine roads and an official transport system (the *cursus publicus*) with inns and posting stations at regular intervals speeded official communications. There were even roadbooks (*itineraria*) for the guidance of the many travellers, including some tourists. Fortified lines, of which Hadrian's Wall is the most celebrated, protected the most vulnerable frontiers. But the empire was not held together only by military force. It was also by judicious extensions of Roman citizenship and other privileges, and by fostering helpful religious practices like emperor worship, that local loyalties were ensured.

The empire's economic and cultural achievements were many. Long-distance trade in such items as grain, metals and luxury goods was important, for example, though there were many hindrances to commerce and most people depended on local agriculture and manufacture. It must also be remembered that there was a heavy dependence on slavery. In the end the Romans could never be assured of political stability, even in their heartlands. Reliance on the army contributed to the empire's eventual undoing – generals and their armies competed for political power, the economy was undermined by overtaxation and other factors, and growing military weakness invited rebellion, and invasion by 'barbarians'.

Thematic Case Study 1.2

European feudalism

- Feudal society was hierarchical – ranked and stratified.
- The majority of the population were subsistence farmers, engaged in extensive forms of agriculture (for example, by cultivating strips in open fields shared with others, raising some livestock), and exploiting local resources like pastureland, woodland and fish.
- Many of the rural population were obligated to their lords and landholders, for example not to move away from their villages, to work the lord's land, to pay the tribute and taxes the lord and the state required, to serve the lord militarily, and to perform other services.
- Lords held their land conditionally from their overlord and ultimately from the king or other ruler. Land could not easily be sold. Lords exercised jurisdiction over those living on their estates.
- Lords owed their overlords (and they in turn their overlords and so on) allegiance or fealty (hence the term feudalism), for example in supporting them with their military forces and performing other services as required.
- Feudal states were quite decentralized, parcelled out into lordships and they in turn into jurisdictions of various kinds.
- Towns were generally small and few and far between. Most were centres of trade and crafts. Their residents (merchants, burgesses) often enjoyed freedoms and privileges ('liberties') denied to most rural dwellers (see Plate 1.3).
- Merchants and itinerant traders joined together networks of local and long-distance trade, linking local fairs and markets with towns, ports and so on. Craft production and trade were generally subject to a variety of restrictions and controls, and often played only a minor role in the lives of rural dwellers.
- The medieval world view was strongly influenced by the Church, whose priests and officials were among the few educated people and were universally present.

hundreds of miles each year on hunting trips, and their geographical knowledge frequently covered an even wider span. Enough has been said above about migrations and the diffusion of artefacts and practices across great distances to suggest how space could be overcome even if individuals were immobile, at least by today's standards.

There was, however, no easily avoiding the problems of communication in pre-modern societies. Most people in agricultural societies lived lives which were bound to their villages and the surrounding regions. Their work consisted essentially in winning the means of subsistence for themselves and their families from their environment, as well as paying the taxes and meeting the other demands which were made upon them. From the time of the appearance of states down until the twentieth century the lives of most people were lived in the context of such peasant communities. As mentioned earlier, such communities are still characteristic of parts of the world today, though increasingly eroded by commercialism and growing contacts with modernity. Certain scholars, impressed by the many different forms that peasant societies have taken across the world and yet also by the things they seem to have in common, have tried to construct a specialized branch of social science out of their study (Shanin 1971).

Scholars researching traditional peasant societies have frequently pointed to the many cultural differences between such rural dwellers and the cities that ruled over them and which they might occasionally visit. A case in point is religion. Religion has been a fundamental factor in social organization and in outlook in virtually every human society down to the twentieth century. Early states were generally associated with an official religion that legitimized the established order and which was often practised in cities but was rarely fully understood by the peasants, most of whom were probably illiterate (see Plate 1.5). Even where the peasants officially followed the same religion as the elite, as in medieval Europe, they almost always interpreted its teachings in their own ways,

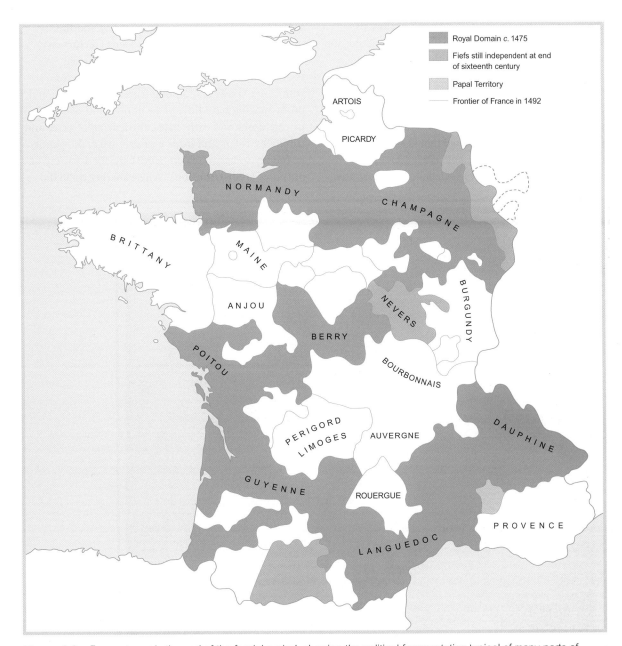

Figure 1.6 France towards the end of the feudal period, showing the political fragmentation typical of many parts of Europe prior to the appearance of modern states.

Source: Based on *Times Atlas of World History* (1989: 15.1). © Collins Bartholomew Ltd 1989, reproduced by permission of HarperCollins Publishers

mixing them with their own superstitions and 'pagan' beliefs. The life and outlook of the ruler and the elite, if not of all cities, were entirely alien to them. And equally their way of life was alien to the cities.

For this reason, it would be completely misleading to think of the subjects of pre-capitalist states as 'citizens' in the modern sense. Traditional rulers knew little of the countryside, where most of their subjects lived, and cared even less (except, perhaps, where they had landed estates, and even then the running of the estate could be left to officials). What mattered to rulers was law and order, and extracting the taxes and tribute the state needed to maintain itself in existence. The idea that rulers should care for the welfare of their ordinary subjects, let alone consult them about their policies, is very modern indeed.

It is difficult for most people living in a world dominated by capitalist relationships to conceive of societies where this was not the case. Before capitalism, most people were engaged in subsistence activities: hunting, gathering, farming, fishing or whatever. They might have to pay taxes or tribute, and they might trade some of what they gathered or produced, but market trade, generally organized by a small minority of merchants and traders, was often quite marginal to them. Moreover, trade, or more properly exchange, frequently took different forms from the market-type exchange that is so familiar today (Polanyi 1968, Dodgshon 1987: 193–224). For example, in many primitive societies exchanges often took the form of ceremonial gift-giving and the mutual exchange of goods and tokens (**reciprocity**). The point was to maintain alliances and social relationships, much as today gifts are exchanged within the family and among friends. Bargaining and profit-making have little place in this kind of activity. In chiefdoms and early states, the levying of taxes and tribute by rulers and lords was also usual, the goods or money so accumulated being used for a variety of purposes (**redistribution**). Where market trade did

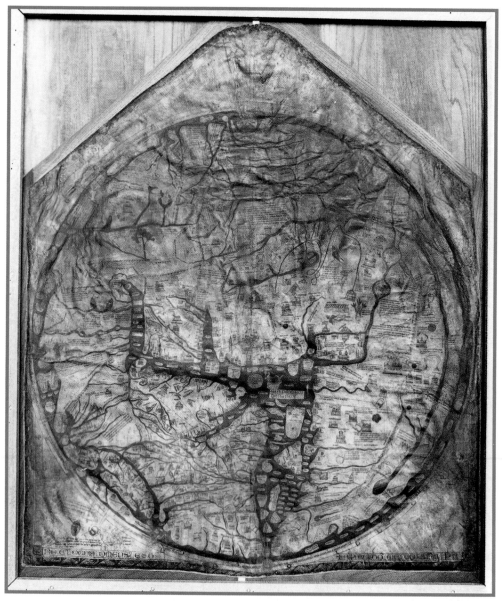

Plate 1.5 The celebrated Mappa Mundi or world map, dating from c. 1300 AD, in Hereford Cathedral, England. Constructed to show Jerusalem in the centre of the world, this map typifies the medieval Christian view of a world explicable only in religious terms and symbols. (By permission of the Dean and Chapter of Hereford and the Hereford Mappa Mundi Trust)

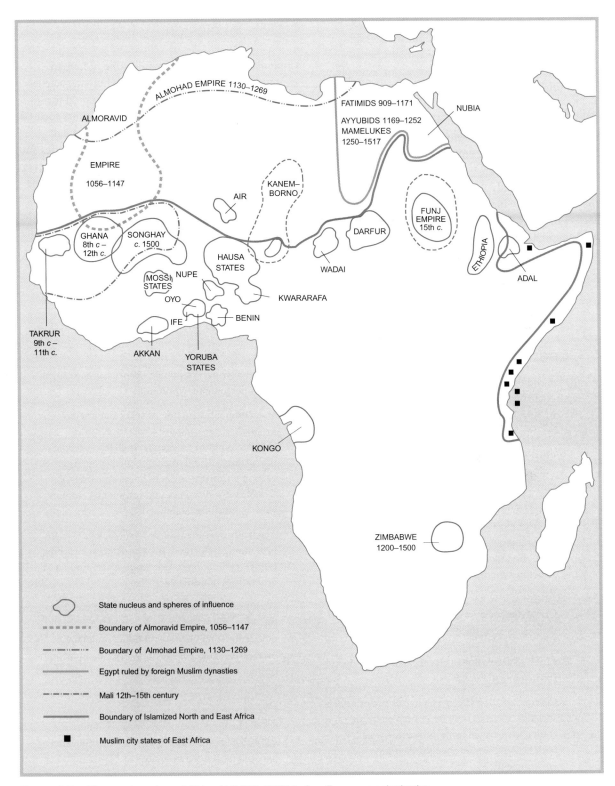

Figure 1.7   The great empires of Africa (AD 900–1500) before European colonization.

Source: Based on *Times Atlas of World History* (1989: 136–7). © Collins Bartholomew Ltd 1989, reproduced by permission of HarperCollins Publishers

exist in early societies (which was quite often the case), it was frequently hedged about by laws and restrictions of various kinds. There might be laws against the taking of interest on loans, for example, regulations on price, or restrictions on where trade might take place or who might engage in it. Only with the rise of modern capitalism did the market become central to the way societies functioned.

It is very difficult to generalize about pre-capitalist societies and almost any generalization is open to objection. Case study 1.2 describes the main features of the feudal society that existed in medieval Europe, the society that was the seedbed for the development of world capitalism (see Chapter 2). Even here, however, one must be careful – the features described would be more or less true, depending on the date and the place being considered.

1.7 Conclusion

Capitalism was preceded by millennia of human existence on the earth during which human societies took many different forms and helped mould the world that can be seen today. This chapter has focused on one way of trying to make sense of those past worlds, but also pointed to some of the difficulties and controversies that surround any attempt to do so. It is clear that human societies have on the whole become bigger and more complex through time, and their environmental impacts greater. What is less clear is the exact nature of the processes that have produced change, and who have been the winners and losers. Change has evidently been the norm rather than the exception, and this became particularly so with the rise of capitalism, the subject of the next chapter.

Learning outcomes

By the end of this chapter, you should begin to appreciate that:

- The surviving evidence necessarily limits knowledge of past societies, since they cannot be observed directly. In general, the further back in time past societies existed, the more uncertain knowledge about them becomes.

- People in the past lived very different lives from those of most people alive today. But it is important not to oversimplify or overgeneralize about past societies that varied enormously in their structures and ways of life and in their degree of complexity.

- Classifying societies into types is a way of making sense of the past and of trying to understand the geographical characteristics of varied societies. But all classifications have their shortcomings.

- Change has always been a characteristic of human societies. Societies have never been static, or completely isolated from the outside world. Societies have always changed their physical environments.

- The greater part of human history has been spent in hunter gatherer communities.

- Overcoming the friction of space was a severe problem in pre-capitalist societies. But it is important not to exaggerate its effects – people often travelled, and often migrated, long distances even in 'primitive' societies, ideas and artefacts travelled vast distances, and continental empires were successfully established.

- The landscapes that can be seen today have been profoundly influenced by past societies, including hunter gatherer ones. Our world is the product of generations of human activity.

- We must not assume that people in the past had the same values and outlook as we do. We should not assume that our ways of doing things are necessarily better (or indeed worse) than past ways, or than the ways of other peoples alive in the world today.

- Change and progress are two different things. Social change almost always means winners and losers. The human record suggests that the latter have often been the majority.

Further reading

Crone, P. (1989) *Pre-Industrial Societies*, Blackwell, Oxford. An excellent introduction to the character of pre-industrial societies 'of the civilized kind', i.e. from the invention of cities and civilization. Written for undergraduates.

Dodgshon, R.A. (1987) *The European Past: Social Evolution and Spatial Order*, Macmillan, Basingstoke. An original examination of European prehistory and history, taken from a geographical perspective. Based on the social classification system highlighted in this chapter. Very detailed, but repays careful study.

Dodgshon, R.A. and **Butlin, R.A.** (eds) (1990) *An Historical Geography of England and Wales*, 2nd edition, Academic Press, London. An historical geography of the region from prehistoric times to 1939. Chapters 4 and 5 illustrate the issue of feudalism.

Hodges, R. (1988) *Primitive and Peasant Markets*, Blackwell, Oxford. A lucid introduction for undergraduates of the intricacies of exchange in pre-capitalist societies and of the development of markets. Contains a considerable amount of geographical material.

Simmons, I.G. (1996) *Changing the Face of the Earth: Culture, Environment, History*, 2nd edition, Blackwell, Oxford. A history of the impact of humankind on the natural environment from the earliest times to the present. Contains excellent chapters on pre-capitalist societies, especially from the environmental point of view, but tends to be a little technical in places.

The Times Archaeology of the World (1999) New edition, Times Books, London. A lavishly illustrated survey of archaeology.

The Times Atlas of World History (1999) New edition, Times Books, London. A highly acclaimed survey of world history beginning with human origins. This is in fact more a history book with maps than a true atlas and fails to provide detailed maps despite its large format. But it is lavishly illustrated and contains a wealth of factual and explanatory material.

Useful web-sites

www.british-museum.ac.uk The British Museum, London. The museum holds a unique collection of art and antiquities from ancient and living societies across the globe. The web-site gives details of the museum's holdings and current exhibitions, hosts discussions on prehistory and past civilizations, and gives guidance on how to find further information.

www.si.edu The Smithsonian Institution, Washington, DC. The Smithsonian is a focus for many kinds of scientific and cultural endeavour in the United States. The web-site contains much that is useful for readers of this chapter, especially in relation to the history and prehistory of North America.

www.jorvik-viking-centre.co.uk The web-site of the Jorvik Viking Centre in York, England, famed for its reconstruction of life in York in the Viking period.

www.channel4.com/history/timeteam Channel 4 (UK)'s popular 'Time Team' programme is devoted to the archaeological investigation of a variety of sites, particularly in Britain.

http://accessibility.english-heritage.org.uk The web-site of English Heritage, which is responsible for the upkeep of many prehistoric and historic sites in England, many of which are open to the public.

www.besthistorysites.net Entitled 'Best of History Websites', a comprehensive guide to history-oriented resources online, starting with prehistory. For teachers, students and others.

 For annotated, clickable weblinks and useful tutorials full of practical advice on how to improve your study skills, visit this book's website at www.pearsoned.co.uk/daniels

CHAPTER 2

The rise and spread of capitalism

TERRY SLATER

Topics covered

- Definitions of capitalism
- The cyclical nature of capitalism
- The transition from feudalism to capitalism
- The beginnings of European imperialism
- Colonial commerce
- Transatlantic migrations and the slave trade
- European colonial empires and racism
- Industrial and agricultural transformations
- New modes of transportation
- Industrial urbanism

Between 1500 and 1900 there was a fundamental change in the way an increasingly large part of the world was organized. Beginning in particular parts of England, and spreading to other parts of Europe, the capitalist economic system was to change, or influence substantially, not only the economies, but also the political, social, and cultural dimensions of newly powerful nation-states. By 1900, capitalism was the dominant socio-economic system over a large part of the world, a by-product of the colonial empires of those nation-states.

2.1 What is capitalism?

A number of writers have provided theoretical frameworks for understanding the workings and ramifications of the capitalist system. The first was devised by the Scottish philosopher, historian, and father of classical economics, Adam Smith, in his book *An Inquiry into the Nature and Causes of the Wealth of Nations* (1776). Like all theoretical models, Smith's is a simplification of reality, but it introduced a terminology and series of conceptualizations that are still familiar to us today. Smith assumed the model was driven by people's selfish desires for gain and self-interest. Thus, production takes place to generate profit; surplus profits are accumulated as capital; and the basic rule of the system is 'accumulate or perish'.

Integral to the system is the determination of prices which, in a free market, said Smith, are determined by the supply of, and demand for, the factors of production. In order to maximize their profits, industrialists will always seek to minimize their costs of production, including the wages that they pay their workforce, so as to out-compete other producers by having lower prices. One way of doing this was through the **division of labour**, dividing manufacturing tasks into simple, repetitive operations that could be performed by unskilled, and therefore cheap, labour (see Chapter 12). Another important way in which manufacturers could seek **competitive advantage** over their fellows was to be first in the use of new technology. The significance of the invention and adoption of new ways of doing things, whether through science and technology, or ideas and organization, has often been crucial to firms, industries and regions in getting ahead of competitors. Most scholars agree that the capitalist system began to cohere into an integrated whole in a particular place (England), at a particular time (the seventeenth century), and that it then took another long period (until the mid-nineteenth century) before it had matured in the states of north-west Europe and in North America (Green and Sutcliffe 1987: 6–7) (Spotlight box 2.1).

Spotlight Box 2.1

Characteristics of states with a mature capitalist economy and society

- The majority of the population in these countries were wage-labourers forming a 'working class' who sold their labour power for wages in cash or kind so as to purchase food and other commodities to survive.
- The majority of these wage-labourers were male and worked outside the home; females increasingly 'worked' in the home, both in the domestic care of their families and by 'selling' surplus labour time for minimal wages.

- Wage-labourers were increasingly closely supervised by managers of the production processes.
- The means of production of wage-labourers was owned by capitalists whose aim was to make a profit on their investment.
- There was an increasing disparity between the 'profit' of capitalists and the 'wages' of workers with an increasing propensity for conflict between the two.
- The vast majority of goods and services, including fixed property like land, were distributed through monetary exchange.

2.1.1 Cyclical characteristics of economic development

Smith went on to model the workings of a national economy over a year (macroeconomics) to elucidate the way in which capitalism works in a series of interconnected cycles. These cycles have been studied subsequently by other economists, most notably the Russian Kondratieff (1925), whose name is now used for the roughly fifty-year-long cycles of boom and depression that have characterized the capitalist world since the mid-eighteenth century (Table 2.1).

In seeking the explanation for these cycles of growth and stagnation, Schumpeter (1939) argued that technical innovation leading to the development of new industries was the key to understanding the growth phase of the cycle. Because technical innovation is spatially uneven, then so too has been the geography of economic development under capitalism. The question then arises as to why the geography of innovation is uneven. Recent research has suggested that the socio-institutional structures of regions of innovation, or of those lacking innovation, are the key to explanation; in other words whether educational, governmental and social organizations encourage or discourage enterprise and change. Others have criticized the technological determinism of these theories (Mahon 1987).

2.2 Other perspectives, other stories

2.2.1 Marxism

The viewpoints of orthodox economics are not the only interpretation of the transformation of large parts of the world over the past three hundred years. Karl Marx (1867) took a very different perspective in his analysis of capitalism in the mid-nineteenth century, *Das Kapital* (see Spotlight box 3.2). He proposed that profit arises out of the way in which capitalists (the bourgeoisie) dominate labour (the workers) in an unequal class-based relationship. This unequal class relationship was perceived by Marx to lead inevitably to class conflict. In his later writing he sketched out ways in which labour could gain control of the means of production and thereby 'throw off their chains'. This was to have dramatic long-term consequences for the socio-political organization of the world through most of the twentieth century between the Russian Revolution and the upheavals of 1989 (see Chapter 3, pp. 73–5). We need to note, however, that Marx's writing is dominated by an historical, but not a geographical, perspective: it traces class relationships in time but not in space. Marxist interpretations of capitalist development have therefore always had difficulty in conceptualizing uneven spatial development.

2.2.2 World systems theory

Another way of conceptualizing the nature of these changes is to be found in the writings of Immanuel Wallerstein (1979), whose 'world-systems theory' provides a threefold categorization of historical socio-economic systems. Wallerstein proceeded to delimit the spatial characteristics of the capitalist world economy into core, periphery and semi-periphery. These terms are not used in the everyday sense but signify areas in which particular processes operate. For Wallerstein, core processes are those characterized by relatively high wages, advanced technology and diversified production; periphery processes are characterized by low wages, simple technology and limited production. Between the two is the semi-periphery: areas that exploit the periphery but which are exploited by the core, so that they exhibit a mixture of both core and peripheral processes (Taylor 1989: 16–17). In the

Table 2.1 Logistic and Kondratieff cycles

Cycle	Period	Growth phase	Stagnation phase
Logistic II	c.1450–c.1750	c.1450–c.1600	c.1600–c.1750
KI	1770/80–1845/50	1770–1815	1815–1845/50
KII	1845/50–1890/96	1845/50–1875	1875–1890/96

period between the seventeenth and the end of the nineteenth century some regions of the world were in the core (north-west Europe, for example) or periphery (central Africa) throughout the period. Other regions moved from the semi-periphery, into either the core (the southern states of the USA, for example) or the periphery (the states emerging from the ruins of the Ottoman/Turkish empire). Wallerstein's theory has been very influential in explanations of the growth of globalization, but it is an explanation only at the 'structural' level. It says little about the complexities of the social and economic networks that enable the system to work, or of the resistance by groups or individuals trying to change it (Ogborn 1999).

2.3 The transition from feudalism to capitalism

The transformation of the European economy into a capitalist one was a long-drawn-out process and historians continue to argue the precise causes of the changes that took place. Writers using an avowedly Marxist frame of reference (Dobb 1946, Kaye 1984) have made many of the most significant contributions in this debate. To them, the development of a class of wage labourers is crucial, together with the assumption of political power by the new class of capitalists. Geographers have been particularly concerned to trace the outworking of these historical processes of social and economic transformation in the particular space economies of local regions (Gregory 1982, Langton 1984).

The disintegration of feudalism (see Case study 1.2), with its carefully controlled market system, was considerably speeded by the after-effects of the Black Death in mid-fourteenth century Europe. The European population was reduced by between one-third and a half, and towns were especially hard hit. The resulting labour shortage meant that enterprise and new ways of doing things were more likely to be rewarded, since the feudal elite required more revenue to maintain their power. At the same time, the slackening of social and cultural controls meant that new ways of thinking could flourish, especially in the growing towns.

By the later fifteenth century, larger European towns were dominated by what has come to be called **merchant capitalism**. Merchants were both the providers of capital and principal traders in a regionally specialized, complex, Europe-wide trading nexus based not on luxury goods as in the medieval period, but on bulky staples like grain and timber, and on an increasing array of manufactured products. Towns that flourished economically in fifteenth- and

---

Spotlight Box 2.2

Characteristics of the period of merchant capitalism in Europe

- Increasing numbers of people sold their labour for money wages. They ceased to work on their own land. This led to increased consumer demand for food, clothing and household goods; in many places there was a notable rise in living standards.
- More producers of both agricultural and craft-manufactured products began to accumulate capital as they produced for this growing market. These prosperous yeoman farmers and manufacturers were the foundation of a new class of capitalists.
- The removal of feudal market restrictions and controls by guilds on production led merchants to

invest in the reorganization of production on a capitalist basis.
- Technical innovations transformed industries; the most significant was probably the development of the printing press by Gutenberg, in Mainz, enabling knowledge to be diffused cheaply and rapidly.
- The rediscovery of Classical knowledge led to new ways of seeing the world and, consequently, to its rebuilding to reflect these new images, especially in towns.
- Ultimately, these changes led to religious upheaval and political revolution whereby the new capitalist class became dominant in the governance of nation-states.

sixteenth-century Europe were those whose inhabitants were able to copy and manufacture imported products more cheaply than the exporting region, and develop a constant stream of new, innovative, marketable products (Knox and Agnew 1994: 154–61) (Spotlight box 2.2). For most commentators, the key question then becomes: why did these processes of transformation coalesce first in sixteenth- and seventeenth-century England?

We can begin to answer this question by suggesting that, since land was one of the key factors of production, critical in the sixteenth century was the enormous transfer of land from conservative ecclesiastical ownership to secular ownership. This was the consequence of the dissolution of the monasteries by Henry VIII in the later 1530s. Even the monastic buildings could be adapted by capitalist manufacturers, as the Wiltshire clothier William Stumpe showed at Malmesbury Abbey where he installed 300 weavers in the 1540s (Chandler 1993: 487–9). Land was also being transformed in the sixteenth and seventeenth centuries

through the process of enclosure. This enabled livestock to be raised more efficiently for the urban meat markets and experimentation in new agricultural techniques to be undertaken by yeoman farmers (Butlin 1993: 178–9).

In the seventeenth century it was the social, religious and political transformation of England that was critical according to Dobb (1946). The struggles between Crown and Parliament during the Civil War and Commonwealth (1642–60) are seen as a conflict between landowners, and capitalist yeomen and manufacturers. Though this struggle was continued through the Restoration (1660) and Glorious Revolution (1688), by the end of the seventeenth century the capitalist bourgeoisie were politically predominant and were able to transform the state to their own advantage.

Elsewhere in Europe landowners remained pre-eminent over a predominantly peasant agricultural workforce and this period is known as the 'Age of Absolutism'. Landowners, especially the rulers of small

Figure 2.1   The palace, gardens, park and town of Karlsruhe, Germany, founded in 1715 by the *Margrave Karl Wilhelm of Baden.*

states, amassed enormous wealth from their control of land. This wealth was expended on increasingly spectacular landscapes of display: huge palaces, filled with works of art and every luxury, set in carefully designed, and often very intricate, landscapes reliant on vast inputs of labour for their maintenance (Figure 2.1). King Louis XIV of France had set the model at his palace of Versailles, outside Paris, using the resources of a much larger state. Such conspicuous displays of wealth and privilege by the few led, ultimately, to a much bloodier revolution than in England. The French Revolution saw the slaughter of not simply the royal house, but of the landowners, intellectuals and the bourgeoisie, since this was the first of the workers' revolutions, with its cry of '*Liberté, egalité, fraternité*'. However, it led rapidly to the totalitarian militarism of Napoleon.

2.4 An expanding world

As the European economic and cultural world was gradually transformed, European travellers began to voyage beyond the shores of Europe. They travelled overland into Asia, drawn both by a thirst for greater knowledge but, more significantly, by the rewards that could be reaped from direct exploitation of scarce commodities for European consumption. These included spices, sugar, silk, muslins, porcelain and the like, which had previously been traded through the eastern Mediterranean. In the late fifteenth century, however, this region was coming under Turkish control. The Portuguese, funded by Italian bankers, were the first to navigate round Africa into the Indian Ocean, with its Arab-dominated sea-based trading networks, whilst soon after Christopher Columbus,

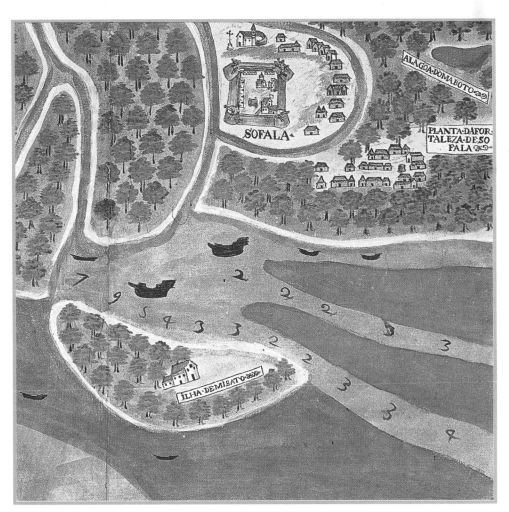

Figure 2.2 The Portuguese fortress and trading station of Sofala, in modern Mozambique, in 1558. For a time it was the administrative centre for all Portuguese trading with India.
Source: Carneiro (1990)

again funded by Italian (Genoese) merchant capital, sailed westwards across the Atlantic (in 1492) to open the Americas to European exploitation.

For Spain and Portugal, this is the heroic 'Age of the Navigators', as they moved rapidly to take control of these new maritime trade routes looking out to the Atlantic. The merchants of Lisbon and Seville grew rich on this trade. For the rest of the world it marks the beginning of the European colonial empires that were to dominate and control almost every aspect of life for the next four hundred years. The Portuguese established a far-flung network of fortified port bases around Africa and the Indian Ocean through which trade in gold, spices and textiles was conducted (Figure 2.2). Equally significant was their use of African slaves to produce sugar on plantation farms in their Brazilian territory. The Spanish founded a more militaristic, oppressive and exploitative empire in central and south America which saw the decimation of local populations through warfare and the catastrophic effects of European diseases, especially smallpox and measles. It also brought enormous inflows of gold and silver bullion into Spain and, ultimately, onwards into the European economy. Blaut (1993) sees this as the event that more than any other kick-started the European capitalist world system.

Fairly quickly, control of the products of this new maritime world economy moved from south-west to north-west Europe. The Dutch began to compete successfully with the Portuguese in the Indian Ocean and the centres of European capital moved from Lisbon and Seville first to Antwerp by the mid-sixteenth century, and then to Amsterdam at the end of the century. The merchants of these ports successfully integrated the Atlantic trade with that flowing from the Baltic and from central Europe via the Rhine valley. It was at Antwerp that the first stock exchange was established, enabling capital raised in one business to fund investment in another. Amsterdam's rise was based on having greater military and naval power than its neighbour and using it to enforce the privileges of a distinctive trading monopoly: the state-licensed joint-stock charter company (the Dutch East India Company) (Dodgshon 1998: 74–83).

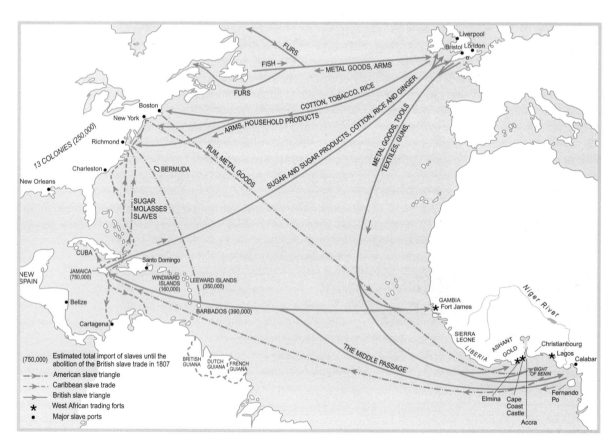

Figure 2.3 The eighteenth-century North Atlantic trading system between Britain, Africa and North and Central America.
Source: Bayley (1989: 46–7)

At the same time, French and English fishermen had followed John Cabot's exploratory voyage across the North Atlantic to begin to exploit the rich fishing grounds off Newfoundland and, subsequently, to trade with native Americans for fur pelts. From these small beginnings French and English merchant influence began to expand through the seventeenth century in the Indian Ocean (using similar company trading monopolies to the Dutch); in the Caribbean, where sugar plantations and slave labour were copied from the Portuguese; and in North America. They superseded the Dutch in the early eighteenth century through superiority of arms, but it was not until the later eighteenth century that the British emerged dominant in the developing world economy. Consequently the primary city of the European capitalist system moved once more, from Amsterdam to London (Dodgshon 1998).

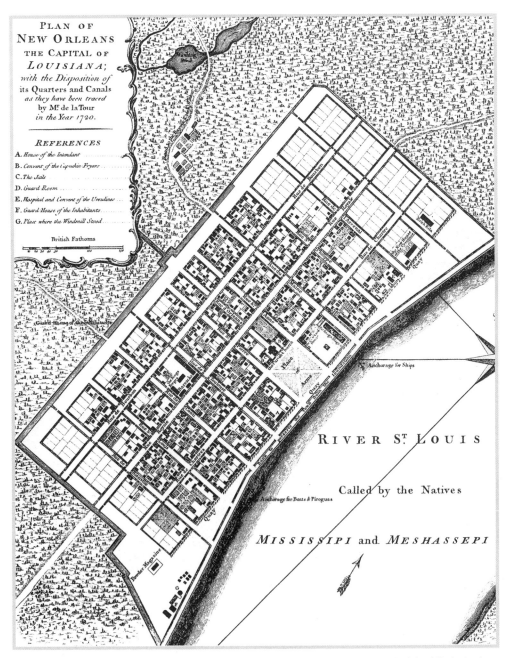

Figure 2.4 Plan of the French colonial entrepôt city of New Orleans at the head of the Mississippi delta in 1720, published in London in 1759 by Thomas Jefferys.

The colonialist trading systems that developed in the period up to 1770 have often been described in simple dualist, oppositional, terms (core–periphery; dominant–subordinate; metropolitan–colonial, for example). Blaut (1993) is especially critical of the Eurocentric character of this type of argument. More recently, a more complex, multi-layered interpretation has been posited. We need to note, first, that different parts of the developing world economy operated in somewhat different ways, were based on different product interactions between raw materials, manufacture and consumption, involved complex transportation flows, and required different politico-military frameworks to make them function properly. Meinig's (1986) model of the Anglo-French mid-eighteenth-century North Atlantic economy uncovers some of these complexities in terms of the commercial, political and social systems that were required for the geography of colonial capitalism, and Ogborn (1999) has added details of networks of individuals and patterns of resistance (Figure 2.3).

2.4.1 Colonial commerce

In Meinig's North Atlantic commercial system, London and Paris acted as the source of finance, commercial intelligence and marketing (London's Royal Exchange was founded in 1566). The expanding industries of particular regions of western Europe were growing, in part because they supplied a developing colonial market with manufactured goods such as textiles, tools, armaments and household necessities. These goods were assembled in Atlantic ports such as Bristol, Liverpool, Bordeaux and Le Havre. After crossing the Atlantic they were stored, disassembled and distributed from colonial ports such as New York, Charleston and New Orleans (Figure 2.4, Plate 2.1) by major traders. They sold on to local traders towards the colonial frontier. At the frontier a barter system of exchange was as likely to be found as a money economy. None the less, fur, fish, timber and agricultural produce flowed in the opposite direction for storage and transhipment back to Europe (Figure 2.3).

Plate 2.1 Madame John's Legacy, c.1788. One of the oldest houses in New Orleans, it was built during the Spanish colonial period (1765–1803). Commercial functions occupy the brick-built ground floor, residence the balcony floor.

In North America, tobacco, rice, sugar and cotton were the major products. All were produced on the plantation system using slave labour. By 1750, some 50,000–60,000 people were being forcibly transported across the Atlantic each year from all parts of the western African seaboard. Many died en route or shortly afterwards but, altogether, some 3.8 million Africans had been transported by 1750, just over half to Latin America, the remainder to English, French and Dutch North America and the West Indies. These staggering figures meant that African Americans were easily the largest new culture group to be established in both North and South America in the colonial period (Meinig 1986: 226–31). By 1750, the trade in slaves was dominated by large European companies using specially constructed and fitted ships, but many smaller traders from the Atlantic ports would sail first to African trading stations to exchange manufactured products for slaves. The slaves would then be taken to the West Indies or Charleston, with ships returning to Liverpool or Bristol with a cargo of sugar, tobacco or rice, and sometimes slaves too. Most of the latter were destined for domestic service in aristocratic households (Figure 2.3).

perceived to be a better new world. Religious faith also inspired another group of settlers who came to convert the native people of the Americas to Christianity. The competition in Europe between Catholic and Protestant versions of the faith – between Reformation and Counter-Reformation – was exported to the colonies since both native Americans and African slaves were perceived as being in need of conversion. Consequently, Jesuit monasteries or Franciscan friaries often stand at the heart of Spanish colonial towns and cities, whilst their dedicatory saints gave name to them: San Francisco, Los Angeles, San Diego and San Antonio, for example (Conzen 1990).

By the beginning of the eighteenth century, significant migrations of Scots, Irish, French, Germans, Dutch, Swiss and Moravians had taken place, but the dominant migrant group were the English. Between 1700 and 1775 the population of the eastern colonies grew tenfold to some 2.5 million and, by 1820, the population of the United States surpassed that of Britain (Lemon 1987: 121). The losers were native Americans who were dispossessed of their lands east of the Appalachians and often taken into slavery.

2.4.2 Colonial society

The slave trade was perhaps the most significant aspect of the cultural transformation effected by the eighteenth-century capitalist world economy in the Americas and, it should not be forgotten, in Africa. One of the distinctive features of this transformation in the Americas was the 'othering' of Africans by Europeans. They were regarded as in every way inferior to Europeans; they were legally and socially defined as different; and they therefore lived their lives separately from white settlers. Where Africans were in the majority, as they were in the West Indies, and in the Carolinas and Louisiana, this enabled distinctive African-American cultures to develop within the confines of political and economic enslavement.

There were, of course, other trans-Atlantic migrations throughout the time period covered in this chapter. The earliest included those seeking to escape religious persecution in Europe, or to establish religious utopias in the new continent. The Puritan 'Founding Fathers' of New England were in this category and, right through to the Mormon migration in the early nineteenth century, such groups continued to leave Europe's shores for what was

2.4.3 Colonial politics

There were major differences in the way European states governed their colonial empires. The Spanish developed a highly centralized system in which all aspects of policy were laid down in Madrid. The Laws of the Indies, which laid down in meticulous detail how colonization should take place and how new towns should be planned, are symptomatic of this centralization (Nostrand 1987) (Case study 2.1). The French governed through a similar centralized structure of military, administrative and ecclesiastical strands of authority linked to government departments and the Crown in Paris. The English, in contrast, allowed each colony to be virtually self-governing and the links to London were many and various. English colonial plantation owners and merchants were not averse to threatening a governor if their capitalist trading relations were threatened. It should occasion no surprise that it was the English colonies that eventually rebelled and fought for their independence. Overlapping these colonial territories were the older territorial patterns of native groupings, those of the Iroquois and Creeks remaining well into the eighteenth century (Meinig 1986: 262).

Thematic Case Study 2.1

Summary of the *Laws of the Indies* by Philip II of Spain, 1573

Laws 1–14

Establish that exploration cannot take place without royal permission; that governors should learn all they can about their territory; they should consult with and negotiate with local 'Indians'. 'Discoverers' should take possession of land and should name rivers, hills and settlements.

Laws 15–31

Lay down that Spaniards should treat 'Indians' in a friendly manner; they should help priests in their work of conversion.

Laws 32–42

Concern the type of region that should be explored, pacified, settled and brought under Spanish mandate; the siting of towns in such regions, and the establishment of local government.

Laws 43–109

Establish the legal and taxation regimes for new colonies and lay down that each town should have at least thirty families. The ideal size of house plots, farms, herds, commons is established.

Laws 110–128

Provide details of the plans of new towns, including the size of the plaza, the orientation and breadth of the streets, the location of churches and public buildings, and the allocation of house plots by lot.

Laws 129–135

Describe provision for common land and farms and the character of domestic buildings.

Laws 136–148

Lay down relationships with local 'Indians' and the way in which they should be converted to Christianity so that they 'can live civilly'.

Source: Crouch *et al.* (1982: 6–19)

2.5 Imperialism and racism

A second phase of colonial expansion, through the nineteenth century, was geographically distinct in that it was focused on Asia and, especially after 1880, on Africa. It is also marked by the development of a much more virulent racism in the 'othering' of all non-Europeans. The consolidation of European states in the nineteenth century, especially of Germany and Italy, and the increasing economic competition they offered to the eighteenth-century leaders of the world economy, Britain and France, led them to seek new opportunities through the colonial exploitation of hitherto economically peripheral lands. Britain seized the Dutch Cape Colony at the tip of southern Africa in 1806 to prevent its use by Napoleon's navy. By 1820, 4,000 British settler graziers were in conflict with

African pastoralists, despite the fact that the low wage-labour of those same Africans was essential to maintain the farms (Lester 1999). At much the same time, the first British farmers were beginning to establish themselves in Australia as its period as a convict colony began to draw to a close. Over extensive parts of the Indian sub-continent the collapse of the Mughal Empire saw its replacement with a variety of British administrations backed by British soldiers. After the 1857–8 Indian Mutiny, India became a Crown Colony and British central administration, schools and economic investments followed.

Meanwhile, in Britain itself, Christian-inspired reformers succeeded in bringing an end to the slave trade, and then to slavery. Sadly, the invective of both populist and bourgeois writing opposed to abolition characterized Africans, Australian Aborigines, Indians

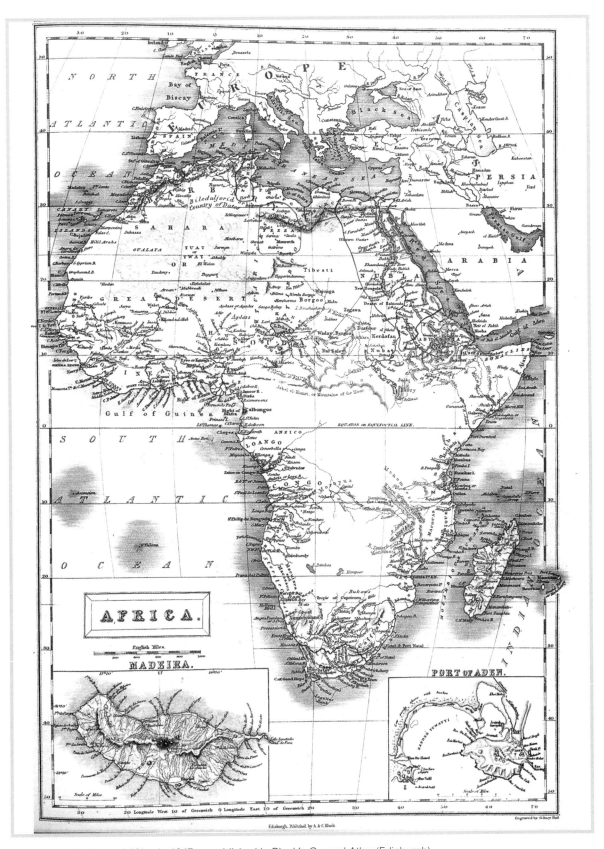

Figure 2.5 'Unknown' Africa in 1847 as published in Black's General Atlas (Edinburgh).

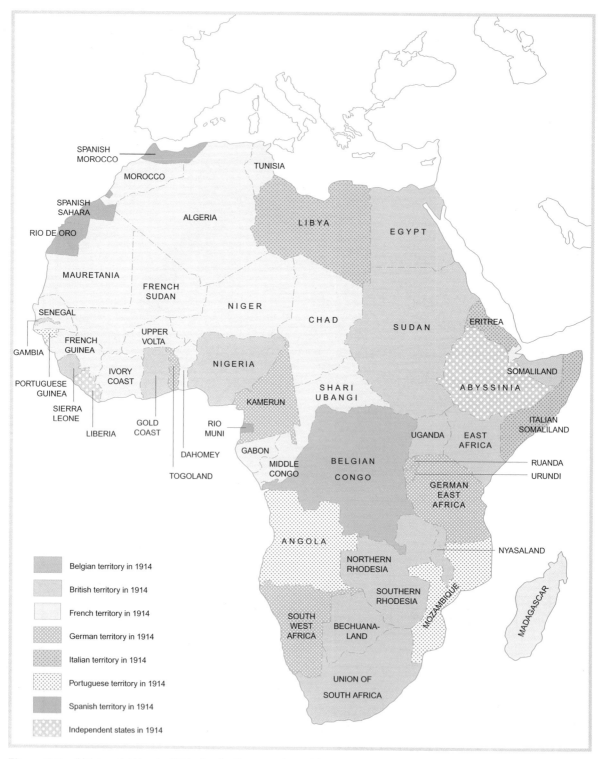

Figure 2.6 'Unknown' Africa in 1914 after the European imperial contest to divide the continent.
Source: Moore (1981: 74)

and native Americans as not only unchristian and therefore 'uncivilized' but as less than human; inferior to Europeans in every way; both savage and yet child-like, and therefore needing to be controlled and disciplined. By the middle of the nineteenth century, these attitudes were widespread amongst all classes in Britain and were fuelled further by popular antagonism to events such as the Indian Mutiny (1857), the Zulu rebellion (1879), and the Maori wars (1860s). By the later nineteenth century, Darwin's ideas on evolution were being misused to give a scientific veneer of respectability to this 'othering' of non-Europeans, to the extent that Australian Aborigines were being shot as 'vermin' rather than seen as fellow humans.

The mid-nineteenth century marked both the deliverance of one continent from colonialism to self-government, as South America broke free from Spanish and Portuguese control, and the colonizing of another, as European powers raced to divide Africa amongst themselves. The interior of Africa remained largely unknown to Europeans until the 1860s except for Christian missionaries. But by the end of the century, borders had been delineated and fought over, natural resources were being rapaciously exploited, and the polities, cultures and economies of native peoples had been disrupted or destroyed (Figures 2.5 and 2.6).

2.6 Industrialization

2.6.1 Proto-industrialization

The early phases of capitalist industrial development in Europe, up to about 1770, are categorized in various ways. To some writers this is the period of merchant capitalism (Spotlight box 2.2), to others this is the period of proto-industrialization (Spotlight box 2.3). Both terms have generated a substantive academic debate as to precisely what processes of transformation were involved.

Textile manufacture is the most studied production process of this period. Its transformation began in several parts of Europe as early as the fifteenth century. By the early eighteenth century, more and more of the production processes were being mechanized and industrialists were concentrating their machinery into single, multi-floor buildings. By the 1770s, the mechanization of spinning had been completed and weaving followed in the 1820s. The logic of this transformation was completed by steam power and the factory was established as both a building (Figure 2.7) and a system of production (Spotlight box 2.4). Different sectors of the textile industry went through this transformation at different times, and some sectors declined absolutely as capital was diverted to the most prosperous sectors. The textile industry, more than any other, had a relentless tendency to geographical specialization and concentration with the result that many towns and regions were dangerously dependent on a single industry (Laxton 1986). By the 1770s in Britain, west Yorkshire was beginning to dominate light worsted woollen cloth, Lancashire and the north Midlands cotton cloth production, the Welsh borders flannel, and the east Midlands hosiery.

The complex interaction of resources, labour skills, technological innovation, and capital circulation in particular regional economies, which increased production and lowered prices, meant that new markets had to be developed and transport improved. Market opportunities were to be found both in the growing urban markets at home, and among the colonial populations overseas. Thus, cotton textiles

---

Spotlight Box 2.3

Proto-industrialization

- Many early industrial processes were located in the countryside.
- Most industries were characterized by some type of 'domestic' production system whereby the capi-talist/merchant provided raw materials/machinery for producers who worked in their own homes.
- Producers generally worked in family units and were paid piece rates according to their production each week or month.
- Machinery was frequently water-powered.

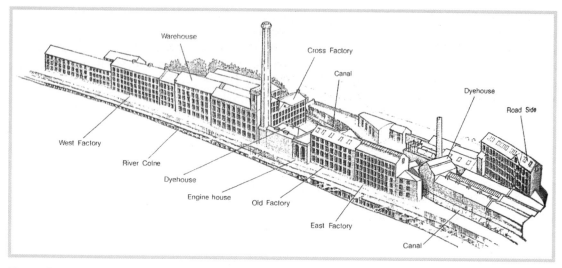

Figure 2.7 Starkey's woollen textile mill, Huddersfield, c.1850. Though this mill was steampowered, by mid-century its powerlooms were still housed in multi-storey buildings rather than a single-storey shed, since the site was a restricted one.

Source: Giles and Goodall (1992: 102)

Spotlight Box 2.4

The factory system of production

Under the factory system the capitalist:

- Had complete control of the production process from receipt of raw materials to finished product.
- Had control of the labour force on whom a new, disciplined time geography could be imposed (early factories often worked day and night).
- Could apply capital to the development of new machinery to simplify processes and reduce labour costs through mechanization; and,
- By dividing the tasks of the production process, could reduce labour costs further by employing women and children.

---

were first imported to Britain from India; the techniques of production were learnt; manufacture commenced; factory production lowered costs; cotton textiles were exported to India; there they undercut and destroyed the Indian industry.

A key transformation of the transport infrastructure, so essential to industrialization, was the improvement of links with growing towns, and with the ports. Canals and turnpike (toll) roads, which were the principal innovations before the 1820s, also lowered the costs of raw materials, especially of coal. This is graphically illustrated by the way in which urban populations celebrated the opening of new canals, not for their industrial potential, but because they dramatically reduced the cost of domestic heating. In Britain, the construction of better roads, and of the

canal network, mostly between 1760 and 1815 (the growth phase of the first Kondratieff, Table 2.1) was financed by private capital. The state was involved only in providing the necessary legislation to enable construction to take place and fees or tolls to be levied. Thus Josiah Wedgwood (china and pottery), Abraham Darby (iron manufacturer) (Figure 2.8), and Matthew Boulton (jewellery and silver plate), as well as the Duke of Bridgewater (landowner and coal owner), all invested some of their capital in canal companies to improve the distribution of their products (Freeman 1986). A similar pattern is observed in New England. In France and Germany, by contrast, new roads were seen to have a military function in the centralization of state control and were therefore largely financed by the state through taxation.

Figure 2.8 The world's first iron bridge, constructed across the River Severn, east Shropshire, in 1788 by Abraham Darby III's Coalbrookedale Company, was part of the transport infrastructure improvements financed by local industrialists.
Source: The Ironbridge Gorge Museum

2.6.2 Agricultural change

It was not just industries such as textiles, metal manufacturing and pottery that were transformed by early capitalist production; food production needed to keep pace with a rising population that was increasingly employed full-time in manufacturing. Consequently, the eighteenth century saw the transformation of agriculture into increasingly capitalistic modes. In parts of Europe, especially in Britain, self-sufficient peasant farming began to come under pressure from landowners who used state legislation to remove common rights to land in favour of private ownership through enclosure. Ownership was increasingly concentrated into fewer hands, and farms were consolidated from scattered strips to single block holdings, again especially in Britain and, rather later, in Scandinavia and northern Germany. Elsewhere in Europe this process was delayed until the twentieth century. In northern and western Europe, landowners invested in transforming production for the new urban

markets, using technology and new techniques and crops first developed in the Low Countries in the seventeenth century. In eastern Europe, the semi-periphery, this period saw a reversion to serfdom and near-feudal relations of production. Capital was also deployed to increase the area of land under intensive, rather than extensive, production. Fenland and coastal marshes were drained; heathland soils improved for cultivation, and the moorland edge of improved pastures pushed higher up hill and mountain slopes.

In the nineteenth century, European agriculture continued to increase production but was unable to keep pace with the demand from the growing urban industrial population. In 1840, Britain repealed the Corn Laws, which had protected local grain producers, and opened its markets to colonial and American producers. New methods of extensive grain farming were used in mid-western North America, and in the new British colonies in Australasia and South Africa. The invention of the steam ship, allowing more rapid transport, and of refrigeration and meat canning from

Table 2.2 European emigration 1881–1910

Source countries	Destination countries	Numbers of migrants
Great Britain	N. America, S. Africa, Australasia	7,144,000
Italy	USA and N. Africa	6,187,000
Germany	North and South America	2,143,000
Austria/Hungary	North and South America	1,799,000
Russia	USA	1,680,000
Scandinavia	North America	1,535,000
Spain	Central and S. America; N. Africa	1,472,000
Ireland	USA	1,414,000
Portugal	South America	775,000
S.E. Europe	USA	c.465,000
France	Central and S. America	223,000
Low Countries	USA	171,000

Source: Moore (1981: 57)

the 1870s, led to similarly extensive ranch grazing of sheep and cattle in these countries. Together with the Pampas grasslands of Argentina, all became an important part of the semi-periphery of Britain's global economy. This new export-dominated agriculture was reliant on overseas investment by British capital in new railway networks to transport these products to slaughterhouses and industrial packing plants at the port cities whilst, in the USA, Chicago was growing rapidly on the same economic foundations (Cronon 1991, Miller 1997). It was accompanied, too, by new waves of migration to the colonies and the mid-west of the USA (Table 2.2), especially from Ireland and Scotland, where the Great Famine and the Highland Clearances, respectively, were other manifestations of capitalist agricultural change in the British Isles. The United States absorbed enormous numbers of European migrants throughout the nineteenth century. Irish migrants predominated in the 1840s and 1850s, followed by Scots and English, Scandinavians and Germans through the second half of the century, with Russians, Italians and south-central Europeans from the 1880s onwards (Table 2.2) (Ward 1987).

This later colonization, and the internal colonization of the United States mid-west in particular, were grandly characterized as a moving frontier of settlement by the American historian, Turner, writing in 1894, who suggested that it also transformed the 'character' of frontier settlers, making them self-reliant, opportunistic, individualistic and democratic. Recent commentators have noted that Turner's hypothesis says little about the continued sweeping aside of the rights of the native populations of these lands (the land was regarded as 'open' or 'free' for white settlement); nor does it give any credit for the settlement process to women, who provided both the domestic labour and, often, especially in the initial phase of settlement, much of the farm labour too. The advancing frontier of capitalist agriculture was as much small-scale, incremental and domestic as it was wide-sweeping and large-scale.

2.6.3 Factories and industrial production

Some commentators have seen the 'Industrial Revolution' in Britain as a short period, between 1770 and 1830, of rapid transformation whereby the country's economy moved from an agricultural to an industrial manufacturing basis. More recent writers see the transformation as much more drawn-out, extending from the seventeenth century and into the early twentieth century. They also see it as much more regionally diversified and geographically uneven, with particular regions specializing in particular products which transformed their production systems at different times. Most attention has been given to the textile and iron and steel industries, both of which are characterized by large factories requiring massive capital investment in buildings and machinery, driven by steam power, with manufacturing processes increasingly vertically integrated, and employing a large, and increasingly disciplined labour force (Figure 2.7). The same forces were at work in New England. By 1855, there were 52 cotton mills in Lowell, Massachusetts, employing more than 13,000 people, two-thirds of them women (Groves 1987). In Europe, Lille (France), Ghent (Belgium) and the Wupper valley

of Germany developed as centres of cotton textile manufacture (Pounds 1990: 402).

Many other industries were developed in the same way as new markets were developed on a national and international basis and production expanded. Food and drink, for example, is rarely thought of as a factory-based industry, but brewing, once the prerogative of almost every village inn, became increasingly an urban, large-scale industry with regional markets for its beers. In England, Burton-upon-Trent developed rapidly as the 'brewing capital' of Britain, with huge factory-scale breweries, on the basis of its colonial contracts to supply bottled 'India Pale Ale' to troops and civilians stationed in the Indian sub-continent as a substitute for local water. For other industries, however, including many manufacturing industries, 'factories' remained small-scale workshops employing fewer than 50 people (the Sheffield cutlery industry in northern England, and wire-drawing in the eastern Ruhr in Germany are good examples). Indeed, many industries remained almost domestic in scale

Figure 2.9 The Great Exhibition, held in Hyde Park, London, in 1851, in Joseph Paxton's Crystal Palace, was intended to display the manufactured products of industrial Britain and its empire to the rest of the world. (Popperfoto)

well into the twentieth century (for example, the Birmingham jewellery and Coventry watch industries, both in central England).

The nineteenth century is characterized not only by this enormous variety in the scale of production, but also by an equally enormous variety of manufactured products. The tag 'The Workshop of the World' is as true of nineteenth-century Britain in general as it is of several of the country's manufacturing cities. London's 1851 Great Exhibition (Figure 2.9) was seen as a showcase of the huge variety of products from Britain's factories and workshops and those of its empire. In other British, European and American regions there was an increasing specialization so that local economies were dependent on maintaining market advantage on a single industry: the Polish textile city of Łódź (Case study 2.2) and Lancashire's cotton mill towns were of this kind. Such places also produced distinctive working-class cultures, which, in the case of Łódź, was also predominantly Jewish (Koter 1990). Again, in those same mill towns much factory employment was female (see the example of Lowell above), whereas in heavy industry centres, such as ship-building in Newcastle and Sunderland, in the north-east of England, waged employment was male-dominated and women's roles were primarily domestic. E.P. Thompson, in his classic Marxist interpretation of *The Making of the English Working Class* (1963), is especially sensitive to the variety of experiences of working people in this period of aggressive industrial capitalism. By the second half of the century British labour had won the right to organize itself into trade unions to improve work and social conditions for members.

There are, however, some common features of nineteenth-century industrial development in Britain and, subsequently, in Europe and North America. First was the fact that the power supply for the majority of industries came to be dominated by steam. There was therefore a move by many industries to the coalfield regions to reduce the cost of fuel (Figure 2.12, see p. 57). The development of the Ruhr region of Germany (Pounds 1990: 412–29), and of western Pennsylvania in the USA (Meyer 1990: 256–8), are good examples of this process. The reason for this change in fuel supply was the invention and commercial development of the rotative steam engine by James Watt and Matthew Boulton in 1769–75.

Secondly this, in its turn, was fundamental to the invention and development of railways that transformed the transport costs of both raw materials

and finished products almost as dramatically as it reduced the friction of distance. The Western and, increasingly, the colonial worlds saw a new time geography from the 1840s onwards, whereby journeys requiring several days to accomplish previously could be undertaken in a few hours. This was especially important in the industrialization of North America with its vast transcontinental distances (the first transcontinental line was completed in 1869). It was also significant in Britain's imperial control of India, and of Russia's colonization of the interior of Siberia and central Asia. In all three cases the railways enabled industrial-scale exploitation of natural resources (from timber to metalliferous ores) to take place, the profits from which could, in turn, be invested in further industrial, urban or railway development. The rapid expansion of San Francisco following the California gold and Nevada silver discoveries in 1849 and 1859 respectively is a good example of this (Walker 1996). Railways were also major industries in their own right, of course; the Indian railways employed three-quarters of a million people by the 1920s, for example.

Third was the fact that, as capitalists sought economies of scale, the transformations of the nineteenth century were largely urban. By 1851, Britain had become the first country anywhere in which a majority of its inhabitants lived and worked in towns. By 1900 there were sixteen cities in the world with a population that exceeded one million, as against one (Peking) in 1800, and another 27 places had half a million or more (Lawton 1989).

2.7 Urbanization

The urbanization of capital has been theorized from a Marxist perspective by Harvey (1985a, 1985b). He notes that, as capitalists over-accumulate, the surplus flows into secondary circuits of capital, of which the principal is the built environment (in its widest sense). As in the primary industrial circuit, these flows are cyclical according to the perceived profitability of such investment. Because of the longevity of the built environment these 'building cycles' (sometimes called Kuznets cycles) are much longer than the business cycles, somewhere between 15 and 25 years. Crises may lead to the loss of profitability of an investment, but they do not often lead to the loss of building fabric. Western cities are therefore made in the image of past capitalist decisions and subsequent adaptations to fit new circumstances. Harvey goes on to suggest ways in

Regional Case Study 2.2

The textile city of Łódź, Poland

- Łódź developed thanks to the political circumstances of the remnant Polish state in the period 1815–31. Rapid industrialization was required to increase export earnings. A new town was founded beside a small market settlement in 1820–1 for 200 migrant drapers (4) with separate gardens (6).

- In 1824, the long straight road to the south, Piotrkowska, was laid out with 272 plots for

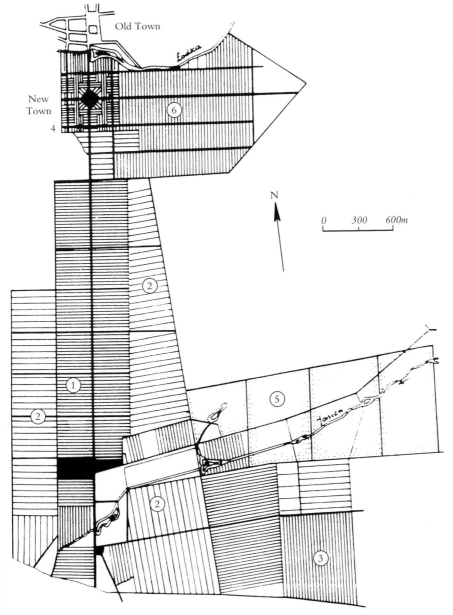

Functional plan-units of Łódź 1827
1. Linen and cotton weavers' colony; 2–3. Linen spinners' colonies; 4. Drapers' colony;
5. Water-powered manufacturing properties; 6. Drapers' garden

Figure 2.10
Plan of Łódź.
Source: Koter
(1990: 111)

Regional Case Study 2.2 continued

weavers and 167 plots for cotton and flax spinners. The long gardens, 1.68 ha in size, encouraged the spinners to grow flax (1 and 2). In 1828 a further area was developed to the south-east for Silesian linen weavers (3) (Figure 2.10).

- One-storey houses were provided by the state, but the tenants could purchase them subsequently. There were water-powered fulling, bleach, dye and print works beside the river (5). The first textile factories were built in 1827 and 1838. By 1860 there were 35,000 inhabitants in the town.

- In the second half of the nineteenth century the population of Łódź increased to 100,000 by 1878; to 314,000 in 1897; and to 506,000 in 1913. This explosive growth was almost entirely supported by the new factory-based textile industry.

- Jewish capitalist entrepreneurs such as Karl Scheibler and Israel Poznański built huge factory complexes. They were steam-powered, with belt-driven machinery, and the buildings were fireproofed. Scheibler and Poznański built grandiose palaces for their own residence beside their factories since they maintained day-to-day management.

- They also built model dwellings for some of their workers. The first were built in 1865. The dwellings were three-storey, one-room apartment blocks with kitchens and toilets in separate blocks in the yards.

- A self-contained workers' settlement was built by Scheibler in front of his Księży Młyn factory in 1875. 144 flats in 18 two-storey brick-built blocks were provided along a private gated street. A school, canteen and hospital were added later (Figure 2.11).

- In the city centre, along Piotrkowska, the long plots of the weavers were developed with tall insanitary tenement blocks on either side of a dark courtyard. Access was through the often grand Neo-Classical street front building by means of a central passageway.

Figure 2.11 Łódź, Scheibler's factory and workers' housing.
Source: Poplawska and Muthesius (1986)

which class struggle is written into the landscape of the Western city (1985a: 27–31), which he explores in more detail in an analysis of Paris in the third quarter of the nineteenth century. More recently, Dodgshon (1998: 148–61) has built on these arguments to show how the built environment of towns and cities is a major source of inertia in capitalist economies and societies.

There are perhaps two major phases of development in Western industrial cities before 1900, and the beginnings of a third. The first phase, which began in Britain in the 1770s (Table 2.1), was characterized by the need to provide homes for workers once the production process was separated from home. Initially, such housing was provided by a process of 'densification' whereby more and more

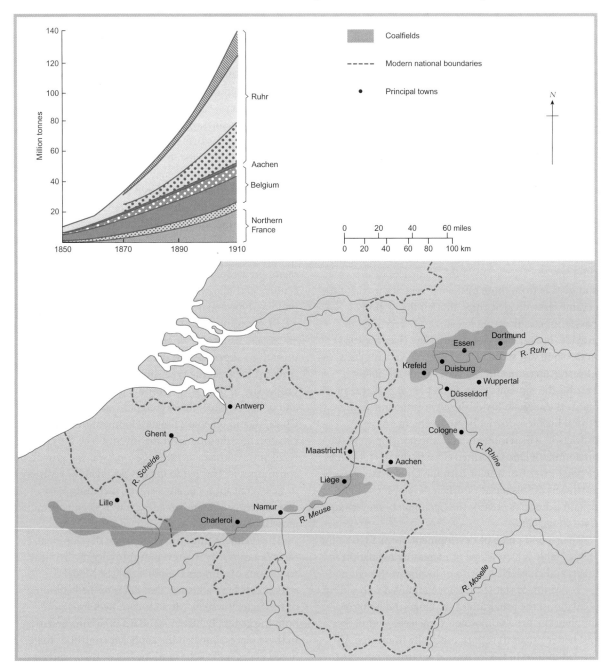

Figure 2.12   The coalfields of north-west Europe saw a concentration of new industrial enterprises after the 1780s as steam-powered machinery became normal in textile manufacturing, iron and steel making and metal goods manufacturing.
Source: Pounds (1990: 44)

living spaces were crammed into the existing built-up area. Gardens and yards were used for increasingly high-density basic accommodation, much of which was multi-occupied. By the 1770s in larger British cities this process could go no further and housing began to be provided on the city fringe where it proved a profitable 'crop' for small landowners. Such housing was brick-built, basic, unplanned and unregulated, and regionally distinctive in its plan forms. The most notorious of these forms was the back-to-back dwelling originating in the Yorkshire textile towns (Beresford 1988). The lack of regulation meant that these housing areas lacked effective water supplies, sewage and waste disposal systems, social and educational facilities, and connected road systems. The consequences, when combined with the poverty of low wages, were ill health, disease and very high urban death rates, especially among children. European and North American cities were also characterized by unregulated slum housing areas as urban populations rose from the 1820s onwards (Case Study 2.3) (Pounds 1990: 368–91, Homburger 1994: 110–11).

The second feature of this first phase of industrial cities was the separation of the residences of the bourgeoisie from those of the workers. Industrial cities became class-divided cities, though the working classes, especially young females, were recruited to provide domestic services by the elite (Dennis 1984). In the southern states of the USA that servant class was African American and still slaves until the 1860s. Thirdly, the capitalist land market began to value more highly the accessibility of the city centre and so the central business district began to emerge, as did districts in which industry and warehousing were the predominant land use, often close to the port facilities of canal, river frontage or harbour. One of the classic portrayals of this phase of urban development in Britain is Engels's (1845) *The Condition of the Working Class in England.*

The second phase of development, from 1840 to the 1890s (Table 2.1), was one of increasing regulation in towns and cities. New forms of local government and the collection of statistical information led to building regulations; the provision of sewerage and better water supplies; the provision of new cemeteries and parks; and better urban transport, especially tramways. All this led in turn to the increasing suburbanization of the better-paid elements of the urban workforce in the last quarter of the nineteenth century. These transformations occurred in most European core

economies, and in North American cities, but with temporal variations. Thus, in England, the characteristic late-nineteenth-century dwelling was a suburban, brick-built, tunnel-back, terraced house. In North America, excepting only the largest cities, the vast majority of urban housing was timber-built; in German and French cities (as well as in Scotland), the apartment or tenement block was the norm (Case study 2.3). The North American city, from the 1840s, was also characterized by the ghettoization of immigrant populations: first Irish and Germans, then freed slaves from the southern states after the Civil War, then Italians, and finally, from the 1890s, Jewish migrants from the Russian empire of eastern Europe (Ward 1987).

Central business districts were modified by two other major transformations in the capitalist system in this second phase. First, banking services were increasingly required to enable surplus capital to be safely stored, efficiently invested and recycled towards new opportunities; such banks were but one aspect of developing office quarters, including, for example, insurance offices, land agents and other property services, and legal services. This tertiary employment sector grew rapidly in the second half of the nineteenth century and 'the office' developed as both a specialized building type and a means of production (Daniels 1975). It was increasingly characterized by female employees using typewriters (the key technological invention) supervised by male managers. In North America the building type showed an increasing propensity to both large floor areas and height at the end of the century (the skyscraper was born in Chicago) and relied on the development of steel-frame construction methods and the invention of the elevator (Gad and Holdsworth 1987). Second was the growth of consumption. This led to the development of a variety of new retail spaces in city centres, including department stores and arcades, as well as an enormous growth in the variety of specialized shops.

The incipient third phase of urban development reflects a minority interest in improving the living conditions of workers by industrialists. Model settlements have a history that dates to the beginning of the factory system. Modern commentators note both the generally higher standard of accommodation in these places, but also the capitalist control of the home life as well as the working life of their workers. At the century's end the experimental garden suburb settlements at Bournville and Port Sunlight in England were widely admired and imitated in Germany and the

Thematic Case Study 2.3

Tenement housing for industrial workers

· The 1865 New York Report on the Sanitary Condition of the City illustrated tenements (Figure 2.13, Plan A) on what had originally been four lots facing First Avenue. The apartments in the five-storey front block were of two or three rooms, only one of which was heated.

SKETCH PLAN A

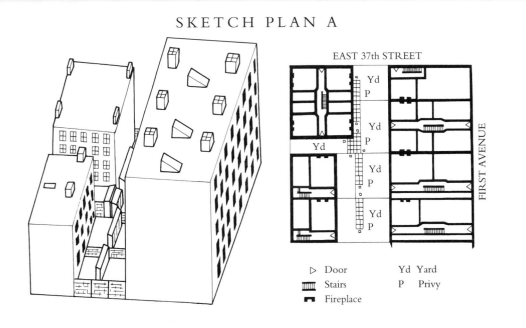

▷ Door   Yd Yard
▥ Stairs   P Privy
▰ Fireplace

SKETCH PLAN B

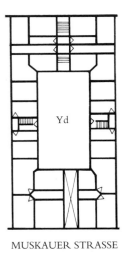

MUSKAUER STRASSE

Figure 2.13 Plans of tenements in New York (A) and Berlin (B).
Sources: Ward (1989: 34), B Borgelt *et al.* (1987: 11)

Thematic Case Study 2.3 continued

- In the southernmost lots dark tunnels provided access to the yards where smaller, cheaper two-room apartments occupied four-storey buildings to the rear of the lots.
- On the northern lots, East 37th Street allowed for crossways sub-division of the property and a block of small two-room apartments had been built.
- Privy blocks occupy the yards which would also normally have been strung with drying washing.
- Berlin was expanding rapidly at the end of the nineteenth century as the capital of newly unified Germany. Workers lived in 'Mietskasernen' (single-room barrack blocks) earlier in the century.
- In the 1890s 'Mietshaüser' (tenement blocks) were the normal housing of workers and artisans.

- Some of these were 'L'-shaped on the lot, some 'C'-shaped, but many were built around a central courtyard, as in the Figure on Muskauer Strasse (Plan B).
- Shops often occupied the ground floor on the street frontages of these blocks; a central passageway gave access to the yard where four stairwells accessed the apartments. Most were five or six storeys, so that the yards were dark and gloomy.
- Apartments were of between one and three rooms, but by the 1890s usually had integral sanitation.

USA. At the same time, the first experiments in the large-scale provision of social housing were underway in London using apartment block housing. These two themes, the garden suburb and social housing, were to combine powerfully in the development of modern planning to shape the urban environment in the first half of the twentieth century in Europe and its empires, but not in North America.

2.8 Conclusion

Within three centuries, Western capitalism had utilized its ill-gotten gains from its first colonial adventures to develop a series of specialized industrial regions supplying worldwide markets. Food production, industrial manufacture, service provision, consumption and transportation were all radically transformed by the capitalist enterprise. Cultures, societies and governments were necessarily impelled to change, too. States became more centralized and powerful. The difference between rich and poor both at the level of individuals and between countries became more marked. For some individuals and groups capitalism brought prosperity, improved living conditions and greater freedom; for others it brought destruction of local cultures, impoverishment, degradation and slavery. At the end of the nineteenth century the South African Boer War gave the first glimpse of the industrialization of warfare which was to scar the twentieth century so deeply.

Learning outcomes

Having read this chapter, you should know that:

- Scholars have interpreted and theorized the development of capitalism in different ways.
- The European colonialist enterprise was critical in the evolution of capitalism.

- Industrial capitalism transformed all sectors of the economy, including the built environment and social relations between individuals and classes.
- Industrial capitalism was essentially urban.

Further reading

Blaut, J.M. (1993) *The Colonizer's Model of the World: Geographical Diffusionism and Eurocentric History*, Guilford Press, New York and London. This is a very readable and thought-provoking polemic of post-colonial writing, providing an alternative explanation of the success of European capitalism.

Dodgshon, R.A. (1998) *Society in Time and Space: A Geographical Perspective on Change*, Cambridge University Press, Cambridge. A more advanced text which provides much more detailed arguments and evidence for the themes of this chapter.

Knox, P. and **Agnew, J.** (1994) *The Geography of the World Economy*, 2nd edition, Edward Arnold, London. A well-written and popular textbook which will give readers another perspective on this period and much else besides.

Langton, J. and **Morris, R.J.** (eds) (1986) *Atlas of Industrializing Britain*, 1780–1914, Methuen, London and New York. Historical atlases usually treat this period very well with innovative cartography and thought-provoking texts. This one deals with Britain in considerable detail.

Michell, R.D. and **Groves, A.** (eds) (1987) *North America, the Historical Geography of a Changing Continent*, Hutchinson, London. There are a number of good historical geographies of North America. This one is well written, thoughtful and copiously illustrated.

Useful web-site

www.besthistorysites.net/20thCentury.shtml Entitled 'Best of History Websites', a comprehensive guide to history-oriented resources online. For teachers, students and others.

For annotated, clickable weblinks and useful tutorials full of practical advice on how to improve your study skills, visit this book's website at www.pearsoned.co.uk/daniels

CHAPTER 3

The making of the twentieth-century world

DENIS SHAW

Topics covered

- The second industrial revolution

- Fordism – new patterns of production and consumption in the twentieth century

- Organized capitalism

- Resistances to Western-style capitalism: Nazism, the Islamic revival

- Resistances to capitalist-type development: Marxism, communism

- The end of European imperialism; informal imperialism

- The Cold War and the collapse of communism

- Disorganized capitalism

The twentieth century could be said to have been the period when capitalism finally triumphed over most of the globe. But it has been neither a straightforward triumph nor an unchallenged one, and capitalism itself has been changed in the process. This chapter is concerned with the various spaces created by and in response to twentieth-century capitalism – spaces of resistance and reinterpretation as well as spaces of adaptation and acceptance. The patchy and unequal world in which we now live reflects the erratic and conflict-laden nature of the processes which have produced it.

3.1 The changing capitalism of the early twentieth century

In the autumn of the year 1933 the writer and journalist J.B. Priestley set out by bus on a trip that was to take him the length and breadth of England and which he later described in his *English Journey* (Priestley 1937). As he left London by the Great West Road, Priestley noted how the road 'looked odd. Being new, it did not look English. We might have suddenly rolled into California.' What struck Priestley as particularly odd was 'the line of new factories on each side' of the road. 'Years of the West Riding', he explained (he was born and raised in Bradford), 'have fixed forever my idea of what a proper factory looks like: a grim, blackened rectangle with a tall chimney at one corner. These decorative little buildings, all glass and concrete and chromium plate, seem to my barbaric mind to be merely playing at being factories.' Armed with a copy of the now celebrated textbook, *The British Isles: Geographic and Economic Survey* by Stamp and Beaver (1933), which was later to be used by generations of geography undergraduates, Priestley went on to make some astute geographical and social points about these factories: 'Actually, I know, they are tangible evidence, most cunningly arranged to take the eye, to prove that the new industries have moved south. You notice them decorating all the western borders of London. At night they look as exciting as Blackpool. But while these new industries look so much prettier than the old, which I remember only too well, they also look far less substantial. Potato crisps, scent, tooth pastes, bathing costumes, fire extinguishers; those are the concerns behind these pleasing facades' (Priestley 1937: 3–5).

In these few words, Priestley summarized some of the major ways in which the industrial world of the twentieth century was to differ from its nineteenth-century predecessor. The fact that he was making his journey by road was itself significant; thirty years before he would have had to go by rail. The new factories he observed were the products of the technological changes that had been transforming industrial capitalism since the late nineteenth century, and many had clearly developed to serve an expanding consumer market (see Plates 3.1 and 3.2). And the location of the factories by the new arterial highway leading westwards out of London was the result not only of a revolution in transport and communications but also of the locational freedom deriving from the availability of electricity and other fuels. The textile industries that Priestley remembered from his childhood in Bradford were tied to the coalfields; the newer industries that were appearing in London by the 1930s no longer needed coal, and were much cleaner and brighter in consequence.

Of course, what Priestley saw along the Great West Road was by no means representative of all the technological changes that had been affecting the industrial economies of Britain and other countries for the previous few decades. What had been happening in these countries was that a whole series of new industrial branches had been developing to supplement, and eventually to eclipse, the traditional activities based on coal, iron and textiles. Not all of these were as pleasing to the eye as those observed by Priestley. In the second half of the nineteenth century, for example, the metallurgical industries had been transformed as a result of a series of inventions allowing the production of cheap steel. Next came the rapid development and proliferation of different branches of the chemical industry (alkalis, dyestuffs, pharmaceuticals, explosives, lacquers, photographic plates and film, celluloid, artificial fibres, plastics). The electricity industry, which began to flourish by the end of the nineteenth century, was dependent on earlier inventions, like the steam turbine. About the same time came the rise of the motor industry, which was in turn associated with other industries like oil and rubber. When Priestley set out on his journey, society was already beginning to adjust to the impact of the many new activities catering to the consumer (most notably, domestic appliances – see Plate 3.2) and to new means of transportation (car, bus, aircraft). Of course the full impact of such developments was to come later, after the Second World War, while some technologies, like regular TV broadcasting, the jet engine, nuclear power and the microchip, still lay in the future.

Plate 3.1 The Hoover factory, Perivale, west London. This splendid example of Art Deco architecture, design by Gil Wallace in 1932, reflects the new consumer industries which were being established in the years after the First World War. (Corbis UK/Angelo Homak)

So profound were the technological and accompanying social changes that affected industrial capitalism from the late nineteenth century that some historians have described them as a 'second industrial revolution' (Landes 1969: 4). But it is important to remember that the older industries – coal, textiles, railways, some forms of engineering – did not die immediately or indeed quickly. One of the features of the changing industrial geography of the late nineteenth and early twentieth centuries was that countries like the USA and Germany, whose industrialization came later than Britain's, now began to forge ahead on the basis of the newer industries described above. Britain remained overdependent on the older and less dynamic branches (Figure 3.1).

For the first half of the twentieth century, the industrial changes described above only directly affected certain parts of the world, notably Western Europe, North America, Japan, and by the 1930s the Soviet Union and some other areas. Much of what was later to become known as the Third World, or the developing world, was still agricultural. Yet, in continuation of earlier processes (see Chapter 2,

pp. 46–9) many colonies and other regions were now being organized commercially to supply the industrial countries with raw materials and tropical products – for example, bananas and sugar from Central America and the Caribbean, Brazilian coffee, Indian tea, Malaysian rubber. They were thus being tied in to the capitalist world economy. Gradually certain of these countries began to adopt the technologies of the industrialized world – the rise of the Indian textile industry is one example – but only later in the twentieth century did industrialization become more widespread.

Thus the foundations of what is now known as a global economy were already being laid in Priestley's day, or even earlier. By the beginning of the twentieth century capitalism had become a world phenomenon, tying far-flung countries together by means of international trade and fostering international capital flows through major financial centres like London. The first multi-national corporations were already appearing. All this was aided and abetted by the new systems of communication and transport – telegraph, telephone, radio (from the 1900s), fast steamships,

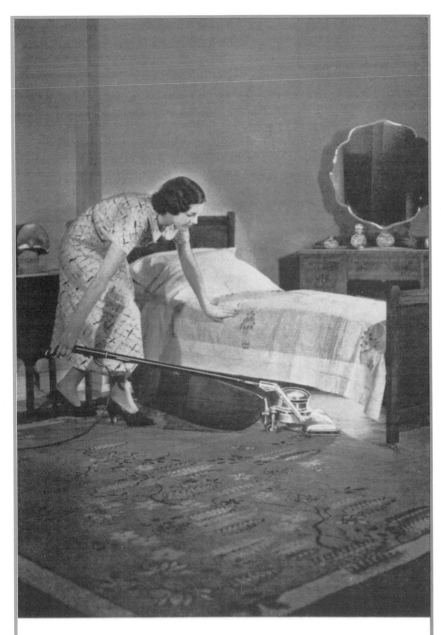

Low furniture does not hinder the Hoover. The handle can be swung down to the ground if necessary.

Plate 3.2 Using a vacuum cleaner or Hoover, one of the new consumer appliances which were becoming widely available by the 1930s. Photos and advertisements of the period almost always depicted women performing such domestic tasks. (Mary Evans Picture Library)

aircraft – which were beginning to span the globe. Of course, none of this bears comparison with the forces of globalization which were to become so significant later in the century. Yet the world was already becoming a smaller place (see Plate 3.8; Case study 3.2). J.B. Priestley himself suggested this when he compared the Great West Road to 'California'. What might this mean to the average English reader in the 1930s? The answer is, a great deal. The Hollywood film industry was in its heyday and the people, homes and landscapes it portrayed were being viewed, and copied, the world over.

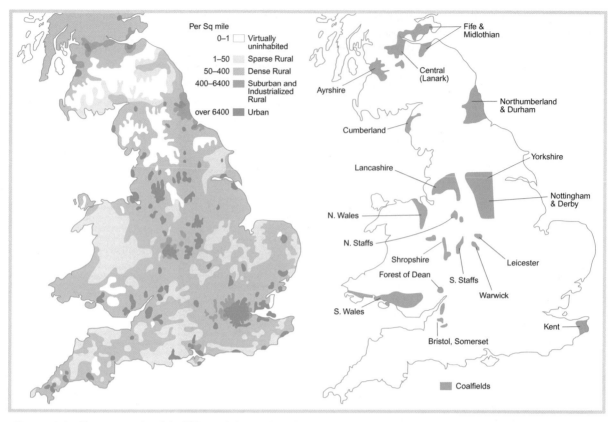

Figure 3.1 The geography of the UK population and coalfields compared. With the significant exception of London, there is a strong correlation between the geography of population and that of coalfields, reflecting both the early stage at which Britain industrialized and urbanized, and the country's long dependence on nineteenth century industries.
Sources: Population, based on Mitchell (1962); coalfields, Stamp and Beaver (1963: 286)

3.2 Organized capitalism

It would be a mistake to suppose that the advance of capitalism in the twentieth century was a story of unmitigated triumph. On the contrary, its fluctuations and misfortunes were such that one historian felt constrained to call the period the 'Age of Extremes' (Hobsbawm 1995). The first half of the century – Hobsbawm's 'Age of Catastrophe' – was particularly disturbed, with two world wars (1914–18, and 1939–45) and a deep world economic depression (1929–33). By contrast, the years between 1945 and 1973 were ones of growing prosperity across much of the globe (Hobsbawm's 'Golden Age') to be followed once more by a disturbed period in the wake of the shock oil price rises of the 1970s.

It was Karl Marx who originally emphasized the unplanned, competitive and even chaotic nature of capitalism's development. But one of the features of the twentieth century has been the attempt, by both

government and private agency, to regulate and even to control it. The reasons for this phenomenon are many, but they are no doubt linked both to the precipitate nature of technological change during this period and to the severe fluctuations mentioned above. Attempts to 'organize' capitalism have taken a number of different forms. For example, already in the late nineteenth century, and especially in the USA and Germany, there were moves towards the formation of inter-firm agreements, cartels and larger companies and corporations. Large corporations could more easily marshal the huge capital resources which modern industry requires, and also influence their markets more effectively. As noted already, the first multinational corporations appeared at this time, those based on the USA being most notable. Nineteenth-century examples include the German electrical firm, Siemens, and the US Singer sewing machine company. Twentieth-century examples include US-based Hoover, Ford, Coca-Cola, Pepsi Cola, Nabisco (Shredded

Wheat), and Kellogg's, all of which invested in Britain in the 1930s. Most of the major international oil firms also date from this period.

A further way in which capitalism became more 'organized' was the phenomenon of mass production, linked especially to the growing consumer market. Here two Americans are regarded as particularly significant. F.W. Taylor (1856–1915) is especially associated with time-and-motion studies, whereby complex tasks on the factory floor could be completed more efficiently and productivity increased. The other is Henry Ford (1863–1947) who organized car production in his Dearborn, Michigan, plant using modern methods like the assembly line and interchangeable parts (see Chapter 14, pp. 320–1). The result of his centralized approach was a significant reduction in the cost of producing cars, which could now be manufactured on a mass basis. The ensuing mass production methods and associated patterns of mass consumption are frequently referred to as Fordism.

Governments were also affected by the desire to 'organize' capitalism and to tackle the many problems to which it seemed to give rise. At the international level, the USA, which emerged after the Second World War as the undisputed leader of world capitalism, took the lead in establishing a series of institutions like the World Bank, the IMF and the General Agreement on Tariffs and Trade (GATT) to ease international monetary payments, promote trade and encourage economic development. At the national level, numerous countries, especially some of the west European ones, pursued democratic agenda of various kinds such as attempts to construct 'welfare states' to tackle such social problems as unemployment, ill health, old age and social inequity. Arguably the experience of government planning and controls in wartime helped pave the way for the optimistic belief in planning and large-scale social engineering that characterized the post-war period. In western Europe especially this was the era of bold experiments in new town and city development, slum clearance and

Plate 3.3   A suburban house built in the Modern style in the 1930s, Essex, England. The suburbanization of the inter-war period was a product of the affluence and increased mobility of some middle-class groups during this period. Suburbanization was to become even more prominent after the Second World War. (Corbis UK/Gillian Darley/Edifice)

Spotlight Box 3.1

Organized capitalism

- Extractive and manufacturing industries are the dominant economic sectors.
- There is an accent on economies of scale, leading to the importance of large industrial plants. Such plants may structure entire regional economies around themselves. Examples might include (in the UK): the West Midlands, based on cars and engineering; Lancashire, based on textiles; the North-East, based on mining, shipbuilding and heavy engineering; and (in the USA): Detroit, based on cars and engineering; and Philadelphia, based on textiles and port-related activities.
- Manufacturing plants are controlled centrally by big industrial corporations – there is an accent on mass production and standardization.

- There is state regulation of the economy to overcome problems generated by the market, for example regional unemployment problems.
- Big industrial cities are the spatial expression of large-scale manufacture.
- There is state-controlled welfare provision to even out social inequalities, address unemployment problems, and raise health standards.
- In culture and social provision, there is an accent on mass provision, for example in housing, consumer goods, TV programming, and newspapers. The emphasis on mass coverage and standardization leaves relatively little choice, reflecting a modernist perspective.

Source: after Lash and Urry (1987)

ambitious social housing schemes, regional planning, and extensive controls over land use (Hall 1996). No doubt the success of these schemes was dependent on the spreading affluence that accompanied Hobsbawm's 'Golden Age'. Across western Europe and North America a tide of suburbanization signalled not only a growing ability to own one's own home in a desirable location but also the availability of the social and physical infrastructure, the private cars and the many new consumer products which now made such a goal possible for many (see Plate 3.3).

The writers Scott Lash and John Urry have described the era ushered in by the methods of Ford and Taylor but reaching its apogee during the period 1945–73 as 'organized capitalism' (Lash and Urry 1987). Some of its principal features (which were especially characteristic of the developed world) are described in accordance with their views in Spotlight box 3.1.

Needless to say, such generalizations would be more or less true, depending on the time and location being considered. Capitalism had different histories in different places, and the exact form it took had much to do with the long-term evolution of each society affected by it.

3.3 Challenges to liberal capitalism

Come, bombs, and blow to smithereens,
Those air-conditioned, bright canteens,
Tinned fruit, tinned meat, tinned milk, tinned beans,
Tinned minds, tinned breath.
Mess up the mess they call a town –
A house for ninety-seven down
And once a week a half-a-crown
For twenty years
From 'Slough' by John Betjeman (1937)

John Betjeman's famous fulminations against the town of Slough, situated just west of London and experiencing developments similar to those observed by J.B. Priestley a few years earlier, are in fact a hymn against modernity. Betjeman was railing against many of the social repercussions of the profound twentieth-century changes noted earlier in this chapter. In this he was by no means alone. Indeed, according to some commentators (e.g. Wiener 1985), Britain has been characterized by an anti-urban and anti-industrial culture that has made it very difficult for the country to adapt to twentieth-century change. However, rather similar problems have also occurred in other countries where modernity (here defined as the spectrum of economic, social, political and cultural changes associated with twentieth-century capitalism) has brought problems of adaptation.

It has already been noted that the twentieth-century cannot be described as a century of uninterrupted progress for capitalism, particularly in its Western 'liberal' form. Capitalism has been subjected to a series of challenges and political struggles that have greatly affected the course of twentieth-century history, and in various parts of the world there have been attempts to create spaces in which alternatives to liberal capitalism can flourish. Interestingly enough, in terms of Wallerstein's world systems theory discussed in Chapter 2 (see pp. 38–9) most of these attempts have been associated with countries outside the core, or with those like Germany after its defeat in the First World War struggling to rejoin the core states. Particularly important for the political geography of the twentieth century were the attempts by Marxists and others to reject the capitalist development model entirely and to reconstruct society on a new basis. This issue will be discussed in the next section. This section will focus on two movements that have also challenged Western-style capitalism in profound ways, but without discarding it entirely: Nazism (National Socialism) with its close relative, Fascism; and the Islamic revival.

Although they had nineteenth-century antecedents, both Fascism and Nazism were essentially products of the inter-war years. Fascism, under its leader Benito Mussolini, ruled in Italy from 1922 until its final defeat in the Second World War in 1945. Nazism under Adolf Hitler ruled Germany from 1933 until it too was defeated in 1945. Various Fascist or neo-Fascist groups ruled or were active elsewhere in Europe, and in some other regions, during this period, and to a lesser extent since.

There is no doubt that Nazism was by far the most influential Fascist movement after 1933, and so the following remarks will be devoted to it (Kershaw 1993). One of the problems of discussing Nazism or other forms of Fascism is their lack of a consistent ideology or philosophy. However, certain general points can be made. Like certain Western intellectuals and others, the Nazis were moved by a dislike of facets of capitalist modernity, such as commercialism, materialism, individualism, threats to the traditional family like the rise of female employment (see Plate 3.4), and similar tendencies that they associated with the 'decadent' Western democracies. They also despised Western-style parliamentary democracy, with its

plurality of political parties and class divisions. In its place, they advocated the concept of a single national community, a *Volksgemeinschaft*, headed by a single Leader or Führer (Hitler), who was regarded as representative of, and chosen by, the people (this was the Nazi concept of 'democracy'). The Leader's power was absolute. Such a creed, however, seemed exceedingly unlikely to come to power in Germany in the early 1930s had it not been for the extreme circumstances reigning there. One was a general sense of resentment at Germany's defeat in the First World War and subsequent national humiliations. As extreme nationalists, the Nazis promised to avenge this defeat. Another was the dire straits to which many of the middle class had been reduced by post-First World War inflation and the Depression that began in 1929. There was also the fear of the many strikes and disorders perpetrated by communists and other left-wing groups (as well as by the Nazis), that were, of course, encouraged by the selfsame economic difficulties.

One of the oddities of Nazism was that it was, at one and the same time, both reactionary and modernizing (Herf 1984). On the one hand, the Nazis looked backwards to an imagined heroic and rustic Germany of the past, to Nordic myths, Germanic landscapes (see Rollins 1995) and happy peasants tilling the fields in traditional costume. They tried to bolster the 'traditional' family: women were to remain at home and raise children for the fatherland. There were even attempts to build villages and garden settlements to reflect such ideals. On the other hand, the Nazis were also modernizers who built the autobahns, fostered industry and spent vast sums on the military. Their aim was to turn Germany into a superpower, able to dominate the European continent and regions beyond (see Chapter 17, pp. 391–5). To this end, the economy was expected to play its full part and, though the Nazis never rejected private capitalism, their economic controls became ever more pervasive, especially with the onset of war in 1939 (many Marxists have argued that Nazism was capitalism's most degenerate form). Similarly, while the Nazis went out of their way to provide for the leisure time and health needs of the German masses, the ultimate goal was to transform them into a race of heroes, ready to fight and die for their Führer.

One of the things that distinguished Nazism and Fascism from other right-wing tendencies was that they were mass movements. The masses were to be organized and captivated by ceremonial and display, by mass rallies and triumphal processions (rather than,

THESE WOMEN ARE DOING THEIR BIT

LEARN TO MAKE MUNITIONS

Plate 3.4 Women workers in armaments plants during the First World War. Although Nazis and some other political groups were later to oppose female employment, the labour shortage during the First World War gave women job opportunities which were increasingly taken as the twentieth century advanced. (Ministry of Munitions)

for example, by rational persuasion) (Mosse 1975; see also Atkinson and Cosgrove 1998). The landscape of Germany was dotted with places where such ceremonials could be performed, the Nuremberg Party Rally grounds being perhaps the most notorious. Had Nazism survived, the whole of central Berlin would have been reconstructed on a grandiose scale to reflect the supposed glories of the Third Reich (Speer 1995:

195–223). Hobsbawm and others have argued that Nazism represented a rejection of the values of the eighteenth-century Enlightenment and the French Revolution, the values upon which modern Western notions of science, rationality and democracy are based. The Nazis wished to replace these values by irrational and Romantic ones, stressing the importance of the will, of faith, even of occult powers. Nazism thus claimed to be a religion in its own right, aiming to compete with and ultimately to replace rival faiths.

Nazism's central and most notorious feature (and one which distinguished it to some extent from Italian Fascism, at least before 1938) was its racism. The German people were imagined to be descendants of an Aryan race, equipped by nature to be rulers of the earth. This belief was related to racist ideas which were all too common in late nineteenth- and early twentieth-century Europe (and which were, no doubt, related in turn to Darwinism). The Nazis took them to their illogical extreme. All non-Aryan peoples were regarded as inferior, particularly the Jews who, because of their culture, religion and cosmopolitan ways, seemed to represent all that the Nazis feared and hated. As time went on, it became clear that the Nazis meant to exterminate the Jews (they killed six million of them) as well as others (gypsies, homosexuals, the insane, certain religious groups) who could have no place in the world they intended to reconstruct. Had Nazi Germany triumphed in the Second World War, it would no doubt have entirely rearranged the ethnic

map of Europe – exterminating some peoples, reducing others to slavery, and resettling yet others. The Nazi contribution to the twentieth century must be regarded as wholly negative and morally repugnant.

Dislike of some of the features associated with Western-style capitalism has also characterized the other movement to be considered here: the Islamic revival. Islam, of course, is quite different from twentieth-century secular (and essentially 'Western') movements like Fascism. Islam is one of the world's ancient religions, with many millions of adherents in a swathe of territory stretching from North Africa, across southern Asia and down to Indonesia, with many others living beyond (Figure 3.2) (Park 1994, Esposito 1995). It has a rich cultural heritage, and indeed for many centuries in the medieval period the Islamic (or Muslim) world easily outstripped Europe in terms of intellectual, scientific and cultural achievement. The Islamic revival is essentially a product of the twentieth century, particularly of its second half, and, in the opinion of many, represents a reaction against the inroads of secularism, Western colonialism, and some forms of

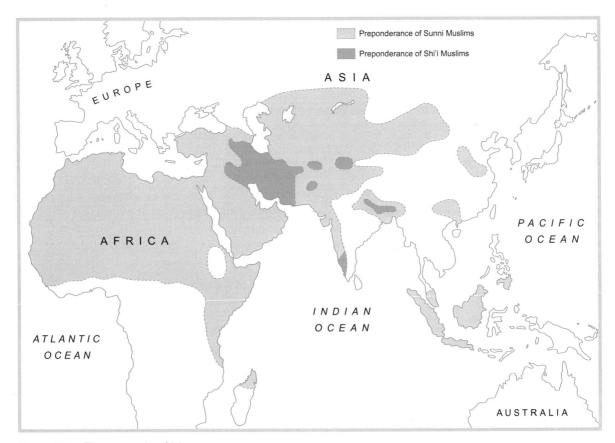

Figure 3.2 The geography of Islam.

Source: Based on *Times Atlas of World History* (1989: 105). © Collins Bartholomew Ltd 1989, reproduced by permission of HarperCollins Publishers

modernization that have affected the Islamic world during this period. Parts of the Muslim world, for example, were colonized by one or more of the European powers several centuries ago, whilst others experienced Western colonialism only during the twentieth century. All such colonies are now politically independent, but the colonial experience naturally had a cultural impact. Furthermore, beginning with Kemal Ataturk's founding of the modern, secular state of Turkey in the early 1920s, many Muslim political leaders have embraced Western-style political, social and economic ideas in the hope of modernizing their countries.

In the opinion of some, the Muslim faith recognizes no divisions between the realms of the sacred and the secular and exhorts its followers to obey its injunctions in every aspect of their lives. The whole of life and society, in other words, is to be regulated in accordance with Islamic teaching (though there is room for debate and disagreement over how far modern social changes can be reconciled with those teachings). Some devout Muslims were particularly offended by the programmes of Westernization and technological development adopted by several Muslim states in the 1960s and 1970s (Scott and Simpson 1991, Park 1994). Secular education, 'immodest' dress and liberal sexual attitudes, the use of alcohol, Western-style entertainment, growing social inequalities and profligate spending by new social elites (offending against Muslim teachings on charity and social justice) have all been subject to criticism. In the minds of many Muslims, such tendencies are associated with the West and its 'degenerate' society. This is hardly surprising in view of the colonial legacy, and the non-core nature of most of the Muslim countries.

Across the Muslim world, there have thus been movements for Islamic revival and renewal, often taking a political form and demanding a return to the original purity of Islam. Sometimes these movements take a particularly militant and anti-Western stance, though this is by no means always the case (Saudi Arabia is an interesting case of a pro-Western state whose society is nevertheless built on strict Islamic principles). Such movements may sometimes threaten existing regimes that are supported by the West. However, the Islamic world is so variable that the political consequences of the revival are very difficult to predict. Perhaps the best known case of a country that experienced an Islamic revolution is Iran, whose previous pro-Western ruler, the Shah, was overthrown in 1979. Here the Islamic clergy took direct control of the government in order to build an Islamic republic

Plate 3.5 Demonstrators carry posters of Ayatollah Khomeini, figurehead of the Iranian Islamic revolution. (Popperfoto)

(see Plate 3.5). This entailed, among other things, the implementation of strict Islamic (Sharia) law that was intended to nullify some of the less acceptable Western cultural influences (Stempel 1981).

One feature of the Islamic revival that has attracted much attention in the West is the veiling of women. This has often been regarded in the West, and in the Islamic world also, as a return to the past when women were exploited and confined to a life of domesticity. An interesting point, however, is that in parts of the Muslim world some educated women have taken the veil voluntarily, seeing it as a protest against the sexual and commercial exploitation of women that is sometimes regarded as a product of Western-style capitalism. Another interesting but less publicized feature of the revival has been the attempt to move to an Islamic system of (zero-interest) banking. This is regarded as not only in accordance with Islamic ethical norms but also much fairer than Western-style banking in spreading risk. It suggests an attempt to construct an economic system based on Islamic rather than Western-style commercial principles.

The Islamic revival can be said to challenge Western-style liberal capitalism both ideologically and geopolitically. Ideologically, it challenges some of the ethical and social values of the West. Geopolitically, it may threaten Western interests in the Middle East and elsewhere in the Islamic world, especially where spaces are created, as in Iran, where Western-style liberal capitalism is unwelcome. Whether a genuine alternative to the latter can be sustained over the long term remains to be seen.

3.4 Communism and the command economy

Perhaps the most significant challenge to capitalism in the twentieth century has come from Marxist-style communism (Calvocoressi 1991). By the 1960s and 70s up to one-third of humanity was living under communist governments which explicitly rejected capitalism as an acceptable way of organizing society. The reasons for that rejection and why it largely failed must now be considered.

An outline of some of the principal features of Marxism is given in Spotlight box 3.2. An important point is that Marx's teachings failed to change those societies at which they were initially aimed – the industrial societies of Western Europe and North America. The point is that Marx himself had expected that communism would find support among the growing industrial working classes of countries like Germany and Britain where factories were bringing such groups together in increasing numbers. However, it was by no means obvious, as the years passed, that the workers of those countries were necessarily being increasingly

exploited, as Marx seems to have expected (later, these countries were accused by Marxists of exporting exploitation to colonies and other less developed regions). Instead, Marxism triumphed in Russia (in 1917), in what was in fact the least industrialized of Europe's great powers. Thus, whereas in terms of Wallerstein's world systems theory Marxism was expected to find favour in the core countries, in fact it initially triumphed in a semi-peripheral one. That it did so changed the character of Marxism, which was now faced with the challenge of building socialism in a peasant society, and in virtual isolation from the rest of the world.

What happened after 1917 in Russia (or the Soviet Union as it was now to be called) was of profound importance for the other countries which later adopted communist systems, if only because Russia was the pioneer. What happened there began to assume something of the character of orthodoxy (Hosking 1992). In view of the difficulties they faced, and in all likelihood because of their own inclinations, the Bolsheviks (as the Russian or Soviet communists were called) adopted a highly centralized political system that brooked no opposition. Eventually, from the late 1920s, they implemented a fully centrally planned economic system. This involved the abolition of private enterprise and virtually all forms of market relations, and the collectivization of agriculture. The purpose of this extraordinary economic system was both to speed the process of economic development and to build up

Spotlight Box 3.2

Marxism

- Marxism, which derives from the teachings of Karl Marx (1818–83), is related to other forms of socialism which seek to moderate or reform the injustices of capitalism.
- Unlike some other forms of socialism, Marxism regards capitalism as an innately unjust and exploitative system.
- According to Marxism, capitalism divides society into two antagonistic classes: those who own the main sources of wealth (the capitalists) and those who must live by selling their labour to the capitalists (the proletariat).

- Marxism teaches that the capitalists are forced by the very nature of capitalism to maximize the profits they obtain at the expense of the workers.
- Marx thought that eventually capitalism would become so exploitative that its downfall was inevitable.
- It would then be replaced by a much more just, classless society (socialism, gradually maturing into communism).
- In the meantime Marx exhorted the world's workers (especially the industrial workers) to organize politically to hasten capitalism's downfall.

the country's military resources. As the 1930s advanced, it became clear that what had earlier seemed a rather vague threat from the outside capitalist world was beginning to take a concrete and menacing form in the guise of Nazi Germany. The centrally planned or command economy (which involved much suffering on the part of the Soviet people) proved equal to this challenge. In the ensuing war with Germany (1941–5), the Soviet Union emerged victorious, but only after sustaining enormous losses.

Soviet victory in the Second World War greatly enhanced that country's prestige. Moreover, because the Soviet armies were now in occupation of much of central and eastern Europe, they were able to ensure that regimes friendly to the Soviet Union (that is, communist regimes) would assume power in those regions. Communism soon spread into other countries, notably China (in 1949), south-east Asia and beyond: in other words, into Wallerstein's 'periphery'. All these countries initially followed the Soviet development model, but soon found that it was necessary to adapt it to their own needs. In the meantime the spread of communism, and Soviet

ambitions, excited the suspicions of the capitalist West. From the 1950s, therefore, the world was split into two armed camps, both equipped with nuclear weapons. The ensuing confrontation, known as the Cold War, profoundly influenced both sides and encouraged their militarization. However, many countries, especially in the developing world, tried to avoid taking sides, while China, though communist, began to pursue its own version of communism outside the Soviet sphere.

While the actual form that communism took in the Soviet Union and other countries may have been a modification of Marx's own ideas, it did represent a radical departure from the capitalist development model. Not only was the command economy a very different, state-centred approach to economic development (and one which was copied in many parts of the developing world after the Second World War, with varying degrees of success), it regarded itself, and was regarded, as a threat to the whole idea of capitalism (Nove 1987). Internally, it led to attempts to reconstruct society on a different basis, particularly in terms of society's spatial structure, but it did in fact give rise to new forms of spatial inequality (Bater 1986). Interestingly enough it

Plate 3.6 Central Moscow in the communist period, showing the open spaces created or enhanced by the Soviet dictator Joseph Stalin for official communist demonstrations and ceremonial display.

also had certain spatial features in common with Nazism and Fascism, such as the emphasis on creating urban spaces specifically for the purpose of mass ceremonial and display (Plate 3.6). The creation of 'spaces of terror', such as concentration camps for the incarceration or elimination of those deemed unacceptable to the regime, was also a feature of the two systems.

In the end, communism failed to prove itself a successful viable challenger to capitalism. Especially from the 1970s, the Soviet Union and its eastern European allies fell behind their capitalist rivals in terms of productivity, flexibility and innovation. Whether this was because of problems inherent to command economies as such, or whether it has more to do with mistakes made by the various political leaderships, is hard to say. Whatever the reasons, by the end of the 1980s practically the whole of the communist world was in a state of economic and political crisis. The subsequent fall of communism in eastern Europe, and the splitting of the Soviet Union into fifteen separate states in 1991, signalled the end of the Cold War and opened up that part of the globe to capitalist penetration (Plate 3.7). Yet the spatial and social consequences of the long years of communist domination are likely to influence the post-socialist states far into the future.

So what was the true character of this most radical challenge to Western-style capitalism? Was it, as many commentators have claimed, merely another form of capitalism that was destined to fail, perhaps because Lenin and the other Soviet leaders should never have tried to build socialism in what at first was still in many ways a peasant society? – or perhaps because socialism is an impossible dream anyway? Or was it a genuine attempt to follow Marx's teachings in difficult circumstances, which was eventually crushed by the forces of international capitalism? Opinions are bound to differ, perhaps above all because the answers we give to such questions will inevitably reflect what we think about the societies in which we ourselves live.

3.5 The end of imperialism?

Father, Mother, and Me
Sister and Auntie say
All the people like us are We,
And everyone else is They.
And They live over the sea,
While We live over the way,
But – would you believe it? – They look upon We
As only a sort of They! –

From 'We and They' by Rudyard Kipling
(1912: 763–4)

Plate 3.7 The fall of the Berlin wall in 1989 signalled the end of communism in the Soviet bloc and the removal of a major twentieth century challenge to world capitalism. (Popperfoto/Reuters)

At the beginning of the twentieth century Britain and several other European powers sat proudly at the centre of a series of empires that spanned the globe (Figure 3.3). As noted in Chapter 2, these empires were the products of a long period of European exploration, settlement, economic exploitation and imperial rivalry. Something of the complacency and condescension with which Europeans commonly regarded their empires at this period is nicely captured by Rudyard Kipling in his comic poem 'We and They', quoted above.

The early years of the twentieth century were not in fact the high points of empire. After the First World War (1914–18), Britain, France and other victorious powers helped themselves to the former German colonies, while Italy invaded the independent African state of Abyssinia (Ethiopia) as late as 1935–6. However, the future of European imperialism was already being questioned even before the war. The English liberal J.A. Hobson, and later the Russian revolutionary V.I. Lenin, popularized the idea that

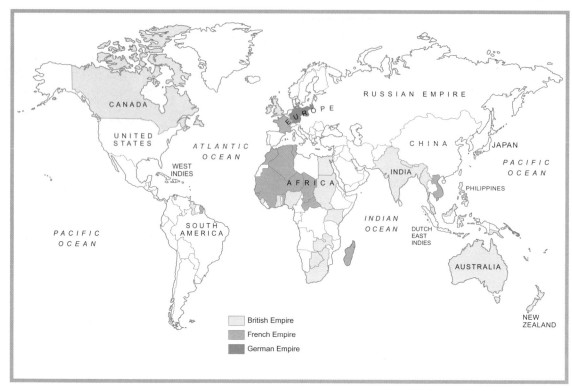

Figure 3.3   The world in 1914 showing British, French and German empires.

Europe's overseas colonies were being economically organized and exploited mainly for the benefit of the European 'mother countries', forming an undeveloped periphery to the European core. Lenin taught that imperialism was an inevitable consequence of capitalism – its 'highest stage'. In the meantime various rumblings of discontent were being felt in various parts of the European empires. Britain, for example, felt constrained to grant home rule to a number of its colonies where white settlement had been significant – Australia, New Zealand, Canada and South Africa. It was also considering doing something similar in Ireland (see Case study 3.1), Egypt and India by 1914. Russia's defeat by Japan in the war of 1904–5, almost the first time that a 'non-white' people had humbled a major European power, suggested that European imperialism might be challenged successfully.

It was, however, the three great episodes of the first half of the twentieth century – the First World War, the Great Depression (1929–33) and the Second World War – that gave conclusive evidence of European weakness and fatally undermined European imperialism. Nationalism, which had had such an impact on the political geography of Europe (see Chapter 19, pp. 424–33), had also influenced the colonial world where 'national liberation movements'

began to demand independence for their countries. Starting with the independence of India in 1947, the next quarter of a century witnessed the break-up of all the European empires (though the Soviet Union, successor to the old Russian empire, finally disappeared only in 1991). Many new, independent states appeared on the map of Africa, Asia and other regions, though not, unfortunately, without considerable turmoil in some cases. The world's political geography was transformed.

In giving (or being forced to give) independence to their colonies, the former imperial powers hoped that they would adopt European-type political systems and capitalist economic systems, partly because these appeared the best basis for future development, and partly because they seemed a reasonable way of upholding European influence. The elites who were now to hold power in the new states were often sympathetic to these aims, since they had frequently been educated by Europeans and wished to see their countries modernized along European or Western lines. However, the adoption of Western models was not always a suitable response to their problems. For example, many of the new states attempted to copy the European idea of the nation-state, hoping to unite their peoples around a common sense of national

Regional Case Study 3.1

Ireland in the twentieth century

Ireland's history in the twentieth century illustrates many of the problems bequeathed by imperialism to newly independent countries.

Although the English had begun to penetrate Ireland as early as the twelfth century, they were never able to assimilate the territory into the English state. Differences between the two countries were exacerbated at the time of the Reformation when England and Scotland became Protestant but most Irish remained Catholic. Sporadic resistance to English/British rule was met by repression. In the nineteenth century the rise of Irish nationalism stimulated several unsuccessful attempts to secure autonomy (home rule). Only after the violence of the Easter Rising in Dublin in 1916 and revolution in 1919–22 did most of Ireland achieve dominion status under the British Commonwealth, becoming the Irish Free State in 1922.

Irish autonomy, however, did not appeal to the Protestants who formed a majority population in the north-eastern part of the island. Most of these people were descendants of settlers who had been deliberately introduced to the region by the English government in the seventeenth century in an attempt to subdue the Irish. Protestants, who had traditionally had a largely privileged status in Ireland, feared incorporation into a Catholic state and demanded continued union with the British crown. As a result of various manoeuvres the northern Protestants were enabled to establish their own statelet of Northern Ireland as a continuing part of the United Kingdom in 1921. However, the six counties of Northern Ireland contained a substantial minority of Catholics, and the partition of Ireland was widely resented by nationalists on both sides of the new border.

After independence governments of the Irish Free State became concerned to secure a greater measure of freedom from Britain and also to assert a post-colonial Irish national identity. Policies such as economic protectionism, attempts to revive the Irish language and Gaelic culture, and imposition of a Catholic moral code and censorship, were thus designed to build a new Ireland, free from the influence of the former imperial master. These policies culminated in Irish neutrality during the Second World War, and declaration of an Irish Republic and secession from the Commonwealth in 1949. However, this also had the effect of strengthening partition, as northern Protestants clung ever more tenaciously to the British connection whilst Northern Irish Catholics often suffered discrimination.

The violence and terrorism that partition has abetted has cast doubt on the extent to which it is a suitable answer to Irish problems. Fortunately recent developments on both sides of the border provide hope of better times in the future. Thus the Irish Republic now seems to be moving away from the narrow nationalism of the past and is evidently seeking a broader and more inclusive identity within the ambit of the European Union and the wider world. Meanwhile Protestant unionists and Catholic nationalists north of the border, divided and segregated for so long, have been encouraged to seek a long-term agreement.

See also Foster (1989)

identity. But the old colonial boundaries had generally been drawn up to suit imperial convenience rather than that of local communities. Thus these boundaries, now the boundaries of independent states, frequently grouped peoples into one state who had no common culture or history while dividing others who did so. This contributed little to the political stability or unity of the new states. Another problem was the feeling among many citizens of the new states that the capitalist economy was responsible for the underdevelopment of their countries. This set off a search for socialist or communist alternatives, much to the annoyance of the West, and encouraged debates about the meaning of 'development' (see Chapter 8, pp. 191–4). Finally, fundamental questions were frequently posed about how far modernization, as generally understood, was compatible with the traditions of the former colonial peoples. The Islamic revival, discussed above, can be seen as one response to this dilemma.

Plate 3.8 Imperial Airways flies to Australia. Imperial Airways was the major British airline company which operated at ever more ambitious international and intercontinental scales during the 1930s. (Vintage Ad Gallery)

Just as imperialism had a profound impact on the political and social geographies of enormous areas in Africa, Asia and the Americas, it equally affected the imperial countries themselves. One of the most important manifestations of this in the twentieth century was the flow of migrants from colonies and former colonies, especially in the tropics and sub-tropics, to take up jobs in the former imperial states. This naturally had a far-reaching cultural impact in cities and regions in western Europe. In fact the whole experience of imperialism led to the mixing of peoples and cultures on a grand scale. It also led to widespread questioning of long-held assumptions about European (and often male) cultural dominance (see Chapter 2, pp. 46–9).

Many scholars have argued that the end of European colonialism did not mean the end of exploitative relationships between the core countries of the world economy (including the former imperial powers) and what was now increasingly referred to as the 'Third World'. According to such thinkers, the formal imperialism of the colonial era had merely been replaced by a more 'informal' alternative, but the basic situation of the core exercising hegemony over the periphery had not really changed (Frank 1969, Wallerstein 1980). From the 1970s, however, some fundamental changes seemed to affect the world economy which, in the opinion of certain scholars, demanded that international relationships be viewed in a new way. Lash and Urry have described these changes as 'the end of organized capitalism' (Lash and Urry 1987).

3.6 The disorganization of capitalism

In 1960 the industrialized areas of Western Europe and North America produced almost 80 per cent of the world's industrial output. Even Japan accounted for only around 4 per cent. Much of the Third World remained agricultural. Only after this time did industrialization spread beyond its traditional centres (which since the 1930s had included the Soviet Union). Meanwhile many of the older industrial countries began to lose industries, even some of those which had arrived with the twentieth century.

Of course, some areas of the 'Third World' benefited far more from this industrial spread than did others. Most spectacularly, the Newly Industrializing

Countries of east Asia, Brazil, Mexico and certain others soon seemed set to join the industrial core. Yet others, like certain Middle Eastern states, earned huge revenues from their energy exports. But there remained many areas which missed out on the new developments (even so the latter part of the twentieth century was a period of unprecedented population growth and urbanization across much of the developing world – see Chapter 9, p. 214, and Chaper 4, pp. 88–93). The term 'Third World', used to group together countries with such disparate economic characters, seemed increasingly redundant, and the world as a whole seemed as unequal as ever.

How is one to explain the changes affecting the world's economic geography in the last third of the twentieth century? Geographers and others argued that such changes are part of the process of globalization. Speedier communications meant that the world was becoming a much smaller place as the twentieth century drew towards its close (Case study 3.2). Capitalism itself was now a truly global phenomenon as markets were internationalized and finance became fully mobile. Before 1960, despite the importance of international trade, the world economy was structured around individual states. After 1960, the world economy became in effect transnational as the boundaries of individual states became ever less important to its functioning. Thus this period witnessed the rise to global importance of the transnational corporations, commercial conglomerates which became major players on a world scale. Because of the wealth and political influence they wielded, such huge companies became increasingly free to switch their operations from country to country as economic circumstances dictated. States, which had previously seemed unchallenged within their own frontiers, found it ever more difficult to control their own economies and began to bid against one another to attract footloose investment and the favours of the transnational corporations. Many industries (and not only industries) now began to locate in parts of the Third World, where costs were cheaper, whilst core industrial countries began to experience deindustrialization and a switch into services and 'control' functions (the headquarters of the transnational corporations still tended to be located in the traditional core countries). A further important result of the development of information technology was that production became much more flexible than before and more geared up to highly specialized markets, changing fashion and the whim of the

Thematic Case Study 3.2

Air travel in the twentieth century

One of the most important agents of globalization has been air travel. In many ways air travel can be said to symbolize the technological achievements of the twentieth century.

While flight by balloon and airship predates the twentieth century, the first successful powered flight is usually taken to be that by the American brothers Wilbur and Orville Wright at Kitty Hawk, North Carolina, in December 1903. Within a few years, in July 1909, Louis Blériot had made the first successful cross-Channel flight between France and England, taking just over 35 minutes to fly the 37 kilometres. The first non-stop transatlantic flight was that by J.W. Alcock and A.W. Brown in June 1919. They took just over 16 hours to fly from St John's, Newfoundland, to Clifden in Ireland. This trip would have taken several days by sea.

In the early days powered flight had a serious competitor in the airship. Zeppelin dirigibles began the first passenger services between German cities in 1910. By the 1930s airships regularly crossed the Atlantic, the ill-fated *Hindenburg* taking some 43 hours at least to complete the trip between Frankfurt and New Jersey. Unfortunately, the hydrogen-filled airships were also very risky and the *Hindenburg's* spectacular destruction at Lakehurst, New Jersey, in May 1937 with the loss of 36 passengers out of 92 largely put an end to this form of passenger travel.

Commercial passenger services by aircraft developed very quickly after the First World War, the regular service between London and Paris inaugurated in 1919 being one of the first. Although flight speeds were slow by today's standards and aircraft range was limited, it soon became possible to cross the world in several days instead of the weeks and months it had taken previously (see Plate 3.8). Thus by the 1930s the Dutch firm KLM was running fortnightly trips between Amsterdam and Jakarta. It took 12 days to complete this journey of 9,500 miles, with 18 stops (including overnight ones). Night flying was dangerous until the development of radar in the Second World War, but overland night flights became possible (as across the USA) with guidance from powerful, land-based beacons. Where water was available, the relatively slow but comfortable flying boats became popular by the late 1930s because of a lack of runways. On the whole, however, air travel remained a luxury enjoyed by the few.

The technological advances of the Second World War were a great boon to air travel. Speeds were greatly increased with the introduction of the jet engine. The first commercial passenger jet was the De Havilland Comet, which began operations in 1952, followed by the Boeing 727 in 1958. Other advances included fully pressurized cabins and advanced instrumentation, allowing aeroplanes to fly higher and faster, the introduction of wide-bodied aircraft from 1970 (the Boeing 747), which carried more passengers, and longer ranges. Air travel thus became possible for many more people and by the 1970s it had replaced sea travel for all but the shortest distances.

International air travel was not just the result of technological advance but also of advances in international co-operation. The International Civil Aviation Authority, established in 1947, sets international standards for safety, reliability and navigation, and facilitates the setting up of routes.

By the 1990s it had become possible to fly by scheduled airliner between Europe and Australia in less than 24 hours with one refuelling stop. The supersonic Concorde flew passengers between London and New York in less than three hours. Quite an advance from Kitty Hawk in under a century!

individual consumer. This new, more flexible approach to production is sometimes known as post-Fordism.

Later chapters of this book will explore some of the economic and social implications of these changes.

However, it is worth stressing here how unsettling such developments were for the late twentieth-century world, and especially for the core countries. The years after 1945 had been ones of unprecedented affluence

Plate 3.9   Deindustrialization in the 1980s afflicted many parts of the world in the wake of the 1970s oil price shocks and of major changes in the world economy. (© Raissa Page)

for these countries as commercial activity boomed, welfare states provided unheard-of security, and leisure opportunities seemed endless. These were the years of rampant suburbanization, ever-rising car ownership levels, international travel and television, as the benefits of mass production truly took effect. They were replaced, from the mid-1970s, by a period of greater uncertainty as deindustrialization set in, unemployment rose, insecurity became more widespread, and social inequality became more apparent. The world was no longer the self-confident, cosy, Eurocentric place it had been at the beginning of the century. Certain writers, notably Francis Fukuyama, celebrated liberal capitalism's victories over several of its twentieth-century rivals, particularly Fascism and communism (Fukuyama 1992). But this was a mood shared by few others, particularly in the wake of the rise of global terrorism towards the century's close. Capitalism had become 'disorganized' and its ability to solve the fundamental problems of our day could still be questioned and challenged as it had been throughout the previous hundred years.

Just as Lash and Urry have used the term 'organized

capitalism' to describe the years down to about 1973 when Fordism reached its apogee, so they have described the last years of the century as those of 'disorganized capitalism'. Some of its more prominent features are listed in Spotlight box 3.3, once again paying particular attention to how the changes have affected the core countries of the world economy.

3.7 Conclusion

The twentieth century was a period of rapid economic and social change over most parts of the world. It was also a period when the world seemed to become smaller and most regions were gradually drawn into an ever more embracing global system. Yet it would be a mistake to imagine that this was an uncontested process, or one that threatens to bring about a global uniformity. The legacy of the twentieth century is a world that is both dynamic and uneven, and therefore very uncertain. The rest of this book tries to grapple with this uncertainty.

Spotlight Box 3.3

Disorganized capitalism

- The onset of 'disorganized capitalism' is marked by a decline in the relative importance of extractive and manufacturing industries.
- There is a relative increase in the importance of service and consumer industries, especially in employment.
- The use of flexible technologies encourages a reduction in the average size of manufacturing plants with more accent on labour-saving investments and more flexible employment processes, all induced by competition.
- Because of the need for flexibility and cost-cutting, industrial firms tend to 'hive off' many of the services and supporting activities they need to other firms and organizations. There are thus more opportunities for small firms, changing the traditionally specialized nature of the regional economy.
- Regional economies are also affected by the greater emphasis on non-standardized production – traditional regional specializations become less marked.
- The global economy reduces the effectiveness of state attempts at economic regulation – from the state's point of view, the economy becomes less predictable.
- Rising costs, demands for reduced taxation, and growing social inequality challenge the idea of a centralized welfare state.
- Smaller, more footloose industries, the rise of services, better communications and other factors reduce the traditional importance of big, industrial cities by comparison with small towns and rural areas.
- There is a rise in importance of the educated social strata needed to work in the new administrative, control, service and related activities – the so-called 'service class' – with more sophisticated and individualized tastes in consumption and other areas. The age of mass cultural provision is replaced by greater cultural fragmentation and pluralism (sometimes referred to as post-modernism). There is a commensurate decline in faith in large-scale planning and similar activities associated with modernism.

Source: after Lash and Urry (1987)

Learning outcomes

Having read this chapter, you should know that:

- Capitalism is inherently dynamic and unstable. It has been so throughout the twentieth century, and is likely to continue to be so in the future.

- The concepts of 'organized' and 'disorganized' capitalism are ways of trying to make sense of the changes which affected capitalist societies during the twentieth century.

- 'Fordism' and 'post-Fordism' (concerning production and consumption) and 'modernism' and 'postmodernism' (concerning culture) similarly try to make sense of twentieth-century change. No concepts, however, can do justice to the complexity of change during this period.

- Western-style liberal capitalism is only one variant of capitalism. It has been challenged in various ways in the twentieth century, some of which have

  had long-term consequences for different parts of the globe. Future challenges might prove more successful than past ones.

- The twentieth century has been an era of nation-states and of nationalism. Towards the end of the twentieth century the role of the nation-state seemed increasingly challenged by globalization. Nationalism, however, may yet flourish as a response to globalization.

- Modernity may not lead to the disappearance of traditional cultural practices. The Islamic revival, and similar religious revivals across the world, are a case in point.

- Despite the view of some that globalization is leading to the emergence of a global culture, it might actually increase the differences between places and hence the importance of geography.

Further reading

Fieldhouse, D.K. (1999) *The West and the Third World*, Blackwell, Oxford. The first attempt to provide a comprehensive survey of the relationship between the West and the Third World, and the debate over its effects, during the twentieth century. Combines theoretical discussion with empirical evidence.

Godlewska, A. and **Smith, N.** (eds) (1994) *Geography and Empire*, Blackwell, Oxford. Discussions of the interplay between the rise of geography as an intellectual discipline and the development of the European empires culminating in the decolonization of the post-war period and the postcolonial experience.

Gwynne, R.N, Klak, T. and **Shaw, D.J.B.** (2003) *Alternative Capitalisms: Geographies of Emerging Regions*, Arnold, London. Examines the effects of globalization and recent economic and political transformations in the world's 'emerging regions', with particular reference to Latin America and the Caribbean, East Central Europe and the former Soviet Union, and East Asia.

Hall, P. (1996) *Cities of Tomorrow,* updated edition, Blackwell, Oxford. A splendid account of urban development in the late nineteenth and twentieth centuries in different parts of the world. The accent is on urban planning and design, but there are many social insights.

Hobsbawm, E. (1995) *Age of Extremes: The Short Twentieth Century 1914–1991,* Abacus, London. What is rapidly becoming a classic account of twentieth-century history, written by a doyen of British Marxist historians. Any chapter is worth reading, but geography students will find those dealing with economic, social and cultural change especially revealing.

Knox, P. and **Agnew, J.** (1998) *The Geography of the World Economy,* 3rd edition, Edward Arnold, London. A survey of geographical change in the world economy which includes an account of twentieth-century developments.

Useful web-sites

www.si.edu/ The Smithsonian Institution, Washington, DC. The Smithsonian is a focus for many kinds of scientific and cultural endeavour in the United States. The web-site is a very useful source for twentieth-century history and developments, with particular emphasis on the United States.

www.besthistorysites.net/20thCentury.shtml Entitled 'Best of History Websites', a comprehensive guide to history-oriented resources online. For teachers, students and others.

For annotated, clickable weblinks and useful tutorials full of practical advice on how to improve your study skills, visit this book's website at www.pearsoned.co.uk/daniels

Population, resources, development and the environment

EDITED BY MICHAEL BRADSHAW

The issues of population growth, resource depletion, food supply and environmental degradation have come to dominate the policy agendas of the international agencies. The eight 'Millennium Goals' to be achieved by all 191 United Nations member states are: the eradication of extreme poverty and hunger; the achievement of universal primary education; the promotion of gender equality and the empowerment of women; the reduction of child mortality; the improvement of maternal health; the combat of HIV/AIDs, malaria and other diseases; the pursuit of environmental stability; and the creation of a global partnership for development (www.un.org/millenniumgoals). Most, if not all, of these issues are examined in this section. The Millennium Goals are a response to a recent world history that has widened the gap between the rich and the poor states. In the 1970s the political manipulation of energy supplies and subsequent recession in Europe, North America and Japan aggravated the economic problems of the so-called developing world. In the 1980s famine struck the

Sahel region of Africa and concerns were raised about the ability of the world's poor states to feed themselves. In the 1990s the mounting evidence that humankind was damaging the earth's ecosystems finally resulted in collective action to reduce greenhouse gases. Unfortunately, despite the proclamations made at the UN, events so far in the first decade of the twenty-first century have continued to compound an already difficult situation. The events of 11 September 2001 and the subsequent US-led invasion of Iraq have resulted in a new phase of economic recession and high energy prices. The plight of Africa, described as the 'Lost Continent', is now recognized, but limited action has so far been taken to address the situation. The 'Kyoto Protocols' that set targets for the reduction of greenhouse gases have yet to be ratified. Clearly, it is one thing to recognize that at the turn of the millennium the world faces a number of major problems, but quite another to bring about the collective action required to address them, especially when some of the world's richest states seem little interested in the plight of those less well off than themselves.

The five chapters in this section examine the interrelationships between population dynamics (Chapter 4), resource production and consumption, (Chapter 5), the supply of food (Chapter 6), environmental degradation (Chapter 7) and development (Chapter 8). They all adopt a global perspective, global in the sense that the processes producing the geographies of population change; energy and food production, habitat destruction, and, ultimately, development are shown to operate and interact at a number of different scales. They stress the importance of the 'history' of human development (discussed in the previous section) in shaping the current geographies of population distribution, resource and food production and consumption, environmental degradation and economic development. Thus, the division of the world between 'developed' and 'developing' countries, or alternatively the 'North' and the 'South', is understood as the product of historical processes that have impacted on different regions in different ways at different times. Such a crude twofold division of the world hides a set of complex and contested issues; the notion of the developing world is defined in terms of its relationship with the developed world.

In assessing the prospects for the future, all of the chapters question the practice of seeing the experience of Europe as a universal model for the evolution of societies elsewhere. Concepts such as the demographic transition and modernization suggest a predetermined progression through a set of stages that describe the European experience. If the world is to manage the problems currently posed by the conflicting geographies of population, resources, development and the environment, it must accept that the European experience does not constitute a universal development process. Furthermore, current dynamics indicate that it will be very difficult to maintain the living standards of the developed world *and* meet the increasing needs of the developing world, unless there are radical changes in the relationships between population, resources, development and environment.

CHAPTER 4

Demographic transformations

TONY CHAMPION

Topics covered

- Scale and distribution of world population growth
- Transitions taking place in life expectancy and fertility
- Developments in international migration
- Population ageing

Though it is theoretically possible for a population to retain a particular size and structure, in practice most populations have been changing substantially over recent decades and seem set to carry on this way for several decades more. This is especially the case at a global scale, now that the total population has surged past the six-billion mark, but it is also true for all major world regions and indeed for most countries around the world. Nor is it just a matter of population size, for also evolving rapidly are age distribution, social structure, ethnic composition, the way people group into households and the life-stage experiences people go through. People born in the first years of the twenty-first century are being exposed to life-chances and behaviour norms that are very different from those of a generation ago, let alone those prevailing in the early years of the twentieth century. Similarly, it is pretty clear that, demographically speaking, the world will be a very different place in another 25 years' time. It is therefore enormously challenging to speculate on the way things might be by the year 2050, let alone the end of the century. In this chapter, the focus is on the scale and distribution of world population growth, the transitions taking place in life expectancy and fertility, the developments in international migration, and the single most important element of demographic restructuring, namely population ageing.

4.1 World population growth

The challenge presented by these transformations can most readily be illustrated by reference to total population numbers. While it is not easy to provide accurate statistics on the world's population for any given year (even with censuses taking place in most countries), we can say for certain that the number of people alive in the world today is greater than it has ever been. It is also undeniable that its growth over the past half-century has been larger than at any previous time. The latter is the case whether considering absolute increments or percentage rates of increase.

Thematic Case Study 4.1

Population 'doubling time'

This refers to the period of time that it would take a population to double in size, given a particular rate of annual growth. The important point to remember is that, as Malthus recognized in his classic (1798) treatise on population, change takes place in the form of a geometric function or – in everyday parlance – at a compound rate, just as money accrues in an investment account if the interest is not paid out. As a result, a growth rate of 1 per cent a year will lead to a doubling of population in roughly 70 years rather than 100. Doubling times can be approximated by dividing 70 by the growth rate, so that at 2 per cent it will take only about 35 years for the population to double, at 3 per cent about 23 years, and so on. The fact that world population doubled from 3 to 6 billion between 1960 and 1999 means that its annual rate of increase over this period averaged just under 2 per cent.

Normally, doubling times are calculated with respect to the rate of natural increase without considering the effect of migration. This is partly because the concept is often used in relation to global population growth and partly because in the past population projections for major regions and countries have been based very largely, if not entirely, on trends in births and deaths. This approach is probably still valid when talking about major regions and where natural increase rates remain very high; for instance, for Africa as a whole, where the late-1990s rate of natural increase is equivalent to a doubling time of barely 27 years. It makes less sense for an individual country like Oman, where the current doubling time is currently just 18 years but where actual population change may be significantly affected by migration. Similarly, it is less relevant in relation to likely population trends for countries with very low rates of natural change; for instance, for the UK where a doubling would take 433 years at the rate prevailing in the mid-1990s but where natural increase was then responsible for only half of overall growth.

Sources: Daugherty and Kammeyer (1995), PRB (2003)

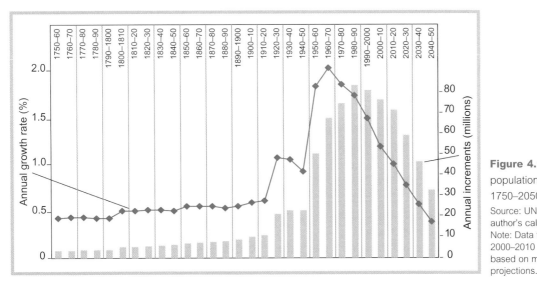

Figure 4.1 World population growth, 1750–2050.
Source: UN (2003) and author's calculations. Note: Data from 2000–2010 onwards are based on medium-variant projections.

The total population of the world is estimated to have passed 6 billion during 1999, representing the latest stage in a relatively short period of generally accelerating absolute growth (Gelbard *et al.* 1999, National Research Council 2000). It was not until the beginning of the nineteenth century that world population attained its first billion (around 1804). Then, while it took some 123 years to add a second billion (1927), the third took only another 33 years, the fourth only 15 years, and the fifth and sixth only 12 years each. Looking at it another way, the first millennium saw virtually no net increase from the 300 million people estimated to have been alive around the time of Christ's birth, whereas the second millennium has witnessed a 20-fold growth of population. Perhaps most dramatically, while the first 1900 years from Christ's birth added 1.3 billion people, the last 100 years have added 4.4 billion.

Moreover, the pace of increase accelerated greatly as the twentieth century proceeded. Whereas world population growth averaged barely 0.5 per cent a year during the first two decades, the rate rose to around 1.0 per cent a year during the 1920s, 1930s and 1940s. It then leapt to around 1.8 per cent in the 1950s and reached 2.0 per cent in the 1960s. Given the 'compound' nature of this growth, the latter figure represents a population 'doubling time' of only 35 years (Case study 4.1). In absolute terms, the annual increment was running at barely 10 million at the start of the century but rose to 20 million in the 1920s, 50 million in the 1950s and almost 83 million during the 1980s (Figure 4.1).

By the close of the century, however, there is plenty of evidence that the rate of world population growth has peaked and is now slowing, though the absolute volume of increase remains very high by long-term standards. The annual rate of increase fell back somewhat during the 1970s and, despite then sticking at around 1.7 per cent for a decade or more, it is reckoned to have fallen to around 1.2 per cent now. This latest reduction in rate has been large enough to offset the effect of the growing population base, so that the size of the annual increments has fallen somewhat in recent years – down to an estimated 77 million a year (Figure 4.1).

These latest trends are expected to continue. According to the 2002-based projections prepared by the United Nations (UN 2003), the growth rate is likely to fall below 1.0 per cent during 2015–20 (almost exactly a hundred years since it first exceeded this rate). It is expected not to be much over 0.3 per cent by mid-century (a level not seen since around 1750). Because of the growing population base, the annual increments seem destined to stay above 70 million till around 2020, but are then expected to more than halve by 2045–50. On this basis, world population would reach 8.9 billion by 2050, almost 3 billion larger than in 2000 but with the majority of this addition occurring in the first quarter of the century and bringing up the 8-billion mark by 2028.

However, if there is one thing that demographers have learnt over the last few decades, it is that projections are never correct and indeed sometimes prove to be wildly inaccurate. Adopting a range of plausible alternative assumptions on fertility, therefore, the UN (2003) demonstrates that world population might be only 7.4 billion in 2050 and already contracting, but equally it could have risen to 10.6 billion, getting on for double the current size. It takes only relatively small changes in basic assumptions to produce very substantial differences over a period of decades.

This review of global trends suggests two major challenges for the new millennium. One is the very practical one of coping with the expected continuation of large-scale growth for the next few decades, further increasing the pressures and problems experienced since 1960 as the world added its latest three billion people. This takes on particular salience when it is recognized that, even more than in the past, the growth will be concentrated in those parts of the world that are economically the least equipped to cope. The other is the challenge posed to people involved in population studies as they try to understand better the factors affecting

population trends and people's behaviour and thereby establish more firmly the likely future course of events.

4.2 Differential population change

The most important feature of the unevenness of population growth across the world is the disproportionate concentration of growth in Less Developed Regions (LDRs). Over the past half-century, these have accounted for almost 90 per cent of the world's 3.8 billion extra people, with their contribution rising steadily to reach 96 per cent by 2000–2005. The UN projections suggest that between 2003 and 2050 the LDRs will grow by 2.6 billion and thereby account for all but 0.6 per cent of world population growth because the population of the More Developed

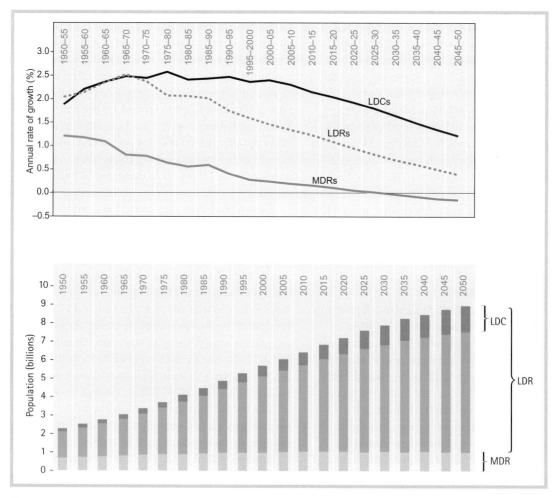

Figure 4.2 Population growth, 1950–2050, for development categories.
Source: Data from UN (2003). Note: Rates of growth are annual average rates for 5-year periods, 1950–55 to 2045–50, estimated up to 1995–2000 and medium variant projections thereafter. The Least Developed Countries (LDCs) are a subset of Less Developed Regions (LDRs).

Regions (MDRs) will be only 16 million higher then than now (Figure 4.2 and Table 4.1).

Within the LDRs, it is the 49 least developed countries that, in aggregate, are now witnessing the greatest upward surge in population (Figure 4.2). Though containing only 8 per cent of world population in 1950, these have contributed almost 14 per cent of world population growth in the past half-century. This proportion is expected to rise to over a third (36.5 per cent) between now and 2050, with their absolute volume of population increase almost doubling to 20 million a year (Table 4.1).

This unevenness in growth rates is producing a substantial shift in the distribution of population (Table 4.1). Whereas the MDRs accounted for almost one-third of all people in 1950 (32.3 per cent), the proportion was under one-fifth by 2003 (19.1 per cent) and seems destined to fall to under one-seventh (13.7 per cent) by 2050. Meanwhile, the share of the 49 least developed countries, while rising relatively slowly between 1950 and 2003, is projected to increase from 10.4 per cent in 2003 to 18.8 per cent in 2050.

This imbalance is also seen clearly in the changes in distribution between the major areas of the world (Table 4.1). Particularly symbolic is Africa's surpassing of Europe's total population size in 1997, in marked contrast to the situation in 1950 when the latter was more than twice the size of the former. By 2050 Europe's share is likely to have fallen to 7 per cent, while Africa's is expected to be nearly three times this. Over the period 2003–2050 Africa's population is projected to grow by just under one billion, representing over one-third of global growth (36.4 per cent) compared to only one-sixth (16.6 per cent) for 1950–2003. Meanwhile, Asia seems destined to remain the main contributor in absolute terms, adding a further 1.4 billion up to 2050, but this is a marked reduction in the pace of growth compared to its 2.4 billion gain in 1950–2003.

At individual country level, China and India continue to dominate, together accounting for over one-third of world population. The number of large-population countries is, however, growing. In 1950 only eight countries (on the basis of 2003 boundaries) contained over 50 million people, roughly the size of

Table 4.1 Population growth, 1950–2050, for development categories and major areas

Category and major area	Population (millions)			Share of world population (%)			Change (millions)		Share of world change (%)	
	1950	2003	2050	1950	2003	2050	1950–2003	2003–2050	1950–2003	2003–2050
World	2519	6302	8919	100.0	100.0	100.0	3783	2617	100.0	100.0
MDR	813	1203	1220	32.3	19.1	13.7	391	16	10.3	0.6
LDR	1706	5098	7699	67.7	80.9	86.3	3392	2601	89.7	99.4
LDC (within LDR)	*200*	*718*	*1675*	*8.0*	*11.4*	*18.8*	*518*	*956*	*13.7*	*36.5*
Africa	221	851	1803	8.8	13.5	20.2	629	953	16.6	36.4
Asia	1399	3823	5222	55.5	60.7	58.6	2425	1399	64.1	53.4
Europe	547	728	632	21.7	11.5	7.1	180	−96	4.8	−3.7
L. America	167	528	768	6.6	8.4	8.6	361	240	9.5	9.2
N. America	172	319	448	6.8	5.1	5.0	148	129	3.9	4.9
Oceania	13	31	46	0.5	0.5	0.5	19	14	0.5	0.6

Note: data for 2050 are medium-variant projections. MDR More Developed Regions, LDR Less Developed Regions, LDC Least Developed Countries (subset of LDR), L. America Latin America and the Caribbean, N. America Northern America. Source: UN (2003) and author's calculations

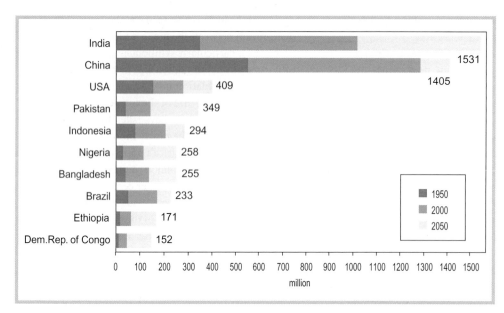

Figure 4.3 The ten countries projected to have the largest populations in 2050, and growth 1950–2050.
Source: Data from UN (2003). Note: 2050 data are medium-variant projections.

Figure 4.4 Annual average rate (%) of population change, 2000–2005.
Source: Data from UN (2003)

the UK at that time. By 2003 this figure had grown to 22 and is projected to rise to 34 by 2050, with the UK dropping to twenty-sixth in the ranking. By that time, India is likely to have overtaken China, with over 1.5 billion people. Figure 4.3 shows what are expected to be the ten largest countries in 2050 and their scale of growth since 1950. One conspicuous feature of the recent past is the rapid growth of countries with a significant Muslim presence, with some of the highest growth rates occurring in the Arabian sub-continent and with the world's main zone of high population growth rates being dominated by the swathe of Islamic countries that stretches from Pakistan westwards to Africa's Atlantic coast (Figure 4.4). This is a trend that seems likely to become even more marked over the next few decades (see Case study 4.2).

Thematic Case Study 4.2

Population growth in the Islamic world

According to Weeks (1988), 'Islamic nations are the world's fastest growing population group'. Only a small part of this growth is coming about through conversions from other faiths. The vast majority is due to the high levels of natural increase, resulting from high birth rate and youthful age structure. Data for 1988 put the fertility of Islamic nations at 6.0 children per woman, a third as much again as the 4.5 figure for other developing nations, while the proportion of people aged under 15 is put at 43 per cent.

Analysis of the latest UN population data indicates the scale of growth over the past half century and how things are expected to change over the next five decades. These data are for the 40 countries where Muslims made up at least half the population in 1988, thus including those of other religions living there but excluding significant numbers of Muslims in other countries like Nigeria, Ethiopia, India, China and former USSR. The population of these 40 countries in 1950 was 302 million, but had more than trebled to 952 million in 1998 and is expected to reach 1,869 million by 2050. The latter increase represents almost one-third of projected world population growth in 1998–2050, compared with their contribution of less than one-fifth in 1950–98. By 2050 these 40 countries would account for 21 per cent of world population, up from 12 per cent in 1950 and 16 per cent in 1998.

These projections, however, assume that the Islamic world moves to replacement-rate fertility (around 2.1 births per woman) before 2050. It is currently very difficult to know whether this is going to be possible. Given that many of these 40 countries are amongst the least developed in the world, much depends on whether they will come to enjoy the economic development normally associated with the

Plate 4.1 The Muslim population is now growing especially rapidly. Muslims are recording more rapid population growth than any other major group in the world, with their high average birth rates producing big school-age populations and the basis for strong future growth. (Panos Pictures/Neil Cooper)

4.3 The progress of the demographic transition

For decades it has been customary to view trends in population growth rates in terms of the **demographic transition** model. The origins and nature of the demographic transition model are outlined in Case study 4.3 and Figure 4.5. While it is an essentially descriptive rather than analytical tool, it provides a simple way of summarizing the state of demographic development reached across the globe. Here, this is done by adopting Chung's (1970) approach to studying the transition's 'diffusion' (see Case study 4.3).

Figure 4.6 presents the 'vital trends' location of the world's major areas and regions in the form of a graph that also shows their relative position within the stages. At this level of geography, it can be seen that by the end of the twentieth century virtually the whole world had reached the third and final stage of the transition process. Only Eastern and Middle Africa are clearly in the first stage with a crude death rate of over 15 per 1,000 and a birth rate well in excess of 30 per 1,000. Southern Africa is also classed as Stage 1, because even though its birth rate is below 30 per thousand, its death rate is above 15. Western Africa, with its crude death rate of 14.9 per thousand, lies just within the second stage, while Other Oceania (Melanesia,

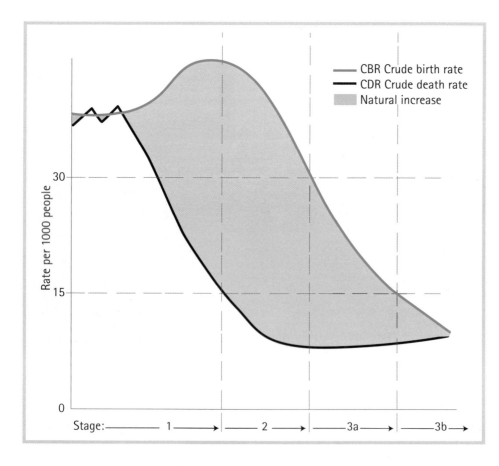

Figure 4.5 The demographic transition model.

Source: Modified from Chung R. Space–time diffusion of the demographic transition model: the twentieth century patterns, in Demko G.J., Ruse H.M. and Schnell G.A. (eds) *Population Geography: A Reader,* 1970, McGraw-Hill. © 1970 The McGraw-Hill Companies Inc. Reprinted by permission of The McGraw-Hill Companies Inc.

Thematic Case Study 4.3

The demographic transition model

The 'demographic transition' concept has its origins in the work of the American demographer Warren Thompson, who in a paper in 1929 identified three broad groups of countries on the basis of birth and death rates. Countries of Western Europe and North America, which he termed Group A, were characterized by death rates that had fallen steadily over the previous few decades and by birth rates that were moving down rapidly towards the level of the death rates, indicating a future of very low population growth. Group B countries, found mostly in eastern and southern Europe at that time, had death rates that were falling quite quickly but birth rates that remained high or were declining only slightly, with the result that their populations were growing rapidly. Finally, Group C countries, accounting for most of Africa, Asia and Latin America, had high rates of both births and deaths and thus quite low overall growth.

Subsequently, mainly through the work of Notestein, these three groups of countries came to be seen as three different stages in a demographic transition. As evidence accumulated on the number of countries that appeared to be following this progression, the model was adopted by Western capitalist agencies as part of their Rostow-based 'stages of economic growth' strategy for Third World development. An early test of this approach was made by Chung (1970), who mapped the diffusion of the demographic transition across the globe from the early years of the century.

Chung's work was based on a very simple framework in which the two criteria for classifying countries were a death rate of 15 per 1000 and a birth rate of 30. His framework of three stages is shown in Figure 4.5, but modified by subdividing the third stage to reflect the substantial post-1960s reduction in birth rates in most MDRs. This forms the basis of an analysis of the situation reached by the first years of the twenty-first century (see Figures 4.6 and 4.7, together with the text commentary).

While the model is still useful as a descriptive device, it has to be admitted that it lacks explanatory power, as it does not explicitly incorporate any causal factors nor even any direct linkage between trends in mortality and fertility. Moreover, its relevance as a descriptive tool is now much weaker. For one thing, a majority of the world's population is now living in countries that have passed through the transition or are at a late stage of it. The model says very little about natural increase in a post-transitional situation (but see Case study 4.5 on p. 102 for a discussion of the 'second demographic transition').

Another criticism is that the model ignores the migration component of population change. According to Zelinsky (1971), it should from the outset have been called the 'vital transition' because it deals only with the vital events of birth and death. International migration has been an extremely important component of population change at particular times and is once again having a substantial impact on many parts of the world.

Sources: Chung (1970), Jones (1990), Daugherty and Kammeyer (1995)

Micronesia and Polynesia combined) is on the verge of moving into the third stage.

At the same time, however, the regions now in the third stage exhibit a considerable degree of diversity (Figure 4.6). While their death rates are closely aligned around 6–8 per 1,000 (except for Europe above this

and Central America just below it), there is a much broader spread of crude birth rates. Some regions lie within 6 points of the 30 per 1,000 threshold, notably Northern Africa, Western and South-central Asia, and Central America. By contrast, Europe's four regions average under 10 per 1,000, while North America,

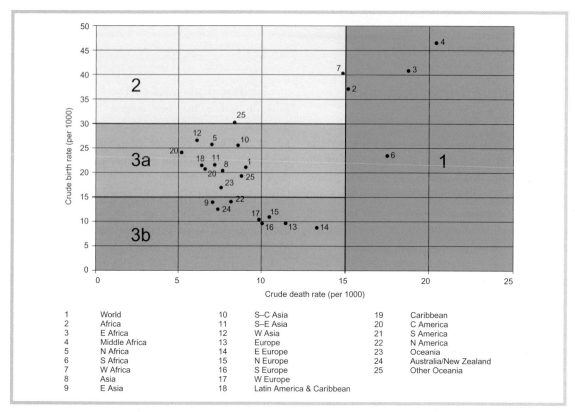

Figure 4.6 Crude death and birth rates, 2000–2005, for major regions and areas.

Source: Data from UN (2003). Note: Shaded panels refer to stages of the demographic transition model (see Case study 4.3 and Figure 4.5)

Figure 4.7 Countries classified on the basis of stage reached in the demographic transition by 2000–2005.

Source: Data from UN (2003)

Australia/New Zealand and East Asia are also relatively low at 13–14. As mentioned in Case study 4.3, Chung's criteria do not specifically indicate what constitutes a post-transitional situation, but on the basis of their distinctive birth rates these seven regions would seem to qualify for this description.

At the level of individual countries, a much more varied pattern is evident (Figure 4.7). In particular, at this scale there are found to be a substantial number of LDR countries placed in Stage 2 on the basis of the criteria used here. These include ten in Asia, four in Latin America and three in Oceania. Africa, too, contains a fair number of the countries with death rates of below 15 per 1,000 but birth rates of 30 or more. Some of these Stage 2 countries can be found within wider regions that remain in Stage 1; namely, Comoros, Eritrea and Madagascar in Eastern Africa and Gabon and São Tomé in Middle Africa. Even so, sub-Saharan Africa is dominated by countries that are still in the first stage, and the only countries there that have moved into the third stage are the small-island territories of Cape Verde, Mauritius and Réunion. The Republic of South Africa, which is also Stage 3a in terms of birth rate, has to be treated as Stage 1 because its death rate has recently moved back up above 15 per thousand – primarily due to the HIV/AIDS epidemic (see the next section). By contrast, Stage 3a dominates North Africa, Latin America and Asia. Clearly, outside sub-Saharan Africa and the Arabian sub-continent, the vast majority of countries have now reached the final stage of the transition (Stage 3a) or moved beyond it (Stage 3b).

4.4 Mortality and life expectancy

During the past half-century enormous strides have been made in reducing mortality and increasing life expectancy. This is the case even in those countries where death rates were already low in 1950. Yet there remains considerable scope for further improvement and new concerns have appeared in recent years. A key question is how quickly the parts of the world where death rates are still high, largely due to the persistence of infectious diseases, will be able to move through the epidemiological transition to a situation of low mortality rates linked mainly to the diseases of old age.

Across the world as whole, crude death rates (deaths per 1,000 people) have more than halved during the past 50 years – from 19.8 in 1950–5 to 9.1 in 2000–5. Even for the least developed countries, rates have

almost halved over this period, dropping from 27.9 to 15.1, the latter being only marginally above Chung's criterion for moving out of the first stage of the demographic transition (see Case study 4.3). The most impressive change has occurred in the more developed parts of the LDRs, because the LDRs in aggregate have seen their death rate plummet by two-thirds, ending up at only 8.8 per 1,000, lower than the average for the MDRs. These trends in crude death rate are, however, partly influenced by differences in age structure. The LDRs' current low rates are helped by their large numbers of younger people who are generally at least risk of dying, while the MDRs' rates are now rising because of their increasing proportions of older people (see below).

In terms of life expectancy, a measure which is not affected by age structure and provides a direct indicator of life-chances, it is found that even the MDRs are continuing to experience gains, though the rate of improvement appears to be slowing (Figure 4.8). Between 1950–5 and 1975–80 life expectancy there rose from 66.1 to 72.3 years for both sexes combined – an increase of one year for every four years of elapsed time. In the quarter of a century since then, however, the rate of progress has slowed, moving up by only 3.5 years of extra life. Part of the explanation for this slowdown lies in the situation in Eastern Europe, which not only emerged from the Communist era with below-average life-chances, but has been hit badly by the effects of the subsequent economic liberalization on employment opportunities and access to health care. Most alarmingly, the Russian Federation's life expectancy for men had slumped to 60.2 by the late 1990s, down from 63.8 ten years earlier. The figure for Eastern Europe as a whole fell from 65.2 in 1985–90 to 63.0 in 1995-2000 – very much lower than the 74 years for each of the other three European regions. More generally, in many MDR countries, the social differential in life expectancy appears to be widening.

For the LDRs, the improvements in life expectancy have been considerably more marked, but here too the pace has been slackening (Figure 4.8). In the 25 years from 1950–5, life expectancy rocketed from 41 to almost 57 years, but over the next quarter of a century average life expectancy advanced by a further 7 years only. The earlier progress can be attributed to the rapid reduction in malaria, principally due to the use of techniques developed to protect soldiers in the early 1940s fighting in south-east Asia, and to sustained programmes of curbing infectious and water-borne diseases like smallpox, tuberculosis and cholera. Within

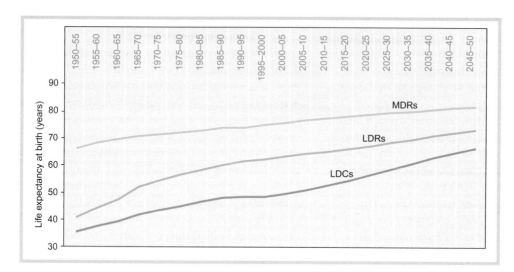

Figure 4.8 Life expectancy at birth (both sexes), 1950–2050, for development categories.

Source: Data from UN (2003). Note: Data from 2000 are projections. Least Developed Countries are a subset of Less Developed Regions.

the space of 20 years, thanks to the importing of mortality control measures from the MDRs, many countries were able to pass through an epidemiological transition that had taken over a century for many European countries. Even so, life expectancy remains stubbornly low in the least developed countries in general and in sub-Saharan Africa in particular.

Thematic Case Study 4.4

The impact of HIV/AIDS

The AIDS virus, unknown to the medical world 30 years ago, has become a major cause of death and is one of the reasons why the UN has recently reduced its projections of future world population numbers. While its impact in the MDCs has been restricted largely to intravenous drug users, recipients of contaminated blood products and gay men, in parts of Africa and Asia its toll is much wider. According to UN estimates, by 1998 some 14 million people had already died from AIDS and a further 30 million were infected by HIV, and by 2002 the toll is reckoned to have risen to 22 million deaths and a further 42 million living with the infection. The 2002-based projections indicate a total death total of 278 million for 2000–50, even assuming a decline in infection rates after 2010 due to changes in sexual behaviour. Altogether, allowing also for babies not born because of the deaths of their potential parents, the world's population in 2050 is expected to be nearly half a billion (479 million) lower than it would have been in the absence of this 'deadliest epidemic of our time'.

The impact of HIV/AIDS on mortality and population growth is, however, so far very uneven across the world. Of the 53 countries classified as 'hardest hit' by the UN (2003), almost three-quarters – 38 – are in sub-Saharan Africa, compared to eight in Latin America and five in Asia, and with the USA and Russian Federation being amongst those included because of having at least one million people infected. The 38 highly-affected African countries are projected to have a 2015 population 91 million, or 10 per cent, lower than it would have been without AIDS. A 19 per cent shortfall in 2015 is projected for the seven countries where HIV prevalence rates are already at or above 20 per cent, namely Botswana, Lesotho, Namibia, South Africa, Swaziland, Zambia and Zimbabwe.

The hardest-hit country at present is Botswana, where one in every three adults is now infected by HIV. Life expectancy there is already below 40 years, down from 65 years in 1990–5, and is expected to fall to just 31.6 years by 2010–15, representing an average loss per person of almost 40 years of life compared to the 70.7 years expected then in the absence of AIDS-

Thematic Case Study 4.4 continued

Plate 4.2 The AIDS epidemic continues to spread in Africa. While AIDS is now a worldwide disease, its greatest incidence is in Africa, where it is already a major cause of death and is bound to affect fertility behaviour in the longer term. (Panos Pictures/Crispin Hughes)

related deaths. Similarly, in Zimbabwe one out of every three adults is already infected and life expectancy will likely be down to 31.8 years by 2010–15, again nearly 40 years below the without-AIDS level. Despite the assumed decline in HIV prevalence after 2010, life expectancy there will still be below 46 years by 2045–50, more than 30 years lower than without AIDS. In South Africa, the epidemic started later and so the main impact is still to come, with projections for its population in 2050 being 9 per cent lower than in 2000 and fully 44 per cent lower than it would have been without AIDS.

Sources: Löytönen (1995), Bongaarts (1996), UN (2003)

Two main challenges face those countries that currently have the lowest life expectancies. One is relatively new, but already appears to be having a significant effect on population growth rates in many countries, notably in Africa: the HIV/AIDS epidemic (see Case study 4.4). The other is infant and childhood mortality. Whereas in the MDRs around four-fifths of deaths are now accounted for by people aged 60 and over, the proportion is only two-fifths in the LDRs. The contrast is even greater for 0–5-year-olds, who make up under 2 per cent of all deaths in the MDRs, over a quarter in the LDRs as a whole and two-fifths in Africa. The main reason for this difference is in the numbers of babies that fail to reach their first birthday. Whereas the infant mortality rate (IMR) is now under 8 in North America, most of

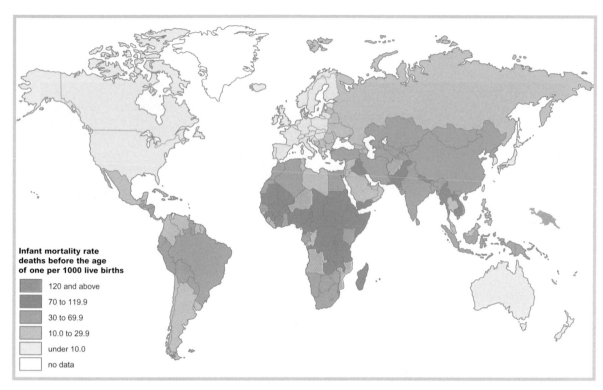

Figure 4.9 Infant mortality rate, 2000–2005.
Source: Data from UN (2003)

Europe and Australia/New Zealand, it averages 32 in Latin America, 68 in South-central Asia and 95 in sub-Saharan Africa. The IMR is particularly diagnostic of progress through the demographic transition, and of the pace of socio-demographic change more generally. Figure 4.9 thus acts as a fairly good delineator of the current extent of the so-called 'Developing World'. Given that the number of children likely to survive into adulthood is a prime consideration in deciding how many children a family should have, the IMR is also a key factor in fertility behaviour.

4.5 Fertility and family size

Despite some uncertainties about the future of mortality, it is primarily fertility behaviour that will determine future population numbers for the world as a whole and most of its major regions. Even quite small changes in fertility rate can produce major differences in population size in the longer term. In looking forward to the middle of the century, the biggest issue concerns whether the least developed countries, especially in Africa, can continue the

reductions in fertility that appear to have got underway over the last couple of decades. In addition, however, it is also not clear whether the well-below-replacement fertility now being recorded by many European countries and some elsewhere will intensify or be subject to a rebound.

On the former issue, it is remarkable how much Total Fertility Rates (TFRs), which are broadly equivalent to the average number of births per woman, have already fallen for most of the LDRs. In 1950–5 their overall TFR was about 6.2 and indeed stayed around the 6 mark for the next 15 years, but since then it has more than halved (Figure 4.10). Asia and Latin America have been the driving force behind this fall, for their TFRs have both fallen from close to 6 to around 2.5. Not all parts of these two regions have, however, been equally involved in this. For one thing, China has made a major contribution to Asia's progress, with its TFR estimated to be down to 1.8, while several countries of Western Asia still have rates above 4. Variation is somewhat less in Latin America, but even so TFR here ranges from 1.5 for some Caribbean states to 4.4 in Guatemala (Figure 4.11). Overall, however, this constitutes nothing less than a **reproduction revolution**.

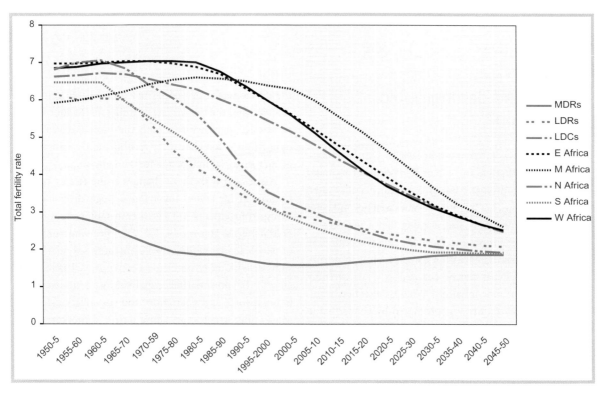

Figure 4.10 Total fertility rate, 1950–2050, for development categories and African regions.
Source: Data from UN (2003). Note: Data from 2000 are projections.

Figure 4.11 Total fertility rate, 2000–2005.
Source: Data from UN (2003)

Thematic Case Study 4.5

A second demographic transition?

This thesis was put forward in 1986 by Dutch demographers Lesthaeghe and van de Kaa primarily to make sense of the dramatic fall in European fertility after the mid-1960s, but its rationale goes well beyond considerations of family size, arguing that a new era in demographic history has dawned. The new regime is seen to be associated with a complete shift in norms and attitudes that can be denoted as a switch from 'altruism' to 'individualism'. Whereas the first transition to low fertility was prompted by concerns for family and offspring, such as providing a high standard of living and ensuring good life chances for the next generation, the second is seen to emphasize the rights and self-fulfilment of individuals.

In particular, it is seen as giving women greater choice of roles. In the past high fertility restricted women's options, while the need for survival limited the variety of family and household forms. Improvements in life expectancy, and especially the reduction in infant and childhood mortality, paved the way to lower fertility, as too did the introduction of pensions to support parents in their old age. Combined with the increasing emancipation of women in higher education and the labour market,

the development of easy-to-use and reliable contraceptives, and the rise of new attitudes about abortion, these have given women both the incentive and the opportunity for limiting fertility and achieving greater financial independence.

Associated with these changes is the rise of non-traditional life-styles, involving a huge diversity of living arrangements and a greater instability in household patterns. This is reflected in the growing number of people living on their own or in same-sex relationships or with other unrelated people, as well as in the growth of non-marital cohabitation, lone-parenthood, divorce and separation, and step-relationships. Most important for long-term demographics, however, are the rise of childlessness and the delay of childbearing into women's later reproductive years.

The key issue for the MDRs is the extent to which these tendencies are influenced by fashion or other factors that may be reversible. For the LDRs the main questions are whether the second demographic transition provides a valid explanation in countries where fertility is already below replacement rate and whether it is transferable to those where it is not yet as low as this.

Sources: van de Kaa (1987), Cliquet (1991), McLoughlin (1991), Hall (1995), Coleman (1996)

The situation in Africa is currently very different, but the fertility transition appears to be underway (Figures 4.10 and 4.11). The majority of countries had TFRs of at least 4.0 in 2000–5, with several of 6.4 or more. As a result, Africa's overall TFR was 4.9 at that time, with Middle Africa being the highest at 6.3, followed by Eastern and Western Africa at 5.6. All three, however, are in the process of moving downwards: Middle Africa from a peak of 6.6 in the 1980s and Eastern and Western Africa from 7.0 in the 1970s (Figure 4.10).

The outlook for fertility, as concluded by the UN's latest reviews of the available information (UN 2003), is a mainly optimistic one. Its medium variant projection suggests that by 2045–50 three out of every

four LDR countries will be experiencing fertility below 2.1 children per woman, the level needed to ensure the long-term replacement of population. On the other hand, the average for the 49 least developed countries is still expected to be well above replacement rate, at 2.5, most notably kept up by Western, Eastern and especially Middle Africa (Figure 4.10). The key questions for Africa, as well as for the two major remaining high-fertility regions of South-central and Western Asia, are how fast the transition will proceed in the future and what will constitute its eventual 'bottom line'.

Progress towards replacement-rate fertility in these parts of the world will no doubt require the same conditions to come into place as happened in other parts of the world. These include the provision of accessible family planning programmes, but before these can be effective, there needs to be the will to limit family size. This, in its turn, depends on a shift from a subsistence life-style to a consumer society, on the widening of education and career opportunities for women and on further progress in tackling disease and other causes of young-age deaths. In short, given their position at the end of the line of the development process and the strength of their Muslim or African traditions, nothing less than a combination of economic and cultural revolutions would appear sufficient to produce a repetition of the Eastern Asian and Latin American demographic 'success stories'.

At the other extreme, the situation in Europe might be expected to shed light on the eventual level of fertility, but in fact there now appears to be greater uncertainty about the significance of recent developments there than in the least developed countries. Following the low fertility of the inter-war period and then the unexpected post-war 'baby boom', it came to be understood that fertility rates would tend to fluctuate in the post-transition regime. Yet, after the mid-1960s, most non-communist European countries saw their fertility plummet to unprecedentedly low levels and stay down there, though there has been some rebound, notably in Scandinavia. Falls were especially marked in Southern Europe, but have been paralleled by the experience of Eastern Europe since the late 1980s. Various epithets have been used to capture the essence of this process, including 'demographic revolution' and 'the twilight of motherhood', but perhaps the most challenging intellectually is the notion of a 'second demographic transition'. The latter suggests that there is taking place a shift to a new societal pattern that is just as fundamental as the original transition (see Case study 4.5).

Whether or not this is the case, it cannot be denied that there are now many countries with fertility levels well below replacement rate (Figure 4.11). Indeed, TFR in a fair number of countries (Austria, Belarus, Bulgaria, Czech Republic, Estonia, Greece, Hungary, Italy, Latvia, Lithuania, Poland, Slovakia, Slovenia, Spain, Ukraine) is now 1.3 children per woman or lower. In 20 of the more developed countries TFR has been below replacement rate for at least two decades. Now that a number of LDR countries (notably in Eastern Asia and the Caribbean) also have TFR of 2.1 or below and there are a great many more where the rates are below 3.0 and still falling rapidly, it is little wonder that great attention is currently being given to whether recent European experience can provide an indication of what is in store for the rest of the world.

4.6 International migration

Migration is an increasingly important component of population change. This is not to say that it has not been significant in the past. In the century leading up to the First World War, upwards of 55 million people moved from Europe to the 'New World', most notably the USA, which gained 30 million between 1860 and 1920 alone. Moreover, over that period migration played a substantial role in relieving population pressures during the long-drawn-out transition process in Europe; for instance, emigration reduced Norway's population growth by over a quarter between 1860 and 1914. Nevertheless, migration has received little attention in demographic growth theory.

There are two reasons for the growing prominence of migration in countries' population trends. One concerns the reduction in natural increase rates in the MDRs. Here migration exchanges are now constituting a larger proportion of their rates of overall population growth than used to be the case and are increasingly determining whether their populations grow or decline. The other is that the volume and nature of migration is changing. The UN (2002) puts the number of people now residing outside their country of birth at an all-time high of about 175 million, more than double the number in 1975. In the words of Castles and Miller (2003), this is 'the age of migration'. While not denying the significance of migration in the past, they identify four ways in which migration is now developing and seems destined to play a major role over the next couple of decades (see Case study 4.6).

The impacts of net migration on population have now become quite significant, especially as far as the MDRs are concerned. On average, almost one in every 10 persons living there is a migrant, compared to one in every 70 in the LDRs (UN 2002). In terms of levels of movement (Table 4.2), there is currently a net flow of over 2 million people a year from the LDRs to the MDRs. For the latter, because of their low rate of natural increase, this is accounting for over seven-tenths of their overall population growth. For North America and Australia the proportion is around two-fifths, while in the case of Europe as a whole net immigration is not enough to produce overall population growth because it is only half the rate of natural decrease. The relative impact on the source regions, by contrast, has generally been much smaller. The only LDRs where migration has removed more than one-twentieth of recent natural increase are the Caribbean and Central America (Table 4.2). For the LDRs combined, the net emigration is equivalent to removing less than 3 per cent of the natural increase.

Over the next few decades to 2050, international migration is expected to remain high, with the MDRs continuing to gain about 2 million people each year (UN 2003). Averaged over the 2000–50 period, the main net receiving countries are projected to be the USA (1.1 million net migrants a year), Germany (211,000), Canada (173,000), the UK (136,000) and Australia (83,000). Meanwhile, the major net senders will likely be China (with an average net outflow of 303,000 a year), Mexico (267,000), India (222,000), the Philippines (187,000) and Indonesia (180,000). By contrast, intra-continental flows are seen as less important in the long term. Past experience has shown that, while sometimes intense at the time, these are not usually sustained in duration. For instance, some of the strong flows seen in the 1990s – from Eastern to Western Europe, from Middle to Eastern Africa, and movement into the Gulf states of Western Asia – have already reduced greatly in magnitude since then or indeed been reversed (Table 4.2).

Within all these changes and especially the more temporary ones, probably the single most important feature has been the increase in numbers of asylum-seekers and refugees. The average level was running higher in the early 1990s than at any time since the first few years after the Second World War. Estimates put the global figure at 18.2 million in 1992, up from 6 million in 1978. Some 3 million asylum applications were lodged in Europe, North America and Australia during 1983–92, with the annual figure rising from just 100,000 to over 700,000 over the period. Europe was most affected by this increase, with the numbers applying for refugee status up from 20,000 in 1980 to 560,000 in 1992. In the following years the numbers of new arrivals in European countries fell in total, but the salience of the refugee problem has been intensified by the fall-out from the ethnic conflicts in former Yugoslavia. Some 340,000 Bosnians fled to Germany in

Thematic Case Study 4.6

The age of migration

While 'migration' can be used to describe any form of residential mobility and the majority of such movement is fairly localized within countries, the epithet 'The age of migration' is used by Castles and Miller (2003) principally with reference to international migration, emphasizing how much this has grown and evolved in recent years.

They point to four general tendencies:

- Migration is becoming more 'global', in the sense that more and more countries are being affected at the same time and that the diversity of areas of origin is also increasing.
- Migration is accelerating, with the number of movements growing in volume in all major regions at the present time.
- Migration is becoming more differentiated, with no one type of movement dominating a country's flows but instead with combinations of permanent settlers, refugees, skilled labour, economic migrants, students, retirees, arranged brides, and so on.
- Migration is being feminized, with women not only moving to join earlier male migrants but now playing a much fuller part in their own right, notably being labour migrants themselves as well as often being dominant in refugee flows.

These trends greatly alter the policy implications of international migration. The growing scale of the process and the increasing involvement of women and families raise new challenges for providing for immigrants and, more generally, serve to increase native hostility in receiving countries. At the same time, increasing globalization and the growing diversity of migrants both make it harder for governments to implement effective policies for curbing migration or influencing its nature.

Sources: Champion (1994), Zlotnik (1998), Castles and Miller (2003)

Table 4.2 Net migration and population growth, 2000–2005, for development categories, major areas and regions

Development category, major area and region	Annual net migration (thousands)	Annual net migration (per 1000)	Annual population growth (per 1000)	Growth due to migration (%)
More Developed Regions	2130	1.8	2.5	71.3
Less Developed Regions	−2130	−0.4	14.6	−2.9
Africa	−168	−0.2	22.0	−0.9
Eastern	−61	−0.2	22.3	−1.0
Middle	17	0.2	26.7	0.6
Northern	−73	−0.4	18.6	−2.2
Southern	−2	0.0	6.2	−0.6
Western	−48	−0.2	25.5	−0.8
Asia	−1293	−0.3	12.5	−2.7
Eastern	−240	−0.2	6.7	−2.4
South-central	−779	−0.5	16.6	−3.0
South-eastern	−281	−0.5	14.0	−3.7
Western	6	0.0	20.7	0.1
Europe	627	0.9	−0.9	n/m
Eastern	−78	−0.3	−4.8	n/m
Northern	172	1.8	2.4	77.5
Southern	181	1.2	0.8	149.6
Western	352	1.9	2.4	78.6
Latin America and Caribbean	−605	−1.1	14.1	−8.0
Caribbean	−79	−2.0	8.7	−23.5
Central	−347	−2.5	16.6	−14.9
South	−179	−0.5	13.7	−3.6
North America	1355	4.2	10.0	41.7
Oceania	84	2.6	12.2	21.5
Australia and New Zealand	95	4.0	9.3	43.8
Other Oceania	−11	−2.8	20.6	−6.4

Source: UN (2003) and author's calculations from rounded figures. Note: n/m no meaningful figure can be calculated.

the mid-1990s, while possibly over half a million ethnic Albanians left Kosovo in 1998–9, though in both cases the majority have subsequently been repatriated. At the same time, in Europe (including 'hot spots' in the former USSR like Nagorno Karabakh in Azerbaijan and Chechnya in the Russian Federation) there were large numbers – estimated at 5 million in the mid-1990s – of 'displaced persons', people forced out of their homes and neighbourhoods but sheltering within their states of origin.

These figures, however, are small by comparison with the upheavals that have been contained within the LDRs. The UN High Commissioner for Refugees estimates that, on average over the past two decades, these have formed the reception areas of almost four-fifths of all international refugees. Through most of the 1980s, Asia hosted almost half of the world's refugees, with Palestinians constituting a long-term problem and with the other major sources at that time being Afghanistan, Vietnam and Cambodia. Another major refugee exodus in Asia was that of Iraqis during the aftermath of the Gulf War in 1991. By the mid-1990s, however, it was Africa where almost half the world's refugees could be found, reflecting the political and ethnic instability in that continent ranging from Sudan to Angola and from Rwanda to Sierra Leone. Not included in these figures are even larger numbers of displaced people. Estimates for the mid-1990s put the number of Africans who were long-term refugees within their own countries at around 16 million, much larger than the 7 million for Asia and 3 million for Latin America. Partly overlapping with these figures no doubt are estimates that put the number of the world's 'environmental refugees' at 25 million for 1998. This is believed to be the first year that this number exceeded those displaced by war and is seen as a worrying portent for the future, whether or not global warming leads to rising sea level and altered patterns of rainfall.

Nevertheless, despite their political salience, even in the 1990s asylum-seekers and refugees made up only around one-fifth of all people living outside their country of citizenship. Moreover, that proportion has

Plate 4.3 Civil war and ethnic cleansing are leading to a rising tide of refugees and displaced people. An ethnic Albanian refugee woman sitting next to a makeshift stove in the Stenovac refugee camp near Skopje, Macedonia, is one of thousands displaced by the conflict in Kosovo in 1999. (Popperfoto/Reuters)

subsequently fallen as restrictive asylum policies helped to reduce the number of refugees to an estimated 12 million by 2001 (UN 2002) and as the number of people living outside their country of birth has kept rising. Increasingly important among the latter are professional, managerial and technical personnel, a group that by contrast causes much less controversy and indeed is often referred to as 'invisible' migrants. Most notably, these include staff transferred between postings by their transnational corporations (especially in manufacturing and financial services) and those recruited to work in MDRs or newly industrializing countries with shortages of medical, construction and other skills. A further group of non-refugee migrants comprises the relatively unskilled labour that is prepared to work in low-wage sectors like agriculture, manufacturing and hotels and catering, often on a transient or clandestine basis. This process is advantageous not only to employers but also to national economies where demographic slowdown or strong unionization would otherwise have led to

higher labour costs and weaker competitiveness. In addition to these main types of international movement, there are a host of others including students, retirees and brides (see Case study 4.6).

In some cases, the impact of these non-refugee migrants on population distribution is relatively small. This may be because many of these people are relatively temporary visitors rather than permanent settlers or because there are flows in both directions between countries. Nevertheless, even in the latter case, this migration can lead to substantial differences in population composition. In the UK, for instance, the continuing pattern is of a net loss of British-born citizens and a substantial net gain of people born in the LDRs. For those countries where either the volume of inflows is considerable (like the USA) or where the rate of natural increase is very low (as across much of Europe), the impact of international migration can over time become very marked. The most conspicuous and politically contentious aspects relate to their economic and cultural effects. While employers tend to

Plate 4.4 Asylum-seekers now form a major component of immigration for many countries. This family waiting with their forms near Heathrow were among 70,000 asylum-seekers filing applications in the UK in 1999, over ten times the average for the late 1980s. (© Karen Robinson/Panos Pictures)

Plate 4.5 International food programmes are proving inadequate at coping with the effects of drought, conflict and famine. (Popperfoto/Reuters)

welcome the arrival of cheaper, malleable labour, the less skilled elements of the indigenous population and the previous cohorts of immigrants may feel their job prospects threatened and the governments are often only too conscious of the extra public spending needed for welfare and education programmes. Even more fundamentally, the arrival of people of different language, religion and especially skin colour may be seen by some as threatening not only the cohesion of local communities but the very integrity of the existing nation-states.

4.7 Population ageing

Current demographic trends are having a profound effect on population composition. As just seen, migration is increasing the ethnic and racial diversity of populations, while the second demographic transition is associated with marked changes in family and household arrangements. Perhaps the most fundamental transformation, however, is in age structure. Across most of the world, population ageing

is now the dominant trend, as people are living longer and – normally the more important factor – falling fertility is reducing the proportion of younger people. Already, many MDR countries have moved a long way from the traditional 'pyramid' age structure and indeed are beginning to look distinctly 'top heavy', but the biggest growth in numbers of older people is now taking place in the newly industrializing countries. For the LDRs as a whole, coping with a 'greying population' constitutes one of the most serious challenges of the new millennium.

The UN's latest figures make extremely impressive reading, as do the simple age-structure diagrams in Figure 4.12. Between 1950 and 2000, the median age (in years) of the world's population rose from 23.6 to 26.4, but by 2050 it is expected to have shot up to 36.8. Much attention has been given to the ageing of the MDRs since 1950, and the late 1990s marked something of a milestone there as the number of older persons (60 and over) exceeded that of children (aged under 15) for the first time. Nevertheless, the rise in median age in the MDRs over the next half-century will be as great as for the last 50 years, rising from 37.3 in 2000 to 45.2 in

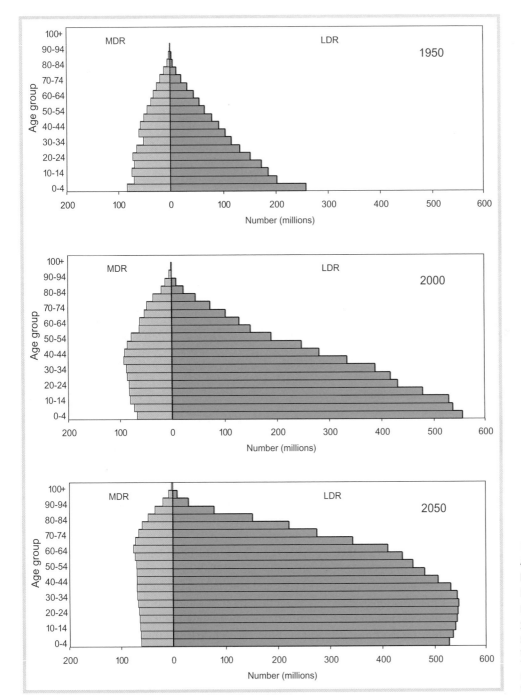

Figure 4.12
Age structure of
More and Less
Developed
Regions, 1950,
2000 and 2050.
Source: Data from
the UN (2003). Note:
The 2050 data are
based on medium-
variant projection.

2050. Europe is, and will remain, the most affected region, with the projections indicating that older people will constitute more than one in three of the population by 2050, outnumbering children by nearly two-and-a-half times. By that time, on current trends, Latvia will be the oldest country in Europe, with half its inhabitants aged 53 or more, second in the world to Japan's expected median age of 53.2 years.

Ageing has been much slower until now in the LDRs in aggregate, with median age rising only from 21.3 to 24.1 between 1950 and 2000, but its impact over the next half century will be far greater than for the MDRs. By 2050 their median age is projected to rise to 35.7, an increase of 11.6 years compared to one of 7.9 years for the MDRs. Over the past half century the LDRs' number of people aged 60 and over grew by

Plate 4.6   The elderly are starting to outnumber children in many parts of the Developed World. This familiar off-peak holiday scene is likely to become commonplace all year round and in all places, as the 'baby boomers' of the 1950s and early 1960s reach retirement age. (Camera Press/Chris Nevaro)

some 264 million, a small fraction of their 3.2 billion total increase. But between 2000 and 2050 the older population of the LDRs is expected to rise by 1,139 million, accounting for two-fifths of their total population increase over the period. By contrast, their number of children under 15 years old in 2050 is expected to be about the same as currently. As a result, the LDRs will have reached the same situation as the MDRs reached in the late 1990s, with the number of older people outnumbering children. Of the three major LDR areas, Asia is the furthest down the ageing road, with half its population aged 26 or over in 2000, while Africa is the youngest, with its median age of 18.1 years being lower than it was in 1950.

These trends constitute a big enough challenge for MDR countries, where governments have for some time been concerned about their ability to fund pensions and care provision through taxes imposed on a shrinking proportion of working-age people. Their recent and future problems, however, pale into insignificance in comparison with the sheer numerical growth of older people expected in the LDRs over the next half century and take on even more frightening

proportions when the fragile economic base of many of these areas is considered. Up till now, the most rapid ageing has taken place in the newly industrialized countries of East Asia, where very rapid economic growth (at least until 1998) helped to ease the burden. If similar growth fails to take root in the rest of the LDRs, and especially in the currently very poor regions of South-central Asia and sub-Saharan Africa, the direct consequences are likely to be tragic and the knock-on effects in relation to other aspects of mortality, fertility and migration are virtually impossible to gauge.

4.8 Conclusion

The world's demographic complexion has clearly undergone some major transformations in the past half-century. As of 2004, there are three times as many people living in LDRs as in 1950. The 'death control' element of the demographic transition, scarcely evident outside the MDRs at mid-century, has now affected all major regions and is largely complete in most of them. The past three decades have seen the fertility transition spread quite rapidly across many LDRs hand in hand with economic growth and family planning programmes. More recently still, international migration has moved to a new plane in terms of numbers of people involved, the range of countries affected and the variety of forms taken. All the time, populations around the world are becoming progressively older, with the traditional 'age pyramid' becoming increasingly obsolete as a description of age structure.

Looking to the future, the most likely scenario is that these trends, linked as they are to the process of demographic transition, will continue and eventually play themselves out. While this is certainly what the latest UN projections suggest, however, this chapter has mentioned a number of ways in which the assumption of continuity might come unstuck. First and foremost, there remain the Malthusian spectres of 'war, pestilence and famine', as exemplified by the many ethnic conflicts, the spread of HIV/AIDS and the latest food emergencies. Secondly, there would seem to be no guarantee that conditions conducive to small family size will appear in the remaining high-fertility countries, given their recent combination of rapid population growth and weak economic progress. Thirdly, while the factors that have led to the new 'age of migration' appear to show little sign of abating, international migration is notoriously volatile and unpredictable and, in addition, does not seem to be readily amenable to management by governments. To cap it all, talk of a 'second demographic transition' in Europe raises new uncertainties about trends there and about the eventual destination of fertility in the LDRs.

Nevertheless, some things seem very certain and raise extremely important issues. Barring a catastrophe, world population is very likely to rise by a further two and a half billion by the middle of the century. Virtually all this growth is destined to take place in the LDRs, leading to further shrinkage in the proportion accounted for by the MDRs. The number of old people is going to continue to increase rapidly, imposing ever greater strains on families and societies in both the MDRs and LDRs. The ethnic balance in the world will increasingly sideline the white, mainly Christian population, as Asian and African populations burgeon and as Islamic and other faiths expand. This is likely to increase tensions both between and within countries, particularly in areas coming up against the limits of their land and water resources. Nor will the MDR countries be immune, as their existing ethnic minority populations grow and as their case for strict immigration controls is subjected to further tests by the international community. Clearly, demographics now constitute a major force for global change and their combination of transformation and uncertainty represents a powerful challenge for current and future generations of decision-makers and academics.

Learning outcomes

Having read this chapter, you should be able to:

- Recognize the dynamic nature of populations, not only in terms of size but also with respect to composition, notably age structure and ethnic make-up.

- Question the inevitability of the demographic transition process for those countries that have entered the mortality control stage fairly recently.

- Appreciate that the completion of the demographic transition no longer appears to herald the end of transformations in the population.

- Acknowledge the increasing importance of migration as a component of national population change alongside fertility and mortality.

- Assess the significance of the challenges which current demographic trends pose for individuals, families, communities and governments.

Further reading

Castles, S. and **Miller, M.J.** (2003) *The Age of Migration: International Population Movements in the Modern World*, Palgrave Macmillan, Basingstoke. This book, now in its third edition, provides a comprehensive survey of past trends and current developments in international migration. Look especially at Chapter 6 on the globalization of international migration and Chapter 12 on the post Cold War era.

Gelbard, A., Haub, C. and **Kent, M.M.** (1999) World population beyond six billion, *Population Bulletin*, **54**, 1–44. This booklet, published by the Population Reference Bureau in Washington DC, reviews the history of world population, the causes and effects of population change, and the prospects for the next 50 years. See also the PRB's annual *World Population Data Sheet*, containing key statistics on population size, structure and change for all geographical entities with populations of 150,000 or more and all members of the UN.

Jones, H. (1990) *Population Geography*, Paul Chapman, London. This is a clear introduction to geographical perspectives on population. Look especially at the chapters on population growth and regulation, international variations in mortality, fertility in developed countries, fertility in less developed countries, and international migration.

National Research Council (2000) *Beyond Six Billion: Forecasting the World's Population,* National Academy Press, Washington, DC. This presents the findings of a US National Academy of Sciences panel on population projections. It contains detailed examinations of transitional and post-transitional fertility, mortality and life expectancy, and international migration, together with assessments of the accuracy of past projections and of the uncertainties in current population forecasts.

United Nations (2003) *World Population Prospects: The 2002 Revision*, United Nations Population Division, New York. This is the latest of the UN's biennial updates on world population, with estimates of population for 228 countries and areas of the world from 1950 to 2003 and projections through to the year 2050. This edition includes special reports on below-replacement fertility, the demographic impact of HIV/AIDS, and population ageing. The UN's *Demographic Yearbook* provides even more detail on the current situation and recent trends.

Useful web-sites

www.un.org/esa/population/unpop.htm Provides access to extracts from the United Nations' *World Population Prospects: The 2002 Revision.*

www.census.gov/ipc/www/idbnew.html US Census Bureau (on line) International Data Base, United States Bureau of the Census, Washington DC. This is a computerized source of demographic and socio-economic statistics for 227 countries and areas of the world from 1950 to the present, and projected demographic data to 2050, freely available on the World Wide Web.

www.prb.org Home page of the Population Reference Bureau, based in Washington, DC. This provides news updates on all aspects of world and US population, plus information about its own publications (including its monthly *Population Today* and *Population Bulletin*) and links to other web-sites.

www.populationaction.org Home of Population Action International, an organization dedicated to advancing policies and programmes that slow population growth in order to improve quality of life for all people.

oprtest.princeton.edu/popindex The primary reference tool to the world's population literature, maintained by the Office of Population Research at Princeton University and with a searchable and browsable database containing over 46,000 abstracts as published in *Population Index 1986–2000*.

For annotated, clickable weblinks and useful tutorials full of practical advice on how to improve your study skills, visit this book's website at www.pearsoned.co.uk/daniels

CHAPTER 5

Resources and development

MICHAEL BRADSHAW

Topics covered

- The nature of natural resources
- Ways of defining and classifying resources
- The factors that determine resource availability
- The changing geography of energy production and consumption
- The relationship between energy consumption and economic growth
- Global energy dilemmas

Resources are defined by society, not by nature.

(Adapted from Rees 1985:11)

Resources are not, they become; they are not static but expand and contract in response to human wants and actions.

Zimmerman (Peach and Constantin 1972:16)

The large-scale exploitation of the planet's resource base may well go down in history as a defining characteristic of the twentieth century. Today the scale of resource consumption threatens the entire global ecosystem (see Chapter 6). Throughout the last century access to, and control over, natural resources was a source of conflict. Equally, patterns of resource consumption remain a major marker between the so-called 'developed' and 'developing' worlds (see Chapter 8 for further discussion). According to the World Bank (2002a), in 1999 the so-called 'high-income countries' accounted for 14.2 per cent of the world's population, but produced 81.6 per cent of global Gross Domestic Product (GDP) and accounted for 50.5 per cent of global commercial energy use. In the early twenty-first century, society is increasingly aware of the finite nature of the planet's resource base. Yet, a seeming relative abundance of supply has made us complacent, as has a blind belief that technological progress will solve the inevitable shortage of non-renewable resources. This chapter explores the nature of natural resources and considers the relationship between resource production and consumption, and economic development. The chapter is divided into three sections, each with a distinct task. The first section examines the meaning of the term 'natural resource', it evaluates the various ways of classifying resources and analyses the diverse factors that influence their availability. The second section analyses the specific case of energy resources and the changing geographies of production and consumption. The third, and final, section considers the interrelationship between energy consumption and economic development. The chapter concludes by discussing the resource dilemmas that currently face the different regions of the world.

5.1 Natural resources

At any moment in time, the planet Earth holds a finite stock of resources, the resource base. However, what human societies have considered a resource has varied through time and across space. Thus, the notion of what constitutes the Earth's resource base also changes. As both Zimmerman and Rees acknowledge above, something is a resource because human society attaches value to it. Many of the things we value today as 'resources' held no value in the past and may hold no value in the future. For example, petroleum was not an important source of energy until relatively recently in human history. Followers of science fiction

programmes, such as Star Trek, see a future based on resources and technologies that have yet to be created.

5.1.1 Defining resources

In his seminal work *World Resources and Industries*, Zimmerman (Peach and Constantin 1972: 9) states: 'The word "resource" does not refer to a substance, but a function that a thing or a substance may perform, or to an operation in which it might take place.' He goes on to note that: 'resources are . . . as dynamic as civilisation itself.' In fact, one can suggest that each major human civilization was sustained by a particular set of resources and technologies for their exploitation (Simmons 1996). Thus, the archaeological record talks of the *Stone* Age, the *Bronze* Age and the *Iron* Age. In reference to the Industrial Revolution of the nineteenth century, Simmons (1996: 208) states that: 'industrialization based on fossil fuel energy represents a turning point in the history of human-nature relations.' If the nineteenth century was based on the exploitation of coal, then the twentieth century will been seen as the era of oil and gas (Hall *et al.* 2003). Thus, the changing notion of what constitutes a 'resource' is an important factor in shaping the relationship between human societies and the natural environment. Undoubtedly, the hunter-gatherer communities living in the Palaeolithic period had a very different relationship with nature from the industrialists of the nineteenth century (see Section 1). As Blunden (1995: 164) observes, 'Because definition as a resource depends on usefulness to human society, natural materials may be required as resources by societies in some times and places but not in others.' At the time that the industrial revolution was taking place in Europe there were societies located elsewhere on the planet that were still at the hunter-gatherer stage in terms of their concept of what constitutes a resource and in terms of their relationship with nature. This example adds a further dimension to the complexity of the term 'resource'. Its definition can vary across space at the same time. While it is still the case today that people in many parts of the world live in what Mather and Chapman (1995: 139) call 'low-energy societies' (dependent upon plants, animals and human labour), one of the defining characteristics of the early twenty-first century, and a consequence of the processes of globalization, is that the majority of societies now have at least some shared notion of what sort of things constitute resources. The Inuit living in northern

Plate 5.1   The use of firewood for cooking in a 'low-energy' society. (Panos Pictures/Sean Sprague)

Canada are as dependent on gasoline for their skidoos as the commuters in their cars in Los Angeles or London are. Furthermore, echoing Lenin's famous edict that 'Communism is electrification plus Soviet labour power', the provision of electricity is seen as a central component of the modernization process in all types of societies. Witness the chaos that ensued when large parts of the US east coast lost all electricity in late 2003. A similar power outage occurred in Italy and there are growing concerns about the ability of national electricity systems to handle ever-increasing demand.

If the notion of resource is dynamic and intimately linked to the evolution of human society, it follows that so-called 'technological progress' both creates and destroys resources. As new technologies emerge, dependent on particular resources, so old technologies and their associated resources become redundant. For example, today we attach little value to flint (it has some value as a building material), yet in the Stone Age it was an essential resource for making tools. Because these resources had 'use value' they were also the subject of trade. On the other hand, many of the resources that were valued in the Bronze Age, such as

copper and zinc, are still valued today but they are put to different uses. Thus, some resources retain their value as new ways of using them are discovered. The use value of resources also means that control over their supply is an important part of political and economic power. In theory at least, resource-rich regions are able to exploit this natural advantage in their dealings with resource-poor regions. Consequently, gaining control over particular resources has been at the heart of many wars and much conflict. The European colonization of what we now call the 'developing world' was in part motivated by a desire to discover and control new sources of resources. It is also no surprise that many of the world's largest corporations, until recently at least, were involved in resource development. Our discussion so far has implied that resources are a homogeneous thing. However, we know that there are many different types of natural resource and the various ways of classifying them is the subject of the next section.

5.1.2 Classifying resources

Natural resources are commonly divided into two types: **non-renewable** or stock resources and renewable or flow resources. Figure 5.1 presents a classification of resources on this basis. Stock resources are those, mainly mineral, that have taken millions of years to form and so their availability is finite. Hence, we also refer to them as non-renewable as there is no possibility of their being replenished on a time scale of relevance to human society. For example, on the basis of the geological time scale, it is possible for new deposits of coal, oil and gas to be created; however, the time this will take means that we only have available to use the stock of already created fuel minerals. Within the category of stock resources, it is also useful to distinguish between those that are consumed by use, such as fuel minerals, those that are theoretically recoverable and those that are recyclable, such as aluminium. A further characteristic of stock resources is that they tend to be highly localized, that is they are found in relative abundance only in specific places, such as ore deposits and coalfields. Some stock resources are more abundant than others and their relative scarcity affects their value. For example, aggregate minerals, like sand and gravel, are relatively abundant, while precious metals, such as gold and silver, are relatively rare.

Renewable or flow resources are those that are naturally renewed within a sufficiently short time span to be of use to human society. Again we can distinguish between different types of stock resource. Figure 5.1 divides flow resources into 'critical zone' and 'non-critical zone' resources. The distinction here is between those flow resources whose continued availability is dependent upon management by society (critical zone) and those that will continue to be available independent of the actions of society (non-critical zone). As indicated by the arrows, it is possible for critical zone flow resources to become stock resources if they are mismanaged and their regenerative capacity is exceeded. Thus, the aim of resource management is to ensure that exploitation of a particular renewable resource does not damage its capacity to replace itself. In recognition of the increasing challenge, and the numerous failures, to manage renewable resources, Rees (1991: 8) has developed an alternative to the conventional two-part typology of natural resources: 'All resources are renewable on some timescale ... what matters for the sustainability of future supplies is the relative rates of replenishment and use ... it seems better ... to think in terms of a "resource continuum" than the conventional two-part typology' (Figure 5.2). The combined impact of population growth and industrialization is placing increasing stress on the planet's finite supply of fresh water and clean air. According to UNEP (1999: xxii), about 20 per cent of the world's population currently lacks safe drinking water, while 50 per cent lacks access to a safe sanitation system. In 2000 the international community committed to the so-called 'Millennium Goals' (see www.developmentgoals.org); Goal 7 to ensure environmental sustainability set a target to halve the

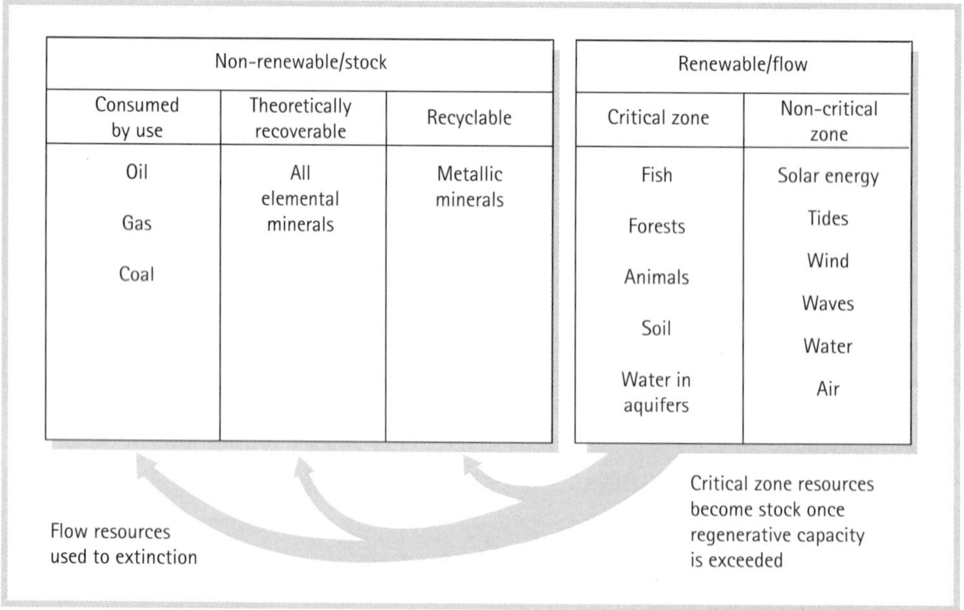

Non-renewable/stock			Renewable/flow	
Consumed by use	Theoretically recoverable	Recyclable	Critical zone	Non-critical zone
Oil	All elemental minerals	Metallic minerals	Fish	Solar energy
Gas			Forests	Tides
Coal			Animals	Wind
				Waves
			Soil	Water
			Water in aquifers	Air

Flow resources used to extinction

Critical zone resources become stock once regenerative capacity is exceeded

Figure 5.1 A classification of resource types. Source: J. Rees, *Natural Resources: Allocation, Economics and Policy*, 2nd edn, Routledge, 1985, p. 15

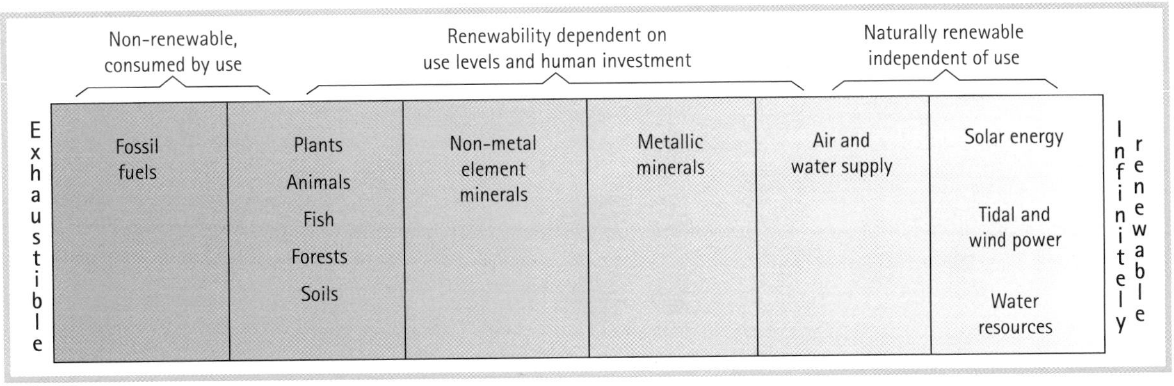

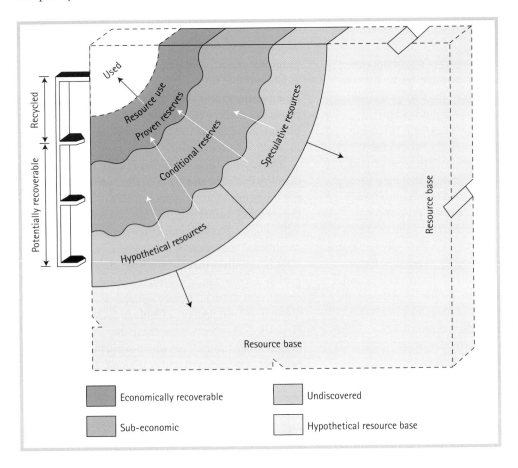

Figure 5.2 The resource continuum.

Source: J. Rees, Resources and environment: scarcity and sustainability in *Global and Challenge: Geography in the 1990s* (Bennett and Estall (eds)), Routledge, 1991, p. 9

proportion of people without sustainable access to safe drinking water and basic sanitation. Meeting the Millennium Development Goals will require providing about 1.5 billion people with access to safe water between 2000 and 2015. The resource continuum recognizes that (clean) air and water supply now have scarcity value and should be considered as finite.

In our discussion so far we have highlighted the complexity of the notion of natural resource and considered the various ways of classifying resources, but what factors actually affect the availability of particular resources?

5.1.3 Resource availability

As was noted earlier, at any moment in time there is a finite stock of natural resources on the planet, the resource base. Each resource has in turn its own

Figure 5.3

Resource availability.

Source: J. Rees, *Natural Resources: Allocation, Economics and Policy*, 2nd edn, Routledge, 1985, p. 20

Plate 5.2   As offshore technology advances, so more remote resources can be exploited. (Popperfoto/Reuters)

resource base, the total quantity of a substance or property on the planet, for example, the total amount of oil in existence today. However, that total resource base is not the amount available for human exploitation. Figure 5.3 illustrates the relationship between the resource base and the various sub-divisions of resource availability. The term **proven reserve** is applied to those deposits that have already been discovered and are known to be economically extractable under current demand, price and technological conditions. Thus, the extent of the proven reserves of a particular resource is highly dynamic and dependent upon a host of partly interlinked factors. These include the availability of the technology and skills to exploit the resource, the level of demand, the cost of production and processing, the price it can command in the market place, the availability and price of substitutes, and the environmental and social costs of developing the resource. These factors determine whether or not a particular resource will be exploited in a particular place and the global level of supply of that resource. The extent to which each of these factors influences resource development also varies across space and time. Today we have the technological ability

to recover resources in geological and environmental conditions that were previously uneconomic. For example, the exploitation of North Sea oil and gas in a physically challenging environment has been made possible by advances in offshore drilling and production technologies.

The category of **conditional reserve** refers to deposits that have already been discovered but that are not economic to work at present-day price levels with the currently available extraction and production technologies. The boundary between proven and conditional reserves is dynamic and bi-directional (that is, resources that change from conditional to proven reserves can, if conditions change, revert to conditional status). The two remaining categories of resource are not readily available to society. **Hypothetical resources** are those that we may expect to find in the future. They are in areas that have only been partially surveyed and developed. They may be in regions, such as Antarctica, where the international community has placed a moratorium on resource development, for the time being at least. **Speculative resources** are those that might be found in unexplored areas which are thought to have favourable geological

Table 5.1 The dimensions of resource scarcity

Type of scarcity	Concern
Physical scarcity	• Exhaustion of minerals and energy. • Human populations exceed the food production capacity of the land. • Depletion of renewable resources such as fish, soils or timber.
Geopolitical scarcity	• Use of minerals exports as a political weapon (e.g. sales embargoes). • Shift in the location of low-cost minerals sources to 'hostile' blocs of nations.
Economic scarcity	• Demand at current price levels exceeds the quantity supplied (therefore shortages). • Needs exceed the ability of individuals or countries to pay for resource supplies. • Rich economies can always outbid the poor for essential resources, creating unequal patterns of resource use. • Economic exhaustion of specific minerals or renewable resources causes economic and social disruption in producer regions or in nations dependent on them.
Renewable and environmental resource scarcity	• Distribution of essential biogeographical cycles (e.g. the carbon dioxide cycle and the greenhouse effect) threatening sustainability of life on earth. • Pollution loads exceeding the 'absorptive' capacity, causing economic health and amenity problems. • Loss of plant and animal species and landscape values, with wide, but poorly understood, long-term consequences.

Source: J. Rees, Resources and environment: scarcity and sustainability in *Global Change and Challenge: Geography in the 1990s* (Bennett and Estall (eds)), Routledge, p. 6

conditions. Finally, there remains a large part of the earth about which we have no information on its potential resource base. The strength of this classification is that it stresses the highly dynamic nature of the concept of resource reserve. The danger is that it leads to the view that society will never run out of resources because there are always more to discover and technological progress will continue to make new resources available for exploitation. Even if the latter were the case, and clearly it is not, the planet now faces the additional problem that consuming resources, such as hydrocarbon fuels, is actually threatening the global ecosystem. Thus, there is a need to rethink the whole notion of resource availability to take into account the ecological cost of our current 'fossil fuels society'.

Given the highly dynamic nature of resources, it is very difficult to estimate at any moment in time the level of resource availability. It is even more difficult to speculate about future levels of production and consumption and, thus, the possibility of resource scarcity. This is not just because extrapolation on the basis of current trends is often misleading, but also because there are a whole variety of factors that can promote resource scarcity. Table 5.1 identifies a variety of different types of scarcity. The resource crisis of the

1970s was motivated by geopolitics, but by increasing the price of energy, it brought about other forms of scarcity. For example, the increased price of oil had a major negative impact upon those countries in the 'developing world' that had embarked upon industrialization and had become increasingly dependent upon imported oil. However, it also shocked the 'developed world' into the realization that energy resources were finite, energy conservation was worthwhile and that alternative sources of energy were required. The 'developed world', and particularly the United States, came to realize the strategic importance of securing access to energy supplies. Critics of the recent war in Iraq say that one of the primary motivations for toppling the regime of Saddam Hussein was the desire on the part of the United States and her allies to secure control of Iraq's substantial oil reserves. The rejuvenation of Iraq's oil industry and subsequent exports are a key component of the country's path to recovery, but it is too soon to tell what impact increased Iraqi oil exports will have on the world price of oil. Today, despite continued geopolitical tensions in the Middle East and elsewhere, we still have a relative abundance of oil and prices are still lower in real terms than after the energy crisis of the 1970s. But, as noted above, society is increasingly

Plate 5.3   The energy crisis in 1973 brought lines at the petrol pumps and a 3-day week. (Getty Images/Evening Standard))

concerned about the ecological consequences of increased energy consumption. The remainder of this chapter looks in detail at the critical question of energy.

5.2 Fuelling the planet

Of all the different types of resource that we have discussed so far, in recent history it is those that provide a source of energy that have been the most sought after. Figure 5.4 provides a simple classification of the different types of energy resource. For most of human history societies have utilized renewable sources of energy; flow resources such as wood that can be depleted, sustained or increased by human activity; and continuous resources, such as water to drive watermills or the wind to turn windmills, that are available irrespective of human activity. They have also used draught animals, horses, buffalo, etc., but these need to be fed and housed. The industrial revolution changed the way in which in certain parts of the world society powered their economies, while the rest of the world remained dependent upon renewable resources.

5.2.1 The dominance of fossil fuels

Since the invention of the steam engine about 200 years ago, much of human society has become ever more dependent upon the exploitation of non-renewable energy resources. Today three fossil fuels account for over 85 per cent of the annual sale of the world's most important minerals. These resources were formed from the decomposition of organic materials millions of years ago and have been transformed by heat and pressure into coal, oil and natural gas. They represent the classic, non-renewable resource, found in specific locations and in a variety of forms and conditions. For example, commercial coal deposits are of three types: lignite (brown coal), bituminous coal and anthracite. There are different grades of oil along a scale from light to heavy and individual refineries are designed to process particular grades of crude oil. Natural gas is often found in association with oil, but is also found on its own. Until recently, the natural gas associated with oil was simply flared off. Today it is often re-injected into the oilfield to enhance oil recovery or used as a hydrocarbon resource in its own right. The mineral

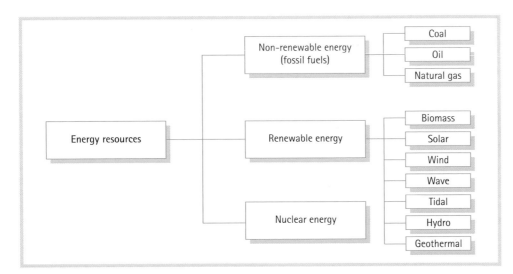

Figure 5.4
Categories of energy resources.
Source: Jones and Hollier (1997: 173)

content of natural gas also varies: for example, so-called 'sour gas' has high sulphur content. This sulphur has to be extracted before the gas can be transported by pipeline. The advantage of fossil fuels as a source of energy is that they are readily accessible, easy to convert using proven technologies, and cost-efficient in production and use.

Of the three types of fossil fuel, oil is the most versatile. Today coal is used in many industrial processes and as a source of heat and remains the most important fuel for electricity generation. However, it is a relatively bulky resource and in recent years, particularly in Western Europe, it has been unable to compete with oil that is more versatile, or with cleaner natural gas. The great advantage of oil is its transportability. It can be transported over long distances in large volumes by pipelines, it can cross oceans in tankers, or it can be moved in smaller

Plate 5.4   Special LNG tankers are used to distribute liquefied natural gas to markets worldwide. (Associated Press/Dita Alangkara)

quantities by rail and road tankers. It has spawned a huge petrochemical industry that now produces a vast range of products. Many of these products, such as plastics and textiles, have substituted products based on renewable resources, such as wood and cotton, for products based on non-renewable resources. At the same time, oil has created many resources for which we have yet to develop commercially viable substitutes, such as jet engine fuel. This means that advanced industrial societies are totally dependent upon the continued supply of oil.

In recent years, natural gas has become an increasingly important source of energy. In many advanced industrialized economies it has even replaced coal and, to a lesser extent, oil as the favoured resource for electricity generation and domestic heating. However, natural gas does not have the same transport flexibility as oil. It is most economic when transported by pipeline, but the infrastructure investment needed requires secure sources of production and stable markets. Natural gas

can also be processed to produce liquefied natural gas (LNG), but the liquefaction process requires substantial investment in plant and consumes large quantities of energy. Nevertheless, LNG is becoming an increasingly important fuel, particularly in the Asia-Pacific region.

5.2.2 Alternative sources of energy

Given the finite nature of hydrocarbon fuels, the problems of scarcity and the environmental consequences of burning fossil fuels, it is no surprise that alternative sources of energy have been sought. As Figure 5.4 shows, these alternatives take two forms: one, the development of new ways of harnessing renewable energy sources, and two, the development of nuclear power. At present all commercial nuclear power stations use so-called fission reactors in which atoms of uranium-235 and plutonium-239 are split by neutrons, thus releasing energy, mainly in the form of

---

Thematic Case Study 5.1

The rise and fall of nuclear power

During the 1950s the peaceful use of nuclear power was heralded as the great hope for the world's energy needs. Harnessing the power of the atom promised a cheap and clean source of electricity. The energy crisis of the 1970s increased interest in a nuclear solution. By developing a nuclear power industry, the developed world hoped that it could reduce its reliance upon OPEC-supplied oil and at the same time reduce emissions of CO_2. The 1970s and 1980s saw a rapid expansion in the generating capacity of the world's nuclear power plants. Then, in the last decade, growth has stalled and there is the real possibility of a decline in total capacity. What went wrong?

In short, as the nuclear power industry developed, expanded and matured it became increasingly apparent that the supposed benefits were far outweighed by the environmental problems it posed. Reddish and Rand (1996: 79) have listed four criti-

cisms levelled at nuclear power by environmental groups:

- That abnormal radiation levels from normal operations will cause cell damage, malignant cancers, genetic diseases, etc.
- That reactor operations and transport of irradiated fuels cannot be guaranteed safe against catastrophic accidents.
- That radioactive waste has to be disposed of safely and reactors have to be decommissioned, all of which is extremely costly.
- That the radioactive materials needed to fuel the reactors need to be kept safe and secure from theft or misappropriation, to make nuclear weapons.

Over the years public trust in nuclear power has been significantly eroded by a number of serious accidents such as those at Windscale, Cumbria, in the UK in 1957, Three Mile Island in the USA in 1979 and Chernobyl in the Ukraine in 1986. Furthermore, as the true cost of building, maintaining and decommissioning nuclear power plants has become apparent many have questioned the economics of nuclear power.

Despite these problems, in 1998 nuclear power accounted for 16 per cent of global electricity generation; but it has been very much an option for the developed world, partly due to cost and technology, but also due to concerns over the non-proliferation of nuclear weapons. In 1996, OECD member states accounted for 86.6 per cent of total nuclear power production. The top three countries were the United

Plate 5.5 The Chernobyl disaster spelled the end for nuclear power as a clean and safe alternative source of energy. (Science Photograph Library/Novosti)

Thematic Case Study 5.1 continued

States 29.6 per cent, France 16.4 per cent and Japan 12.5 per cent. In the same year, nuclear power accounted for 78.2 per cent of total domestic electricity generation in France and 52.5 per cent in Sweden. But Sweden has decided to phase out its nuclear power stations and to switch to conventional thermal power stations. The United Kingdom's energy strategy envisages that no new nuclear power stations will be built in the foreseeable future. Prior to the Asian financial crisis, nuclear power was seen as a solution to Asia's shortage of indigenous energy resources. However, the high cost of constructing nuclear power stations, rising public opinion against them and the increasing availability of cheaper alternatives, such as LNG, means that substantial expansion is now unlikely. Following a series of safety problems and scandals, Japan is reconsidering the role of nuclear power in its energy mix. In North America, western Europe and the former Soviet Union, decommissioning of old reactors is not being mirrored by construction of new capacity. Thus, while nuclear power will continue to provide an important source of electricity, it is no longer seen as the obvious solution to our energy problems.

---

heat, which is then used to generate electricity. In the 1950s nuclear energy was heralded as a cheap and clean alternative to fossil fuels. However, the nuclear industry has failed to expand as expected (see Case study 5.1). Today, few countries are expanding their nuclear programmes and many now face the problem of decommissioning the older generation of reactors. Furthermore, the collapse of the Soviet Union has left the crisis-ridden post-Soviet republics with a nuclear energy system they cannot afford to maintain. In retrospect, the current nuclear energy industry may be seen to have created as many problems as it has solved. There is still the possibility that a new generation of fusion or breeder reactors can fulfil that early promise, but few forecasters see a significant increase in the role of nuclear power in the short term.

The other alternative to fossil fuels is the development of 'new' renewable sources of energy. Many of these renewable sources involve applying new technologies to historic sources of energy supply (see Elliott 2003 for further discussion of renewable energy sources and their prospects):

- *Biomass energy* involves the burning of plant and animal residue to produce heat. In much of the world this is still the traditional source of energy. According to the IEA's Biomass Project, the share of biomass energy in Africa, Asia and Latin America averages 30 per cent. However, in some of the poorest countries of Asia and Africa it can be as high as 80–90 per cent. It is estimated that for some 2 billion people biomass is the only available and affordable source of energy (see Case study 5.2).

- *Solar energy* usually involves capturing the sun's rays in order to warm buildings and heat hot water. As the efficiency of solar panels increases and their cost declines, solar energy is becoming a viable small-scale local solution. However, it is subject to daily or seasonal variations and is weather-dependent, but it can provide an efficient supplemental source of energy for heating domestic hot water, etc. (visit the web-site of the International Solar Energy Society at www.ises.org).

- *Wind power* is of historical significance and has now made a return in the form of 'wind farms'. Improved design has increased the efficiency of modern windmills and they are becoming an increasingly common sight in exposed coastal and upland locations in the developed world. Wind farms can generate electricity for local use or for the power grid. However, some would argue that modern wind farms lack the aesthetic qualities of traditional windmills and therefore many see them as a source of both visual and noise pollution (for more information on wind power see www.windpower.org). In 2003, the UK government announced plans for huge offshore wind farms to be built off the UK coast in the Thames Estuary, the Greater Wash and the North-west. The government is aiming to source 10 per cent of UK electricity from renewable power by 2010 and the energy from these new wind farms is expected to power 15 per cent of UK households.

- *Hydropower generation* also pre-dates the industrial revolution, but the watermills that powered the

Thematic Case Study 5.2

Biomass energy: problem or solution?

By its very nature, the contribution of biomass energy to global energy consumption is very difficult to measure. It largely takes place outside the market economy and is the result of unpaid (female and child) labour. Because it is not part of the commercial energy mix, it seldom figures in official statistics. Yet, the IEA has estimated that it currently represents 14 per cent of world final energy consumption, a higher share than that of coal (12 per cent) and comparable to those of gas (15 per cent) and electricity (14 per cent) (www.iea.org/ead/biomass.htm). Biomass energy in the form of fuelwood for cooking and heating is particularly important in countries in Asia and Africa, but there are considerable variations between countries in the two regions. Take Africa as an example. According to the UNEP (1999: 64) in Africa as a whole 'traditional' sources of energy account for 36.9 per cent of total consumption and 'commercial' energy 63.1 per cent. However, in Eastern Africa and the Indian Ocean islands traditional sources account for 83.9 per cent of energy use and in Western and Central Africa 74.2 per cent. By contrast, in Southern Africa commercial energy accounts for 79.2 per cent of energy use and in North Africa 96.8 per cent. Thus, problems related to fuelwood are concentrated in the Sahel regions and the tropical rainforest of equatorial Africa. In the 1970s and 1980s the international agencies were concerned with the so-called 'fuelwood crisis'. They saw the collection of fuelwood in the Sahel as a major cause of deforestation, desertification and hunger. Subsequent research has shown the situation to be far more complex. Undoubtedly, the collection of firewood is a contributory factor, but the clearing of land for agriculture and the encroachment of urban areas are just as, if not more, problematic and are increasing as the population continues to grow. Thus, biomass energy places added stress on an already threatened and diminishing forest ecosystem. The problem is that there are so many conflicting demands being placed on the forest. In Africa, commercial forestry is of limited significance at present, but in Latin America and Asia it is a major threat. Everywhere, the pressure to clear land for agriculture and settlement is ever increasing. Ultimately, removal of the forest means a reduction in the supply of fuelwood.

The UNEP (1999: 57), reviewing the environmental situation in Africa, has concluded: 'Africa's forests are threatened by a combination of factors including agricultural expansion, commercial harvesting, increased firewood collection, inappropriate land and tree tenure regimes, heavy livestock grazing, and accelerated urbanization and industrialization.' The same report notes that the consumption of fuelwood nearly doubled during the 1970–94 period and that consumption could rise by 5 per cent by 2010. This pressure on the forest resource is unsustainable, but what should be done? One solution would be to encourage the use of commercial sources of energy based on fossil fuels, but this would require people to earn money and would likely increase deforestation to expand agricultural production; it would also promote further urbanization. An alternative approach would be to manage the development of firewood plantations, on both a commercial and non-commercial basis. Middleton (2003: 306) advocates the development of dedicated energy plantations on land that has been deforested. This has the added benefit of arresting problems caused by the removal of the forest. He also notes that such a use of biomass fuels produces no net emissions of carbon dioxide since the amount released into the atmosphere by burning is then taken up by growing plants. Thus, in the future sustainable development of biomass energy could offer a solution to the problem of deforestation, rather than contribute to the problem, as well as meet the energy needs of the poorest inhabitants of the developing world.

early machines have now been replaced by dams and hydro-electric schemes. In many regions of the world hydro-electricity has become an important source of energy, but, in all but a few cases, it cannot offer a large-scale solution to a country's energy needs. The building of dams to store water and build a 'head' to drive the turbines that generate the electricity floods large areas of land; it also starves river systems of flow and damages deltaic and riverine environments. The construction of Three Gorges dam on the Yangtze river in China involves planning the displacement of 1.2 million people and the creation of a reservoir occupying 111,000 hectares. Other forms of waterpower, such as wave and tidal power, are also being developed; again, they cannot provide a viable alternative to fossil fuel power generation.

- The final form of renewable energy is *geothermal energy*. As the name suggests, this form of energy uses naturally occurring heat in the earth's crust and is only really viable in areas of volcanic activity. There are two forms of geothermal energy. One is wet-rock geothermal energy where steam or hot water is trapped from boreholes or surface vents

and used to heat buildings or generate electricity. The other is hot-dry geothermal energy, which is accessed by boreholes drilled into hot dry rocks. Water is then forced down and the steam used to generate electricity. Geothermal energy, by its very nature, is a highly localized energy source and requires careful management. For example, there is now evidence from New Zealand that excessive extraction of geothermal energy has reduced the level of geyser activity, creating a conflict between the desire to generate power and heat houses and the need to maintain the revenues from tourists attracted by the volcanic springs. In 1998 geothermal energy accounted for 15 per cent of New Zealand's energy production.

In nearly all of the above cases, renewable energy seems capable of providing localized, small-scale solutions to energy needs. At present, renewable energy sources do not provide a viable alternative to fossil fuels. Furthermore, it could be argued that most of these alternatives require considerable capital investment and access to advanced technology; therefore, they do not offer a solution to the fast-growing energy needs of the 'developing world'.

Plate 5.6   Windfarms: a sustainable energy source for some, an eyesore for others. (Popperfoto)

5.2.3 The changing energy mix

It follows from the discussion above that the relative importance of the various sources of energy has varied across time and space. The balance between the various sources of energy is known as the energy mix. During the last century, there was a substantial change in the global energy mix, but as Table 5.2 shows, there still remains substantial regional variation. The evolution of the energy mix tends to be linked to the development of the industrialized world, simply because it accounts for most of the world's energy consumption (Table 5.3). At the beginning of the twentieth century, coal was by far the most important energy resource in the industrialized world. The oil industry was still in its early stages; the first modern oil well was drilled in Pennsylvania in the United States in 1859. Natural gas was not recognized as a resource at all. It was not until after the Second World War that the dominance of coal was threatened. Advances in transportation technology, such as the invention of the jet engine, and the emergence of a petrochemicals industry led to a rapid increase in the demand for oil in the post-war boom years. In 1950 coal accounted for 61 per cent of the world's commercial energy consumption and oil 27 per cent. By 1970 coal had fallen to 30 per cent, oil had increased to 44 per cent and natural gas to 20 per cent. The absolute level of

coal production did not decline (in fact at a global scale coal production continues to increase). The rapid economic growth experienced by the developed world was instead increasingly underpinned by oil and, more recently, by natural gas. The recent 'dash-for-gas' in the developed world is being driven by a desire to reduce dependence on oil imported from the Middle East and by the fact that using natural gas to produce electricity produces less greenhouse gases than coal or oil. Thus, in the last hundred years, the energy mix in the industrialized world has gone through two major transitions, from coal to oil and from oil to gas.

These energy transitions have not been a universal experience. As Table 5.2 illustrates, the energy mix in the developing world is somewhat different. For the most part the developing world remains more reliant upon oil for its commercial energy and in some regions coal remains the dominant resource. For example, in China coal still accounts for 75 per cent of commercial energy production and China accounts for 30 per cent of global coal consumption. Elsewhere dependence upon oil has made the developing world particularly vulnerable to the sudden increases in oil price that have been experienced at times over the last three decades. But, unlike the developed world, these economies lack the capital and technology to develop their own energy potential or to diversify their energy mix. However, as the developed world has depleted its

Table 5.2 Regional variations in energy balance, 2002 (per cent of total consumption)*

	Oil	Natural gas	Coal	Nuclear	HEP	Total (mtoe)
OECD	40.8	23.1	20.9	9.8	5.4	5,346.1
EU 15	43.2	23.6	14.8	13.7	4.7	1,468.9
North America	39.2	26.2	21.8	7.5	5.2	2,715.4
Europe & Eurasia	32.7	33.2	17.9	9.9	6.3	2,829.5
S. & C. America	47.9	19.7	4.0	1.0	27.4	448.2
Africa	40.8	20.9	31.1	1.0	6.4	291.0
Middle East	51.5	45.9	2.1	0.0	0.5	403.1
Asia Pacific	36.5	10.9	43.5	4.3	4.7	2,717.8
World	37.5	24.3	25.5	6.5	6.3	9,405.0

* Includes commercially traded fuels only. Therefore excludes wood, peat and animal waste as fuels.

Source: BP Statistical Review of World Energy (2003: 38). (Also available at: www.bp.com)

own sources of oil and gas it has become increasingly dependent upon supplies from the developing world. Thus there has been a marked change in the geography of energy production and, to a lesser extent, consumption. The transnational oil companies have been key actors in developing the energy resources of the developing world, but increasingly the governments of the oil-producing states in the developing world have seized control over their oil and gas industries. Now the leading oil majors, such as BP, ExxonMobil, ChevronTexaco and Shell, find themselves working in partnership with regimes whose actions in relation to human rights and environmental protection are condemned by governments and NGOs in the developed world.

5.2.4 The changing geography of energy production and consumption

Whereas in 1950 North America, Europe and the Soviet Union accounted for nearly 90 per cent of world energy demand, by 1990 their share had fallen to two-thirds (Jones and Hollier 1997: 181). In 1997 the member states of the Organization for Economic Cooperation and Development (OECD) accounted for 58.4 per cent of global energy consumption (Table 5.3). The redistribution of the global pattern of energy consumption has not been due to decline in the developed world, although during the 1990s this has

been the case in the so-called 'transition economies', but to increased demand in the developing world. Table 5.4 provides information on the average annual growth in commercial energy use between 1980 and 1995. During this period global energy consumption grew by 3.2 per cent, but growth in the high income economies (the developed world) was only 1.7 per cent a year, while growth in the low- and middle-income countries (the developing world) was 5.6 per cent a year. Thus, in relative terms, demand for energy in the developed world has been slowing, while it has been increasing in the developing world (compare this pattern with that of population growth discussed in the previous chapter). This differential growth rate, together with the decline in indigenous production in the developed world, has major implications for the geography of energy production and consumption and for global geopolitics.

There are at least three elements to the geography of energy production: the global distribution of reserves, the distribution of energy production and consumption and the resultant pattern of trade between energy surplus and energy deficit regions. As we noted earlier, the potential to produce coal, oil and natural gas is geologically determined, but whether or not a given deposit is exploited is the consequence of a whole host of factors. The regional distributions of global reserves of coal, oil and natural gas are presented in Tables 5.5, 5.6 and 5.7. Each table also includes the reserve to production ratio at the end of

Table 5.3 Global distribution of primary energy consumption, 2002 (per cent of total world consumption)*

	Oil	Natural gas	Coal	Nuclear	HEP	Total
OECD	61.9	54.1	46.6	85.6	48.9	5,346.1
EU 15	18.0	15.2	9.0	33.0	11.7	1,468.9
North America	30.2	31.2	24.7	33.6	24.1	2,715.4
Europe & Eurasia	26.3	41.2	21.1	45.8	30.2	2,829.5
S. & C. America	6.1	3.9	0.7	0.8	20.7	448.2
Africa	3.4	2.7	3.8	0.5	3.1	291.0
Middle East	5.9	8.1	0.3	0.0	0.3	403.1
Asia Pacific	28.1	13.0	49.4	19.3	21.6	2,717.8

* Includes commercially traded fuels only. Therefore excludes wood, peat and animal waste as fuels.

Source: BP Statistical Review of World Energy (2003: 38). (Also available at: www.bp.com)

Table 5.4   Energy production and efficiency

	Commercial energy use, average annual % growth 1980–95	Commercial energy use, per capita kg of oil equivalent 1995	GDP per unit of energy use (1987 $ per kg oil equivalent)	
			1980	1995
World	3.2	1,474	2.2	2.4
Low income	5.5	393	0.9	1.1
Excl. China & India	5.8	132	3.4	2.7
Middle income	5.8	1,488	1.2	1.1
Lower middle income	9.0	1,426	1.0	1.0
Upper middle income	1.9	1,633	1.7	1.5
Low & middle income	5.6	751	1.1	1.1
East Asia & Pacific	5.3	657	–	0.9
Europe & Central Asia	8.8	2,690	–	0.6
Latin America & Caribbean	2.7	969	2.2	2.0
Middle East & N. Africa	5.2	1,178	3.3	1.8
South Asia	6.6	231	2.0	1.7
Sub-Saharan Africa	2.0	238	2.1	1.9
High income	1.7	5,123	2.9	3.4

Source: World Bank (1998: 144–8) © The International Bank for Reconstruction and Development/The World Bank.

1997. This is a measure of how many years the current proven reserves would last if current levels of consumption were maintained. From a comparison of the reserve to production ratios for the three fossil fuels, it is quite clear that coal is the most abundant fossil fuel, followed by natural gas and then oil. At the end of 1997 the world had sufficient reserves of coal to last 219 years at current rates of consumption, while current oil reserves would last 40 years. These tables also reveal some interesting regional variations in the level of reserves. While the OECD countries have more than sufficient coal to meet their needs, it is these economies that have reduced their reliance upon coal the most. Thus, the developed world is dependent upon the very fuels for which it has the lowest reserves to production ratios. What this means is that the developed world is likely to become even more dependent upon imports of energy from the developing world. In the case of oil reserves, the Middle East still accounts for the vast majority of the

world's oil. Natural gas reserves are provided by two dominant regions, the Middle East and the former Soviet Union (predominantly Russia). Thus, the majority of the world's oil and gas reserves are in regions that are presently politically and/or economically unstable.

The relationship between production and consumption is illustrated in Figures 5.5 and 5.6. These diagrams compare the share of world production and consumption by major region. The comparative distributions of oil production and consumption show that Asia-Pacific, Europe and North America are the major deficit regions with Africa, the Middle East, the former Soviet Union and South and Central America as the major surplus regions. The individual countries involved in the resultant trade are identified in Table 5.8 and the movements of oil are shown in Figure 5.7. Despite being the second largest producer in 1997, the United States still imports almost as much oil as it produces. Likewise, North Sea production falls well

Table 5.5   Coal reserves at end of 2002 (million tonnes)

	Anthracite and bituminous	Sub-bituminous and lignite	Total	Share of total %	R/P ratio[1]
North America	120,222	137,561	257,783	26.2	240
S. & C. America	7,738	14,014	21,752	2.2	404
Europe & Eurasia	144,874	210,496	355,370	36.1	306
Africa & Middle East	56,881	196	57,077	5.8	247
Asia Pacific	189,347	103,124	292,471	29.7	126
Total World	519,062	465,391	984,453	100.0	204
Of which OECD	211,084	234,686	445,770	45.3	217

[1]Reserve to production ratio

Source: BP Statistical Review of World Energy (2003: 30). (Also available at: www.bp.com)

Table 5.6   Proved oil reserves at end of 2002

	Thousand million tonnes	Share of total %	R/P ratio[1]
North America	6.4	4.8	10.3
S. & C. America	14.1	9.4	42.0
Europe & Eurasia	13.3	9.3	17.0
Middle East	93.4	65.4	92.0
Africa	10.3	7.4	27.3
Asia Pacific	5.2	3.7	13.7
Total World	142.7	100.0	40.6
Of which OECD	9.4	6.9	9.7
OPEC	111.9	78.2	82.0
Non-OPEC[2]	20.1	14.4	11.9
Former Soviet Union	10.7	7.4	22.9

[1]Reserve to production ratio
[2]Excludes Former Soviet Union

Source: BP Statistical Review of World Energy (2003: 4). (Also available at: www.bp.com)

Table 5.7 Proved natural gas reserves at end of 2002

	Trillion cubic metres	Share of total %	R/P ratio[1]
North America	7.15	4.6	9.4
S. & C. America	7.08	4.5	68.8
Europe & Eurasia	61.04	39.2	58.9
Middle East	56.06	36.0	Over 100 years
Africa	11.84	7.6	88.9
Asia Pacific	12.61	8.1	41.8
Total World	155.78	100.0	60.7
Of which OECD	15.38	9.9	14.1
European Union 15	3.14	2.0	14.4
Former Soviet Union	55.29	35.5	75.5

[1] Reserve to production ratio

Source: BP Statistical Review of World Energy (2003: 20). (Also available at: www.bp.com)

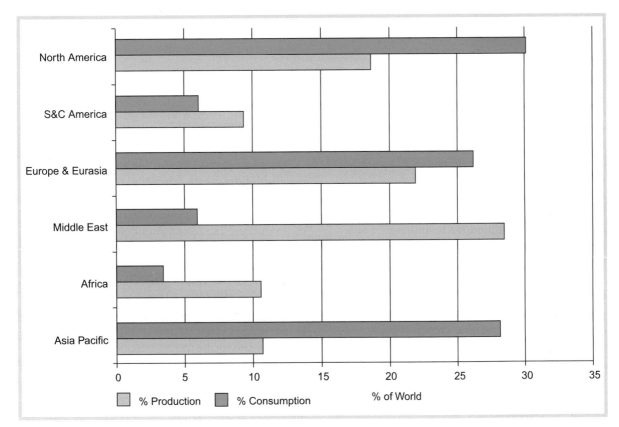

Figure 5.5 Distribution of oil production and consumption in 2002.

Source: BP Statistical Review of World Energy (2003: 6, 9). (Also available at: www.bp.com)

Table 5.8 Producers, exporters and importers of crude oil

Producers in 2002	Mt	% of world	Exporters in 2001	Mt	Importers in 2001	Mt
Saudi Arabia	409	11.5	Saudi Arabia	303	United States	426
Russia	378	10.7	Russia	162	Japan	209
United States	350	9.9	Norway	151	Korea	119
Mexico	178	5.0	Iran	119	Germany	105
Iran	176	5.0	Venezuela	109	Italy	91
China	169	4.8	Nigeria	105	France	86
Norway	156	4.4	Mexico	93	India	79
Venezuela	153	4.3	Iraq	90	Netherlands	61
Canada	133	3.7	United Kingdom	87	China	60
United Kingdom	115	3.2	UAE	79	Spain	58
Rest of World	1,331	37.5	Rest of World	663	Rest of world	663
World	3,548	100	World	1,961	World	2,057

Source: OECD/IEA, 2003, Key World Energy Statistics (Also available at: www.iea.org)

short of meeting Europe's demand for oil and is now past its peak production. The result is a movement of oil production dominated by Middle Eastern supply to Europe and North America. Figure 5.7 also reveals substantial movements to Singapore and Japan.

The major exporters and importers of natural gas are listed in Table 5.9, while Figure 5.8 shows the movements of natural gas. A comparison of the two figures reveals the less 'transportable' nature of natural gas; for the most part there is a balance within each region. The notable exceptions are Africa and the former Soviet Union (Russia and Turkmenistan), which have a surplus of natural gas, and Europe, which has a deficit. Natural gas supply into Europe is via pipeline from West Siberia (Russia) and North Africa or as LNG from North Africa. There is also a growing movement of natural gas in the form of LNG to Japan and South Korea. In 1995 LNG accounted for 14.3 per cent of Japan's mineral fuel imports (Japan Institute for Social and Economic Affairs 1996: 74–7). At present most of this LNG is supplied from within the Asia-Pacific region, but there is likely to be an increase in supply from the Middle East. For example, in 1995

the top three suppliers of LNG to Japan were Indonesia (41.2 per cent), Malaysia (18.4 per cent) and Australia (16.5 per cent). There is also the possibility of pipeline supply of natural gas from Russia to China, Japan and South Korea.

The geographies of production and consumption described above are the result of a complex interaction of economic and political factors. Not surprisingly, there are substantial regional differences in the cost of oil production. The record-low oil prices of the late 1990s focused attention upon the actual cost of producing oil. A report by Cambridge Energy Research Associates puts the cost of production in the Middle East as low as US$2/barrel (*Economist*, 6 March 1999: 29). That same report estimated the cost of production in Indonesia at US$6, Venezuela and Nigeria at US$7, Mexico at US$10, the United States and the North Sea at US$11 and Russia at US$14. These production costs pose two questions: why develop high-cost fields when the Middle East has massive reserves and very low production costs, and what will happen to the high-cost producers if oil prices fall?

The answers lie in the realms of geopolitics as much

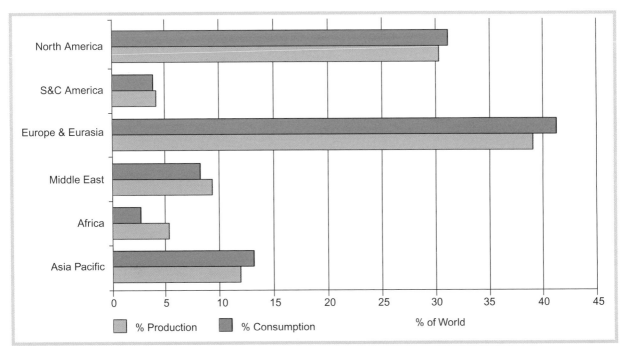

Figure 5.6 Distribution of natural gas production and consumption in 2002.
Source: BP Statistical Review of World Energy (2003: 22, 25). (Also available at: www.bp.com)

as economics. Prior to the first oil shock of 1973–4 the industrialized world had become increasingly dependent upon supplies of cheap oil from the Middle East (Odell 1989). When the Middle Eastern oil producers formed OPEC (the Organization of Petroleum Exporting Countries, see: www.OPEC.org) and used oil as a political weapon to punish the industrialized West for supporting Israel in the Arab-Israeli war this brought about a rapid increase in the price of oil. The first oil shock was followed by a further shock in 1979–80 following the Iranian Revolution and a mini-shock in 1990–1 because of the Gulf War (see Figure 5.9). Following the second shock prices started to decline. There were at least three reasons: firstly, high energy costs promoted conservation; secondly, the high price of oil and the actions of OPEC promoted production by high-cost non-OPEC producers; and thirdly, recession and economic restructuring reduced the growth of demand in the developed world. The economic crisis in Asia further dampened demand for energy. As a response to very low prices, OPEC members agreed to cut back production and this action, combined with global economic recovery, saw prices recover during 1999. However, subsequent events dramatically illustrate the volatile nature of oil and gas markets.

In the aftermath of 11 September 2001, a global economic downturn kept oil prices low. Then the global geopolitical situation changed dramatically with the war in Afghanistan and then the second Iraq War. As this book goes to press, a combination of growing demand for oil, particularly in the US and China, uncertainty over the future status of Iraqi oil production and conflict between the Russian government and Russia's largest oil company Yukos, have conspired to push oil prices well above $40 a barrel. There is even media speculation that we could see oil prices over $50 a barrel for some time to come, threatening global recession. In response, OPEC, particularly Saudi Arabia, has agreed to increase production, but the market remains unconvinced. High oil prices have now resulted in renewed interest in non-renewable sources of energy and even a growing enthusiasm for nuclear power. Furthermore, continued instability in the Middle East is encouraging western governments and the multinational oil companies to look for oil and gas elsewhere, such as off the west coast of Africa. All of these developments illustrate the complexity of the notion of resource scarcity. But what does the current situation mean for the developing world? Much depends on whether or not the developing world follows the same pattern of economic development as the developed world; this is the subject of the final part of this chapter.

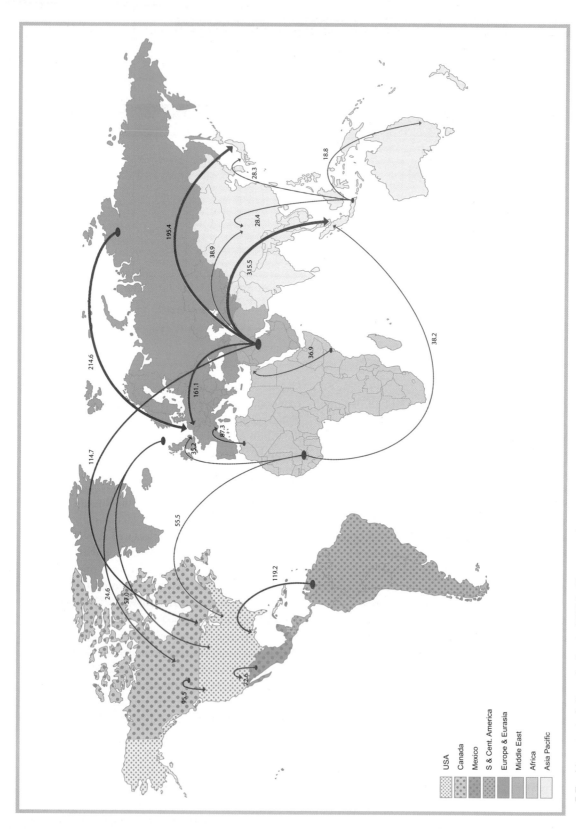

Figure 5.7 Movement of oil production in 2002 (million tonnes).

Source: BP Statistical Review of World Energy (2003: 19). (Also available at: www.bp.com)

Table 5.9 Producers, exporters and importers[1] of natural gas

Producers in 2002	Mm^3	% of world	Exporters in 2002	Mm^3	Importers In 2002	Mm^3
Russia	595,000	22.7	Russia	190,000	United States	113,480
United States	539,349	20.6	Canada	106,232	Germany	81,341
Canada	182,075	7.3	Algeria	59,980	Japan	72,637
United Kingdom	108,204	4.1	Norway	56,260	Italy	59,291
Algeria	82,554	3.2	Netherlands	52,529	Ukraine	55,519
Netherlands	75,315	2.9	Turkmenistan	39,391	France	45,271
Indonesia	70,816	2.7	Indonesia	35,147	Netherlands	26,771
Norway	67,627	2.6	Malaysia	18,574	Korea	23,280
Iran	66,320	2.0	Qatar	18,429	Spain	20,725
Saudi Arabia	60,570	2.3	United States	14,618	Turkey	17,113
Rest of world	770,023	29.4	Rest of world	104,297	Rest of world	188,338
World	2,617,853	100.0	World[2]	695,457	World	703,766

[1] Exports and imports include pipeline gas and LNG
[2] World trade includes intra trade of former USSR

Source: OECD/IEA, 2003, Key World Energy Statistics. (Also available at: www.iea.org)

5.3 Energy and development

What is the relationship between energy consumption and economic development? The experience of the developed world suggests that in the initial phases of industrialization there is a direct link between increased energy consumption and economic development. That is, as industrial activity grows it consumes more and more energy. However, in the last thirty years it has become increasingly apparent that in the developed world the relationship between economic development and energy consumption has changed. This was a direct consequence of the OPEC-inspired energy crisis of the 1970s and is largely due to the processes of de-industrialization and economic restructuring that have seen the less energy-intensive service sector replace heavy industry and, to a lesser degree, manufacturing as the major generators of wealth. At the same time, conservation and technological change has made industry more energy-efficient. Similarly, a change in transportation technology has increased energy efficiency. Thus, today's post-industrial societies have decoupled the link between economic growth and increased energy consumption. Nevertheless, it is still the case that the developed world consumes the lion's share of the world's energy. Indeed, just one country, the United States, which accounts for only 5 per cent of the world's population, consumes almost 25 per cent of total world energy consumption (Mather and Chapman 1995: 140).

5.3.1 Energy consumption and economic development

The relationship between energy consumption and economic development is usually depicted as a scatter plot between the level of per capita energy consumption on the one hand, and GNP per capita on the other hand (Figure 5.10). The data do indeed suggest a clear relationship between energy consumption and GNP per capita; the higher the level of energy consumption the higher the level of GNP per capita. There are obvious outliers, but these are easily explained away. Some countries traditionally have higher levels of energy consumption than GNP. This was the case in the Soviet-type economies that had a bias towards heavy industry

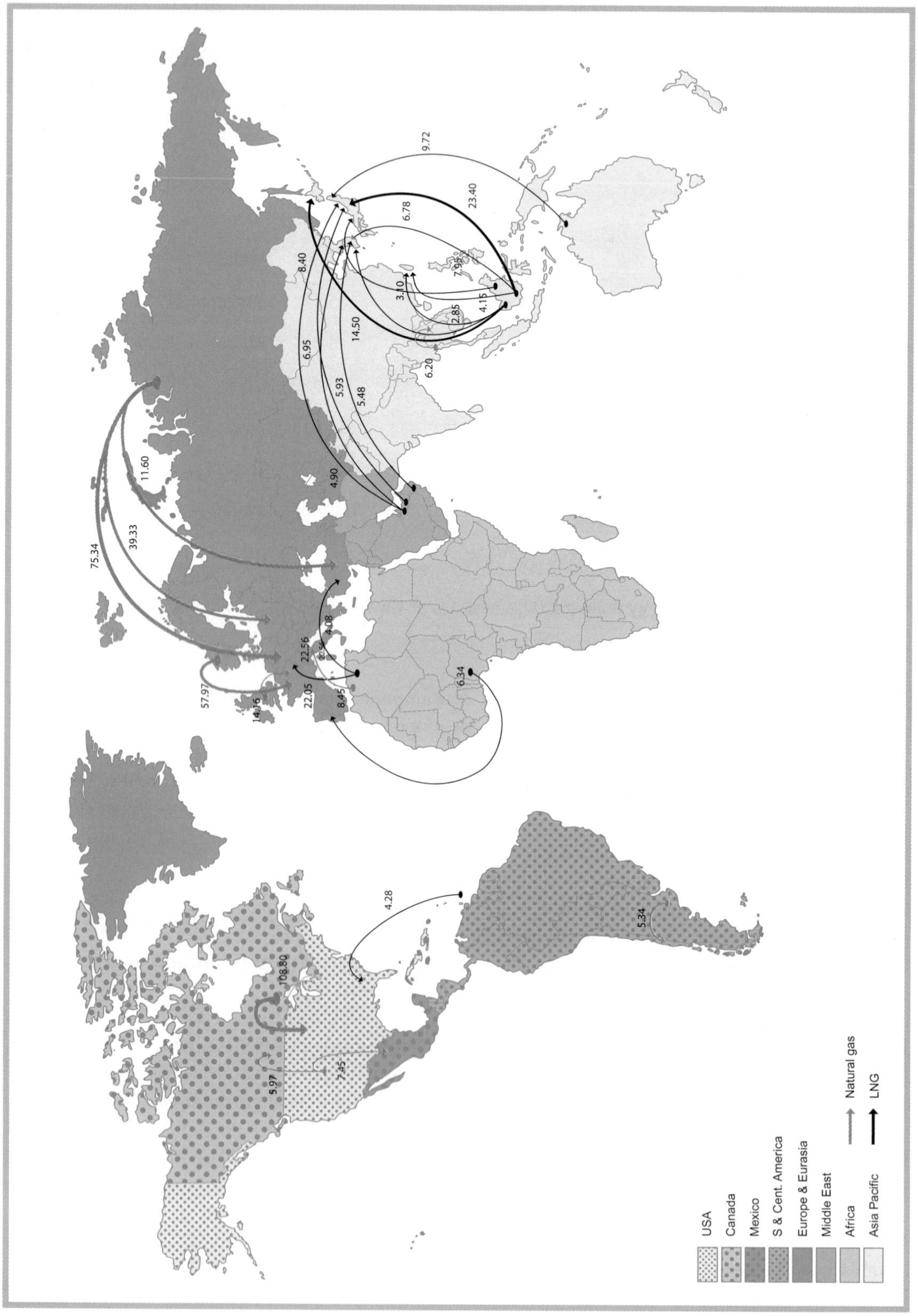

Figure 5.8 Movement of natural gas in 2002 (billion cubic metres).
Source: BP Statistical Review of World Energy (2003: 29). Also available at: www.bp.com)

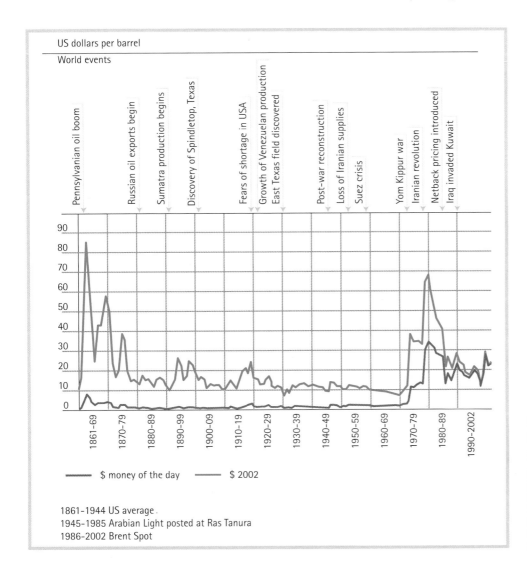

US dollars per barrel

World events

Pennsylvanian oil boom
Russian oil exports begin
Sumatra production begins
Discovery of Spindletop, Texas
Fears of shortage in USA
Growth of Venezuelan production
East Texas field discovered
Post-war reconstruction
Loss of Iranian supplies
Suez crisis
Yom Kippur war
Iranian revolution
Netback pricing introduced
Iraq invaded Kuwait

90
80
70
60
50
40
30
20
10
0

1861–69
1870–79
1880–89
1890–99
1900–09
1910–19
1920–29
1930–39
1940–49
1950–59
1960–69
1970–79
1980–89
1990–2002

—— $ money of the day   —— $ 2002

1861–1944 US average
1945–1985 Arabian Light posted at Ras Tanura
1986–2002 Brent Spot

Figure 5.9
Historical trends in the world oil prices.
Source: BP Statistical Review of World Energy (2003: 14). (Also available at: www.bp.com)

and were notoriously wasteful in their use of energy. Others, such as Japan, have low levels of energy consumption given their high GNP. This is because they have introduced energy-saving technologies and have exported energy-intensive industry offshore. Finally, there are some countries, such as Canada, where climatic extremes (both hot and cold) require a large amount of energy to be used for heating or for air conditioning. There are also other factors that affect the level of energy consumption: for example, the level of urbanization. But it has also been suggested that there is a similar relationship between economic development and urbanization as there is between the former and energy consumption. Finally, the size of a country may also affect the level of energy consumption, as larger amounts of energy are required to move between places. Such an argument could be marshalled to explain the

very high levels of energy consumption in the United States, as could climatic factors.

On the face of it the relationship between energy consumption and economic development seems unproblematic. However, as Chapter 10 reveals, the whole question of what stands for 'development' is problematic and contested. The measurements used in Figure 5.10 systematically understate the relative position of the developing world. First, because the use of commercial energy as a measure of energy consumption ignores the role of non-commercial biomass energy sources, as a consequence it only measures the 'modern' sectors of the economy. Second, the monetary measure of GNP per capita is an inadequate measure of the 'human condition' in many of the world's poorest countries. That said, a re-evaluation of the relative position of the poorest

Plate 5.7   When 'muck was money'. (© Sheila Gray)

Plate 5.8   Not a smoke stack in sigh on this post-industrial estate. (Popperfoto)

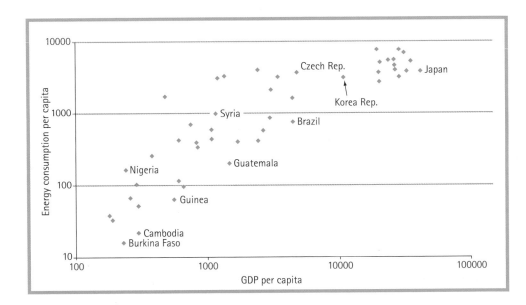

Figure 5.10
Relationship between energy consumption and economic growth.

countries would only move them slightly 'up the curve'. It would not alter the fact that there is a clear relationship between industrialization and energy consumption. Even so, industrialization does not necessarily equate to development.

The relationship between industrialization and energy consumption may seem to suggest that all the countries of the developing world will eventually follow the same energy and development trajectory as that experienced by the developed world. Thus, the patterns of energy and development can be equated to a stage model of the kind proposed by W.W. Rostow (this is discussed further in Chapter 8). Such a model is presented in Figure 5.11. This model examines the change in energy ratios over time as an economy develops. The energy ratio is the relationship between the rate of change in energy consumption and the rate of change in economic growth (Mather and Chapman 1995: 154). In the pre-industrial phase, the energy ratio is less than one as these are 'low-energy' societies, mainly dependent on subsistence agriculture. As discussed earlier (also see Chapter 2), the industrial revolution resulted in a change in the relationship between society and resource consumption, and the harnessing of new sources of energy was at the heart of this process. As economies industrialize, the energy ratio begins to exceed one. In the early phases of industrialization energy efficiency is low and economies are dominated by energy-intensive industries. As the industrial economy matures so the energy ratio declines. This is due to increased

efficiency, increased energy costs or the actual decline in the level of economic activity. This stage is typified by the developed world during the 1970s. Eventually, the economy moves into a post-industrial phase and the ratio falls below one. If the model seems very familiar, it is much like the model of the demographic transition (see Chapter 4) because it describes the evolution of the energy economy as it actually occurred in Western Europe and the United States.

What evidence is there that the rest of the world will inevitably have to go through the same stages? It could be argued, with some justification, that the Newly Industrialized Countries (NICs) of Asia have followed this pattern of industrialization; however, the time taken to move through the phases here seems to have been compressed as the NICs have tried to diversify their economies and promote service sector growth. They have also tried to industrialize without the advantage of substantial indigenous energy resources, a fact that has made them vulnerable to the effects of price increases. Somewhat ironically, the financial crisis in Asia was largely responsible for the fall in energy prices in the late 1990s. By contrast, the transition economies of the post-socialist world (East-Central Europe and the former Soviet Union) pose a different challenge to the model. Prior to 1989–90 these economies were undoubtedly on the steepest part of the curve, consuming large amounts of energy and producing a relatively low level of economic growth. Isolated from the oil price shock by cheap supplies of

Thematic Case Study 5.3

Resource abundance: blessing or curse?

From all that has been said in this chapter, one would assume that having a relative abundance of natural resources would convey an advantage in terms of prospects for economic development. However, the reality is somewhat different. There is now substantial evidence to suggest that resource-abundant economies have actually performed worse in terms of rates of economic development than resource-poor economies. The geographer Richard Auty (1993, 2001) was the first to term this phenomenon the 'resource curse thesis'. Sachs and Warner (2001) have weighed up the evidence and conclude that there is sufficient evidence to suggest that 'high resource intensity tends to correlate with slow growth.' In other words, economies that have a high degree of dependence on the resource sector tend to grow more slowly than resource-poor economies. While the 'resource curse thesis' is now generally accepted, there are exceptions to the rule, those usually cited being Botswana, Chile and Malaysia, and there is no single explanation for this under-performance. Sachs and Warner suggest that most forms of explanation follow some form of 'crowding out' logic, whereby the dominant resource sector inhibits the development of the non-resource sector. Resource economies are particularly susceptible to fluctuations in income due to the volatility of resource prices, added to which, as we know from earlier discussions, the resource base itself is often soon depleted. Once the 'resource boom' is over, the economy is not sufficiently developed or diversified to sustain living standards and a period of 'bust' often follows. Such a 'boom and bust' cycle is by no means inevitable: effective government policy can use resource income to promote a more diversified economy that can sustain living standards once the boom has passed. A review of the literature (Stevens 2003) suggests that there are number of dimensions to the 'resource curse' (though Stevens favours the term 'resource impact'):

- In many resource-rich economies there has been a failure to save income during boom periods to cover periods of bust (some states have created so-called 'stabilization funds' to save for a rainy day), plus a tendency to spend income on consumption (usually through increased imports) and on prestige projects.

- There is also a failure to redeploy income from the resource sector to promote a more sustainable pattern of economic development. In some instances there is also a tendency to use resource income to subsidize and protect the activity of inefficient producers in the non-resource sector. Later, when the resource income dwindles and the economy is opened up to international competition, these inefficient producers then fail.

- In many instances, often as a consequence of a colonial heritage, the resource economy remains relatively isolated from the rest of the economy. This minimizes the multiplier impact of large-scale resource-based investment projects, beyond their payment of taxes. One solution is to impose a 'local content' requirement on the resources companies, forcing them to use local suppliers of goods and services. However, given the relative lack of 'economic development' in many resource economies it is often difficult to source goods and services locally.

- The increase in export income associated with a resource boom tends to result in a strengthening of the value of the domestic currency of the resource economy. This can have the effect of making the cost of domestic production in the agricultural and manufacturing sectors higher than the cost of imports. This is known as 'Dutch Disease' following the experience of the Netherlands; the net result is a decline in the competitiveness of the non-resource sector, which aggravates the problems discussed above.

- Finally, there is increasing evidence that suggests that a sudden influx of resource income tends to promote crime and corruption, armed conflict and an abuse of human rights. Such problems not only have a direct impact on the welfare of individuals, they also promote increased social inequality and

Thematic Case Study 5.3 continued

undermine the effectiveness of the state (see Ross 2001, Renner 2002).

The continuing under-performance of resource-rich economies has led the World Bank and the major resource companies to reassess the impacts of resource development. The World Bank through its Extractive Industries Review (www.eireview.org) is reconsidering whether it should be promoting resource-based development as a means of improving living standards and promoting sustainable development. The EIR's final report, called *Striking a Better Balance*, concludes: 'the Extractive Industries Review believes that there is still a role for the World Bank Group in the oil, gas and mining sectors – but only if its intervention allows extractive industries to contribute to poverty alleviation through sustainable development and that can only happen if the right conditions are in place'. The EIR was prompted by the World Bank's decision to finance the construction of the Baku–Tbilisi–Ceyhan pipeline from landlocked Azerbaijan to the Mediterranean coast of Turkey. The collapse of the Soviet Union has prompted a dash to gain access to new energy resources in the Caspian and Central Asia, and more recently in Siberia and the Russian Far East. At the same time, there is also increased oil and gas activity off the shores of West Africa (Gary and Karl 2003) as the developed world seeks to gain control of new, non-OPEC, sources of energy. These developments will create a new set of resource abundant economies and regions; NGOs such as Revenue Watch (www.revenuewatch.org) are now closely monitoring the flow of revenues surrounding these new resource projects (see Caspian Revenue Watch 2003). The dangers of the resource curse are now well recognized, but only time will tell if the international community can assist the newly resource-rich economies to avoid the pitfalls of the past.

Soviet oil, these economies had failed to increase their energy efficiency during the 1970s and 1980s. This failure was a major reason for the collapse of the Soviet-type economic system. Transitional recession has followed the collapse of the socialist economies and this has depressed energy consumption. The question now remains: what will the relationship be between energy consumption and economic recovery in the transition economies? Will they follow the path of post-industrialization (see Chapter 13) or will they re-industrialize? The likely answer is that individual economies will follow their own paths (OECD 1999). A much bigger question mark hangs over the developing world. Is it inevitable that it will have to industrialize to improve living standards? If it is, then the availability of energy resources will be very much centre stage for world economic development in the twenty-first century. Furthermore, those same countries are likely to be the major suppliers of oil and gas and other minerals and there is plenty of evidence to suggest that such resource-based development often fails to provide a sustainable basis for improving living standards (see Case study 5.3).

5.4 Conclusions: global resource dilemmas

This chapter has considered the nature of resources, through a detailed case study of the relationship between energy and economic development. The chapter concludes by considering the resource dilemmas that now confront the world. It is no longer the relatively simple matter of whether or not the world has sufficient resources to meet demand. In the short term, there is little danger of physical scarcity at a global scale. Rather the essential dilemma is whether the global ecosystem can absorb the consequences of continued increases in resource production and consumption. Furthermore, the exact nature of the resource dilemma varies across the globe. Therefore, there is no single resource problem, nor a universal solution.

In the advanced post-industrial societies of the developed world, the dilemmas relate to the economic and political costs of geopolitical scarcity (see Table 5.1). Increasingly, the economies of the developed

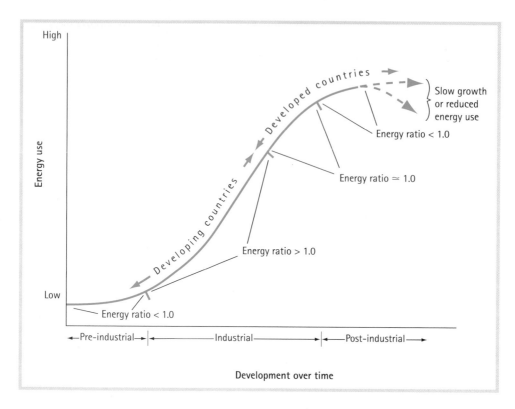

High

Energy use

Developed countries

Slow growth or reduced energy use

Energy ratio < 1.0

Energy ratio ≈ 1.0

Developing countries

Energy ratio > 1.0

Low

Energy ratio < 1.0

←Pre-industrial→ |——Industrial——| ←Post-industrial→

Development over time

Figure 5.11
Energy ratios and economic development.
Source: A.S. Mather and K. Chapman, *Environmental Resources*, Pearson Education Limited, 1995

world are finding themselves dependent upon what they perceive to be 'hostile' and 'unstable' sources of supply in the developing world. The response has been to develop high-cost resources close to home, to promote increased efficiency and conservation and to seek alternatives, such as renewable energy. This strategy imposes an economic cost as it forces the developed world to pay more for its energy than it needs to. However, decoupling economic growth from energy consumption, increasing energy efficiency and reducing carbon dioxide emissions compensate for the increased cost of energy. The developed world can afford to seek technological solutions to its resources problems, particularly in the area of energy supply.

In the transition economies of Central Europe and the former Soviet Union resource dilemmas are somewhat different. The Soviet economic development strategy imposed a resource-intensive form of industrialization upon these countries that made them dependent upon cheap supplies from the Soviet Union (predominantly Russia); it also caused widespread environmental degradation. The economic transformation now taking place in the post-socialist world has closed many of the smoke-stack industries; however, the newly independent states in this region now face the problem of having to secure access to resources. The resource-rich states, such as Russia,

hope that economic recovery will bring a resource bonanza as regional demand rebounds and prices increase. It remains to be seen if economic recovery in Central Europe and the former Soviet Union will create increased demand or whether these economies will follow the path of the developed world and achieve economic growth without an equivalent increase in resource demand. There is already evidence that the fast-reforming economies of Central Europe and the Baltic States are likely to follow the model of Western Europe; while the member states of the Commonwealth of Independent States will follow a different, and as yet undetermined, path. Russia's reluctance to sign the Kyoto Protocols (see Chapter 6) is partly due to uncertainty over the future structure of the domestic economy and thus its environmental footprint.

In the developing world there is considerable diversity in terms of resource dilemmas. The NICs of south-east Asia have succeeding in industrializing, but in doing so they have become increasingly dependent upon resource imports. Most recently, economic crisis in Asia has severely depressed resource demand. Elsewhere in the developing world a dualistic pattern of resource consumption seems to have developed. On the one hand, there remains a 'low-energy' society dependent upon biomass energy and subsistence agriculture, in which the majority of people exist. Here population pressure is leading to environmental problems as forests are stripped for fuelwood and land ploughed up or over-grazed. At the same time, there exists a growing 'high-energy' society linked to increasing commercial, agricultural (see the next chapter for more on this) and industrial activity and the spread of urbanization. Much of this activity is aimed at supplying resources to the developed world. The two sectors of society combine to create an increasing demand for resources and pressure on the environment. The balance between the sectors also varies greatly among the countries of the developing world.

Given that the majority of the world's population lives in the developing world, this growth in demand will be the major factor contributing to increases in global resource demand. In its *Global Environment Outlook 2000*, the United Nations Environment Programme (UNEP 1999: 2) concluded that: 'A tenfold reduction in resource consumption in the industrialized countries is a necessary long-term target if adequate resources are to be released for the needs of developing countries.' It may be that in the future it is increasing demand in the developing world that raises the spectre of resource scarcity in any number of forms. However, the developed world already recognizes this threat and is using the issue of global warming to try to curb increased resource consumption in the developing world. For this to happen without curtailing the life chances of the people of the developing world, the developed world must help to shape an entirely new relationship between resources and development.

Learning outcomes

Having read this chapter, you should understand that:

- The notion of what is a resource is socially constructed and varies through time and across space.

- There are a variety of different ways of classifying resources.

- There are a complex set of factors that influence the availability of resources.

- There are substantial regional variations in energy production and consumption.

- There is a complex and changing relationship between energy consumption and economic development.

- Different types of resource dilemmas confront the different regions of the world.

Further reading

Elliott, D. (2003) *Energy, Environment and Society*, 2nd edition, Routledge, London. Provides a good overview of the key issues and energy alternatives.

Jones, G. and **Hollier, G.** (1997) *Resources, Society and Environmental Management*, Paul Chapman, London. Provides more detailed discussion of the issues raised by this chapter and deals with some essential issues, such as water resources, which have not been discussed in any detail.

Mather, A.S. and **Chapman, K.** (1995) *Environmental Resources*, Longman, London. Provides a detailed, technical analysis of the resource issues raised by this chapter with numerous supporting examples.

Middleton, M. (2003) *The Global Casino: An Introduction to Environmental Issues*, 3rd edition, Arnold, London. Provides concise analyses of numerous environmental issues; each chapter contains a list of suggested readings and recommended web-sites.

O'Riordan, T. (ed.) (2000) *Environmental Science for Environmental Management*, 3rd edition, Prentice Hall, Harlow. An essential guide to the science behind the issues, plus up-to-date discussion of key challenges such as sustainable development and climate change.

Sarre, P. and **Blunden, J.** (eds) (1995) *An Overcrowded World? Population, Resources and the Environment*, Open University and Oxford University Press, Oxford. One of a series of books produced by the course team at the Open University in the UK. This volume deals with many of the issues raised in this section. Chapter 4 'Sustainable resources?' and Chapter 6 'Uneven development and sustainability' are of particular relevance to this chapter.

Useful web-sites

There is a huge amount of material available on the web that relates to the issues discussed in this chapter. Some of the key reference sources are listed below; additional web-sites have been referenced in the text.

www.bp.com/centres/energy Home of the BP *Statistical Review of World Energy*. This web-site, which is updated on an annual basis, has a wealth of statistical information on the energy sector, much of which is down-loadable in Excel format or as PowerPoint slides.

www.iea.org The official site of the International Energy Agency. The site contains information on the Agency's operations and publications, as well as statistics on energy production and consumption. Also has links to other international organizations.

www.eia.doe.org The official site of the US Government's Energy Information Administration. In addition to information on US energy matters, it contains reports on individual countries and a massive list of links to other US government agencies, international agencies, foreign governments and commercial company sites.

www.worldbank.org The official site of the World Bank. The site contains information on World Bank operations and publications, as well downloadable statistics and briefing documents.

www.wri.org The World Resources Institute is an independent centre for policy research and technical assistance on global environmental and development issues. Its web-site provides information on its own activities and publications, some of which are downloadable, as well as links to other organizations in the same area. The Institute, together with the UNEP, UNDP and World Bank, produces the biennial *Resource Report* that is an indispensable reference guide to the state of the earth's resources.

 For annotated, clickable weblinks and useful tutorials full of practical advice on how to improve your study skills, visit this book's website at www.pearsoned.co.uk/daniels

CHAPTER 6

Environment and environmentalism

ANDREW MILLINGTON AND JENNY PICKERILL

Topics covered

- Why and in what ways environment is an issue for society at the beginning of the twenty-first century

- The local and global dimensions of environmental issues

- The evolution of modern environmentalism from the early 1960s, through the Rio Earth Summit in 1992, to the World Summit on Sustainable Development in 2002

6.1 Environment matters

Open your local newspaper. If this happens to be in the UK, you will probably read about flood alleviation, fly (illegal) tipping of household and business waste, field trials of GM crops, and new laws about the storage of garden chemicals. Read also about the opinions of local politicians, pressure groups and local residents on these and other issues. Subscribe to the popular science magazine *New Scientist*, and in the last eight months you would have read about the results of research and the views of commentators on issues such as GM crops in the developing world; GM crop trials in the UK; how the 'green revolution' has made Africa's farmers poorer; how civil engineers plan to build huge dams and alter the courses of some of Asia's biggest rivers; flaws and failures in nature conservation; and the probable collapse of the Kyoto Protocol. Here are two types of journalism, two scales of journalism – but what lessons can be learnt about the environment from them?

Firstly, environmental issues affect all places and societies – the urban and the rural, the rich and the poor, and the developed and the developing – nobody is unaffected. Secondly, different groups of people (let's use the term 'stakeholders') hold strong views and different opinions on the same environmental issue: views and opinions that reflect their status in society, political persuasion and nationality. Thirdly, all environmental issues affect people, but many are also directly or indirectly caused by people's actions. Fourthly, genetic modification of crops and the engineering of major rivers provide examples of our continued reliance on advances in technology to control the environment. Finally, the examples illustrate that environmental issues range in scale from the local to the global.

6.2 The local to the global

Archaeologists, such as those researching the copper mining area at Wadi Faynan, Jordan, which dates back 4,500 years (Grattan *et al.* 2003), routinely show that humans have impacted on their local environments for millennia. The archaeological and geographical literatures are replete with examples of human impacts on the environment through prehistorical, historical

Plate 6.1   A report, released by the University of Sydney's Environmental Geology Group, found that the world-famous Sydney Harbour sea-floor is one of the most polluted urban marine environments in Australia. (Corbis/David Gray)

Table 6.1   Main global environmental issues at the start of the twenty-first century

Cause(s)	Main effects
Intensification of farming and aquaculture systems	Excessive chemical use (e.g. fertilizers, herbicides, pesticides, antibiotics) contaminates (pollutes) soils, surface waters, groundwater and coastal waters which alter ecosystems, leading to declining water quality (which can affect human health), contaminate foods (which can also affect human health) and impact on biodiversity.
Inappropriate agricultural management practices (can be caused by changes in rural population densities, or the economics of agricultural production)	Soil erosion by water and wind create on-site effects such as decreased crop yields which increase poverty and may lead to migration from economically-marginal farmlands. Off-site effects with high economic costs include siltation of irrigation systems, dams and harbours.
Poorly designed irrigation systems	Irrigation often leads to soil salinization and waterlogging which depress crop yields. Ultimately the land can be lost to cultivation permanently. Over-extraction of groundwater can lead to shortage of water supplies and, in coastal areas, the influx of seawater into the groundwater. This also compromises agricultural productivity and farms can be abandoned.
Biotechnology (particularly genetic modification)	Concerns over loss of biodiversity and uncontrolled interbreeding between GMOs and unmodified organisms, though largely unproven at present time.
Deforestation	Changes to livelihoods of forest people, loss of biodiversity, changes to hydrological cycle and increased soil erosion, impacts on climate change through release of carbon to atmosphere and reduced carbon storage. Mainly concentrated in tropical and boreal forests.
Overfishing	Inadequate or non-existent controls on fishing in many of the world's oceans and seas have led to declining fish stocks (which have knock-on effects on marine ecosystems), which directly affects people's livelihoods and national economies where fishing is a major element.
Industrialization	Fossil-fuel based industrialization without clean technology (especially in Asia) is leading to increased greenhouse gas emissions thereby elevating global temperatures and, more generally, climate change. Transboundary air pollution leading to environmental acidification of environments downwind: this affects farming systems and aquaculture. Localized contamination of air, soil and water affecting health of local people. At the global scale this is offset by the decline in, and the clean-up of, heavy industries in North America, Europe and the CIS.
Urbanization	Loss of agricultural and other types of land use causes declines in agricultural production and loss of biodiversity. The local and regional hydrological cycles are disrupted: flooding is often more frequent. Urban heat islands develop. There are often transport- and industrialization-related issues (see Transportation and Industrialization, this table).
Major land use changes	1 Loss of agricultural land to urban and industrial development reduces agricultural production and displaces people. 2 Loss of wetlands (marshes, lakes, etc.) reduces biodiversity and affects the local and regional hydrological cycles. The environmental cleansing function of wetlands is lost leading to increased pollution. Livelihoods that depend on wetlands (e.g. farmers in semi-arid areas) are compromised. 3 Conversion of forest (see Deforestation, this table).
Transportation (mainly increased private car ownership, and high volumes of air transport)	Increased greenhouse gas emissions leading to elevated global temperatures and, more generally, climate change. CFCs and HCFCs also deplete the ozone layer. Localized effects: loss of land to new roads and airports, noise pollution and respiratory health effects affect people's livelihoods and their health.
Tourism	Tourism increased rapidly in the late twentieth century and some analysts predict that it will be the largest industry globally within a few decades. Impacts include increased fossil fuel use (see Transportation, this table), land use conversion (especially in coastal regions) (with local impacts on ecology and hydrology), and specific types of damage (e.g. to coral reefs). The impacts are offset by increased employment and, in some areas, by sustainable tourism.

Table 6.1   continued

Cause(s)	Main effects
Climate change	Climate change is considered the most important environmental issue at the present time as it impacts on most issues directly or indirectly. For example, biodiversity is lost (e.g. coral bleaching), increased storminess increases damage to coastal communities in particular, changes in crop production and cropping patterns directly affect farming communities.
Drought	Major changes to livelihoods through reduction in crop and animal productivity, or total loss of agricultural production. Increased levels of poverty in, and migration from, drought-prone areas. At the nation state level droughts affect national food security.
AIDS	Reduction in labour in labour-intensive agricultural and aquaculture systems leads to decline in food production and environmental damage because of lack of maintenance.
Poverty	Increased levels of poverty, particularly amongst rural populations, are forcing them to fall back on natural resources and use them more intensively. The urban poor are vulnerable to enhanced health risks through, for example, working in unsafe conditions in dangerous industries, scavenging on refuse dumps, and living in low-quality housing.
Terrorism	Disruption to normal patterns of life and commerce. Threats of toxic releases (chemical and biological agents).

and recent times (many syntheses – e.g. Turner *et al.* (1990), Goudie (1993), Hannah *et al.* (1994), Meyer (1996), UNEP (1997, 2000, 2003) – are available). What is new, however, is that we currently live in an era of unprecedented environmental concern. We identify a series of environmental problems that encompass the entire planet, and impact on most of the world's population and economies – these problems are called global environmental issues. Reasonable consensus has been reached concerning the scope and importance of such issues (Table 6.1); but what makes them global? Environmental scientists identify two types of global environmental change. On the one hand, systemic change occurs when there is a direct impact on a physically interconnected, global system, i.e. the atmosphere or the oceans: for example, the affects that increased emissions of greenhouse gases have on global climate. Cumulative change, on the other hand, occurs when many discrete events become significant because their distribution is global or because, added together, their impact is felt across a large proportion of the globe. For example, across the Americas, Africa and Asia-Pacific, the relatively small amounts of tropical forest lost in each area being converted from forest to farmland, or being lost to a mining operation, add up to a global-scale problem affecting humid tropical forests, one of the most biodiverse biomes found on Earth: forests of vital significance to the global water and carbon balances.

Recognition of this duality is important in formulating policies and strategies. Those addressing systemic issues need to be global, with all countries agreeing and adhering to a particular policy or strategy. Those focusing on cumulative issues require both global approaches (e.g. in the fields of data exchange and comparative research) and local approaches (to reduce the impacts of a problem because slightly different causes exist in different geographical locations) (Middleton, 2003).

6.3 Understanding causes

How have global environmental issues arisen? Whilst the Earth's environment has changed throughout geological time, it is only recently that we have been able to (i) demonstrate that the impact of humankind on many aspects of the environment is greater than previous natural change, and (ii) understand the complexity of society–environment links.

For some global environmental issues, evidence clearly points to the current changes having a human origin. Three examples are:

- The exponential rise in carbon dioxide levels and global temperatures after the industrial revolution

- Presence of halocarbons (e.g. chlorofluorocarbons and hydrochlorofluorocarbons) in the atmosphere as they do not occur naturally, i.e. they must be the result of manufacturing industries

- The rate of extinction of plant and animal species is greater than that in any of the previous five episodes of biological extinction that geologists recognize from the fossil record, all of which occurred before humans had evolved.

For many, evidence like that listed above is unequivocal proof coupling human actions and environmental damage. However, it does not explain, though it may imply, cause. Moreover, these examples only reach the outer strands of a complex web of relationships between humans and the environment. Middleton (2003), for example, develops a fourfold typology of society–environment relationships (Table 6.2). Explanations by scientists of the human causes of environmental degradation have, until recently, almost entirely focused on direct causes. For example, there is general agreement amongst biologists that **biodiversity** loss has five direct causes:

- destruction or degradation of habitat

- the effects of pollution

- the introduction of new species into an area

- **climate change**

- over-harvesting or hunting.

But isn't this list overly simplistic? Surely, it fails to ask (at least) five pertinent questions:

- Why is habitat being destroyed?

- What is causing pollution?

- Why are new species introduced into an area?

- Why is climate changing?

- Why are plants over-harvested and animals hunted to extinction?

A group of economists working for the World Wildlife Fund tackled questions such as these by identifying proximate causes (those in the first list above) and root causes (the underlying socio-economic and political drivers of environmental change – the answers to the questions in the second list) (Figure 6.1). This is a very useful, if somewhat deterministic, way of examining society–environment links because it links the 'local' causes-and-effects that are relatively easily attributed to people to the economic, political and social factors that drive human actions, but which often emanate far away – sometimes on other continents. This is a useful framework in which to explain environmental issues.

Table 6.2 A typology of society–environment relationships

Type of relationship	Examples
Deliberate environmental manipulation	1 Diversion of watercourses for irrigation which leads to changes in river flows. Positive benefits include flood control and water availability for agricultural production over a long period, e.g. the Indus Irrigation project in Pakistan. 2 Construction of large dams which leads to forced relocation of people, loss of productive land and wetland habitats, and changes in water resources and ecology downstream, e.g. Three Gorges Project in China.
Unanticipated impacts of activities that have been carried out to improve society	1 Irrigation in former Soviet Union led to the dessication of the Aral Sea due to diversion of too much water from the rivers that flowed into it. This led to the loss of the Aral Sea commercial fishery and health problems related to wind erosion of agrochemicals trapped in former lake bed sediments. 2 Expansion of cereal cultivation onto semi-arid grasslands has led to unexpected episodes of severe soil erosion (e.g. the US 'dustbowl' in the 1930s).
Complex problems in which cause and effect are not obvious. These are often characterized by human activities and natural forces combining synergistically	This applies to climate change, where there is still debate as to what proportion of elevated temperatures is due to natural climate change as against the effects of burning fossil fuels.
Linkages characterized by overlap and interaction	Contemporary deforestation is causing changes to people's livelihoods, losses of biodiversity, changes in the regional water balance, and changes in global climate. All of these impacts overlap at scales ranging from local to global.

Source: After Middleton (2003)

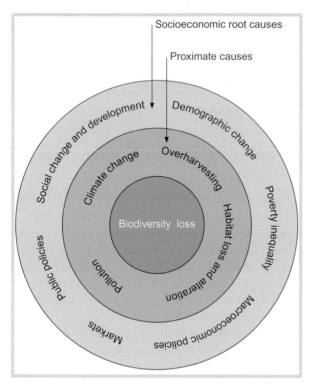

Figure 6.1   Biodiversity loss: proximate and socioeconomic root causes.

Source: Wood *et al.* (2000)

6.4 Obtaining solutions?

Solutions to the perceived obstacles that the environment places in the way of progress had, until the late 1950s and early 1960s, drawn broadly on science and technology. The period of post-Second World War expansion of the world economy and attendant modernizations and the colonial era prior to the Second World War are full of examples: 'solutions' often related to flood control, irrigation and hydroelectric power generation. For example, the world's largest irrigation system, in the Indus Valley Project in Pakistan, was started during Queen Victoria's reign and has continued to expand to the present day. Its history is one of advances in civil engineering, agronomy and soil science and it contains major feats of engineering such as the world's largest earth-filled dam – the Tarbela. Its objectives are to overcome seasonal droughts and floods, and thereby increase both the overall output and reliability of crop production. Economic expansion in the western USA was based on similar technological feats such as the construction of the Hoover Dam between 1933 and 1935 on the Colorado

Plate 6.2   The Aswan Dam controls the River Nile, Egypt and provides water that is vital for the irrigation of surrounding areas. (Corbis/Lloyd Cluff)

River, and Qwens Aqueduct system that still delivers water from the eastern Sierra Nevada to Greater Los Angeles and generates hydroelectric power for the city en route.

In their broadest sense technological advances have accompanied human activity from prehistoric times (e.g. early crop breeding, water harvesting for agriculture, the invention of the wheel) and we still rely on science and technology today (e.g. harnessing genetics to improve crop production, the installation of catalytic converters to reduce pollution emissions from car exhausts). However, acceptance of science and technology as a solution to environmental problems is now tempered by the knowledge that they can also lead to environmental problems. Solutions depend on using scientific and technological advances in the right socio-economic and political milieus. Indeed, some solutions only depend on adjusting social, economic or political factors – we do not have to search for new technical fixes. Why and when did our awe of science and technology diminish?

6.5 The rise of environmentalism

We now know that many of today's most pressing environmental issues were first identified in the nineteenth and early twentieth centuries. Significant time lags exist between many 'issues' first being identified and their recognition by politicians (Table 6.3). This table opens a window onto the evolution of society–environment relationships that allows us to ask and answer questions such as:

- If George Perkins Marsh was able to present evidence for consequences of tropical deforestation

in 1864 in his book *Man and Nature*, why was action not taken until the late 1980s?

- Why did the British government 'discover' the causes and consequences of tropical deforestation all over again in its colonies in the first half of the twentieth century?

- Why did we 'discover' the consequences of deforestation yet again in the Amazon Basin in the 1980s and 1990s?

The answer to these questions lies, in part, with the ways in which societies have viewed the environment in the past. Barrow (1995: 3–16) provides an accessible review of human attitudes towards the environment and development from prehistoric times to the present day (Pepper 1996 provides a more detailed synthesis). However, in this chapter we are concerned with the modern environmental movement (or **environmentalism**) because it is in this period that we have become aware that:

- many environmental problems are caused by scientific and technological advances

- environmentalism has provided the impetus to bring about changes in society to resolve environmental issues

- many issues have become global in their scope.

During the late 1950s concerns about the impacts of economic development on the environment gained prominence more or less simultaneously in North America, Europe and Japan and continued to grow during the 1960s. Sandbach (1980: 2–6) notes that the number of articles in newspapers and popular journals

Table 6.3 Identification of some key environmental issues and decade of their political acceptance

Issue	First identified	Decade of political acceptance
Air and water pollution	Carson (1962)	1960s
Tropical deforestation	Marsh (1874)	1960s
Acid precipitation	Smith (1852)	1960s
Stratospheric ozone depletion	Molina and Rowland (1974)	1970s
Carbon dioxide-induced climate change	Wilson (1858)	1970s
The effect of other greenhouse gases on climate change	Wang *et al.* (1976)	1980s

Source: After Doös (1994)

Plate 6.3   Farm workers set Amazonian rainforest ablaze to clear land to be used for farming and ranching in Rondônia state, Brazil. (Getty Images/Stephen Ferry)

rose during the 1960s and early 1970s, indicating the growing awareness of public interest.

Of particular importance to this embryonic movement was California where public interest law firms began to fight for environmental causes (Harvey and Hallett 1977: 62). Barrow (1995: 10) attributes the growth in environmentalism in North America during the 1960s and 1970s to:

- media coverage

- increased amounts of leisure time

- growing affluence

- a growing awareness of environmental problems and dangers associated with the arms race between the USA and the Soviet Union and political instability in the Middle East

- environmental knowledge gained from a suite of global scientific programmes such as the International Geophysical Year (1957–8), the International Biosphere Programme (1964–75) and the International Hydrological Decade (1965–74).

The early environmental movement had three areas of concern, namely the pollution impacts of technology, nature conservation and concerns about so-called '**overpopulation**'. Air and water pollution drew comment from a wide range of people and occasionally prompted government action, e.g. the Clean Air Act in the UK in 1956 (although almost a century earlier the UK government had introduced the Alkali Act (1863) to curb air pollution). Long-standing concerns about wildlife protection and nature conservation, which date back to the late 1800s in both North America and the British Isles (Evans 1991), were successfully taken up by people influential in the media and beyond. In the UK, establishment figures like the Duke of Edinburgh (who became President of the World Wildlife Fund), Sir Peter Scott (a noted wildlife artist who founded the Wildfowl and Wetlands Trust) and Sir David Attenborough (who began a highly successful media career based around natural history) began to take up the cause at this time. The third theme at this time was the predicted effect overpopulation would have on the future global

Thematic Case Study 6.1

The rise of environmental activism

Environmental NGOs (ENGOs) and related groups serve two key roles in society: they often identify environmental issues and, importantly, they seek to translate environmental concern into practical strategies for change.

Dissatisfied with the progress, important new environmental groups formed in the 1970s. In 1971 Friends of the Earth (FOE) was launched in London; it has become one of the biggest and most influential of the British environmental NGOs, and is part of an international FOE network. Its formation reflected a frustration with the staid conservation movement's lack of action against those perpetrators of environmental degradation (Rawcliffe 1998). They have a five-pronged strategy:

- to lobby those in political power and industry, often using legislative activity to press for change
- to generate scientific research and publish it in accessible formats
- to employ the media to attract attention to particular issues
- to mobilize the public through local groups
- to coordinate and cooperate with other groups to run large-scale campaigns.

In the early 1990s there was dissatisfaction with the operations of some ENGOs. FOE, for example, had evolved from a radical group to one that seemed less keen to press for far-reaching change. In effect it had become reformist – pragmatically seeking inclusion of environmental concern within the current political and social system. Many saw this reformist approach as being ineffectual, slow and hierarchical. Drawing inspiration from other movements and other countries, smaller, more radical environmental groups (or informal networks) began to proliferate. These groups are radical because they argue that environmental protection will only occur if we instigate major changes to our existing ways of life. The visions they espouse are often idealistic and the steps to realizing their dreams often challenging, such as dramatically reducing consumption, abandoning the use of cars, or changing our eating habits.

Such groups have chosen to focus on particular issues, but they are not single-issue activists. Instead they pose a systematic challenge to existing societal practices, often rejecting 'representative' democracy and formal politics and promoting grassroots participation in environmental decision-making and a 'do-it-yourself' (DIY) approach. The tactics employed reflect their belief in the need for radical change and often involve the use of non-violent direct action. Direct action is 'intended to have an immediate effect on a situation, as distinct from political activity which might have a roundabout effect through representatives, or demonstrative activity whose effect was to get publicity' (Rooum 1995: 27). Examples include destroying GM crops, road blockades and office occupations.

Earth First! (EF!) formed in the US in 1979 and spread to the UK in 1991. Based upon anarchist ideology, it is a network of autonomous groups that eschew formal membership and hierarchy (and thus leadership) and espouse consensus decision-making structures. It is based on the belief that actions speak louder than words, and has become synonymous with (often illegal) non-violent direct action and the DIY approach: one of their slogans is 'If not you, then who?' (Wall 1999, Doherty 2002). EF! has been influential in British anti-roads protests, e.g. Twyford Down (1992), Newbury (1995–6) and A30 Devon at Allercombe and Fairmile (1994–97). They used protest camps with tree-sits, tunnels and lock-ons with the aims of (i) physically preventing road construction; (ii) generating media publicity; (iii) educating the public; and (iv) acting as a catalyst for mass mobilization. EF! activists have expanded actions against issues such as GM crops and the arms trade. The defining characteristics of such activism are the continuously evolving creativity in tactics coupled with a broad concern for a variety of issues, for example 'they see exploitation of the Third World, the global poor, women, animals and the environment as a product of hierarchy, patriarchy, anthropocentrism, racism and, most prominently, capitalist economic relations' (Seel and Plows 2000: 114).

Thematic Case Study 6.1 continued

In Australia, environmentalists have employed direct action since the late 1970s. Actions at Terania Creek (New South Wales), Daintree (Queensland) and surrounding the Franklin Dam (Tasmania) all focused on preventing the logging of old-growth trees. Activists set up protest camps and used tree-sits and road blocks to deter environmental destruction (Cohen 1997, Doyle 2000). Such tactics are still employed today at anti-logging protests at East Gippsland (Victoria) and Styx Valley (Tasmania). India has an even more exten-sive history of *satyagraha* (non-violent action) inspired by Mahatma Gandhi. In the early 1970s vil-lagers (mostly women) from Uttar Pradesh began to physically try to prevent local logging – resulting in the Chipko Andalan movement (McCormick 1989). More recently, villagers threatened with losing their land, homes and livelihoods by the construction of the Narmada Dam have committed to drowning rather than forced resettlement (Shiva 2002). Such examples clearly illustrate the implicit link between environ-mental concern and social justice.

The rise of environmental activism since the 1970s has highlighted the importance of environmental protection with some notable successes. Britain's road-building programme of the 1990s was curtailed, and the logging of old-growth forests has been all but stopped in Western Australia. Yet, as Carter (2001: 2) surmises, 'the frequency with which governments adopt a business-as-usual response to environmental problems raises the cynical thought that perhaps nothing much has really changed'. The dilemma remains how best to enact change, be it through ENGOs that are able to lever access to the halls of power but in doing so can appear elitist, or through radical direct action which encourages grassroots par-ticipation and individual responsibility but which in turn can be small in scale and effect. For now the answer appears to lie in the vibrancy of environ-mental movements composed of a variety of contesting, challenging and supportive groups, incor-porating international ENGOs alongside more radical small-scale groups.

---

environment. North American commentators like Paul Ehrlich, Garrett Hardin and Barry Commoner predicted impending crises. They were so dogmatic that they became known as the 'Prophets of Doom' or 'Ecocatastrophists'. They were considered neo-Malthusians because they saw environmental problems as a consequence of population growth, following the arguments set out by Thomas Malthus in the late eighteenth century.

One book, written by the biologist Rachel Carson (1962), distilled the essence of the broad concerns of the era, and is generally regarded as a milestone in modern environmentalism. Carson pointed out the long-term ecological consequences of intensive agriculture, in particular the ways in which synthetic pesticides (notably DDT) persisted in the food chain and poisoned birds and mammals (including humans). The publication of *Silent Spring*, in 1962, signalled the start of the modern era of environmental concern for many. Social scientists and concerned scientists in universities and research institutes began to speak with a strong voice on environmental problems at this time. Environmental non-government organizations

(ENGOs) in North America, Western Europe and Japan expanded in number and influence. Though a few, like the North American Sierra Club, had been in existence since the late nineteenth century, new and more radical ENGOs like Greenpeace established themselves at this time. These ENGOs and their allies would later become central to environmental debates and protests across the world (see Case study 6.1).

Unlike the early concerns over technology and pollution, and wildlife loss, both of which for the most part have worsened since the 1960s, the predictions of neo-Malthusians have generally not materialized. Hardin (1968, 1972) introduced the 'tragedy of the commons' in which he argued that common pool resources (i.e. resources not under private ownership such as many grazing lands, oceans and fisheries) would be grossly overexploited because maximization of short-term economic gains would outweigh long-term conservation goals. Land resource 'free for alls' have been extremely rare in the last 30 years, and have drawn criticism for being too simplistic and lacking an evidential base (Harrison 1973, Boserüp 1990: 41). But the state of many of the world's fisheries, e.g. the North

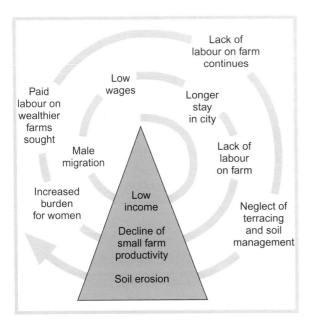

Figure 6.2 The implications of male migration to towns and cities for subsistence farming.
Source: Millington *et al.* (1989)

Atlantic fishery at the start of the twenty-first century, suggests that Hardin's arguments have some validity. The misguided simplicity of neo-Malthusian explanations of environmental problems is shown in studies that have tried to explain land degradation in terms of population dynamics. Not only is the evidence contradictory, for example population growth in northern China has been linked to increased land degradation (Takeuchi *et al.* 1995), whereas in northern Yemen (Carapico 1985, Hurni 1989) and

Kenya's Machakos Hills (Tiffen *et al.* 1994) it has been related to population decline, but it is clear from these studies that linking land degradation to one explanatory parameter – population size – masks the complexity behind population dynamics and labour–land relationships (Figure 6.2) (Millington *et al.* 1989). The neo-Malthusian argument is only one strand of environmentalism. Chris Barrow (1995: 12–13), in summarizing the literature on environmentalism, identifies 13 other strands (Table 6.4).

6.6 The road to Rio

In the previous section we identified two groups of people whose voices were heard loud and clear in the early stages of environmentalism: researchers (both scientists and social scientists) and ENGOs. During the 1970s, politicians and bureaucrats began making meaningful contributions to environmental debates around the world, especially bureaucrats in the United Nations (UN) system and Green Party politicians in Germany, New Zealand and Switzerland where they gained seats in national parliaments. By analyzing all of these contributions, it is possible to identify an environmental road map, with a series of milestones to the UN Conference on Environment and Development held in Rio de Janeiro in 1992 and beyond (Figure 6.3).

The first milestone – Rachel Carson's *Silent Spring* – was followed by the publication of *Limits to Growth* by the Club of Rome in 1972 (Meadows *et al.* 1972), which, along with other publications in the

Table 6.4 Types of environmentalism

Moderate environmentalism, weak sustainability (light to medium greens*)	Early conservation/nature protection Modernism Ecocentric and biocentric environmentalism Economic environmentalism Technocentric environmentalism Ecological scientific rationalism
Radical environmentalism, also known as 'deep greens'*, strong sustainability	The deep ecology movement Social ecology (ecosocialism) Shallow ecology Radical ecology Postmodern environmentalism Anti-establishment environmentalism Ecofeminism The New Age movement

* The 'greens' classification: light greens generally accept the current relationship(s) between people and the environment, but often seek changes to the Western approach to development; medium greens seek revisions to economics/politics to improve people–environment relationships; and deep greens seek drastic changes to people–environment relationships.

Source: After Barrow (1995: 12–13)

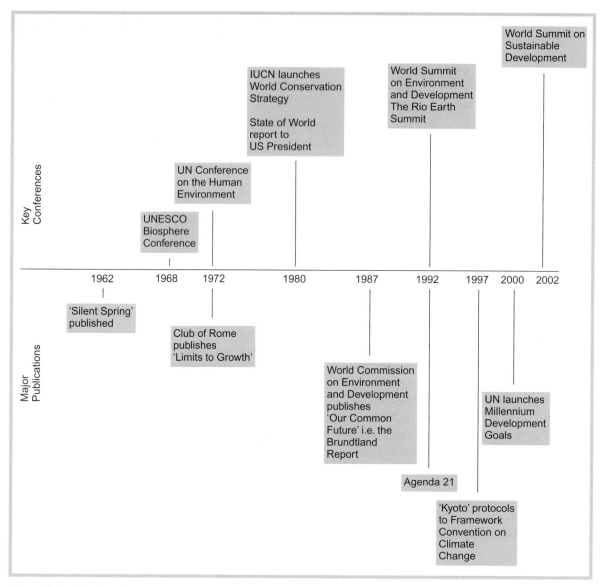

Figure 6.3 Timeline of key environmental conferences and publications.

ecocatastrophic genre, predicted doomsday scenarios. We know those predictions to have been overly pessimistic (the authors updated their predictions in *Beyond the Limits* in 1992, but still retained the overall message), but this report and other literature rang alarm bells and heightened awareness of the critical importance of the interrelationships between environment and economy. The energy crisis of the early 1970s, and the subsequent global recession that followed, promoted increased concerns for overall efficiency and conservation in resource use (see Chapter 5). It became apparent, for people in the developed world at least, that it was possible to maintain economic growth and reduce resource

consumption per unit of wealth created. This argument saw the emergence of an alternative view to ecocatastrophism: a view that would grow into the concept of sustainable development. This view was codified in the Declaration of the United Nations Conference on the Human Environment held in Stockholm in 1972 (this declaration and related conference documents can be read at www.unep.org/Documents/Default.asp?Document ID=97&ArticleID=1503).

Although it was possible at this time, very few states passed significant amounts of legislation to reduce environmental impacts while maintaining economic growth: West Germany was a notable exception. For most countries urbanization and population growth, and the growing pressures from the aspirations for 'development' (see Chapter 5), meant that the environment was subject to increasing stresses.

Such concerns are flagged in milestones in the early 1980s such as the IUCN's World Conservation Strategy. Designed to address global environmental problems, it had three main, ecologically focused objectives:

- to maintain essential ecological processes and life support systems

- to preserve genetic diversity

- to ensure the sustainable utilization of species and ecosystems.

The strategy built on the notion of sustainability. However, by placing the needs of a relatively narrow focus on the environment (a concern that would later

Table 6.5 The main outcomes from the 1992 Earth Summit

Outcome	Salient points and official web-sites
The Rio Declaration	Set out principles of sustainable development in a format acceptable to all governments. www.unep.org/Documents/Default.asp?DocumentID=78&ArticleID=1163
Agenda 21	A 41-chapter document that reported the state of the world environment in the early 1990s, and indicated where actions were needed. www.un.org/esa/sustdev/documents/agenda21/index.htm
UN Commission on Sustainable Development (CSD)	Establishment of CSD which has become the umbrella agency for monitoring and promoting the agreements reached at Rio. www.un.org/esa/sustdev/index.html
Framework Convention on Climate Change (UNFCC)	Convention designed to address issues of contemporary climate change identified by the Intergovernmental Panel on Climate Change (IPCC). unfccc.int/
Convention on Biological Diversity (CBD)	Convention designed to address the issues of biodiversity loss that had been identified in the 1980s, and build on the three existing wildlife conventions.[1] www.biodiv.org/index.html
Convention to Combat Desertification (CCD)	Convention designed to build on and improve international (UN) actions to halt dryland degradation, which had begun in the 1970s. www.unccd.int/main.php
Statement of Forest Principles	A 'regressive document' (Saint-Laurent 1996: 68) that exposed the lack of agreement on forests at Rio. www.un.org/documents/ga/conf151/aconf15126-3annex3.htm
The Inter-Governmental Panel on Forests (IPF)	Established by the CSD in 1995 to take forward work on forests and sustainable development. www.un.org/esa/forests/ipf_iff.html
The Conference on Small Island States	Conference of Small Island Developing States (SIDS), low-lying coastal countries that share similar sustainable development challenges. Its deliberations are reported at: www.un.org/esa/sustdev/sids/sids.htm
The Conference on Straddling and Highly Migratory Fish Stocks	Conference designed to discuss common-pool fish stocks: www.un.org/Depts/los/fish_stocks_conference/fish_stocks_conference.htm. A Straddling Fish Stocks Agreement was negotiated by 1995.
Local Agenda 21 Initiative	Actually launched by International Council for Local Environmental Initiatives (ICLEI) in 1991, but usually associated with a 'launch' at Rio. Progress on Local Agenda 21s has been good. www.iclei.org/

[1] The three conventions were: CITES – the Convention on International Trade in Endangered Species, the Bonn Convention on Migratory Species, and the Ramsar Convention on Wetlands.

be articulated as biodiversity – a word that, coincidentally, was first introduced in 1980) ahead of those of human society, it failed to garner the political support necessary to realize its objectives. Nonetheless, it was an important contribution to the ongoing global debate about economic growth and the environment. The debate led to the publication of *Our Common Future* – more commonly known as the Brundtland Report – in 1987.

The Brundtland Report adopted the position that it was possible to pursue economic growth without compromising the environment, and it is generally accepted as introducing the concept of sustainable development (the seeds of which had been sown some 15 years earlier). It also provided the first, and still widely used, definition of sustainable development: 'development which meets the needs of the present without compromising the ability of future generations to meet their own needs.' While few people disagree with this as an aspiration, as a definition it is too vague: for example, how can present and future needs be determined, and whose needs does it refer to? Many attempts have been made to refine the concept, which has led to numerous, often contradictory, definitions. The result is that 'Sustainable Development' is often little more than a soundbite signalling recognition that the environment matters but failing to provide substance. The real success of the Brundtland Report was the emphasis it placed on the relationships between economic growth, social conditions and environmental degradation, an emphasis that placed sustainable development firmly on the global political agenda, and that over the next five years would see a concerted effort by UN agencies and many governments towards a global summit on environment and development. In 1992 – after five years of negotiations and preparatory meetings – the UN Conference on Environment and Development (also known as the Earth Summit) was held in Rio de Janeiro.

Much has been written about the Earth Summit (two of the most accessible accounts are provided by Dodds 1996 and O'Riordan 2000). It brought together 178 government delegations, over 120 heads of government, over 50,000 NGO representatives and a press corps of around 5,000; but even before this, the largest global gathering to discuss environmental issues had been written off by some environmentalists as a

publicity stunt. There were also strong geopolitical tensions. A rift opened up between the representatives of the developing world and the developed countries, the former being worried that environmental concerns would be used by the developed world to 'limit [their] development for the sake of the Earth'. Shifting tensions between geopolitical blocs developed over issues such as biodiversity, climate change and desertification. In the arena of biodiversity, geopolitical alliances gelled around one issue but dissolved around another. For example, in the negotiations towards the Convention on Biological Diversity the 'Global North' battled the 'Global South' over finance, technology transfer and conservation; the G77 and the EU opposed the USA over biosafety; Latin America and Africa were not entirely united as developing world blocs; and even within EU member states there were disagreements (Dodds 1996: 49).

Whatever its failings, the key point is that the Earth Summit resulted in a number of specific agreements that now form the basis of a global strategy for sustainable development (Table 6.5) and it had a wider, less quantifiable impact on environmental issues worldwide, particularly in that:

- it elevated the importance of the environment on the political agendas of most countries (most countries now participate in the conventions that were agreed to at Rio, and many have used these conventions to pass new environmental legislation)

- the multilateral negotiations led to some issues gaining internationally important status (perhaps most notably biodiversity loss, which had only been recognized by scientists some 12 years earlier), and which changed long-standing, often ineffective attitudes to nature conservation (see Case study 6.2)

- legislation at the international scale was negotiated

- there was recognition that many stakeholders had legitimate voices in environmental policy making (the recognition of the stakeholder principle was vital in allowing actors outside government to become effective in bringing about changes in environment and development, a point that proponents of the bottom-up approach to development had long argued for).

Thematic Case Study 6.2

The changing nature of conservation

Whether directly, through hunting for example, or indirectly through their development activities, people are usually held responsible for losses of wildlife. Until the 1980s conservationists' solutions relied on controlling people's actions through strategies like hunting bans and establishing protected areas. These strategies have a long history. The East Riding Association for the Protection of Seabirds – the UK's first nature conservation organization – was founded in 1867; a year later the first international convention on nature conservation was held in Vienna; and the late nineteenth century witnessed the creation of the first protected areas in Africa and North America.

These strategies – collectively known as protectionism – are overtly neo-Malthusian and have dominated nature conservation. Without the relevant ecological tools and theories (which were not available until the middle of the twentieth century) protected areas were established in areas considered wilderness. North American conservationists like John Muir, Aldo Leopold and Henry Thoreau introduced the wilderness concept to describe pristine landscapes untouched by humankind. The concept has been largely discredited for ignoring 'Others' (in the case of North America, the Amerindians) and the concept of climax vegetation (now rejected by ecologists) (Gomez Pompa and Kaus 1992). Conservationists, armed with the flawed wilderness concept and a paucity of scientific theory, continued to annex land from local communities to create nature reserves through much of the twentieth century.

We now acknowledge that most protected areas were created in the absence of consultation with local communities and maintained at their expense. In fact, the welfare and development of these people were considered to be in direct conflict with biodiversity conservation goals, as these characteristics of protectionism show:

* conservation that works against human welfare

* exclusion of local people, sometimes by expelling them from areas where they had lived for generations
* ignoring long-standing rights of access to natural resources
* association with colonial administrations (especially in Africa)
* the economic marginalization of those excluded, often on the boundaries of their former lands.

The exclusion of local people by conservation organizations led to this being called fortress conservation or the fences-and-fines approach. Some African national parks, e.g. Kruger in South Africa, had boundary fences, but for reserves the fences were metaphorical though the boundaries were strictly enforced (wherever finances allowed). Protectionism received global support. The World Conservation Strategy (1980) had, as one of its aims, 'the *preservation* of genetic resources', and the World Charter for Nature, passed by the UN General Assembly in 1982, called for, amongst other things, 'the *conservation* of unique areas and representative samples of ecosystems containing rare or endangered species'.

By the early 1980s, however, the foundations of fortress conservation were being undermined. The World Conservation Strategy had also called for '*sustainable use* of species and ecosystems', and during the 1980s, development experts, NGOs and indigenous peoples began to question protectionist conservation strategies on two fronts: namely to (i) counter the problems posed for local communities (see above) and (ii) promote social equity. This movement reflected a shift in the power base of many societies from national to local; the acceptance of new thinking in ecology, which espoused dynamics and disequilibria rather than stasis and equilibria; and the rise of neoliberalism (Hulme and Murphee 1999). Brown (2002) identifies three types of new conservation:

* integrated conservation and development programmes, which were prevalent in the 1980s
* community conservation initiatives
* wildlife utilization programmes, which came to the fore in the 1990s.

Thematic Case Study 6.2 continued

New models of conservation were even accepted by some conservationists who realized that many protected areas were too small. Conservation had to reach into the community, and the local people, once the enemies of conservation, now became partners.

The trade-off between social and economic equity, and conservation, was underpinned by the Convention on Biological Diversity. Its primary objectives – the conservation of biodiversity and the sustainable use of its components, and the fair and equitable sharing of the benefits arising out of the utilization of genetic resources – make this clear. Here then is global legislation which, unlike previous conventions, is not preservationist and assumes human use of biodiversity at sustainable levels. What of these new partnerships – is it possible to avoid species loss whilst at the same time allowing relatively low levels of economic development in the same area? – and what happens in the future when people's economic aspirations grow? We simply do not know the answers.

6.7 Crossing the South Atlantic

The Rio Conference was itself a milestone, for it not only signalled the start of a new leg from Rio along the road map to sustainable development, a route that took us to Johannesburg for the World Summit on Sustainable Development (WSSD) in 2002, but it also heralded an era of global environmental policy-making and legislation.

Each of the three main conventions agreed at Rio has been followed by annual or biannual conferences of the parties (COPs) (i.e. the countries that have signed a convention) at which progress towards ratification and implementation of the conventions has been discussed. Whilst the Convention to Combat Desertification and the Convention on Biological Diversity proved relatively straightforward to ratify and implement (these are well documented on their respective web-sites – www.unccd.int/main.php and www.biodiv.org/default.aspx), the Framework Convention on Climate Change and its most important protocol – the Kyoto Protocol – have yet to be ratified. We discuss this in some detail in Case study 6.3.

In addition to the progress made on the main conventions, there are other positive elements on the post-Rio balance sheet. These mainly concern changes in overall attitudes towards environmental and natural resources issues, and their integration with globally agreed development objectives. Firstly, it is clear that the fears of physical scarcity that prompted the resource crisis of the 1970s have now been replaced by environmental scarcity (see Table 5.1). O'Riordan (2000: 32) observes that the world is not running out of non-renewable resources; rather it is the renewable resource base that is besieged due mainly to our inability to balance the desire for economic growth with environmental needs. Secondly, although a few governments have found it difficult to ratify conventions and make efforts to achieve targets (see for example Case study 6.3), the implementation of sustainable development actions by local administrations – Local Agenda 21s – has been far more successful. The International Council for Local Environment Initiatives (2002) reported that 6,416 LA21s were underway or committed to in 113 countries, that national campaigns were underway in 18 countries, that formal stakeholder groups had been established in 73 per cent of administrative units with LA21s and that, universally, water resource management was the prime issue. Equally universal were the facts that a lack of sufficient political commitment by national governments and inadequate financing were major obstacles to progress. Elsewhere, however, progress has been made on financing environment and development. It is clear that the costs of the agreements made at Rio for mainstreaming sustainable development and environment exceeded the means of most developing countries and the budgets of many post-communist states of the former USSR and Eastern Europe. Before Rio it was already evident that funding from the developed world for remedial environmental actions was unreliable. For example, it is generally agreed that actions under the umbrella of the UNEP to halt desertification in the world's drylands prior to the Earth Summit were compromised, in part, by the failure of countries in the

Thematic Case Study 6.3

Implementing a convention? The obstacles faced in tackling climate change

Although there is residual controversy surrounding the issue of global warming, three basic propositions can be accepted:

* That there is a natural greenhouse effect that warms the Earth's surface by some 33°C. Without this human habitation would be impossible
* Since the onset of industrialization, some 150 years ago, the amount of CO_2 released into the atmosphere has increased markedly. Carbon dioxide is a greenhouse gas – that is, it traps the long-wave radiation emitted by the warming of the Earth's surface by the sun and, in doing so, heats up the atmosphere. Industrialization has also increased the amounts of other greenhouse gases, e.g. methane and nitrous oxide, and has introduced new compounds such as chlorofluorocarbons (CFCs), which promote global warming and, in some instances, deplete the ozone layer.
* Land use change, particularly deforestation, has reduced the Earth's ability to absorb CO_2 through the process of photosynthesis.

These processes are promoting an increase in the level of greenhouse gases in the atmosphere at a time when the Earth is experiencing a natural warming trend. While natural processes account for some of the warming trend, it is now accepted that elevated levels of CO_2 in the atmosphere are a major cause of global warming. The single largest anthropogenic source of CO_2 is the burning of fossil fuels (for details on global warming and guidance on further reading, see O'Riordan 2000, Chapter 7).

Solutions to the problem of global warming lie in reducing fossil fuel combustion in transportation, industry and power generation. The question is how?

The United Nations Framework Convention on Climate Change (UNFCC) was signed at Rio in 1992 (visit the UNFCC web-site at www.unfcc.de for further details). The ultimate objective of the UNFCC was to 'stabilize greenhouse gas concentrations in the atmosphere at a level that would prevent dangerous anthropogenic interference with the climate system. Such a level should be achieved within a time-frame sufficient to allow ecosystems to adapt naturally to climate change, to ensure that food production is not threatened and to enable economic development to proceed in a sustainable manner'. Conferences of the Parties (COP) are attended by all the states that have ratified or acceded to the Convention (over 194 by February 2004) every two years. The third COP held in Kyoto, Japan, in 1997 saw agreement on the 'Kyoto Protocols', the first actions to address greenhouse gas emissions. The breakthrough at Kyoto was due to a flexible approach that allowed individual countries to set their own targets, while aiming at a 5 per cent global reduction in CO_2 over 1990 levels. The target date for individual countries was set as some time between 2008 and 2012. In addition, the developed countries agreed to reduce emissions in 2000 to their 1990 level, a target few have met. The European Union has an emission reductions target of 28 per cent, the USA 27 per cent and Japan 26 per cent. The transition economies of central Europe and the former Soviet Union were granted concessions; for example, Russia has a zero change target. Countries seeking admission into the EU – such as Poland – accepted reductions. Not surprisingly, a complex set of rules and regulations surrounds the targets, making it possible to trade 'emissions reductions' between states. Thus, for example, if Japan helped Russia to reduce its emissions below the 1990 level, that reduction would be set against Japan's target.

For the protocol to come into force, the countries responsible for 55 per cent of the greenhouse gas emissions in 1990 have to ratify it. And herein lies the problem, for although 120 countries had ratified the Kyoto Protocols in February 2004, their total emissions only amounted to 44% of 1990 emissions. Only two countries can project the cumulative total over the threshold: the USA, which refused to ratify in March 2002, and Russia, which is prevaricating. Fuelled by cheap energy and increased oil and gas production, Russia's economy has grown at rates of

Thematic Case Study 6.3 continued

up to 10 per cent since 1999. Will ratifying Kyoto stifle Russia's growth, which analysts suggest will continue for another two decades at least? There is another reason behind the choice of 1990 as the baseline: 1990 was the last year of a fully functioning Soviet economy, and Russian emissions fell by 39 per cent from 1991 to 1999 (Webster 2003).

At the same time as the developed world committed itself to emissions reductions, the developing world was persuaded to accept the concept of the clean development mechanism (CDM). The developing countries feel that emissions reductions in the developed world should be on such a scale that would compensate for increased future emissions due to economic growth in the developing world. CDM is aimed at enabling economic growth without a commensurate increase in greenhouse gas emissions. However, this will require transfer of considerable amounts of capital and technology. In other words, the developed world will have to help the developing world implement the CDM. It is unclear how this will happen. It is proposed that developed countries can earn credits towards their own reduction targets by assisting in the implementation of CDM in the developing world. O'Riordan (2000: 202) notes that: 'the CDM mechanism denies the right of Third World nations to select their own CO_2 future. This is "eco-logical colonialism" by another name. While the developed world has access to the capital and technology to achieve the Kyoto targets, it still requires the political will. In the developing world, political will amounts to little without the material requirements to balance the economic growth with environmental needs.'

There is an alternative on the table: this is known as contraction and convergence (C&C). At COP9 in Milan many representatives admitted privately that 'C&C [was] what [they] had been waiting for' (Pearce 2003: 6). Contraction means that greenhouse gas emissions would be reduced globally, resulting in dramatic cuts during the next half century. Convergence would see each country's emissions reduced, until by 2050, according to authorities in the UK and Germany, everybody in the world would have an equal right to pollute – the amount being 0.3 tonnes C per person. Carbon trading permits would help the heaviest polluters to reduce their emissions rapidly. C&C also overcomes the USA's objection to Kyoto, which is that their Asian economic competitors such as China have no emissions reductions targets. Some environmentalists and politicians are now beginning to regard Kyoto as an obstacle, not a solution; it remains to be seen whether they say the same about C&C if the world adopts that route.

developed world to meet financial commitments they had made at the UN Conference on Desertification in 1977. Many countries fell back on established bilateral aid agreements (Thomas and Middleton 1994). Therefore, a very important outcome of Rio was the establishment of the Global Environmental Facility (GEF), an independent clearing house funding organization managed by the World Bank, UNDP and UNEP that disburses funds to developing countries for projects that benefit the global environment and promote sustainable livelihoods. GEF is the financial mechanism for four international environmental conventions:

- the Convention on Biological Diversity

- the Framework Convention on Climate Change

- the Convention to Combat Desertification

- the Stockholm Convention on Persistent Organic Pollutants (signed in Stockholm in 2002, details of which can be found at www.pops.int).

The GEF has raised US$4.5 billion from developed world economies and US$14.5 billion of co-funding between 1991 and 2003 for projects on biodiversity, climate change, ozone depletion and international waters in the developing world and countries in transition (Figure 6.4). This amounts to over 1,300 projects in 140 countries. Since 2003, GEF funds are also being disbursed to solve problems of land degradation and persistent organic chemicals.

Although the focus at Rio was on sustainable development, there was a clear dichotomy between

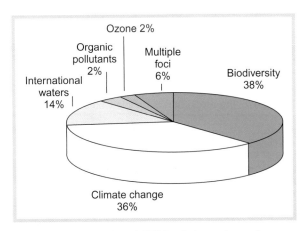

Figure 6.4 Allocation of GEF funds by sector up to 31 December 2002.

environment and development in many areas, with the environment often establishing the priorities. The chapter headings of Agenda 21 (which can be read at www.un.org/esa/sustdev/documents/agenda21/index.htm) reinforce this point. Since Rio, particularly in the late 1990s and the run-up to the WSSD, the emphasis has changed to one that seeks to integrate environment and development more closely and for the environmental agenda to be led by concerns about poverty and development. These new lines of thinking are underpinned by the World Bank's Poverty Alleviation Strategies, the Millennium Development Goals which were agreed by the UN General Assembly in September 2000 (Table 6.6), and as a general response to lack of progress on poverty and debt in the Third World.

The WSSD was, like Rio, a mega-conference attracting parallel events such as the People's, Children's and Indigenous People's Summits amongst other events throughout South Africa (Bigg 2003). Vaguely equivalent to the Rio Principles in 1992 was a political statement, the Johannesburg Declaration. The main outcome of the WSSD, however, was the Plan of Implementation (for Agenda 21). This had been negotiated at three preparatory conferences, yet was still incomplete at the start of the summit. Its structure bears little resemblance to that of Agenda 21. Whilst this might at first appear odd, it can be regarded as positive because, in the dynamic arena of environment and development, we should not base new goals and plans on agendas which have moved on. Bigg (2003) provides one of the earliest reviews of the WSSD and, in identifying 30 targets in the Plan of

Implementation, recognizes that most simply reiterated the UN's Millennium Development Goals and other already agreed global commitments. However, he identified six new targets:

- Halve the number of people without access to basic sanitation by 2015
- Introduce new measures to regulate toxic chemicals by 2008
- A commitment to restore fish stocks by 2015, and new marine protected areas by 2012
- Achieve a 'significant reduction' in species loss by 2010
- Establish a stronger regime to share benefits from bioprospecting and biopatenting for developing countries
- Improve the access of developing countries to alternatives to ozone-depleting substances.

In time, the Global Deal agreed at WSSD (i.e. a set of commitments and responsibilities to implement sustainable development, led by governments and heads of state but also incorporating business and civil society at all levels) may become a landmark in implementing sustainable development. As part of this deal, two types of agreements (known as outcomes) were negotiated:

- Type 1 outcomes are commitments and agreements negotiated by governments during preparatory meetings which are included in the plan of implementation.
- Type 2 outcomes are implementation partnerships and commitments between groups of stakeholders to translate agreed outcomes into actions by governments, international organizations and other major groups.

Type 2 outcomes are a potentially interesting, yet controversial development. Whilst expressing market-led and neoliberal approaches (these are discussed further in Chapter 8 on development), are they a signal that some governments are backing out of their moral obligations and financial commitments to the global environment good? Do they also signal that global environmental issues are businesses for sale, and if so, where will the contracts go – to the developed or the developing world? The answers to these questions are unclear – only actions over the next decade will yield the answers to these and similar questions.

163

Table 6.6 United Nations' Millennium Development Goals

Goal	Targets associated with goal
1. Eradicate extreme poverty and hunger	1. Halve, between 1990 and 2015, the proportion of people whose income is less than $1 a day. 2. Halve, between 1990 and 2015, the proportion of people who suffer from hunger.
2. Achieve universal primary education	3. Ensure that, by 2015, children everywhere, boys and girls alike, will be able to complete a full course of primary schooling.
3. Promote gender equality and empower women	4. Eliminate gender disparity in primary and secondary education, preferably by 2005 and in all levels of education no later than 2015.
4. Reduce child mortality	5. Reduce by two-thirds, between 1990 and 2015, the under-five mortality rate.
5. Improve maternal health	6. Reduce by three-quarters, between 1990 and 2015, the maternal mortality ratio.
6. Combat HIV/AIDS, malaria, and other diseases	7. Have halted by 2015 and begun to reverse the spread of HIV/AIDS.
7. Ensure environmental sustainability	8. Integrate the principles of sustainable development into country policies and programmes and reverse the loss of environmental resources. 9. Halve, by 2015, the proportion of people without sustainable access to safe drinking water and basic sanitation. 10. Have achieved, by 2020, a significant improvement in the lives of at least 100 million slum dwellers.
8. Develop a global partnership for development	11. Develop further an open, rule-based, predictable, non-discriminatory trading and financial system (includes a commitment to good governance, development, and poverty reduction – both nationally and internationally). 12. Address the special needs of the least developed countries (includes tariff- and quota-free access for exports enhanced programme of debt relief for HIPC and cancellation of official bilateral debt, and more generous ODA for countries committed to poverty reduction). 13. Address the special needs of landlocked countries and small island developing states. 14. Deal comprehensively with the debt problems of developing countries through national and international measures in order to make debt sustainable in the long term. 15. In cooperation with developing countries, develop and implement strategies for decent and productive work for youth. 16. In cooperation with pharmaceutical companies, provide access to affordable, essential drugs in developing countries. 17. In cooperation with the private sector, make available the benefits of new technologies, especially information and communications.

Source: www.un.org/millenniumgoals

6.8 Constructing futures

Environmental issues doggedly remain with us – now not only the domain of scientists (if they ever were), environmentalists or politicians, but of the entire world population. Current efforts to reduce global warming, if successful, will not have much of an impact before *circa* 2050; implementation of recovery plans for fish stocks in the North-west Atlantic are underway and the International Council for the Exploration of the Sea recommends zero catches for commercial fish like cod, hake, whiting and capelin in parts of the region (ICES 2003); and despite the establishment of hundreds of protected areas in the 1980s and 1990s we continue to lose habitat and species in some of the Earth's most biologically diverse

Plate 6.4 Greenpeace activists, dressed as polar bears, protest about BP's offshore drilling in the Arctic Ocean. (Corbis/Scott Olson)

places. The list is long and – to environmentalists – depressing.

So have the efforts of environmentalists and policy-makers outlined in this chapter been in vain? Well, no, society is taking action: policies are being designed and implemented at global, regional and local scales. But there are issues that still dog progress:

- Remediation and restoration of many types of environmental damage takes years.

- The costs of environmental remediation and protecting environments are often enormous.

- Our incomplete knowledge about many aspects of the environment, and how environment–society linkages work, can act as a barrier to finding effective solutions.

- The speed of economic and cultural change is rapid, the desire and drive for economic development strong; in contrast, the negotiation and implementation of global and regional environmental policies are slow.

Local initiatives are more dynamic and may be the key

to a solution; however, they present problems in themselves. Each initiative may be relatively easy to negotiate, implement and finance, but replicating them across thousands (maybe millions) of communities globally (as is required to tackle cumulative global issues) raises concerns about economies of scale, and how to share knowledge and good practice.

So, some of the environmental issues we read about in our local newspapers are being addressed by local administrations and politicians. In the UK these are often paid for out of local and national taxes, and sometimes by EU funds: usually under the umbrella of new environmental legislation and, often, as part of the establishment of a Local Agenda 21. But what of solutions to global environmental issues – cumulative issues such as the potential loss of biodiversity because of the introduction of GM crops, or systemic issues like the impact that the imminent failure to implement the Kyoto Protocol will have on global warming? Here local actions often appear futile. What can one person, one household, one community or one administrative unit do when governments of some of the world's most economically powerful states cannot be brought onside? The environmentalist's answer is to be a good global citizen, to put pressure on administrations, governments, corporations and supra-national organizations through political and non-political means, and to educate others. We have gone some way down this route in the last three decades, but significant problems remain. Problems, to paraphrase words of Roger Blench (1998: 2), exist at the two extremes of an economic axis: '...*the environment* [our emphasis] is being destroyed ... by individuals too absorbed by financial gain to notice or households too poor to care'. Problems are also found along the same economic axis but at different geographical scales, and lead to questions such as:

- Is it morally justifiable and politically plausible for the developed world to try to dictate the terms of sustainable development and good environmental management to the rest of the world?

- What will it take for all developed world countries and elites everywhere to cast aside their domestic concerns and privileges in favour of the global and collective good?

- How do these questions relate to wider human geographies of power, politics and ideology?

Learning outcomes

Having read this chapter, you should understand:

- The differences between systemic and cumulative global environmental issues.

- The linkages between proximate and root causes, and how they relate to human impact on the environment.

- How thinking about the environment and environmentalism has evolved since the early 1960s.

- How global policies that have developed around sustainable development have incorporated the environment.

Further reading

Barrow, C.J. (1995) *Developing the Environment: Problems and Management*, Longman, Harlow. Chris Barrow provides a nice balance of examples of environmental problems (e.g. climate change, desertification) and how they can be managed with more conceptual material such as reviews of changing attitudes towards environment and development over time, a discussion of what makes a crisis, and a chapter on the greening of development.

Dodds, F. (ed.) (1996) *The Way Forward: Beyond Agenda 21*, Earthscan, London. Felix Dodds gathered together a group of knowledgeable authors to write this 22-chapter book on Agenda 21. It provides a very readable account of the lead-up to Rio, the main agreements made and progress in the years immediately after this important milestone. Of course, further progress has been made in many of these sectors since 1996.

Hannah, L., Lohse, D., Hutchinson, C., Carr, J.L. and **Lankerani, A.** (1994) A global assessment of global impact, *Ambio*, **23**, 246–50. Possibly the most succinct introduction to human impacts on the environment, this should be read before moving onto books such as Meyer (1996), Middleton (2003) and Turner II *et al.* (1990).

Meyer, W.B. (1996) *Human Impact on the Earth*, Cambridge University Press, Cambridge. One of a number of books in this list which review a wide range of human impacts.

Middleton, N. (2003) *The Global Casino*, 3rd edition, Arnold, London. Nick Middleton provides an up-to-date introduction to a wide range of environmental issues in 22 chapters – from tropical deforestation to climate change, from urban environments to food production – as well as chapters on sustainable development and war. It provides an excellent starting point for delving deeper into individual issues.

Pepper, D. (1996) *Modern Environmentalism: An Introduction*, Routledge, London. A comprehensive introduction to the modern environmental movement.

Turner II, B.L., Clark, W.C., Kates, R.W., Richards, J.F., Matthews, J.T. and **Meyer, W.B.** (1990) *The Earth as Transformed by Human Action*, Cambridge University Press, Cambridge. Another book providing a comprehensive review of human impacts on Planet Earth.

UNEP (2003) *GEO-3: Global Environment Outlook 2003*, Earthscan, London. The third book in a series of three-yearly syntheses of the global environment produced by the United Nations Environment Programme. The book is also available online at www.unep.org/GEO/geo3/, as are GEO2 and GEO1.

Vitousek, P.M., Mooney, H.A., Lubchenco, J. and **Melillo, J.M.** (1997) Human domination of Earth's Ecosystems, *Science*, **277**, 494–9. A scientific analysis of how human actions are affecting global (planetary) scale physical and ecological processes and phenomena.

World Resources Institute (WRI) (2003) *World Resources 2002–2004*, WRI, Washington, DC. Like UNEP's GEO series, WRI has published biannual reviews of the world's environment in its 'World Resources' series since the mid 1980s. The latest book is also available online at: http://pubs.wri.org/pubs_description.cfm?PubID=3764.

Worldwatch Institute (2003) *State of the World 2003*, Earthscan, London. Another book in a series of more-or-less annual reports of the state of the world's environment and related development issues.

Useful web-sites

www.un.org/millenniumgoals/ The United Nations web-site dedicated to the Millennium Development Goals. Not only are the goals outlined, but there is more detailed information on each goal and target. This information is provided in a slightly different format at the World Bank's web-site (www.developmentgoals.org).

www.unep.org/GEO/ The information portal of the United Nations Environment Programme, containing statistical information and online publications.

 For annotated, clickable weblinks and useful tutorials full of practical advice on how to improve your study skills, visit this book's website at www.pearsoned.co.uk/daniels

CHAPTER 7

Changing geographies of global food production

BRIAN ILBERY

Topics covered

- Population and food supply
- Globalization and food regimes
- Agricultural industrialization
- Regional speciality food products
- Future challenges for the supply of food

Food is a basic necessity for human survival and shapes people's lives in profound cultural, ideological and economic ways (Goodman and Watts 1997). Yet, rapid population growth, especially in the least developed countries (LDCs), creates a major challenge for the production of food on a global scale (see Chapter 4). Rising from just 1 billion people in 1800 to 4 billion in 1975 and a projected 8.3 billion by 2030 (Bruinsma 2003), population growth is having a major impact on levels of food self-sufficiency around the world. Despite major technological advances in food production and the emergence of global agri-food systems, the reality is that the developed world has too much food whilst many parts of the developing world have too little.

7.1 Population growth and food supply

Although food supply has increased in most major world regions since 1950, too many of the world's population are still experiencing hunger. Defining hunger primarily in terms of undernutrition, the Food and Agriculture Organization (FAO) has shown that the number of undernourished people fell from 786 million in 1990 to 693 million in 2001 (Table 7.1); this

is projected to fall to around 440 million by 2030 (Bruinsma 2003). However, Table 7.1 also indicates that the world's worst-fed people are growing in numbers the fastest, with sub-Saharan Africa having one-third of their population classed as undernourished. Whilst heavily concentrated in the LDCs, it is important to emphasize that certain groups within society in developed countries also go hungry. In California, for example, up to 5 million people are short of food whilst others are consuming 'designer' organic vegetables that have been shuttled around the world in sophisticated and refrigerated forms of transport (Buck *et al.* 1997, Goodman and Watts 1997).

Hunger and food insecurity are the result of population growth, conflict, poverty and environmental instability (Grigg 1985). While Bruinsma (2003) suggests that per capita food supplies will have increased by 2015/2030, the output of food per capita has been falling in the LDCs since the 1960s. Indeed, they now account for approximately 30 per cent of world food imports by value, a significant increase on the 1970 figure. Their falling levels of food self-sufficiency can, in part, be explained by three factors (Ilbery and Bowler 1996):

* Population growth has outpaced the rate of increase in food production in many LDCs; often combined

Table 7.1 Population growth and undernutrition, around 2000

World region*	Annual population growth rate 1975–2001(%)	Total fertility rate (per woman) 2000–5	Number under-nourished (millions) 2001	Percent population under-nourished 1998–2000
Sub-Saharan Africa	2.8	5.4	206.7	33
Arab States	2.7	3.8	37.7	13
South Asia	2.1	3.3	349.2	24
East Asia & Pacific	1.4	2.0	–	–
Latin America & Caribbean	1.9	2.5	62.7	12
Central/Eastern Europe & CIS	0.5	1.4	36.9	9
OECD	0.8	1.8	–	–
World	1.6	2.7	693.2	11.3

* Seven world regions based on standard United Nations regional classification of countries, 1999.

CIS: Commonwealth of Independent States

Source: UNDP (2003)

with environmental disturbances, this has meant that imports have been necessary just to maintain per capita food consumption levels.

- Many countries, for example Brazil and Kenya, have expanded their agricultural exports whilst neglecting the domestic food staples. This has been necessary to earn foreign currency to repay debts acquired to fund development programmes.

- Incomes have risen in some LDCs, for example in oil-exporting countries such as Saudi Arabia, Venezuela and Libya. Thus, the wealthier sections of the population have increased their consumption of imported wheat, rice and livestock products.

Consequently, many LDCs have become net food importers, increasing their dependence on developed world exports. In the 1950s and 1960s, considerable proportions of these exports were in the form of food aid (Tarrant 1980). Although falling significantly by the 1980s, food aid is still important to many low-income countries in sub-Saharan Africa and south Asia. For example, Asia and sub-Saharan Africa are the largest recipients of cereal aid, accounting for 73% of the global figure in 2002. However, such aid has been criticized on numerous grounds, not least because it increases dependency, distorts local markets and stifles domestic production (Maxwell and Singer 1979). By the 1990s, there had been a shift in emphasis, away from food aid and towards food trade. This has been encouraged by multinational organizations such as the World Trade Organization and the World Bank, and the increasing liberalization of world trade in agricultural products. Indeed, the United Nations' World Food Programme accounted for 43% of global food aid and 45% of cereal aid in 2002, with 89% of their cereal aid being concentrated in sub-Saharan Africa and Asia.

Considerable debate has developed since the 1960s over the extent of the 'world food problem' (Dyson 1996). Pessimists from the neo-Malthusian school of thought predicted a 7 per cent fall in per capita food production in the 1990s, leading to an absolute global shortage of food and widespread starvation and famine in developing countries (Myers 1991, Ehrlich *et al.* 1993). In support of their claims, they point to the fall in the global output of cereals per person since the mid-1980s (cereals are often used as a proxy for general food trends), the failure of the 1960s and 1970s 'green revolution' to continue to develop new and higher-yielding varieties of wheat and rice into the 1980s and

1990s, and the real likelihood of future global harvest failures due to drought, leading to a repeat of the 1972–4 'world food crisis' (Dyson 1996). Of real concern is the increasing reliance by LDCs on the import of especially cereals, meat and milk from North America, Europe and Australia.

Although accepting that global food prospects are mixed and that there remain too many undernourished people in sub-Saharan Africa, the more optimistic anti-Malthusian school of thought believes that the world food problem is not deteriorating (Dyson 1996). The number of undernourished people actually fell by 201 million between 1980 and 2001, despite a significant increase in population during that period. Indeed, Dyson (1996: 75) suggests that

> the frequency and demographic impact of famines has declined greatly during recent decades. Modern food crises have had a diminishing and increasingly short-lived impact on population growth. The phenomenon of famine has become more and more restricted to Sub-Sarahan Africa.

Whilst accepting the general decline in per capita cereal production, optimists argue that population growth has not outstripped cereal production in all world regions. Indeed, cereal yields are not levelling off, neither are the effects of the green revolution, which in any case was not a revolution in the sense of creating a major upsurge in global cereal yields (except possibly in Pakistan). Any slowdown in cereal yields has occurred in developed rather than developing countries, in response to production control measures and prevailing international cereal prices. Indeed, Bruinsma (2003) predicts that the arable area in LDCs will increase by approximately 120 million hectares by 2030, mainly in sub-Saharan Africa (60 million) and Latin America (40 million). Finally, as will be demonstrated later in this chapter, food production and consumption patterns are becoming more diverse and so cereals may no longer be an adequate proxy for general food trends.

The aim of this chapter is not to explore the relationships between population and food (see Dyson (1996) and Bruinsma (2003) for excellent accounts), but to examine the changing geographies of global food production and their consequences for both developed and developing economies. More specifically, it describes the process of globalization and the 'industrialization' of the food supply system. In so doing, it emphasizes the dominant role played by transnational agribusinesses in developing globalized agro-food systems and in concentrating their activities in particular agricultural sectors and regions. The chapter then focuses on both the transfer of industrialized farming systems to particular regions in the LDCs and the 'relocalization' of food systems in the developed world. Case studies are included to illustrate the key points. Finally, the implications of these trends for global food production are considered.

7.2 Globalization and food regimes

Defined as the 'degree of purposive functional integration among geographically dispersed activities' (Robinson 1997: 44), globalization coincided with the emergence of global money and markets, the increase in transnational economic diplomacy, and global cultural flows in terms of tourism, guestworkers and consumerism (Amin and Thrift 1994). Global integration has been guided by powerful transnational corporations (TNCs), institutions and actors and has led to a new political economy of agriculture in which agro-food industries are epitomized by the mass production of manufactured food (for more on globalization and its many facets see Section 4).

Whilst there remain large inequalities in food supply between developed and developing countries, they are increasingly linked together in 'highly industrialized and increasingly globalized networks of institutions and products, constituting an agro-food system' (Whatmore 1995: 37). McDonald's is the classic agro-food expression of globalization (Ritzer 2000), and Friedmann (1982) used the term **agro-food complex** to describe the relations of production and consumption in particular food products, especially for highly industrialized agricultural commodities like salad crops (Friedland 1984, FitzSimmons 1986). Indeed, Friedmann and McMichael (1989) argued that relations between agriculture and industry have

Spotlight Box 7.1

Three food regimes since the 1870s

First regime: pre-industrial (1870s–1920s)

This involved settler colonies supplying unprocessed and semi-processed foods and materials to the metropolitan core of North America and Western Europe. Characterized by *extensive* forms of capital accumulation, the main products involved were grains and meat. The regime gradually disintegrated as agricultural production in developed countries competed with cheap imports and trade barriers were erected.

Second regime: industrial (1920s–1970s)

This regime relates to the productivist phase of agricultural change, focused on North America and the development of agro-industrial complexes based on grain-fed livestock production and fats/durable foods. Characterized by *intensive* forms of capital accumu-lation, the second regime incorporated developed and developing nations into commodity production systems. The resulting agricultural surpluses and environmental disbenefits began to undermine pro-ductivist agriculture in the 1970s.

Third regime: post-industrial (1980s onwards)

This developing regime relates to the crisis surrounding productivist, industrialized farming systems and involves the production of fresh fruit and vegetables for the global market, the continued reconstitution of food through industrial and biotechnological processes, and the supply of inputs for 'elite' consumption in developed countries. Characterized by a *flexible* form of capital accumulation, the third regime is dominated by the restructuring activities of agribusiness TNCs and corporate retailers.

Sources: Based on Le Heron (1993), Robinson (1997) and Ilbery and Bowler (1998)

historically been more global than generally thought. They used the concept of **food regimes** to link international patterns of food production and consumption to the development of the capitalist system since the 1870s. It is possible to identify three food regimes (Spotlight box 7.1), with each representing for its time the modern food system. The first involves settler colonies supplying unprocessed and semi-processed foods and materials to the metropolitan core of Western Europe and North America. The second relates to the 'productivist' phase of agricultural change, focused on North America, and the development of agro-industrial complexes and grain-fed livestock. Finally, the third involves greater flexibility, including the production of fresh fruit and vegetables for the global market, the continued reconstitution of food through industrial and bio-industrial processes, and the supply of inputs for 'elite' consumption in the DCs. Unlike the second regime, TNCs have been dominant in the evolution of the third food regime.

The globalization of agriculture is thus presented by these authors as a logical progression, when in reality it is a contested, partial, uneven and unstable process (Whatmore and Thorne 1997). The global presence of McDonald's and Pepsi Cola, for example, is not typical of the complex and highly uneven process of globalization that has reshaped food production and consumption in the post-war period (Pritchard 1998). Indeed, Le Heron (1993) highlights the importance of global–local interactions, leading to uneven development. Thus the role of locality is still important and 'globalization does not imply the sameness between places, but a continuation of the significance of territorial diversity and difference' (Amin and Thrift 1994: 6).

Globalization tendencies are, therefore, mediated by regional/local relationships and TNCs can play an increasing role in the social and political development of individual regions and nation-states (Marsden 1997). Indeed, giant TNC food companies and retailers are transforming the global agro-food economy; they have become dominant actors in the reconstruction of international and national food systems, based around the increasing demands of groups of consumers in the developed world for high quality, natural foods. As FitzSimmons (1997) suggests, TNCs have become the primary agents of globalization in the agro-food sector; they sit at the centre of webs of relationships that link farming, processing and marketing. Indeed, corporate retailers play an essential role in developing

regulatory systems that ensure their dominance over the supply of major food products. Not surprisingly, food retailing is becoming increasingly concentrated in the hands of a small number of supermarkets; this is particularly pronounced in a number of European countries (Poole *et al.* 2002).

7.3 Industrialization of agro-food systems in the developed world

Agricultural production in the developed world has become increasingly concentrated on a limited number of large-scale farms (Ilbery and Bowler 1998). At the same time, there has been an increase in capital expenditure on major agricultural inputs such as chemicals and machinery, and a growth in the processing and manufacturing of food. As a consequence, it becomes vital to view farming in the wider context of an **agro-food system**, in which the production sector itself is inextricably linked to various 'upstream' (input supplies) and 'downstream' (processing, distribution and marketing) industries (Atkins and Bowler 2001). The food supply system is dominated by large **agribusinesses** which, according to Davis and Goldberg (1957) and Wallace (1985), are the sum of all operations involved in the manufacture and distribution of farm supplies, the production operation of the farm, storage, processing and distribution of farm commodities and items made from them. These agribusinesses often develop agricultural commodity chains beyond national boundaries.

As Le Heron (1993: 45) suggested, 'industrialisation is progressively obliterating the farming base'. However, it is especially the non-farm sectors of the agro-food system that have become highly industrialized and dominated by TNCs. This has occurred through two discontinuous processes (Goodman *et al.* 1987):

- **Appropriationism**, whereby certain agricultural *inputs* are appropriated and replaced by 'industrial' alternatives. Good examples would include synthetic chemicals replacing manure and tractors replacing animal power.

- **Substitutionism**, which focuses on *outputs* rather than inputs and is concerned with the increased utilization of non-agricultural raw materials and the creation of industrial substitutes for food and fibre. Examples would include the substitution of sweeteners for sugar and nylon for cotton.

The rate and direction of appropriationism and substitutionism are determined largely by prevailing levels of technological change (e.g. witness the continued controversies over genetically modified crops), state support policies for agriculture, and consumer tastes and fashions. Thus the industrialization of agriculture proceeds in successive rounds of appropriation and substitution. This leads to increasing concentration in most segments of the agro-food system, from input suppliers to supermarkets. For example, in the UK over 70 per cent of the agro-machinery market is concentrated in just four companies, as is over 50 per cent of turnover in food retailing (Whatmore 1995, Poole *et al.* 2002).

However, agro-food systems are different from other production systems because agriculture is bound by biological processes and cycles. So, agribusiness TNCs attempt to reduce their dependence on 'nature' and increase their influence over farming indirectly through a process of *formal subsumption*, where arrangements or contracts are made with farmers to provide 'raw materials' for their value-adding food manufacturing activities. However, farmers are not 'passive in this logical progression of ... market forces' (Whatmore 1995: 45) and the family farm has a capacity to survive through various forms of self-exploitation. This hinders the industrialization process and the farming sector remains quite diverse, predominantly family-based, and geographically uneven.

The restructuring of farming, therefore, is dominated by geographical differences in the integration of agriculture into a global and industrialized agro-food system. Indeed, differentiation and diversity are key features of capitalist agriculture. Thus agricultural industrialization becomes concentrated in particular sectors and regions characterized by large farm businesses that have adopted intensive farming methods and integrated into global networks. In the European Union (EU), for example, 80 per cent of agricultural production is concentrated on less than 20 per cent of farms in particular 'hot spots', including East Anglia, the Paris Basin, southern Netherlands, and Emilia Romagna in north-east Italy (Whatmore 1995). The remaining 80 per cent of farms are marginal to the global agro-food system and are located in 'lagging' rural regions, where farming is less market-dependent.

Overall, agricultural industrialization is increasingly global in scale, where food commodity chains lengthen and producers become 'distanced' from consumers.

Such a system is unstable and uneven. It leads to increasing differentiation over space, overproduction, increased indebtedness as farmers invest in more technology, environmental disbenefits, and concerns over food security and safety. One reaction of TNCs and major retailers is to relocate the production function of the agro-food system to new agricultural spaces, often in LDCs. This is usually associated with the intensive production of high-quality horticultural products where little consideration is given to environmental degradation.

7.4 Reorientation of developing world agriculture

For many years, food production in the developing world was often of a subsistence nature. Early attempts in the late eighteenth and early nineteenth centuries to transform agriculture were through the creation of foreign-controlled and operated plantations, which provided raw and semi-processed materials for industries in the developed world. Thus many European settlers in such African colonial countries as Kenya, Mozambique, Angola and Southern Rhodesia began exporting crops such as tea, coffee and tobacco to the developed world, especially to European imperial states. Encouraged by colonial governments, African smallholder farmers also grew cash crops for sale to European traders at major ports, even though local people often had too little protein. A dependency culture thus developed on the basis that foreign earnings from export crops could be used by government agencies for development programmes. By the time these countries had become independent in the post-Second World War period, agricultural policy was focused on exports, with a neglect of food staples. Consequently, LDCs became net importers of staples and processed foodstuffs.

In more recent years, many former plantations, often run by agribusiness TNCs, have applied intensive and industrialized systems of farming, including the employment of wage labour and the application of capital and advanced technology to their production activities (Loewenson 1992). Many of these estates specialize in one crop, such as tea, coffee, sugar, bananas, green beans or flowers, destined for markets in the developed world. Estate managers often put in place infrastructure and marketing arrangements; these can attract government support because of perceived local development advantages. Indeed, estates, as earners of export income, importers of technology and employers of labour, are seen as one answer to the challenge of poverty and population growth in many LDCs.

As Goodman and Watts (1997) demonstrate, the classical export commodities (coffee, tea, sugar, cocoa) have now been complemented by high-value foods, including fruits, vegetables, poultry and shellfish. Throughout the 1980s, the value of world trade in high-value foods increased by 8 per cent per year. By 1990 they accounted for 5 per cent of global commodity trade, one-third of which came from the LDCs (far exceeding the value of LDC exports of coffee, tea, sugar, cotton, tobacco and cocoa) (Goodman and Watts 1997). By the early 1990s, 24 low–middle income countries annually exported over US$500 million of high-value foods; indeed, just five countries accounted for 40 per cent of such exports from the LDCs. Friedmann (1993) labelled these the newly agriculturalizing countries, namely Brazil, Mexico, China, Argentina and Kenya.

The major growth in high-value food exports in the 1980s and 1990s reflects many factors, including technical changes in the food industry and the liberalization of world agricultural trade. Significantly, however, dietary changes in the developed world have been instrumental and Goodman and Watts (1997) highlight how high-value foods are produced to satisfy the tastes and life-styles of particular groups of consumers in developed countries. Agribusiness TNCs are attracted to high-value food production in newly agriculturalizing countries because of low-cost labour, government support (e.g. through Structural Adjustment Programmes), good international communication links, and the ability to produce high-quality, fresh products for niche markets in the developed world.

Marsden (1997) explains how recent agricultural development in the newly agriculturalizing countries is based on the use of intensive production techniques, which are underpinned by quality conditions and control (from planting through to consumption). Design and quality are vital ingredients in high-value food production. Responding to consumer demand and tastes for size, shape, colour and content of products, the concern for 'quality' emanates from the strategies of corporate retailers. Storage and transhipment are other important functions, often involving exporters and importers, acting under instructions from the retailers. Produce is rejected if quality and appearance are not good. In turn,

exporters have to satisfy quality control procedures specified by importers and retailers. As Marsden (1997: 177) states, 'those actors and agencies who are closest to the definition and implementation of quality conditions begin to accumulate power in the food network. This leads to a growing social and economic differentiation in the region, with smaller producers prone to exclusion in globalised food networks.'

So, corporate retailers in the developed world play a key role in translating quality definitions of foods globally. Indeed, they develop their own regulatory systems that ensure a dominance over the supply of main food products; it is highly competitive and vulnerable to consumer reactions. Production for export markets leads to a decline in production for the local population that are reliant on increasing amounts of food imports. There is little doubt that large-scale agribusinesses and corporate retailers have successfully linked regional economies and crop sectors of LDCs to a global system of food production and consumption. However, their behaviour has attracted a number of criticisms as regards their economic, social and environmental impacts on host countries (Espiritu 1986, Ilbery and Bowler 1996):

- They contribute to sharp inequalities in income, productivity and technology compared to the sector producing domestic staples.

- Labour-saving innovations by agribusiness mean that numbers employed are reduced, with many becoming deskilled, part-time or seasonal. Specialist export experts are often brought in.

- The benefits of intensive agriculture accrue disproportionately to foreign investors and, in the case of nationally owned estates, to urban-based people and local elite groups.

- Large estates, especially those of agribusiness TNCs, often take the best land for export crops. This land absorbs the most inputs, investment and expenditure, even though it occupies only a small proportion of the total agricultural area. Unable to compete, localized agricultural production often begins to break down.

- Many of the agricultural practices of agribusiness TNCs are unsustainable. Environmental degradation is a real problem and there are reported instances of groundwater depletion and agribusinesses moving on to new areas when soils have become depleted.

7.4.1 A Kenyan case study

Since 1990, an increasing number of African countries have entered the export trade in high-value foods and cut flowers (Barrett and Browne 1996, Hughes 2000). The competitiveness of African suppliers in this sector reflects low costs of production, complementarity to European seasons, relatively short flight times and the ability to supply produce of the quality and quantity required in international markets all year round. Kenya is the only newly agriculturalizing country in Africa; it is also an established exporter of horticultural products to the UK market.

Between 1991 and 1996, Kenyan exports of horticultural produce increased 58 per cent to reach a total output of 84,824 tonnes; by the year 2003, this was expected to rise to over 120,000 tonnes (HCDA 1997). During this period of economic liberalization in Kenya, huge investments have been made in the industry, especially by the ten largest producers in the cut flower and pre-packed vegetable sectors. This has mainly been to attract and keep lucrative contracts

with large British supermarket chains. Kenya specializes in the export of high-quality and often pre-packed green beans, mange touts, Asian vegetables (mainly okra, chillies and aubergines), avocados and cut flowers. The export of cut flowers, especially roses and carnations, has been increasing at about 20 per cent per annum since the early 1990s (Hughes 2000). Europe is the main customer for Kenyan fresh horticultural produce; in 1998, 85 per cent went to just four EU countries: UK (30 per cent), The Netherlands (29 per cent), France (16 per cent) and Germany (10 per cent).

New marketing chains have evolved to facilitate the Kenyan export trade in high-value foods and cut flowers (Barrett *et al.* 1999, Hughes 2000). First, there is an older, more fragmented *wholesale chain*, linking small- and medium-scale growers into the export market through a series of agreements and contracts between growers, agents, exporters, freight agents, importers and retailers (Figure 7.1). Many of the links in this lengthening chain are based around complex family-based connections where, for example, Kenyan-

Plate 7.1 High value foods and cut flower exports from Kenya.

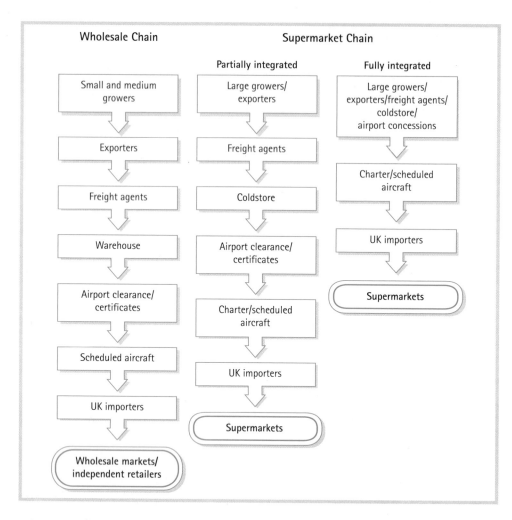

Figure 7.1
Marketing chains and the export of high-value foods and cut flowers from Kenya into the UK.
Source: Based on Barrett *et al.* (1999:165)

Asian exporters deal with British-Asian importers and retailers. The produce for this market is generally imported and sold loose; significantly, it does not have to meet supermarket specifications. Consumers are interested in variety, quality, flavour and value for money rather than presentation and packaging. Although still accounting for approximately 25 per cent of Kenya's horticultural exports to the UK, the wholesale trade is declining. This reflects the growing importance of a second chain – the *supermarket chain* – which accounts for at least 70 per cent of horticultural exports. High-value fruit, vegetables and cut flowers require speed in harvesting and packaging, as well as in transportation. Quality, and thus price, is dependent on the environment in which the produce is stored and shipped. Certain very delicate crops such as green beans, mange touts and cut flowers are better preserved if travelling along a 'cool chain', where most or all stages of transit between harvest and retail point are temperature-controlled. Not surprisingly, this chain mainly comprises large-scale commercial growers,

geared to supplying the expanding supermarket trade in trimmed and packaged products.

The supermarket chain consists of *partially integrated* and *fully integrated* systems (Figure 7.1). Whilst the former have developed primarily for large-scale producers who are also exporters, the latter see production, exportation, freight handling and often importing being controlled by one company. The fully integrated system is heavily associated with the cut flower and supermarket trades; the large-scale farms involved use modern technologies, mostly imported, to achieve high yields of uniform-quality produce and high standards of presentation and packaging. Seed stock, irrigation, floodlighting and other technologies encourage intensive, 'productivist' farming systems, with which small and medium growers cannot compete. The partially integrated chain supplies about 25 per cent of exports, but differs from the wholesale chain in that a small number of large-scale growers have seen the benefits of integrating production and exporting aspects of the business, thus bypassing agents in Kenya. This integration allows supermarket demands for quality and traceability to be satisfied. To remain competitive, the producers have started to trim and package vegetables and to diversify into baby vegetable production.

Developing since the early 1990s, the fully integrated chain is a response to the specific needs of the cut flower industry and supermarket demands (Hughes 2000). Quality, freshness and speed of delivery are paramount. In order to achieve this and maintain control of the 'cool chain', the three largest producer/exporters of horticultural produce in Kenya have set up their own export and freight companies; they proudly claim that within 48 hours of being harvested, produce can be on the supermarket shelves in Britain. They now account for almost 50 per cent of Kenya's horticultural exports. Most of the produce is sold in pre-packs, which are bar-coded and priced in Kenya as directed by the supermarkets who provide the stickers. To maintain the loyalty of the supermarkets, and the quality requirements of customers, producers have invested heavily in EU-standard packing stations, refrigerated trucks and coldstores (including at Nairobi airport). The 'farms' have research and development laboratories where quality is monitored, packaging tested and new products screened. Once inspected by customs and excise officials at the airport, the produce is loaded directly onto specially chartered planes. The produce hardly ever leaves the 'cool chain'.

EU importers sourcing from Kenya are a buffer between exporter and supermarket. In the fully integrated system, they are often part of the same group that includes producers, exporters and freight agents. Importers are responsible for forwarding the produce to warehouses and undertaking necessary health, safety and quality checks as required by regulators. Most importers deal with the major supermarkets through informal rather than formal contracts, which are 'worked out' on an annual or six-month cycle. These are then confirmed on a monthly, weekly or even daily basis, with trade increasingly moving towards a more flexible 'just-in-time' principle. It is thus the importers who take the greatest risk in the marketing system. The supermarkets' demands for a high standard of packaging and presentation, including a significant movement towards pre-packs in the country of origin, favour the sourcing of produce from large-scale 'productivist' agri-business companies.

It is quite clear that the supermarket retailers are the most powerful 'actors' in this lengthening agro-food system. They have flexible arrangements with different importers, and exert their power through 'remote control' (Whatmore and Thorne 1997). Only when inspected and approved by the retailers' quality control systems will produce be accepted and paid for. Overall, therefore, horticultural exports from Kenya are increasingly concentrated in the hands of large, highly capitalized producers. The requirements for documentation about sourcing and traceability militate against small-scale producers who are likely to become increasingly marginalized. Yet, rather than produce much-needed staple products for the indigenous population, they attempt to participate in the export sector. The real paradox of this whole system is that it has embraced intensive 'productivist' methods to satisfy the demands from 'new consumers' in regions of the developed world rather than internal demands for more staple foods.

7.5 Relocalization of developed world agriculture

The 'productivist' model of agriculture in developed market economies, based on the principles of efficiency, rationality and high levels of government support, has been under challenge since the early 1980s (Ilbery and Holloway 1997). This challenge has been stimulated by the environmental and socio-economic disbenefits of agricultural industrialization and, as a consequence, alternative systems associated with environmental protection, sustainability, food quality, reduced price support, and the integration of agriculture within broader rural economic, social and environmental objectives have arisen.

The movement away from an essentially productivist ethos in developed world agriculture has been conceptualized as the **post-productivist transition** (Shucksmith 1993, Lowe *et al.* 1993, Ilbery and Bowler 1998, Walford 2003). Although criticized as a theoretical device (Wilson 2001, Evans *et al.* 2002), a number of known characteristics are associated with the post-productivist transition (Spotlight box 7.2). It embraces the whole of the agro-food system, from production and processing to consumer choices and the marketing systems that link them. Government policy is central to the post-productivist transition, but this time responding to rather than stimulating change. So, for example, the gradual withdrawal of

Spotlight Box 7.2

The post-productivist transition in agriculture

Major characteristics

- A reduction in food output and greater emphasis on food quality
- Progressive withdrawal of state subsidies for agriculture
- The production of food within an increasingly competitive and liberalized international market
- The growing environmental regulation of agriculture
- The creation of a more sustainable agricultural system

Pathways of farm business development

- Pathway 1: Extension of the industrial model of productivist agriculture, based on traditional farm products
- Pathway 2: Redeployment of farm resources into new agricultural products on the farm (agricultural diversification)
- Pathway 3: Redeployment of farm resources into new non-agricultural products on the farm (business structure diversification)
- Pathway 4: Redeployment of farm resources into employment off the farm (other gainful activity)
- Pathway 5: Maintaining the traditional model of conventional farm production (extensification and/or adoption of agri-environmental programmes)
- Pathway 6: Winding down to hobby or semi-retired farming
- Pathway 7: Retirement from farming

subsidies for farmers under the Common Agricultural Policy (CAP) of the EU is a response to processes such as the globalization of the agro-food system, trade liberalization and the growing international significance of the environmental movement. Further reforms are an almost inevitable consequence of continuing negotiations with the World Trade Organization. Continued liberalization of world trade in agricultural products and moves towards market orientation are key issues.

Similarly, and in response to the environmental disbenefits of 'productivist' agriculture, there has been an increasing environmental regulation of agriculture and a willingness to embrace the principles of sustainable agriculture. Again within the context of the EU, agri-environmental programmes were introduced as part of the CAP reforms in 1992 and further strengthened under the Agenda 2000 reforms (Robinson and Ilbery 1993, Lowe *et al.* 2002). These enabled member countries to introduce a series of voluntary schemes to reduce the application of chemicals and fertilizers in specified areas and to encourage the development of, for example,

environmentally sensitive areas, countryside stewardship, extensification, organic farming, moorland restoration and habitat improvement. Although a large number of agri-environmental schemes now exist in the EU, there is considerable debate as to whether or not such programmes provide real environmental benefits (Morris and Potter 1995, Hart and Wilson 1998, Wilson and Hart 2001).

However, one of the key drivers of the post-productivist transition has been a shift in consumer demand and the increasing interest in, and knowledge of, food (Coakley 1998). In response to concerns over food safety, epitomized by the major outbreak of foot and mouth disease in the UK in 2001, there has been a movement towards 'healthy eating' and fitness, and a growing demand for high-quality regional and local food products with authenticity of geographical origin and *traceability* (Gilg and Battershill 1998, Ilbery and Kneafsey 1998, Hinrichs 2003, Winter 2003). Bell and Valentine (1997) argue that food is an indicator of socio-cultural status and Chaney (1996) sees the consumption of food as part of a life-style image. Thus some consumers (the 'new consumer') are giving more emphasis to aspects of quality and convenience than to price and quantity (Barrett *et al.* 1999).

As suggested earlier in this chapter, the shift from productivist to post-productivist farming systems does not signal the end of agricultural industrialization. Indeed, both types of farming system are likely to coexist and become more spatially differentiated. So, on the one hand, an intensive, high-input, high-output system emphasizing food *quantity* and continued integration into the agro-food system may continue in 'hot spot' regions in both developed and developing countries. On the other hand, a more extensive, low-input, low-output system which emphasizes *quality* and sustainability may develop in more marginal agricultural areas in developed countries (those left out of the globalized agro-food system). The latter may present opportunities for a relocalization of agricultural production, whereby farmers in such areas develop alternative strategies by, for example, concentrating on niche markets and responding to consumer demand for either high-quality regional and local food products and/or ecologically sound farming practices. Such relocalization also encourages the shortening of food chains and Marsden *et al.* (2000) classify short food supply chains into three main types:

- *Face-to-face*, where the consumer buys direct from the producer or processor.

- *Spatial proximity*, where products are sold through retail outlets in the region, locality or place of production.

- *Spatially extended*, where information about those producing the food and the place of production is translated (through product labelling/branding) to consumers who are located outside the region of production.

Undoubtedly, the post-productivist transition will lead to the emergence of new agricultural land uses. However, which ones emerge in which areas will depend on the different adjustment strategies adopted by farm households. Bowler (1992) and Bowler *et al.* (1996) referred to these as *pathways of farm business development* (Case study 7.2). The different pathways fall neatly into three categories:

- Maintaining a full-time, profitable food production element to the farm business (pathways 1 and 2).

- Diversifying the income base of the farm business by redirecting resources into non-farm enterprises and occupations (pathways 3 and 4).

- Surviving as a marginalized farm business at a lower level of income, perhaps supported by investment income, pensions and agri-environmental payments (pathways 5 and 6), or retiring from farming altogether (pathway 7).

The pathways are not mutually exclusive and farm households may combine different elements, through for example organic farming and 'adding value' by processing and marketing. However, if households begin to make the transition from pathway 1 towards pathway 7, it is highly likely that they will become marginalized for ever. Much will depend on the nature of the farm household and whether its members can be described as *accumulators* (risk-taking, venturesome, and able to redeploy resources) or *survivors* (conservative, traditional, with some diversification and possible 'passive' adoption of agri-environmental programmes).

7.5.1 Niche markets and regional speciality food products

With food becoming a statement of taste, sophistication and fashion in particular niche markets (i.e. sub-groups of mainly middle- and high-income consumers with particular tastes, life-styles and aspirations), there are opportunities for a (re)localization and reintegration of agriculture within the rural development strategies of developed market economies. In the UK, for example, the Policy Commission on the Future of Farming and Food (2002), published in the wake of the foot and mouth epidemic, sees locally produced food as an important means of reconnecting both farming with its market and the rest of the food chain, and consumers with what they eat and how it is produced. Indeed, certain food markets are increasingly being shaped by the demands of particular groups of consumers. A life-style image is related to class and income, and more affluent groups are able to perceive food as not only good to eat but also good to think about (Beardsworth and Keil 1992). Thus for the 'new consumer' a relationship emerges between food and personal identity.

The growing demand by such groups for high-quality, wholesome and traceable food raises issues over just what is meant by *quality* and how regional/local food products can be linked to particular places. Quality is a contested concept, but relates to the

Table 7.2 Indicators of quality in speciality food and drink products

1 Achieving *certification* by a professional organization, government or other external bodies:
- appellation contrôlée
- organic symbol
- quality mark
- self-regulation

2 Establishing *association*, either geographically within a region or historically with a tradition or culture:
- regional designation
- traceability
- historic/cultural importance
- local environment

3 Ensuring *specification* of:
- production method (e.g. small-scale workshops, 'authentic' recipes)
- raw materials (e.g. local milk, water from a particular spring)
- ownership

4 Generating *attraction* by tapping into the subliminal wants of consumers:
- designs
- texture, flavour and taste
- freshness/appearance
- premium price
- consumer perception

Source: Ilbery and Kneafsey (1998: 331)

satisfaction of consumer needs (Vastoia 1997) and a consistent level of performance, taste and so forth provided by the product (O'Neill and Black 1996). It has connotations of being able to command higher prices and is a positional characteristic – something that is above minimum standards. A quality product is 'differentiated' in some positive manner that is recognized by consumers. It is possible, therefore, to identify a number of indicators of quality (Ilbery and Kneafsey 1998, 2000a). These range from ensuring *specification* of production method and raw materials used, establishing an *association* with a region, local environment or cultural heritage, to generating consumer *attraction* through such attributes as taste, texture and appearance, and achieving some form of official *certification* for the speciality food product (Table 7.2). Thus quality implies a number of 'actors' in the agro-food system, from initial producer to final consumer. While the consumer may perceive quality in terms of food safety, appearance, taste and traceability, producers often see quality as a market device to increase sales, differentiate their product and gain premium prices.

The linking of regional speciality food products to particular places is a necessary prerequisite for the relocalization of food production. Bell and Valentine (1997: 155) argue that there are already strong links between products and places, so that 'almost any product which has some tie to place, no matter how invented this may be, can be sold as embodying that place'. The creation of a strong regional identity is thus vital and the 'fixing' of local food products to place can be achieved through marketing and place promotion (Ilbery and Kneafsey 1998, Ray 1998, Kneafsey and Ilbery 2001). Emphasis is placed on *selling places*, but the image created must be distinctive to the region and not related to all rural areas. So, marketing becomes particularly important because constructions of quality will be interpreted, understood and consumed differently by individuals in different places and at different times.

With the growing demand for regional speciality food products, it is important that the sector is properly regulated. In the UK, for example, the Food Safety Act (1990) has effectively handed over the regulatory power for consumer protection to corporate retailers. This requires supermarkets to show 'due diligence' by demanding 'traceability' for all the products they sell. Not surprisingly, they have taken the initiative on such issues as food quality, authenticity and convenience. Supermarkets have also seen the potential of using regional imagery and marketing to exploit the quality food market, often creating 'false' images of traditional fare and wholesome food (Gilg and Battershill 1998). Most **quality assurance schemes** have been initiated by retailers and by supermarkets;

Plate 7.2 Speciality food and drink products in the UK. Wallace and Gromit courtesy of Aardman.

very few have been started by local speciality food producers (Morris 2000, Morris and Young 2000). Traditional retailers are in danger of losing their market share, partly because they cannot compete with the supermarkets and partly because consumers demonstrate less confidence in their claims of quality. Thus, there is a real possibility of agribusiness TNCs and corporate retailers appropriating the speciality food market and integrating it into their industrialized agro-food systems. Buck *et al.* (1997) have produced early evidence of such a process in relation to organic farming in northern California.

One response to such a process is for the state to regulate regional speciality food products through a formal system of quality 'labels'; another is for small producers to establish regional speciality food groups. As an example of the latter, the government quango 'Food from Britain', with financial support from the Ministry of Agriculture, has overseen the establishment of regional speciality food and drink groups in England (Figure 7.2a). These regional groups are able to offer a broad range of products to attract buyers, arrange trade fairs and exhibitions, provide training on

marketing and food promotion, and gain access to markets and funding. The largest regional group, Taste of the West, has over 200 producer members, and survey evidence suggests that they define quality in terms of specification and attraction rather than official certification schemes and association with a regional environment, culture and tradition (Ilbery and Kneafsey 2000a).

It is for this reason that official certification of 'quality' regional speciality food products, where a product can be related specifically to place, is required. In this context, the EU has recognized the potential of differentiating speciality food products on a regional basis. Two related regulations were passed in 1992, one of which (Regulation 2081/92) is concerned with speciality food products from specific areas. In an attempt to harmonize and develop Community-wide quality labels for particular food and drink products, producers can apply for either a PDO (Protected Designation of Origin) or a PGI (Protected Geographical Indication). A PDO relates to a product originating in a specific region where the quality of that product is due exclusively to a particular

181

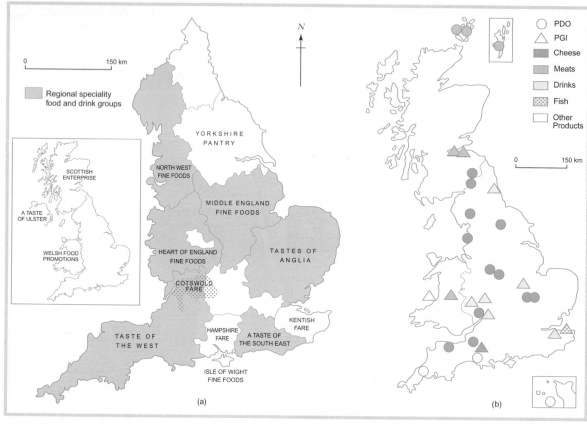

Figure 7.2   Regional speciality food and drink products in the UK: (a) regional speciality food and drink groups in England; (b) PDOs and PGIs in the UK.

Source: B. Ilbery and M. Kneafsey. Producer constructions of quality in regional speciality food production: a case study from southwest England. *Journal of Rural Studies*, **16**, 2000, pp. 217–30. Copyright 2000, with permission from Elsevier

geographical environment with its inherent human and natural factors. In contrast, a PGI relates to a product originating in a particular region that possesses a specific quality, reputation or other characteristics attributable to that geographical origin (but not necessarily due to its natural environment or raw materials). Such quality 'marks' protect producers from attempted copies of their product and act as a marketing device to attract customers. A scheme of this nature aims to replace national quality labels with an international system and thus ensures fair competition between EU producers; it also helps to ease the transfer of locally produced speciality food products to regional, national and international markets.

The United Kingdom already has over 30 PDO/PGI designations (Figure 7.2b). Specialist cheeses dominate the PDOs, indicating that these are made exclusively from local raw materials (e.g. Stilton cheeses). In contrast, drinks dominate the PGIs. Thus Newcastle Brown (a beer) is a PGI because, although associated with the city of Newcastle, not all of the ingredients

come from that region. There are now over 350 PDO and 200 PGI designations in the EU (Parrott *et al.* 2002). Some member countries have very few (e.g. Finland, Sweden, Denmark, Ireland, the Netherlands), while others have a large number (e.g. France and Italy with over 100 each and Greece and Portugal with over 80 each). These differences suggest that food has more cultural significance in some European societies than in others (Parrott *et al.* 2002).

7.6 Meeting the challenge of food production

This chapter has examined various dimensions of the globalized agro-food system. As noted earlier, some commentators (e.g. Le Heron 1993, Pritchard 1998) suggest that a third food regime (Case study 7.1) is emerging which, whilst being quite flexible in nature, is focused on the supply of fresh fruit and vegetables for 'elite' consumption in the developed world.

Unfortunately, such a situation is likely to exaggerate further the huge inequalities that already exist in global patterns of food supply and consumption, thus condemning large, but declining numbers of people to continued poverty and hunger.

Such inequalities are, of course, a function of the prevailing capitalist mode of production, in which agribusiness TNCs and corporate retailers have developed considerable power and so control the global food supply system. They have targeted their activities on particular 'hot spots' around the world, where 'productivist' farming methods are practised and integrated into essentially industrialized systems. Indeed, farming is just a small cog in large agro-commodity chains. These hot spot regions are not confined to the developed world and can now be found in a number of 'newly agriculturalizing countries', as demonstrated in the Kenyan case study. A majority of farmers and thus regions, in both developed and developing countries, have been left out of these agro-food chains and so become marginalized. Nevertheless, small-scale producers in some LDCs often attempt to supply food for what is perceived to be lucrative export markets rather than focus on the production of much-needed staple products for the indigenous population. Continued liberalization of world trade in agricultural products and the movement towards greater market orientation in agricultural policy can only increase international competition in food production and further strengthen the power of agribusiness TNCs and corporate retailers.

These developments present a major challenge for the future supply of food to the most needy groups in the world, and especially to regions of high population growth in LDCs. Such a challenge cannot be resolved without state regulation. However, such regulation must not be imposed on different regions in a 'top down'

fashion. Instead, endogenous 'bottom up' programmes of integrated rural development are required, which are multisectoral, sustainable and based on local needs through the principle of subsidiarity. Thus, local self-sufficiency becomes the first priority, rather than satisfying the demands of 'elite' groups from developed countries. In the case of LDCs, this means a refocusing of attention on the production of basic staple foods for the local population. This requires the dependency culture on export crops to be broken, which in turn necessitates changes in international and national policy. The benefits of relocalized production must also be spread to all sectors of society and not confined to local elite groups. In the more marginal agricultural areas of developed countries, farmers need to adjust to the continued globalization of the agro-food system by embarking upon more sustainable and 'alternative' forms of agriculture, based on local knowledge and customs. Indeed, a focus on high-quality, authentic and traceable regional/local food products for niche markets is one possible pathway. As shown in the second case study of regional speciality food products, this is especially so if they are 'fixed' to territory, so that there is potential for the local area to keep more of the economic benefit and, at the same time, retain some control over the type of economic activity that occurs there (Ray 1998).

In the final instance, global–local interactions are vital when attempting to meet the challenge of feeding the world's growing population. Thus the encouragement of 'bottom up' forms of rural development must be complemented by a much fairer global system of food trade, in which the vast food-growing potential of regions such as Eastern Europe can be utilized to feed those millions of people still experiencing hunger.

Learning outcomes

Having read this chapter, you should understand:

- The different schools of thought concerning the complex relationships that exist between global population growth and food production.

- The continued globalization of the agro-food system in terms of the concept of food regimes.

- The dominance and spatially uneven penetration of the global agro-food system by transnational agribusinesses.

- The adoption of 'productivist' farming techniques in parts of the developing world to satisfy the tastes and life-styles of particular groups of consumers in developed countries.

- The increasing demand for, and renewed significance of, high-quality regional/local (niche market) food products from specific places within the developed world.

- The main issues facing the changing geographies of global food production and consumption in the twenty-first century.

Further reading

Atkins, P. and **Bowler, I.** (2001) *Food in Society: Economy, Culture and Geography*, Arnold, London. A collection of fairly short chapters on many of the global food production and population growth issues raised in this chapter.

Dyson, T. (1996) *Population and Food: Global Trends and Future Prospects*, Routledge, London. An excellent exposition of the complex relationships between global trends in both population growth and food production.

Goodman, D. and **Watts, M.** (eds) (1997) *Globalising Food: Agrarian Questions and Global Restructuring*, Routledge, London. A diverse collection of essays on aspects of the globalization of food production and consumption.

Ilbery, B. and **Bowler, I.** (1998) From agricultural productivism to post-productivism, in Ilbery, B. (ed.) *The Geography of Rural Change*, Longman, London. A review of conceptual and empirical research on the main dimensions of 'productivist' and 'post-productivist' farming systems in the developed world.

Robinson, G. (2003) *Geographies of Agriculture*, Pearson, London. The most recent and up-to-date textbook on geographical aspects of agriculture and food supply systems.

Useful web-sites

www.europa.eu.int   The official site of the European Union. Within this very large site, one Directorate General has responsibility for Agriculture. This DG site contains information on the EU's Common Agricultural Policy, key speeches and policy papers, statistics, links to member-state sites and an Information Resource Centre.

www.fao.org   The official site of the United Nations Food and Agriculture Organization. The site contains information on FAO programmes and FAOSTAT, an online multilingual database with information on food production and trade.

www.defra.gov.uk   The site of the UK government's Department of Environment, Food and Rural Affairs (formerly the Ministry of Agriculture, Fisheries and Food). The site contains information on UK agricultural policy, statistics and links to other government departments, non-ministerial departments and NGOs. Similar sites exist for other national governments, for example **www.usda.gov** is home of the US Department of Agriculture.

www.oxfam.org   Home of Oxfam International, which is dedicated to addressing the structural causes of poverty and related injustices. The site includes Oxfam International's Position Papers that address some of the issues raised by this, and other, chapters.

www.soilassociation.org   Home to the UK organization dedicated to promoting organic food and farming. The site contains an online library with a variety of information from local food links to policy papers. The site also contains information on regional food groups in England (such as **www.heff.co.uk** for the Heart of England Fine Foods Group).

For annotated, clickable weblinks and useful tutorials full of practical advice on how to improve your study skills, visit this book's website at www.pearsoned.co.uk/daniels

CHAPTER 8

Worlds apart: global difference and inequality

MARCUS POWER

Topics covered

- The importance of the 'three worlds' schema
- Development as knowledge and power
- The view from 'the South' and the view 'from below'
- Alternative geographies of global development and inequality

What is the geography of the Third World? Certain common features come to mind: poverty, famine, environmental disaster and degradation, political instability, regional inequalities and so on. A powerful and negative image is created that has coherence, resolution and definition. But behind this tragic stereotype there is an alternative geography, one which demonstrates that the introduction of development into the countries of the Third World has been a protracted, painstaking and fiercely contested process.

(Bell 1994: 175)

Perhaps one of the most interesting features of the word 'development' is that it has produced a bewildering array of labels for people and places that are not considered 'developed'. Thus countries are variously seen as belonging to the 'Third World' or the 'Global South', to be somehow 'lesser developed', underdeveloped and behind or even backward in some way. In an interesting study of student travellers to the 'Third World', Desforges (1998) illustrates the links between travel and identity for many students who sought to 'collect places' that offered authentic individual knowledges and personal experiences that could be gained through travel. This involved a framing of 'Third World' others and places as different and the assumption that it was possible to 'collect' experiences of Third World places. In this way, travel can be understood as one way in which 'youth identities "stretch out" beyond the local to draw in places from around the globe' (Desforges 1998: 176). As Morag Bell suggests in the above quote, the space of the 'Third World' has lots of tragic stereotypes of famine, poverty, drought, etc., to animate it and make it seem coherent and defined. Learning to understand and appreciate social or economic differences relative to the area of the world in which we live is a difficult and value-laden process. The term development is notoriously hard to define and often refers simply to 'good change', a positive word that in everyday parlance is practically synonymous with 'progress' and is typically viewed in terms of increased living standards, better health and well-being and other forms of common good that are seen to benefit society at large (Case study 8.1). Conditions of poverty clearly vary between different areas as does the way poverty is experienced, so simplistic stereotypes of 'Third World poverty' are of little use. In some ways the lack of an agreed set of international development indicators and measures or of common systems of data collection tells its own story of the failure of international development since 1945. A major problem with the maps produced from these statistics is that they have allowed some observers to label whole areas as 'Third World' or 'lesser developed' as if the same could be said of all its constituents (Wood 1985). As Morag Bell puts it:

> In emphasising what people are deprived of (as is implied by poverty), they [statistics] impose a negative uniformity upon non-western societies. As objects of study poor people become categories and are labelled as a homogeneous group. (Bell 1994: 184)

This chapter seeks to explore the notion of 'Third World' development and outlines some of the ways in which this has been theorized and practised. In order to do this, the chapter examines the historical and geopolitical dimensions of development and focuses on the need to contextualize our studies of development. Rather than seeing poor people as 'objects of study' we need to attend to the ways in which poor people themselves understand and make sense of their experiences of poverty.

Moving beyond the labelling of 'Third World' peoples and places as a homogeneous group, it is important to grasp how places and peoples are *differentiated* through development, experiencing progress and 'good change' in a variety of ways. The key question here is how to listen to and understand the differences that define and distinguish how development is lived in different sorts of places and amongst different groups of people. This chapter is divided into four further sections. The first discusses two of the most important conceptual perspectives that have been formulated on the relations between development and inequality: the *modernization* and *dependency* schools. Although there have been many other different strands of development thinking, both these perspectives have been widely influential and remain relevant to an understanding of theory and practice today. Section 8.3 outlines the need to view development historically and formulate a sense of how it has been redefined through time. How have historical forces shaped our understanding of the geography of development and in what ways are the legacies of the past important to understanding contemporary global economic difference and inequality? The next section focuses attention on people and places and explores how the subjects of development experience developmental interventions. Three major 'brokers of development' (non-governmental organizations (NGOs), states and international institutions) dominate the literature on development, but how are the needs and aspirations of intended beneficiaries understood by these agents of the development process? The final concluding section of the chapter poses questions about the political, cultural and economic implications of globalization for the future of development theory and practice and examines possible alternatives to development based around local cultures and knowledges.

Regional Case Study 8.1

The UK and Uganda

A comparison of the UK and Uganda is useful here in allowing us to evaluate the relative value of global indicators of inequality (Figure 8.1). Both countries are roughly the same size but their populations clearly do not have the same levels of access to water, health services or education. The UK can even afford US$586 per head on defence expenditure compared with US$7 per head in Uganda. These statistics clearly suggest massive inequalities between the two countries such that UK citizens can apparently expect to live 37 years longer than their Ugandan counterparts. In understanding how this is possible and how to respond to this unevenness there must be agreement on the significance of these inequalities and their causes or origins. There must also be an appreciation of history (Uganda was a British colony) and of the fact that not every Ugandan citizen has a life expectancy of 40 years or earns US$190. Ultimately simple measures like these tell us very little about wider social and economic differences or about how these variations are produced.

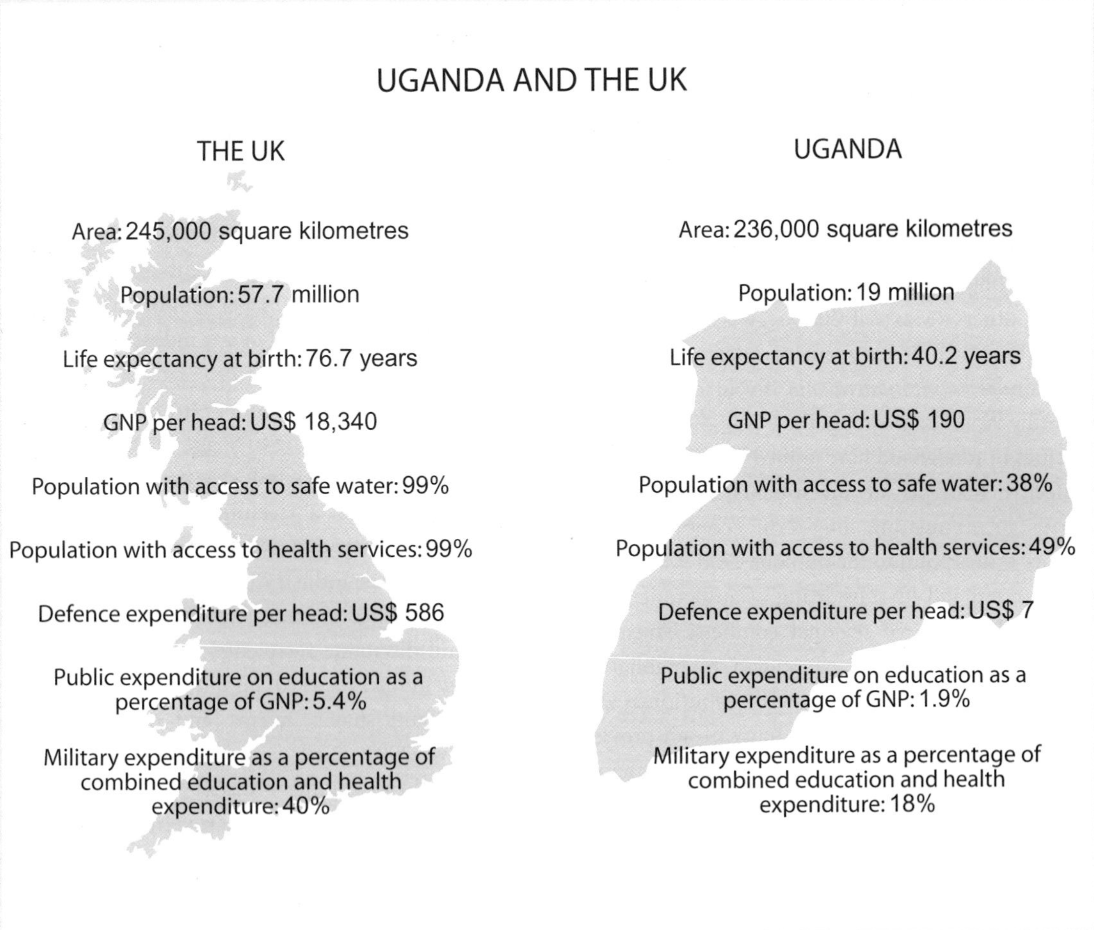

UGANDA AND THE UK

THE UK

Area: 245,000 square kilometres

Population: 57.7 million

Life expectancy at birth: 76.7 years

GNP per head: US$ 18,340

Population with access to safe water: 99%

Population with access to health services: 99%

Defence expenditure per head: US$ 586

Public expenditure on education as a percentage of GNP: 5.4%

Military expenditure as a percentage of combined education and health expenditure: 40%

UGANDA

Area: 236,000 square kilometres

Population: 19 million

Life expectancy at birth: 40.2 years

GNP per head: US$ 190

Population with access to safe water: 38%

Population with access to health services: 49%

Defence expenditure per head: US$ 7

Public expenditure on education as a percentage of GNP: 1.9%

Military expenditure as a percentage of combined education and health expenditure: 18%

Figure 8.1 A comparison of the UK and Uganda: some key statistics.
Source: International Broadcasting Trust (1998)

8.1 Geography and 'Third World' development

By whom is development being done? To whom? These are potent questions. They should particularly be asked when 'solutions' are put forward that begin 'We should . . .', without making clear who 'We' are and what interests 'We' represent (Allan and Thomas 2000: 4).

The problem with labelling vast regions as 'lesser developed' or 'Third World' is that a set of negative assumptions are often made and a 'negative uniformity' is imposed. Thus the picture of unevenness and injustice in the contemporary world that comes to us through these labels is not always a sharp, coherent and precise one and often this unevenness is not effectively conveyed in the statistical measures that are taken as indices of what constitutes 'development' (see Case study 8.2). Crucial then to the construction of a 'Third World' is a process of setting worlds apart and a politics of labelling that it is necessary to understand and be critically aware of. All too often the 'developing world' has been defined as a 'problem' for Western governments that can only be resolved with the

---

Thematic Case Study 8.2

The 'strategic interests' of foreign aid

In many ways examining the case for aid can tell us a great deal about the history of development theory and practice and provides a useful opening onto wider discussions of North–South relations (see Figure 8.2). In 2002 total aid amounted to about US$56 billion from 21 major OECD countries and some others (Maxwell 2002). Two-thirds of this is government to government or *bilateral aid* and the remainder is *multilateral aid* disbursed by agencies like the World Bank group, the UN bodies and the EU. Critics of foreign aid have argued that it has been less effective than private investments and commercial loans in stimulating long-term economic growth. Other critics point to the dubious Cold War record of foreign aid and its subsidizing of autocratic regimes and inflamation of regional conflicts. Thus, many critics of foreign aid have sought to highlight the 'strategic interests' at work in its distribution as well as the inequality and unpredictability of aid provision. Strangely, except for a brief period during the mid-1970s, anti-poverty measures have not been an important focal point of foreign aid, whilst aid has led to many reversals as well as to advances. Seen as a (simultaneous) remedy to problems of growth, governance, poverty and inequality, it has become (not unlike the idea of development) overburdened with expectations (Sogge 2002) and an overambitious enterprise (Rist 1997). Successive UN 'Development Decades' have recognized that economically advanced states like the US have a responsibility to contribute to the financing of development and have specified that each 'developed' country should provide *at least* 0.7 per cent of its annual GNP as official development assistance. While still the world's biggest aid donor, the US has slashed overseas development assistance since 1990 by some 34 per cent and foreign aid accounts today for just 0.1 per cent of gross domestic product – the lowest level of any industrialized nation (Engardio, 2001). There are also plans to further limit foreign aid in order to minimize the US budget deficit (Zouza 2002). According to Sogge (2002) aid involves (1) a financial services industry promoting loans on easy terms, (2) a technical services industry improving infrastructure and know-how, (3) a 'feel-good' and image industry that can relieve guilt, (4) a political toolshed stocked with carrots and sticks to train and discipline recipients, (5) a knowledge and ideology industry and (6) an industry that sets policy agendas and shapes norms and aspirations. The expensive 2002 Earth Summit (RIO+10) held in Johannesburg (South Africa) in August and September 2002 has been criticized by many for being no more effective at pioneering new and alternative perspectives on development aid than its predecessors (except through the social/protest movements that mobilized around it).

intervention of Western 'experts', donors, technology, expertise or ideology (see Case study 8.2).

Definitions of the 'Third World' have been contested, as have the origins of the phrase (Mountjoy 1976, O'Connor 1976, Auty 1979, Pletsch 1981, Wolfe-Phillips 1987), yet the concept of three worlds can hardly be said to convey a precise meaning. The geographical boundaries of the 'Third World' are thus incredibly vague and open-ended (Chaliand and Rageau 1985, Griffiths 1994, *New Internationalist* 1999) and the idea has different meanings in different cultures. The 'three worlds' schema is very much a Cold War conceptualization of space and is strongly associated with the global social and political conflict between capitalism and communism, between the US and USSR, in the second half of the twentieth century. The term posited a first world of advanced capitalism in Europe, the USA, Australia and Japan, a second world of the socialist bloc (China's position within this has been much debated) and a third world made up of the countries that remained when the supposedly significant spaces of the world had been accounted for. This extremely rudimentary schema of classification has provided many observers with a convenient shorthand expression and is undoubtedly linked to the

formation of 'mental maps' and 'imagined geographies' of inequality (Bell 1994). Given that definitions of development have often been contested, these 'imagined geographies' often overstate a set of (tragic) stereotypical common features of the so-called 'Third World' such as overpopulation or political chaos. In discussing unevenness and inequality therefore a representation or an imagination of the people and places subject to the development process is constructed which is only partial and provisional. An 'alternative geography of the Third World' is proposed here, based on a recognition that development has been a 'protracted, painstaking and fiercely contested process' (Bell 1994: 175). This chapter tries to offer such a critical 'alternative geography' that aims to challenge some of the 'representations of peoples and places upon which grand [development] theories are based' (Bell 1994: 193–4).

It is extremely difficult to generalize about the people and places of the 'Third World' since relations between different regions of the world are imagined and simplified by geographical expressions like 'North' and 'South', 'rich' and 'poor', 'developed' and 'underdeveloped', 'First World' and 'Third World'. These terms have to be approached with some caution.

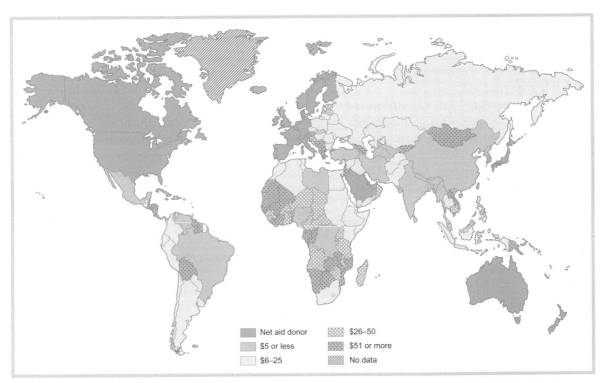

Figure 8.2 Flows of multilateral and bilateral aid between 'North' and 'South' give a useful indication of patterns of wealth and inequality at the global level.
Source: World Bank (1998b). © The International Bank for Reconstruction and Development/The World Bank

Becoming particularly popular after the publication of the Brandt Report in 1980 (Brandt 1980), the idea of 'North' and 'South' is widely seen as preferable to the three worlds scheme. However, no definitions were mentioned in the report and some of the boundaries it delineated between 'North' and 'South' were considered erroneous. Subscribers to this, and the three worlds scheme, have been criticized for the simplicity of these divisions and their failure to recognize diversity and difference within these spaces; the world does not consist of a series of individual national economies in the way often suggested in United Nations and World Bank reports. Many commentators have argued instead that there is a single global economic system with ever-closer degrees of interdependency (Allen 1995). Our task is to think critically about the uneven and unequal geography of the world economy and the ways in which the global capitalist economy *actively produces* inequality and uneven development. For some observers, a big part of the economic development and wealth of the rich countries is wealth that has been directly imported from the poor countries. The key question here then is to what extent does the world economic system *generate* inequality and run on inequality? Is it potentially fraudulent and misleading to hold up an image of wealthy Western living conditions as something that is available to all and a realistic or attainable objective? If the global economy is a pyramid and everyone is encouraged to stand on top, how is this to be arranged?

Development equality – 'catching up' with the wealthy through economic activity – increasingly seems less likely at the start of the twenty-first century and as evidence grows of widening and deepening global inequality. 'Development' has served then in part as a kind of 'lighthouse' (Sachs 1992) or as a 'lodestar' (Wallerstein 1991) into which several different movements, governments and institutions have invested faith and meaning. Like the term 'Third World', development partly represents a geographical imagining, a representation of a better world, a belief in the idea of correctable inequalities/injustices between nations, states and regions and within the existing global economic orders. Escobar (1995) argues that the idea of the 'Third World' also gives enormous power to Western development institutions to shape popular perceptions of Africa, Asia or Latin America. The Third World thus is defined by and becomes intelligible through the languages and representations of the agencies and institutions of global development.

In a famous speech in 1949 US President Harry Truman spoke of the emergence of an 'underdeveloped' world that presented a 'handicap and threat both to them and the more prosperous areas' (Truman 1949). Born of the ashes of colonialism and of the fires of anti-communism in the United States, 'development' occupied centre-stage as an ideal in the global arena after the Second World War and in particular during the Cold War. Truman went on to explain the need for 'modern, scientific and technical knowledge' and announced the beginning of a 'bold new program' within the 'developed world' to resolve 'underdevelopment' and poverty in 'backward' areas (Sachs 1992). Since then 'development' has become:

> an amoeba-like concept, shapeless but ineradicable [which] spreads everywhere because it connotes the best of intentions [creating] a common ground in which right and left, elites and grass-roots fight their battles.
>
> (Sachs 1992: 4)

Ideological interpretations of development have varied but they have usually always claimed 'the best of intentions'. The idea of development is often discussed in relation to 'developing countries' (itself a value-laden expression) but development has taken numerous forms in numerous locations and is actually still relevant in societies which proclaim themselves to be 'developed'. A sense of development being simultaneously important in both 'developing' and 'developed' societies was recently formulated in a Human Development Index (HDI) (UNDP 1998). Again assuming that 'development' can be depicted in a single measure, the UNDP Index combines data (by country) on life expectancy, literacy, income, environmental quality and political freedom. Development is defined here as a series of numerical stages with a linear progression indicated (statistically) as degrees of departure from Western norms and ideals. Human development is represented as a series of league tables that measure progress with the unquestioned underlying assumption that every country can reach the highest levels of performance set in 'the West' or have access to the 'purchasing power' of the 'upper income' world. As in the case of GNP per capita, the HDI does in many ways (albeit at a very crude level) point to growing gaps between different areas of the world. Its significance is undermined by the failure of its 'generalized' indicators. However effective statistical measurement of development can become, it is necessary to be critical of the way in which statistics are collated and definitions formulated and to be careful in the assumptions and prescriptions made as a result. It is through the amassing of statistics and surveillance by global development agencies that the image of an underdeveloped, primordial, traditional and war-ravaged 'third world' is (re)produced (Ahluwalia 2001).

8.2 Conceptualizing development

The idea of progress has become an article of faith in definitions of 'development', which have themselves taken on a 'quasi-mystical connotation' in the sense that seeking to become 'developed' has been constructed as an objective that is unquestionably positive and beyond reproach. 'Development thinking' (Hettne 1995), or the sum total of ideas about development theory, ideology and strategy, has however often been caught in a 'Western' perception of reality. Conceptualizing development is thus partly about the negotiation of what constitutes progress and improvement and the definition of what constitutes 'appropriate' intervention in the affairs of 'poor' of 'lesser developed' countries. Since all-encompassing

---

Regional Case Study 8.3

Bandung, non-alignment and the 'Third World'

The Bandung conference was a meeting of representatives from 29 African and Asian nations, held in Bandung (Indonesia) in 1955, which aimed to promote economic and political cooperation within the 'Third World' and to oppose colonialism. The conference was sponsored by Burma, India, Indonesia, Ceylon (Sri Lanka) and Pakistan and tried to cut through the layers of social, political and economic difference that separated nations of the 'Third World' in order to think about the possibility of *common agendas and actions*. The aims of the 29 nations that attended included a desire to promote goodwill and cooperation among Third World nations, to explore and advance their mutual as well as common interests. Bandung was in many ways the 'launching pad for Third World demands' where countries distanced themselves from the 'big powers seeking to lay down the law' (Rist 1997: 86). It was not hard for countries with shared histories of colonial exploitation to find something in common, since the 'agenda and subject matter had been written for centuries in the blood and bones of participants' (Wright 1995: 14). In his opening speech to the conference on 18 April 1955 President Sukarno of Indonesia urged participants to remember that they were all united by a common 'detestation' of colonialism and racism (Sukarno 1955: 1) and pointed out that colonialism was not dead or in the past but also had its 'modern'

(neo-colonial) forms. The conference was especially successful in hastening the arrival of new international institutions explicitly dealing with 'development' (Rist 1997). The term 'Bandung Era' was coined by Samir Amin in a discussion of the national approaches to development that emerged from this period (Amin 1994, Berger 1994, 2001). The Non-Aligned Movement (NAM) was then founded (among 29 states most of which were former colonies) in 1961 on a principle of non-alignment with either the US or the USSR in the Cold War. A year after the NAM was set up, the 'Group of 77' was then established (mostly made up of the attendees of previous conferences) to press for changes from UN agencies in the ways they viewed 'developing countries'. The NAM has continued to function as a space through which nations of the so-called 'Third World' meet and discuss common issues and objectives and now has some 115 member nations. The organization has been restructured since the end of the Cold War and continues its commitment to founding principles whilst also now incorporating a focus on reforming the international economy. South Africa took over as Chair in 1992 when the organization met in Jakarta to decide on future policy directions, whilst NAM today cooperates with NGOs, the UN, the EU and the G8. At its 40th anniversary in New York in November 2001 NAM restated its commitment to work with the major global development agencies to help bring about a reform of the HIPC initiative concerning debt.

definitions have been contested and controversial, little consensus exists today but some core conceptions have emerged, many of which have continued relevance in the contemporary world. Although there are many different strands of development thinking to explore, the modernization and dependency approaches have been two of the most influential in the twentieth century. In discussing these different conceptions it is important to think about where and when they have emerged (Slater 1993). Most reflect some of the priorities of development thinking characteristic of their era. The formation of development theories therefore depends on different perceptions of 'development challenges' at different times (see Case study 8.3).

8.2.1 The modernization school: an anti-communist manifesto

With the end of the Second World War and after the United Nations was established, conceptualizations of development received a decisive stimulus. Between 1945 and 1981, UN membership rose from 51 to 156 nation-states (Berger 2001). With many new states being formed after the end of colonialism and in the context of the Cold War between the US and the USSR, theorizing development became a more complicated and contested enterprise. In some ways the task of development was seen by some as seeking to provide 'an ethos and system of values which can compete successfully with the attraction exercised by Communism' (Watnick 1952–3: 37). In this regard many observers called for the modernization of 'underdeveloped areas' and painted a picture (following President Truman) of 'underdeveloped peoples' confined to 'backwardness' but torn between the appeal of communism and the prospect of Western modernization. This was an essential characteristic of the *modernization school*, which was often dualistic, opposing 'traditional' to 'modern' life-styles, 'indigenous' to 'Westernized', as if no country or citizen could belong to both categories.

Hirschmann (1958) was a key proponent of modernization ideas, voicing the optimistic view that the forces of concentration ('polarization') will 'trickle down' from the core to the periphery at national, regional and global scales. Hirschmann explained how development economics might 'slay the dragon of backwardness' (quoted in Rist 1997: 219). Friedmann's (1966) core–periphery model adopts many of the same

assumptions about the polarization of development in 'transitional societies' and the 'trickle down' effect of development. This model suggests that a number of stages exist in the national development of countries or cities leading to a final stage that represents the culmination of the development process. Geographers at the time sought to contribute to the mapping of modernization geography, seeking to look at how progress trickled down along urban hierarchies, through transport systems or with the introduction of modern technology. The message from these modernization geographies was that underdeveloped countries could move briskly into the modern tempo of life within a few years, whilst the state (concerned with macro issues and the national economy) would be the key monitor and broker of development.

One of the modernization theorists who identified stages of growth in the development process was the anti-communist writer W.W. Rostow (1960). Again the focus was on a top-down 'trickling' of capitalist development from urban-industrial areas to other regions (Stöhr and Taylor 1981). Rostow (1960) predicted that nations would 'take off' into development, having gone through five stages, which he likened to the stages an aeroplane goes through before take-off, from taxiing on the runway to mid-flight. Ranging from stage one, 'traditional society', to stage five, the 'age of high mass consumption', the theory takes our faith in the capitalist system for granted since Rostow assumes that all countries will be in a position to 'take off' into development. Further, in common with other modernization approaches, Rostow's model devalues and misinterprets 'traditional societies' which represent the 'lowest' form or stage of development. The advanced state of modernization is always represented as 'Western modernization'; traditional societies seem like distant, poor relations.

Geographies of inequality and development cannot be neatly summarized as a set of prescriptive stages. Urban areas, for example, are themselves subject to uneven development and inequality. Modernization approaches have also largely failed to address the importance of gender, assuming that men and women occupied equal positions in terms of power relations and decision-making. The 'trickle-down' effect has often failed to materialize among those who have been the subjects of modernization projects. This approach seemed to suggest that development could be mimicked, copied and replicated and that underdeveloped countries should try to reproduce the development paths of richer developed nations like

Britain or the USA. As Gunder Frank has argued, this approach can also be understood as reflections of the 'Sinatra Doctrine':

> Do it my way, what is good for General Motors is good for the country, and what is good for the United States is good for the world, and especially for those who wish to 'develop like we did'
> (Frank 1997: 13).

As Abrahamsen (2000) argues, in the modernization schema, there is nothing before development that is seen as worth retaining or recalling, only a series of deficiencies and absences. The approach was also in a sense very much based around a 'top-down' rather than a 'bottom-up' approach, implying that the process could be brokered by states or development institutions rather than emerge from the 'grass-roots' struggles of 'Third World peoples' as had been called for in some more radical approaches. In terms of common criticisms, the division of the world into modern and traditional has often been seen as problematic (Pletsch 1981). Modern societies were much more fractured and were divided by race, class and politics and were not as united and responsive to the blueprints of planners as was often assumed. The scale of modernization programmes was also often a problem in that they assumed that 'big is beautiful' (involving large dam-building and irrigation projects, for example). Furthermore, the school and its practitioners often de-politicized development, making few if any references to history and culture. Rostow was able to say little about the 'final stage' of his organic model of development since his underlying principle was that growth had no limits (Rist 1997) and instead he simply discussed a number of very generalized scenarios. Like so many theorizations of development that followed, it ended with a creed, a set of principles about what was to be done, and heavily invested faith in the goals of mass consumption and Westernization.

8.2.2 The dependency school: beyond 'core' and 'periphery'?

One of the major weaknesses of the modernization approach was that the notion of a 'trickle-down' diffusion of development implied that there were precise demarcations available of where the 'core' ends and the 'periphery' begins when this has never really been the case. Radical *dependency approaches* that

emerged in the 1960s and 1970s challenged this notion of positive core–periphery relations, identifying instead exploitation between 'satellites' and 'metropoles'. Whilst the modernization approach was taken up by many international institutions and bilateral donors, dependency approaches comprised all those opposed to US post-war imperialism and allied in some way to the movement of 'third worldism' (see Case study 8.3). The Dependency school is most commonly associated with Latin America but also emerged in Africa, the Caribbean and the Middle East. Drawing upon Marx's writings about the unevenness of capitalist development, dependency scholars such as Celso Furtado (Furtado 1964) and Milton Santos (Santos 1974) drew attention to the mode of incorporation of each country into the world capitalist system, identifying this as a key cause of exploitation. The dependency school is most commonly associated with the work of André Gunder Frank who published a number of seminal pieces on 'The development of underdevelopment' in the late 1960s (Frank 1967). This thesis is relatively uncomplicated in that it views development and underdevelopment as opposite sides of the same coin; the development of one area often necessitates the underdevelopment of another. Dependency on a metropolitan 'core' (e.g. Europe, North America) increases the 'underdevelopment' of satellites in the 'periphery' (e.g. Latin America, Africa). Unlike many modernization approaches, dependency theorists (*dependistas*) sought to view development in historical contexts, and argued, for example, that colonialism helped to put in place a set of dependent relations between core and periphery.

These peripheral satellites, it was argued, were encouraged to produce what they did not consume (e.g. primary products) and consume what they did not produce (e.g. manufactured/industrial goods). Thus, rather than likening the process of economic growth to an aeroplane setting off into a blue sky of urban-based, Western life-styles and consumption, the dependency school was arguing that many 'underdeveloped' areas had been stalled on the runway by unequal relations and a history of colonialism, denying them a chance of ever being airborne, 'modern' or 'industrialized'.

In many ways, dependency approaches were so directly opposed to modernization approaches (almost point by point) to such an extent that eventually both 'seemed to checkmate each other' (Schuurman 2001: 6). Dependency even seemed to preserve the dualistic and binary classification of the world into developed/underdeveloped, First World and Third World, core and periphery, and also lacked a clear statement of what 'development' actually is. Key criticisms directed at the *dependistas* were that the theory represents a form of 'economic determinism' and also overlooks social and cultural variation within developed and underdeveloped regions. The dependency framework seemed to leave the simplistic impression of an 'evil genie who organizes the system, loading the dice and making sure the same people win all the time' (Rist 1997: 122). Another point of contention was that the dependency theorists seemed to be calling for a de-linking from the world capitalist economy at a time when it is undergoing further globalization and economic integration. Elements of the dependency writings were, however, quite thought-provoking and remain relevant, particularly their contention that the obstacles to development equality were *structural*, arising not from a lack of will or poor weather conditions but from entrenched patterns of global inequality and 'dependent' relationships. A brief comparison of the two schools of development thinking is useful here, particularly given that dependency was primarily a critique of the classical traditions of modernization. According to Bell (1994) both perspectives deal in dualistic either/or scenarios:

> these contrasting schools of thought share a common starting point – poverty is viewed in deprivationist terms while their different views of development have a positive connotation.
>
> (Bell 1994: 184)

Concepts of development like the modernization and dependency perspectives are far from being static, uniform or unified, however. In this way it is sometimes easier to refer to modernization and dependency 'schools' since neither represents a singular commonly agreed approach and both are complex and varied. In some ways it is the ability of each approach to capture and accommodate difference and diversity between states and within regions that is most important here. As we shall see in the following section, in order to do this it is important to further consider theories and practices of development in their specific historical and geographical contexts.

8.3 Development practice (1): the historical geography of development

Whilst both modernization and dependency approaches alluded to the importance of 'tradition', historical perspectives were often lacking in many early writings about development (Rist 1997). As Crush (1995) points out, development is primarily 'forward looking', imagining a better world, and does not always examine issues of historical and geographical context. It is particularly important to examine the significance of Empire in the making of international development. Between 1800 and 1878, European rule, including former colonies in North and South America, increased from 35 per cent to 67 per cent of the earth's land surface, adding another 18 per cent between 1875 and 1914, the period of 'formal colonialism' (Hoogevelt 1997: 18). In the last three decades of the nineteenth century, European states thus added 10 million square miles of territory and 150 million people to their areas of control or 'one fifth of the earth's land surface and one tenth of its people' (Peet with Hartwick 1999: 105). Colonialism has been variously interpreted as an economic process of unequal exchange, as a political process aimed at administration and subordination of indigenous peoples, and as a cultural process of imposing European superiority. According to the dependency theorists it was in this period that the periphery was inserted and brought into an expanding network of economic exchanges with the core of the world system. A new sense of responsibility for distant human suffering also first emerged during this time as the societies of Europe and North America became entwined within global networks of exchange and exploitation in the late eighteenth and early nineteenth centuries (Haskell 1985a, 1985b). Thus the origins of a humanitarian concern to come to the aid of 'distant others' lay partly in response to the practices of slavery in the transatlantic world and to the expansion of colonial settlement in the 'age of empire':

> Not only did colonisation carry a metropolitan sense of responsibility into new Asian, North American, African and Australasian terrains, it also prompted humanitarians to formulate new antidotes, new 'cures' for the ills of the world.
>
> (Lester, 2002: 278)

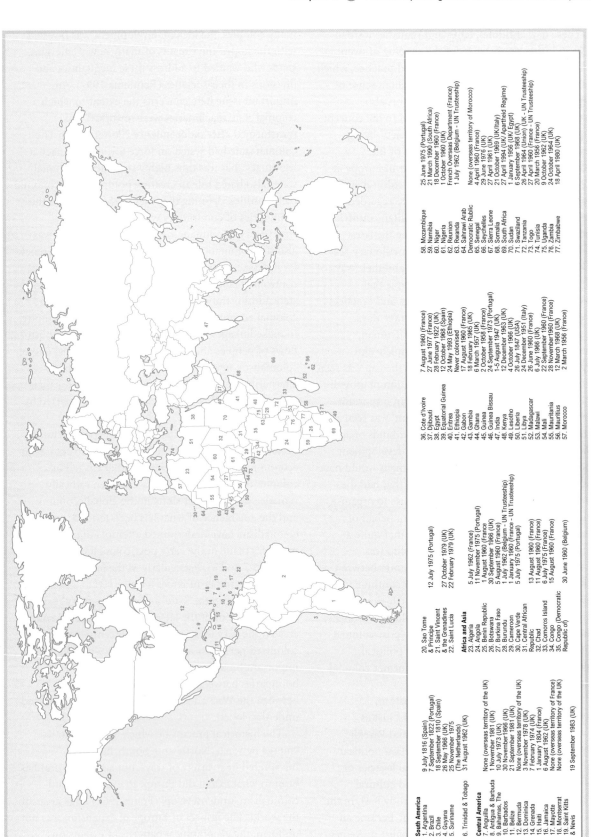

Figure 8.3 Map of the world showing dates of independence from colonial rule and the European colonizing country.

Source: M. Power, *Rethinking Development Geographies*, 2003, Routledge

South America
1. Argentina   9 July 1816 (Spain)
2. Brazil   7 September 1822 (Portugal)
3. Chile   18 September 1810 (Spain)
4. Guyana   25 May 1966 (UK)
5. Suriname   25 November 1975
                       (The Netherlands)
6. Trinidad & Tobago   31 August 1962 (UK)

Central America
7. Anguilla   None (overseas territory of the UK)
8. Antigua & Barbuda   1 November 1981 (UK)
9. Bahamas, The   10 July 1973 (UK)
10. Barbados   30 November 1966 (UK)
11. Belize   21 September 1981 (UK)
12. Bermuda   None (overseas territory of the UK)
13. Dominica   3 November 1978 (UK)
14. Grenada   7 February 1974 (UK)
15. Haiti   1 January 1804 (France)
16. Jamaica   6 August 1962 (UK)
17. Mayotte   None (overseas territory of France)
18. Montserrat   None (overseas territory of the UK)
19. Saint Kitts   19 September 1983 (UK)
& Nevis

20. Sao Tome   12 July 1975 (Portugal)
& Principe
21. Saint Vincent   27 October 1979 (UK)
& the Grenadines
22. Saint Lucia   22 February 1979 (UK)

Africa and Asia
23. Algeria   5 July 1962 (France)
24. Angola   11 November 1975 (Portugal)
25. Benin Republic   1 August 1960 (France)
26. Botswana   30 September 1966 (UK)
27. Burkina Faso   5 August 1960 (France)
28. Burundi   1 July 1962 (Belgium – UN Trusteeship)
29. Cameroon   1 January 1960 (France – UN Trusteeship)
30. Cape Verde   5 July 1975 (Portugal)
31. Central African   13 August 1960 (France)
Republic
32. Chad   11 August 1960 (France)
33. Comoros Island   6 July 1975 (France)
34. Congo   15 August 1960 (France)
35. Congo (Democratic 30 June 1960 (Belgium)
Republic of)

36. Cote d'Ivoire   7 August 1960 (France)
37. Djibouti   27 June 1977 (France)
38. Egypt   28 February 1922 (UK)
39. Equatorial Guinea 12 October 1968 (Spain)
40. Eritrea   24 May 1993 (Ethiopia)
41. Ethiopia   Never colonised
42. Gabon   17 August 1960 (France)
43. Gambia   18 February 1965 (UK)
44. Ghana   6 March 1957 (UK)
45. Guinea   2 October 1958 (France)
46. Guinea Bissau   24 September 1973 (Portugal)
47. India   1–5 August 1947 (UK)
48. Kenya   12 December 1963 (UK)
49. Lesotho   4 October 1966 (UK)
50. Liberia   26 July 1847 (USA)
51. Libya   24 December 1951 (Italy)
52. Madagascar   26 June 1960 (France)
53. Malawi   6 July 1966 (UK)
54. Mali   22 September 1960 (France)
55. Mauritania   28 November 1960 (France)
56. Mauritius   12 March 1968 (UK)
57. Morocco   2 March 1956 (France)

58. Mozambique   25 June 1975 (Portugal)
59. Namibia   21 March 1990 (South Africa)
60. Niger   18 December 1960 (France)
61. Nigeria   1 October 1960 (UK)
62. Reunion   French Overseas Department (France)
63. Rwanda   1 July 1962 (Belgium – UN Trusteeship)
64. Sahrawi Arab   None (overseas territory of Morocco)
Democratic Rublic
65. Senegal   4 April 1960 (France)
66. Seychelles   29 June 1976 (UK)
67. Sierra Leone   27 April 1961 (UK)
68. Somalia   21 October 1969 (UK/Italy)
69. South Africa   27 April 1994 (UK/Apartheid Regime)
70. Sudan   1 January 1956 (UK/Egypt)
71. Swaziland   6 September 1968 (UK)
72. Tanzania   26 April 1964 (Union) (UK – UN Trusteeship)
73. Togo   27 April 1960 (France – UN Trusteeship)
74. Tunisia   20 March 1956 (France)
75. Uganda   9 October 1962 (UK)
76. Zambia   24 October 1964 (UK)
77. Zimbabwe   18 April 1980 (UK)

These new antidotes, cures and remedies were to have enduring significance for the shaping of twentieth century global development theory and practice, which also often carried an implicit 'metropolitan sense of responsibility'. Colonial development was also associated with an unconditional belief in the concept of progress and the 'makeability' of society, being heavily conditioned by the dominance of the evolutionary thinking that was popular in Europe at the time. Imperialism was viewed as a cultural and economic necessity where colonies were regarded as the national 'property' of the metropolitan countries and thus needed to be 'developed' using the latest methods and ideas. With this came a missionary zeal to 'civilize' and modernize the colonized and their ways of life. An important contention here then is that colonialism 'conditioned' the meanings and practices of development in a number of important ways.

After 1945 and under US President Truman, 'underdevelopment' became the incomplete and 'embryonic' form of development and the gap was seen as bridgeable only through an acceleration of growth (Rist 1997). As we shall see, there was an important sense in which, after 1945, 'development' was seen as a process that was entrusted to particular kinds of agents who performed the acts and power of development upon distant others. Globally, development would have its 'trustees', guiding 'civilized' nations that had the 'capacity' and the knowledge or expertise to organise land, labour and capital in the South on behalf of others. This is an important idea then that quite a paternal and parental style of relationship was established through the imperial encounter between colonizer and colonized in ways which continue to have a bearing on the definition of North–South partnerships in the 'post-colonial' world. Additionally, what is also relevant here is that many 'post-colonial' states continue to maintain important political, cultural and economic ties with their former colonial rulers.

Decolonization is thus simultaneously an ideological, material and spatial process, just as complicated as colonization (Pieterse and Parekh 1995). Colonialism put in place important political and economic relations but the cultural legacies of colonialism bequeathed deep social and cultural divisions in many societies (see Case study 8.4). In the process of decolonization 'development' became an over-arching objective for many nationalist movements and the independent states they tried to form. Although experiments with development were tried in many colonies, the idea of development was invested with the hopes and dreams of many newly emerging states who wanted to address these inequalities and divisions in their societies (Rahnema 1997). An important issue here concerns the extent to which colonial state machineries were reworked and transformed after independence (Power 2003). The colonial state had rested on force for its legitimacy, a legitimacy that was thus highly superficial. Colonial states also had a role in creating political and economic communities, defining the rules of the game and the boundaries of community whilst creating power structures to dominate them. The colonial state was also the dominant economic actor, creating a currency, levying taxes, introducing crops, developing markets, controlling labour and production. Above all, colonial state administrations sought the integration of the colonial economy into the wider economies of empire, to make linkages with the metropole and to establish flows of peoples and resources. After the formal end of colonialism, new states have had to formulate alternative methods of garnering legitimacy for their authority (i.e. other than the use of force preferred by the colonists).

As the European capitalist system expanded and became ever more global in its reach, the structures of economic, social and political life that existed in colonies (before colonialism) were often radically remade. Several recent shifts in the global economy and the emergence of what has been termed 'neo-colonialism' have undermined attempts to meet promises made in the early post-colonial era (see Chapters 7 and 8). Particularly in the last three decades, a number of issues have served to sustain the linkages between the 'core' and the 'periphery', usually to the detriment of the latter. In a way imperial rule created the very idea of an imperial centre and a colonized periphery and established a whole variety of important binaries, divisions and constructed boundaries between civilized and non-civilized, between the West and the rest, between developed and underdeveloped.

The historical process by which 'gaps' began to emerge between 'North' and 'South' has been interpreted in a variety of ways, but a key question has been to what extent did European expansion and colonialism 'underdevelop' (Frank 1967) large areas of the world. The impact of European expansion was not uniform; the geographical patterns of expansion varied, as did the motivations for it. Hall (1992) argues that an important divide was put in place between 'the

Regional Case Study 8.4

Rwanda: looking beyond genocide

For those with limited background knowledge of the history and geography of Central Africa, the genocide in Rwanda in 1994 can seem very difficult to interpret and understand. Media representations relying upon the image and myth of Africa as the 'Dark Continent' (Jarosz 1992) explained the tragedy in terms of 'ethnic cleansing' and deployed myths and images of tribal savagery (Prunier 1995). In fact, Hutu and Tutsi cannot really be described as 'ethnic groups' in that both speak the same language, inhabit the same territory, and respect similar traditions and taboos (Destexhe 1995). It is important to view the Rwanda genocide in historical context, looking for example at the specific forms of German and later Belgian colonialism. Successive colonial governments accentuated differences between these supposedly distinct 'ethnic groups' and aggravated tensions between them, fuelling racial hatred. Rwanda became independent in 1962 and only one year later several massacres of Tutsi people began, led by the new Hutu government administration. Appeals to ethnic identity after independence became a way of winning support and in 1972 President Kayibanda imposed a quota on the Tutsi people allowing them only 10 per cent of the places in schools, universities and civil service posts (Prunier 1995). Successive presidents further fractured the country along the lines of identity depending on which region of the country people came from until when in 1990 the Tutsi-led Rwanda Patriotic Front (RPF) launched a major offensive against Hutu people, invading from neighbouring Uganda. In July 1994, the UN made the fateful decision to withdraw its peace-keeping troops from Rwanda and by the end of that year half the population had been killed, been displaced within the country or were in exile. The roots of the Rwandan genocide, for example, can also be linked directly to Africa's colonial past but also to its post-colonial present and it is this interaction between past and present that post-colonialism is fundamentally about. This is relevant because it recognizes that Africa 'has to deal with its past in order to understand its present and confront its future' (Ahluwalia 2001: 133).

West' and 'the rest' as a direct result of this imperial expansion, reminding us that 'the West' is much more of an idea than a geographical reality. By accelerating contact between cultures and economies 'the West' was presented as 'the best' and most advanced or 'civilized' of all humanity. Many accounts of the history of European expansion are thus dominated by the presumed supremacy of 'Europe' and 'the West' with only limited references to the complex histories and cultures of the areas that were colonized. Similar criticisms have been made of some development theories (e.g. modernization) that have also often been based around 'Western' philosophies, experiences and histories (Hettne 1995, Power 2003).

Many recent histories of development have dated its beginnings as an area of theory and state practice to President Truman's speech of 1949. The idea of development is, however, much older than this (Cowen and Shenton 1996) and has much more diverse geographical origins. Development was not a simple 'gift' following contact with Europeans but predates the 'age of discovery' (1400–1550) and the 'age of empire' (1875–1914) (see Section 1). Indigenous peoples in Africa, the Americas, Asia and Australasia had highly developed and sophisticated cultures and technologies prior to colonization (Dickenson *et al.* 1996).

8.4 Development practice (2): subjects of development

In addition to thinking historically about the production of global difference and inequality, it is also necessary to listen to what we might call the *other side of the development story*, or the views of women, the views of people in the South, the views of people that are excluded socially, and the view from 'below'. People are central to the development process and an integral

Thematic Case Study 8.5

States and the geopolitics of development

At the time of the terrorist attacks in Washington, DC and New York city in 2001, world leaders made numerous comments about the links between international poverty and terrorism, suggesting that this was to become *the* key issue of the twenty-first century. The ensuing *Global War on Terrorism* led by states defining themselves as the 'free' or 'civilized' world began in Afghanistan and has shaped contemporary development thinking and practice in many important ways. These recent events highlight the need to more fully understand the geopolitics of international development and in particular the way in which many states in the South are seen to lack necessary degrees of 'order', hierarchy and societal organization. In thinking about geopolitics it is important to explore the rights and abilities of each state to control the territory encompassed by its boundaries or to gain international recognition for its territorial authority. This also means thinking about the way in which foreign aid is used by Western donors to solicit support for the war on terrorism

from states in poorer countries. The state in Angola, for example, was offered a wide range of incentives in 2003 in order to gain Angolan support for the US when the African country had a temporary seat on the UN Security Council. Foreign aid has sometimes been seen as a kind of political narcotic, fostering addictive behaviour among states that receive it and thus come to depend on it. States are thought to exhibit the symptoms of dependence – a short-run 'fix' or benefit from aid, but external support sometimes does lasting damage to the country. During the Cold War, the USSR in particular played on its ambiguous position as both 'inside' the international system and 'outside' it, presenting itself as the natural 'midwife' for completing the independence of new-born states (Laïdi 1988). With the accelerating pace of decolonization and the creation of independent states in the South, geopolitical questions were addressed from a set of new or 'Third World' perspectives, new forms of North–South geopolitics which had emerged from the legacies of colonialism and the perception that 'underdeveloped' countries had distinct geopolitical considerations from those of Western societies (see also Chapter 15).

element in all development strategies. However, their needs have often been ignored and there has been a general failure to consider the implications of development policies for individuals, households and communities. Development strategies have also been driven by economic objectives, relegating provision for people's basic needs, their empowerment or efforts to raise their living standards to second place (Potter *et al.* 1999). Development strategies often assume that 'people' and 'communities' are homogeneous and passive rather than differentiated and dynamic. The knowledge and skills of people in particular communities have often been overlooked in favour of broader and less sensitive approaches to 'modernization' that bypass local cultural traditions and histories. Three major agents of development have dominated the literature on development: nation-states, international development institutions and

NGOs. It is difficult, however, to view any one of these three 'brokers of development' in isolation since, for example, nation-states are the primary focus of the United Nations while NGOs often try to work at the interface between national governments and international governmental institutions.

Many international development organizations have often made erroneous assumptions about the distribution of power and decision-making within the nations and communities where they operate. Consider the level of the household, for example. It is taken for granted that an equal division of labour exists between men and women within households and communities. As we have already noted, national statistics often conceal significant variations within and between specific regions, between rural and urban areas or between men and women. These aggregations of household labour and resources often fail to capture

Plate 8.1   The environmental pollution left by Shell's oil refineries in Nigeria has had particular implications for the people of Ogoniland who have seen very little of the profits generated by oil reinvested in their state. Some observers have argued that Shell's relationship with the Ogoni people and the Nigerian state is one of 'neocolonialism'. (Still Pictures/Detlee Pypke)

the changing and important contributions of household members such as children (see Case study 8.6). The contributions of women to the productive and reproductive activities of the household are still widely overlooked in national 'official' development statistics (Momsen 1991, Østergaard 1992). Nonetheless, the household has been and is likely to remain the basic living unit in many parts of the global South, the key to controlling and organizing production, consumption and decision-making. Household sizes and structures differ considerably from those of the global North, often including an extended family of grandparents, aunts, uncles and other more distant relatives. Unequal access to resources means that all these households do not have the same level of access to income, assets (e.g. land) or local and national power structures. It is important for international development organizations to understand the different roles and responsibilities of men and women within these diverse households. The contribution of women in particular to household income and welfare needs to be much more carefully

understood (Moser 1993, Kabeer 1994, Sweetman 2000, Perry and Schneck 2001).

In development policy and practice it has often been assumed that the household in the 'global South' corresponds to the Western model of the nuclear household, yet these households are far from homogeneous in their structure and in the amount of control over decision-making and resources that women enjoy. The contribution of women to the development process, of their knowledge, time, labour and capital, has often been underestimated by development 'experts'. Mohanty (1991) rejects the idea of the stereotypical 'Third World woman' as the passive recipient of social services that has often been constructed by development planners and theorists alike. Similarly, the visibility of women's productive, reproductive and community management roles in the recording of national development statistics has been widely criticized by gender and development (GAD) researchers (Rathgeber 1990, Moser 1993). Women were often not defined as 'productive' contributors to their societies, as it was abstractly assumed that women

Thematic Case Study 8.6

Children and development

One of the clearest examples of global inequalities comes from a comparison of the relative position of children in societies of the global 'North' and 'South'. In wealthier societies children are often well-fed, have clean clothes and access to an education until the age of 16 or older and have good protection against ill-health. Their counterparts in some of the world's poorest countries have few of these assurances. Child poverty is also common in 'developed' countries as well, however: UNICEF estimates that one in six children in rich nations is poor (UNICEF 2000). A report from ActionAid (1995) clearly shows that children and children's work have been neglected research areas despite the fact that, as Robson (cited in Earthscan 1998) has argued, '[b]y the age of 10–12 years, some children may contribute as much to household sustenance as adults'. Again, gender is crucial in determining the roles and responsibilities of children, with girls often expected to provide assistance with household duties and boys involved in

more public forms of labour. Access to education for boys and girls is also often unequal, with the continuation of girls' education being seen as much less of a priority for many poorer households (Momsen 1991). Children have also often been caught up in the political conflicts that divide many societies and make development a more complicated proposition. Graça Machel's (1996) study on the impacts of armed conflict on children makes clear that 'in the course of displacement, millions of children have been separated from their families, physically abused, exploited and abducted into military groups, or they have perished from hunger and disease'. The 2003 *State of the World's Children Report* (UNICEF 2003) focuses on the importance of children's participation. The report argues that child participation is the 'right' of every child at every age, the responsibility of governments, organizations and families, and a way to promote tolerance, respect for human rights and an appreciation of diversity and peace. Greater investments to protect the human rights and well-being of children and young people are key to eradicating world poverty.

were largely confined to the private, reproductive or domestic sphere (Marchand and Parpart 1995). In agriculture, for example, access to credit, rural extension services, land ownership and farm technology has often been denied to women, despite the fact that women produce more than half the world's food (Koczberski 1998). In this regard it is important to note that many countries in the South have attempted to create indigenous versions of European welfare states, which has proven economically very difficult for many states, leading to the uneven development of citizenship benefits amongst different socio-economic groups and between men and women (see Chapter 17). Again, however, we see the problems created by trying to draw on a universal set of standards based on the experiences of Western liberal democracies and the problems created by ignoring or overlooking the particularity and diversity of peoples and places. Development planners have often neglected the importance of women's

'double day' (Moser 1993) or the limitations placed on their time by the need to combine reproductive (e.g. child care) with productive (e.g. farming) activities. In addition to these two important roles, women often participate in the management of their communities. They organize political, social and economic structures and play a crucial role in the management of the environment and natural resources (Dankelman and Davidson 1988, Sondheimer 1991, Shiva and Moser 1995, Sittirak 1998).

Although the importance of women's participation in development planning has increasingly been recognized by international institutions, the capacity of national governments to respond effectively to the needs of their societies at the 'grass roots' has been severely curtailed by indebtedness, reduced commodity prices, dependence on international donors, recession and protectionism in the global economy. The state has typically been seen as the primary agent of development, particularly since the 1960s when many countries attempted ambitious large-scale modernization projects with the backing of international development institutions and commercial lenders. The formation of new states, particularly in sub-Saharan Africa, followed at a time when development ideals were becoming increasingly popular worldwide. The 'age of great optimism' imbued these emerging states with a sense of the importance of development in the post-colonial era. Again, as with so many other aspects of 'development' there are no universal definitions of the state and states do not behave in any predetermined way. States in the South have become subject to enormous pressures from the societies within their borders, becoming sites of protest and contention when their activities impinge on the lives of their subjects and citizens.

Ideology also has an important bearing on the way the role of the state in development is viewed. For example, reflecting the World Bank's 'neoliberal' framework, the annual World Development Reports (WDRs) often view an effective state almost exclusively 'through the lens of economic efficiency' with states seen as providing either 'barriers' or 'lubricants' to free-market economic reforms (see Case study 8.7). Neoliberalism promotes and normalizes an economic growth-first strategy where the social and welfare concerns of citizens are seen as less of a priority. Neoliberal development models also naturalize an image of international markets as fair and efficient, privileging 'lean government', deregulation and the removal of state subsidies. More importantly, in many ways neoliberalism forecloses alternative paths of development, narrowing the ideological space in which it is possible to think outside the 'development box' (Power 2003). Behind neoliberalism is the core assumption that the economy should dictate its rules to society rather than the other way around.

Not surprisingly, the World Bank often 'locks society out' (Hildyard 1997: 43) in its definition of the state (Figure 8.4). Although the WDR calls for greater participation in decision-making in order to bring 'government closer to the people', a number of key policy areas are seen as 'off-limits'. According to Chussudovsky (1997) the World Bank has claimed to have shifted policy focus towards 'sustainable development' and 'poverty alleviation', but neoliberal policy prescriptions continue their dominance within the international development industry.

The term 'empowerment' is often used by the Bank and other sections of the 'development industry' (Ferguson 1990) to describe their commitment to

popular participation in decision-making for development. Facilitating 'people's participation' is now on the agenda of most international development institutions, including the United Nations Development Programme (UNDP) which produced a major review of participation and democracy in 1993 (UNDP 1993). For Rahnema (1997), the increasingly widespread acceptance of the idea of participation suggests that it has been severely diluted and has lost some of its more radical potential:

> governments and development institutions are no longer scared by the outcome of people's participation ... there is little evidence to indicate that the participatory approach, as it evolved, did, as a rule, succeed in bringing about new forms of people's power.
>
> (Rahnema 1997: 118–23)

Escobar (1995) also seems somewhat disillusioned with the development concepts of 'participation' and 'empowerment', pointing out that development planners and politicians have often tried to manipulate experiences of participation to suit their own ends. Munslow and Ekoko (1995) similarly argue that empowerment has been stronger on rhetoric than in reality and that advances in widening political participation have been slow, particularly in Africa, where the 'democracy and development' agenda has been manipulated by political movements across the continent (Baker 1998). Empowerment speaks of making people the agents of their own development while doing little about the causes of inequality, and without transforming the nature of ties with international development institutions and their ideologies.

The idea of empowerment and its increasing acceptance in development circles is partly a consequence of the rising influence of non-governmental organizations (NGOs), which have often been highly localized and focused on particular issues. NGOs have been defined as a 'residual' category that includes a wide variety of formal associations that are not government-led, even though definitions vary enormously and many NGOs do receive finance and support from their home governments. They can operate at a local, national or international level, but there is often little overall coordination between different NGOs even though they usually work on similar issues, sometimes in the same country or city. A key issue that has been raised in critiques of NGO activity is the closeness of their contact with local

Thematic Case Study 8.7

The World Bank and the IMF

Both the International Monetary Fund (IMF) and the World Bank are based in Washington, DC. The Fund was set up in the 1940s as a result of the Bretton Woods conference to maintain currency stability and develop world trade by establishing a multilateral system of payments between countries based on fixed exchange rates and complete convertibility from one currency to another. The IMF is not controlled by the people or the governments of low-income countries and is not like other UN agencies where every government has a vote (Wood and Welch 1998). Instead the fund is run by a select band of high-income countries that fund its operations (the US has an 18 per cent share of votes and has *de facto* control of the organization). The World Bank is now increasingly involved in macro-economic issues and is always chaired by an American, while the President of the IMF is usually European. The latter is dominated by OECD countries and in particular by the US and is organized in quite a similar way to the Bank.

The International Bank for Reconstruction and Development (better known as the World Bank) was founded as a means of reviving war-damaged European economies, a mandate that was later extended to developing nations. The Bank is funded by dues from members and by money borrowed in international markets. It makes loans to member nations at rates below those of commercial banks to finance development 'infrastructure' projects (e.g. power plants, roads, hydro-dams) and to help countries 'adjust' their economies to globalization. Like the IMF, the Bank was an early and enthusiastic supporter of the neoliberal agenda. According to a Brazilian delegate at the Bretton Woods conference in 1944 these agencies were inspired by a desire to establish a central lending source to which governments would con-

tribute such that 'happiness be distributed throughout the world' (IBT 1998: 3). In 1990 the Bank's then President Barber Conable remained confident of alleviating world poverty: 'we know a great deal about who the poor are, where they are and how they live. We understand what keeps them poor and what must be done to improve their lives'. The 2000/2001 World Development Report prepared by the Bank was entitled *Attacking World Poverty* (World Bank 2001a) and promised 'the most detailed-ever investigation of global poverty' (World Bank 2001b). The result of more than two years of research, the WDR draws on a 'background study' that sought the 'personal accounts' of more than 60,000 men and women living in poverty in 60 countries around the world. This approach allows the Bank to claim that their vision of development is based upon 'the testimony of poor people themselves'. The 2002 WDR *Building Institutions for Markets* (World Bank 2002b) focuses on weak domestic institutions 'which hurt poor people and hinder development'. Instead, the Bank argues, institutions suitable for (supporting) markets should be promoted so that all poor people can feel the benefits of the free market and free trade. Out of an annual budget of US$15 billion the Bank reserves about US$12 billion for project grants with adjustment lending coming to an estimated $3 billion. Additionally, the Bank annually concludes about 40,000 contracts in projects it finances in 95 countries worldwide (*Geobusiness* 2001: 61). The Bank has contacts and business dealings with some 300 businesses across Europe, representing 47 per cent of its contracts. Consultants in borrowing countries cream off another US$1.2 billion a year (*Geobusiness* 2001). While the agency listens to the 'voice of the poor', consultants are paid handsomely for their work and perspectives on development projects.

cultures and knowledges. Although many have argued that the work of NGOs is often poorly monitored and sometimes imposes foreign solutions in delicate situations, many observers have argued that they can provide an effective link between state and society,

often working in 'very practical ways' (Edwards and Hulme 1992: 14).

The relationship between the state and NGOs is crucial to the prospects of inclusive democratic development. Particularly in situations of political

Figure 8.4 This illustration depicts the idea that World Bank 'experts' view the role of the state in development as comprising a simple formula or recipe, with several distinct 'ingredients'.

Source: *The World Bank and the State: a Recipe for Change?* by Nicolas Hildyard, Bretton Woods Project, www.brettonwoodsproject.org

instability or conflict NGOs can provide critical linkages in the absence of effective or disputed state administration. Many international development institutions, however, fail to recognize and understand the causes and conditions of conflict and its wider impact on development. According to modernization theories, ethnic identities were seen as irrelevant and as something which belong to a distant 'tribal' past. However, ethnicity remains a very important, if neglected, feature of the 'development process' and has been at the heart of a number of recent post-colonial conflicts in Africa (e.g. Angola and Rwanda) (see Case study 8.4).

Most recently, debates about 'globalization' and development (which have often been abstract and lacking in clarity) have highlighted how ethnicities have come to be more hybrid as a result of globalization and cultural homogenization. Again, however, we need to avoid falling into the trap of viewing the Third World and the South as a uniform space of sameness, a space without difference and diversity. Globalization is a key means through which many people seek to understand transitions amongst human societies at the start of the third millennium (see Section 4). Much has been made of the impact of

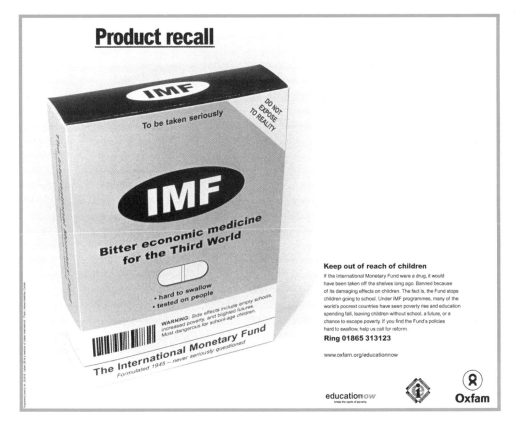

Plate 8.2 The IMF and its hard to swallow 'bitter economic medicine' has had major economic implications for 'Third World' countries depicted in this Oxfam campaign. Its side effects include empty schools, increased poverty and 'blighted futures'. (Oxfam)

globalization, the consequent 'annihilation of space by time' and the increased ease of travel and communication that characterize the contemporary world. In some ways, however, globalization has been 'fetishized' to the extent that as an idea it now has 'an existence independent of the will of human beings, inevitable and irresistible' (Marcuse 2000: 1). There has also been a tendency to speak of the 'end of geography', envisioning a time when all places will have similar social and cultural characteristics as global corporations and as media agencies spread similar products and images across the globe (Holloway and Hubbard 2001). It is necessary, however, to avoid the view that globalization is a relatively new and recent phenomenon, led by the 'unstoppable' juggernaut of global changes in information, knowledge and technology. What is often not clear in these debates is what role there is for the people of the South to shape their own destinies and to what extent borders and geographical divides are becoming less important in a global context.

8.5 Conclusions: geography, unevenness and inequality

Radical development geography must adopt a framework that is liberated from the tyranny of dualism and allows for changes in the world economy and variations between states – one that accommodates alternative historical and contemporary geographies, cutting across the North–South divide.

(Bell 1994: 188)

Development has nearly always been seen as something that is possible, if only people or countries follow through a series of stages or carry out a series of prescribed remedies and recipes. It is also often assumed that an organic and inherently positive and 'natural' process of evolution is somehow at work, one that is both progressive and forward moving: 'from little acorns do giant oaks grow'. Development also

Plate 8.3 Campaigners in London demonstrate the need to 'drop the debt now!' through the Jubilee 2000 campaign which has attempted to raise awareness about the scale of the debt problem and the need for Tony Blair's New Labour government to act. (Photofusion/Ulrike Preuss)

Thematic Case Study 8.8

Globalization, agriculture and development

The key political and economic trends that together summarize what we usually take the term 'globalization' to mean include: (1) the increasing importance of the financial structure and the global creation of credit, leading to the dominance of finance over production; (2) the growing importance of the 'knowledge structure', with knowledge defined as a significant factor of production; (3) the transnationalization of technology and the increasing rapidity with which technologies become redundant, with increasing emphasis on 'knowledge industries'; (4) the rise of global oligopolies in the form of TNCs: corporations must 'go global', acting simultaneously in a number of different contexts; (5) the globalization of production, knowledge and finance leading to a decline in the regulative power of nation states; and (6) the new 'freedom' of capital from national regulative control and democratic accountability, which is said to have led to increasing poverty, environmental destruction and social fragmentation. In thinking about how globalization has impacted upon countries of the South we can take the example of the world food economy, which is very much a global reality that shapes our lives in profound cultural, ideological and economic ways (Goodman and Watts 1997: 3).

Since the UN staged the first World Food Conference in Rome in 1975, amidst growing anxiety about food availability and food prices, the world food economy has changed in many different ways. At the level of *production and distribution*, there has been a radical restructuring. These areas have been internationalized or become more global in scope and character, particularly since the emergence of transnational corporations (TNCs). By the second World Food Conference in Rome in 1996 UN agen-cies were becoming increasingly aware of the ways in which TNCs are (re)shaping the world food economy in ever more global ways, fundamentally reorganizing the division of labour in the global agro-food system over the last 25 years (see Chapter 9).

Transnational biotechnology corporations like Monsanto/Pharmacia, Hoechst, Pannar and Calgene have claimed that the recent emergence of GM or genetically modified crops has the potential to eradicate global food shortages and malnourishment, offering sustainable food security for the poorest parts of the world. These claims have, however, been widely contested in the 'global south'. In March 1999, leading scientists and activists from 12 countries (including Malaysia, India and Bangladesh) met at a conference in Delhi to discuss the prospect of 'biodevastation' associated with the introduction of GM crops and vowed to bring down Monsanto and other biotech TNCs and to build a global mass movement for sustainable and organic agriculture (BBC 1999).

In April 1999 an alternative year-long seed *Satyagraha* was launched in India in direct opposition to Monsanto's GM seeds. Vandana Shiva, Director of the Research Foundation for Science, Technology and Ecology that helped to launch *Satyagraha*, described Monsanto's seeds as 'terminator technology' and has promised that '[w]e will drive Monsanto out of this country. It will be destroyed. Free trade in food is a recipe for famine and farmer's suicide' (Shiva 1999: 9). Almost 1,500 organizations, many of them established by women, have come together across India to reject the new GM seeds like Monsanto's Bollgard Cotton, which can destroy beneficial species and can create superior pests. Many Indian environmentalists described the recent patent crisis as a 'neo-colonial' ploy by the West and the World Trade Organization to (re)colonize the 'developing world' (Shiva 1999: 5).

Plate 8.4   An anti-GM protestor demonstrates outside the Houses of Parliament in London. (Corbis/James Darling)

often means simply 'more': whatever we might have some of today we might or should have more of tomorrow (Wallerstein 1994). After nearly six decades of global development the promise that tomorrow will bring something 'more' is wearing very thin for many poor people.

As Radcliffe (1999: 84) has argued, '[r]ethinking development means reconsidering the categories we use in development geography, and unpacking the power relations that shape them'. This question of the power of development and the unpacking of the power relations that are inscribed in development practices is a critical one. A rethinking of development is also necessary because the lines that have so far divided North and South are now present within *every nation-state* and are making ever less appropriate the conventional language used to interpret the geography of development in the world economy (rich/poor, North/South, First World/Third World, developed/developing) (Power 2003). It is also necessary to (re)formulate a view of development that focuses on the relationships to households and communities and not just (as often happens) on 'formal' institutions like the state, the transnational corporation, the international development agencies or NGOs. This is an incredibly difficult balance to maintain consistently since development operates across so many spatial scales.

At the very core of development studies and development geography there has been a normative preoccupation with the poor (saying what 'ought' to be *done*) and with the 'marginalized and exploited people of the South' (Schuurman 2001: 9). Perhaps we should also ask ourselves what ought to be done about wealth creation and the wealthy? As Hart (2001) suggests, a distinction can usefully be made between 'big D' Development and 'little d' development. The former refers to a post-Second World War project of intervention in the 'Third World' that emerged during the time of decolonization and the Cold War. The phrase 'little d' development, however, points to the development of capitalism as a 'geographically uneven, profoundly contradictory set of historical processes' (Hart 2001: 650). D/development can thus be viewed simultaneously as both a project and a process, with overt and covert forms, shot through with contradiction and unevenness.

In recent decades one of the major trends has been that the world in which we live has become ever more global in character and orientation (globalization is the subject of Section 4). Global corporations and global

marketing activities have resulted in the availability of standardized projects and products, and communications have significantly improved, creating new regimes of 'interconnectedness' (Allen 1995) (Case study 8.8). In studying globalization and its impacts it is necessary to consider its economic, political and cultural strands (Allen 1995).

Globalization provides important historical contexts and frameworks through which development has been understood and experienced. Similarly, globalization has to be viewed in relation to the conditions of 'unequal development' that characterize the contemporary world. The notion of a 'shrinking world' has been important in discussions of the impacts of globalization, but patterns of global inequality have not shrunk after 50 years of development theory and practice. Ake (1996) reminds us that for Africa and other 'peripheral world regions', globalization is a highly uneven process that has not affected all areas of the so-called 'Third World' in the same way. Development in Africa, for example, has never failed; it just never really got started (Ake 1996). In addition, although development thinking has itself been far from monolithic, many conceptions had the habit of 'robbing people of different cultures of the opportunity to define the terms of their social life' (Esteva 1992: 9). It is crucial then to avoid painting a picture of development (and its subjects) as monolithic and to avoid repeating this habit of robbing people of their cultural and historical difference and diversity. The *Dictionary of Human Geography* (Johnson *et al.* 2000) defines development as a process of 'becoming' and as a 'potential state of being', but also mentions how:

> The achievement of a state of development would enable individuals to make their own histories and geographies *under conditions of their own choosing*.
> (Johnston *et al.* 2000: 103; emphasis added)

Local perspectives and knowledges of development have been often overlooked by international development institutions and state approaches have often been heavily centralized and 'top-down'. Whilst many sections of the 'development industry' have in recent years come to acknowledge the importance of people power and popular participation in development (Friedmann 1992), 'empowerment' has been co-opted by the mainstream of development theory and practice and has been stripped of its radical potential. UN agencies in particular have been criticized for their neglect of participatory approaches and their excessive bureaucracy, but prospects for reform look bleak. In some ways, the policy prescriptions of macroeconomic global institutions like the World Bank and the IMF claim to be leading the campaign to eradicate world poverty, but a wide variety of 'anti-capitalist' and 'anti-globalization' protests in recent years have shown that these prescriptions are inflexible and ignorant of the particularities of place and country-specific contexts. Many recent critiques of development appear disillusioned with the future of the development industry and its capacity to understand and alleviate world poverty. Escobar (1995), for example, has argued that development has 'created abnormalities' such as poverty, underdevelopment, backwardness and landlessness and then proceeded to address them in a way that denies value or initiative to local cultures. Escobar (1995) calls the whole process of 'development thinking' into question, showing how the globalization of development ideas and categories has been mediated by historical forces and by local peoples and social identities.

Many 'Western' politicians are only gradually beginning to wake up to the realities of these contemporary global inequalities. A number of them see these concerns as those of distant geographies, a world of problems pushed and worlded beyond the universe of immediate moral concern:

> Elite western policy-makers seem to regard the growing income equality gap as they do global warming. Its effects are diffuse and long term and fears of political instability, unchecked migration flows and social disruption are regarded as alarmist.
>
> (Wade 2001: 80)

Thus poverty and inequality are seen as too diffuse and so long-term as to be of immediate concern. Additionally, rather than thinking of a single, interconnected world economic system, this impoverishment and inequity is constructed as somehow unique or exclusive to the peoples of the 'Third World'. If our concern then is to build a more radical development geography it needs to be understood that poverty also occurs in 'developed' countries and that the aid and 'development' policies of such countries, far from being a part of the solution, may actually be considered a part of the problem. Marginality and deprivation (or for that matter, excessive consumption amongst the affluent) in Europe, North America or Russia and other post-communist 'transition economies' should also be seen foremost as issues of 'development' (Woods 1998, Jones 2000). Radical development geography must be 'liberated from the tyranny of dualism' and the constant stratification of the globe into three distinct worlds. This also means moving away from the developed/developing or lesser/more developed dualisms, which see a simplistic bipolar world of 'haves' and 'have nots'. It is important therefore to allow for variations between states (historically and geographically) and to move beyond the tragic stereotypes of a single condition of 'third world poverty' and a single 'geography of the third world'. This could make a world of difference.

Learning outcomes

Having read this chapter, you should understand:

- The importance of critical and alternative approaches to the study of global development.

- The historical forces and pressures that shape development theories and practices.

- The role of some of the key brokers of development such as states, United Nations organizations, communities and NGOs.

- The differentiation and diversity of peoples, places and spaces of development at the beginning of the twenty-first century.

- The importance of a people-centred view of development and the need to understand the *other side of the development story*, or the views of women, the views of people in the South, the views of people who are excluded socially, and the view from 'below'.

Further reading

Desai, V. and **Potter, R.** (eds) (2001) *The Arnold Companion to Development Studies*, Arnold, London. Contains short and accessible chapters on a wide variety of development themes. A useful companion text in studying development with suggestions for further reading on each topic.

Kothari, U. and **Minogue, M.** (eds) (2002) *Development Theory and Practice: Critical Perspectives*, Macmillan, London. Explores a wide range of perspectives on critical approaches to the study of development, looking at globalization, governance, social development, participation, feminism and post-colonialism.

Potter, R., Binns, T., Elliott, J.A. and **Smith, D.** (2003) *Geographies of Development*, 2nd edition, Longman, London. Updated and revised in a second edition, this book offers a wide-ranging discussion of theories of development, urban/rural spaces and the important institutions of global development.

Power, M. (2003) *Rethinking Development Geographies*, Routledge, London. Draws out the spatial dimensions of development and outlines how the discipline of geography has been implicated and involved in the theory and practice of 'development'. The books offers a critical and stimulating introduction to the imperial and geopolitical dimensions of development, looking at cold war and colonial constructions of 'The Tropics' and the 'Third World'.

Radcliffe, S.A. (1999) Re-thinking development, in Cloke, P., Crang, P. and Goodwin, M. (eds) *Introducing Human Geographies*, Arnold, London, 84–92. Presents a critical discussion of the role of state power in development and of the importance of gender relations. Forms part of a larger three-chapter section, 'On Development Geographies' with contributions from P. Routledge and S. Corbridge.

Useful web-sites

www.id21.org or **www.ids.ac.uk/eldis/** The first site is a useful development research gateway with regular updates and short summaries of recent research topics; the second site is prepared by the Institute of Development Studies (IDS) and offers an indispensable gateway to a range of relevant development resources. The site allows detailed searches (by country or theme).

www.globalizethis.org/ or **www.challengeglobalization.org/** For examples of sites that issue a call for globalizing resistance and challenging globalization.

www.eldis.org/ The ELDIS Development information research gateway site with a useful country search facility and good weblinks.

www.Jubileesouth.net/ or **www.studentsforglobaljustice.org/** The first site concerns the global debt crisis and the second links student resistance to a variety of anti-war, anti-capitalist and global economic justice campaigns.

www.twnside.org.sg or **www.anotherworldispossible.com/** The first site is the home of the Third World Network (TWN) and the second relates to the 'Another World is Possible' campaign.

www.globalissues.org/ or **www.focusontheglobal-south.org** The first site contains useful material on global issues and has introductory material on geopolitics; the second is the home of 'Focus on the Global South'.

www.nam.gov.za/ The home of the Non-Alignment Movement (NAM).

 For annotated, clickable weblinks and useful tutorials full of practical advice on how to improve your study skills, visit this book's website at www.pearsoned.co.uk/daniels

Society, settlement and culture

EDITED BY DENIS SHAW

Geographers are interested in the ways places differ from one another and also in the similarities and interconnections between places. But the complexities of the changes that human societies have experienced since the late twentieth century render the geographer's task of making sense of the world increasingly difficult. Old dichotomies, like that between 'urban' and 'rural', and the lines which geographers have customarily drawn on their maps to divide country from country, region from region, and culture from culture, seem to make less and less sense in a globalizing environment. Even so the results of such processes seem not to be a world that is more uniform. While distant places interconnect and interact with increasing intimacy, new spaces appear where people are excluded from such developments, and almost everywhere some individuals who share spaces with others are denied the rights and advantages that the latter enjoy.

It is such complexities, making themselves felt in urban and rural spaces and in the sphere of culture, that form the subject of this section. The emphasis is on the multifaceted effects of globalization on human societies and the way people think. Thus the world has become more urban – most of the world's population now lives in cities and only three major regions of the world (sub-Saharan Africa, south and much of south-east Asia, and China) remain predominantly rural (though this still represents a substantial proportion of the world's population). But the experience of urbanization differs radically between cities, spaces within cities, and individuals. The corollary is that the world has become less rural – rural spaces may be urbanized, linked into global networks, or lose population to the cities – but again those spaces, and the individuals within them, are affected in very different ways by such processes. And the world has become more 'cultural' in the sense that present-day changes are as significant for their cultural effects as for any others. Issues like identity and the meanings and ways of life that are shared by, or that divide, individuals and groups become matters of concern, contestation and even conflict.

Chapter 9 on cities considers how the term 'urban' is to be defined and the implications of urbanism for the ways in which people live and interact. The significant issue of the role cities now play in global and regional economic and power networks is addressed before the chapter turns to examine the issue of urban governance and how urbanization processes are and can be regulated. Chapter 10, entitled 'Rural Alternatives', is similarly concerned with definititions of the 'rural', and analyses the alternative consequences for both rural spaces and individuals of the processes of urbanization and globalization. Chapter 11 on social inequalities argues that all modern societies are riddled in inequalities, some of which are more fundamental than others. The chapter demonstrates that one of the clearest ways in which social disadvantage is reflected is through its geographical expression, particularly in the existence of spaces described as being 'on the margins'. Chapter 12 on culture defines the latter as a process and as a system of shared meanings. The argument is that globalization is not producing a single global culture although the idea of a global culture, is becoming important. Global cultural processes interact with processes occurring at national and regional levels to produce a complex world in which it is necessary to foster approaches that will welcome and celebrate cultural differences rather than shun and fear them.

This section makes clear the fact that the geography of a globalizing world is as much a geography of exclusion as it is one of interconnection and interlinkage. It shows that it is a dynamic world where social and cultural change is very much the order of the day, and yet it is a world punctuated by continuities. Finally it suggests that it is a world in which people can make a difference, where they can be more than mere ciphers dominated by impersonal forces like global financial and power networks, and where ideas, attitudes and relationships matter.

CHAPTER 9

Cities

ALLAN COCHRANE

Topics covered

- The present-day importance of cities
- Defining urbanism
- Wirth's definition of urbanism
- The significance of the suburbs
- Urban heterogeneity and its implications
- Social polarization and social diversity
- Cities of connection and disconnection
- City cultures

Thematic Case Study 9.1

The growth of cities

Patterns of global population growth and migration are discussed in Chapter 4 of this book, and it is important to remember that: 'The steepest growth rate has been in cities. In 1950 29 per cent of the world's population was urban. In 1965 it was 35 per cent, in 1990 50 per cent, and by 2025 it could be at least 60 per cent. The world annual growth rate of urban population between 1965 and 1980 was 2.6 per cent; but between 1980 and 1990 it was 4.5 per cent. Nearly all the current increase is in poor countries' (Tickell, in Rogers and Gumuchdijan 1997: vii).

As Massey (1999: 114) notes: 'about 3,000 million people are living in urban areas, each of which is home to many millions of people. In the late 1990s between 20 and 30 million people were leaving the countryside every year and moving into towns and cities. We have never faced a situation like this before. We are becoming, increasingly, an urbanized world.' Between 1900 and 1990 the number of cities with populations of over half a million grew from 20 to nearly 600 (Short 1996: 37). And according to UN-HABITAT (2003) we can expect the number of people living in cities across the world to rise to 6,000 million by 2050. They go further to estimate that nearly 1,000 million people (mainly in poorer countries) currently live in urban slums and project a doubling of these numbers over the next 30 years.

Most of the world's population now lives in cities (Case study 9.1).

It is no longer possible – if it ever was – to divide the world into a developed world, with big cities, and an 'underdeveloped' or 'Third' world, without cities. A neat line can no longer be drawn on the globe that divides a world of urbanites from a world of peasants. Some of the biggest cities in the world are in countries that might previously have been described as 'underdeveloped'. São Paulo, Mexico City, Shanghai, Mumbai (Bombay), Buenos Aires and Beijing are all among the ten cities with the biggest populations in the world, while Rio de Janeiro and Calcutta are not far behind (see Table 9.1).

Table 9.1 Cities with the largest populations in the world (millions of people)

1	Tokyo/Yokohama	26.84
2	São Paulo	16.40
3	New York	16.33
4	Mexico City	15.64
5	Mumbai (Bombay)	15.09
6	Shanghai	15.08
7	Los Angeles	12.41
8	Beijing	12.36
9	Calcutta	11.67
10	Seoul	11.64

Source: *Philip's Atlas of the World. Comprehensive Edition*, 8th edn, London: George Philip, in association with the RGS with the IBG, p. 26

The dramatic global growth of cities makes it very important for us to understand them and the ways in which they work. It represents a dramatic shift in the spatial organization of society – in the geographies of everyday life. As Harvey (1997: 23) emphasizes: 'The urban and city are not simply constituted by social processes, they are constitutive of them.' In other words, the changing spatial organization of our lives represented by global urbanization also fundamentally changes the ways in which we live our lives and the social relations of which we are a part. But, in what ways does it do so?

This chapter starts out by looking for ways of defining cities. It explores the features of urban life that distinguish it from other forms of human settlement, and other forms of socio-spatial organization. It considers the implications of some of these features for the ways in which people live their lives in cities, before going on to discuss the key role of cities within wider global and regional networks of social and economic power. Finally, the chapter turns briefly to the challenge of urban governance, asking how the complex and fluid social relationships of urban life are governed and regulated in practice.

9.1 Trying to define cities

Recognizing the importance of urban development in

Plates 9.1–9.4 Images of 1930s Chicago: the slums (Corbis UK/Bettman); the stockyards (Corbis UK/Hulton-Deutsch Collection); Gangsters (Corbis UK/Underwood & Underwood); the Wrigley Building (Corbis UK/Hulton-Deutsch Collection)

general terms is, however, only a starting point. Most of us probably feel that we have a reasonable grasp of what makes some places cities, while others are towns, villages or rural areas. But it is important to explore rather more systematically what it means to call some places cities. This should help us to understand what is shared between cities (and the social relationships associated with them) as well as being able to assess how they differ, for example, playing different roles in global systems of power, offering different opportunities to and placing different constraints upon their residents. So, for example, a large population may

make a place a city (even a 'megacity' in Castells's terms, Castells 1996: 126–7), but only some cities (such as Tokyo, New York and London) are key nodes of financial and economic power within the global economic system, to the extent that they are often referred to as global cities (or world cities) (see, for example, Sassen 1991). As a result, one might also expect a higher concentration of wealth and of members of global elites in those cities, so that their social structures are likely to be quite different from those of other large cities, such as Mexico City or Mumbai.

So, what are the key – defining – features of cities?

In the 1930s, Lewis Wirth – a sociologist based at the University of Chicago (see Plates 9.1–9.4) – set out to specify them systematically, in ways that incorporate what are probably still among the dominant common-sense understandings of urban life. Wirth's definition was timeless and could, apparently, be applied to all societies. For him the city was: 'a relatively large, dense and permanent settlement of socially heterogeneous individuals' (Wirth 1938: 50). In other words cities could be defined through four main characteristics. They are:

- large

- densely populated

- permanent

- socially heterogeneous.

But what are the implications of such a definition? Wirth himself believed that a series of more or less logical consequences flowed inexorably from these characteristics.

- *Size.* The large population size of cities was held to mean that contact between urban residents would generally be impersonal, superficial and transitory, since no one person could know all the others in the city. Wirth's arguments build on a longer tradition that sought to analyse the rise of industrial society. Toennies (1887), for example, distinguished between what he called traditional and industrial society by associating *Gemeinschaft* (community) with traditional society and *Gesellschaft* (association or society) with industrial society. Closeness and stability, in which there was a high degree of informal personal contact within primary groups, characterized the former and everyone understood their responsibilities within a stable society. By contrast, the latter was characterized by transitory and superficial relationships, in which individuals only come together in groups to achieve particular ends (or defend particular interests). Pre-urban (rural or peasant) society has often tended to be associated with *Gemeinschaft*, while urban society has tended to be seen as the prime example of *Gesellschaft*. According to Wirth, the size of cities meant that contact between individuals was usually restricted to those occasions in which there was some mutual advantage to be gained. As a result, he argued, human relationships were fragmented along the lines of the services that each individual might perform for another.

- *Density.* Similarly, the density of population was said to produce increased differentiation between individuals, an increasingly complex social structure, social segregation and a dissociation of workplace from residence. It fostered a spirit of competition, aggrandizement and mutual exploitation, requiring complex rules to manage the resulting friction.

- *Permanence.* It is the permanence of cities that creates the spaces within which the social processes identified by Wirth work themselves out. In a sense it is also that permanence which confirms the importance of 'urbanism as a way of life' (the title of Wirth's seminal article). Urbanism gains its status because it represents a particular stage in the development of human society, as the dominant form of social organization in the industrial (or modern) era.

- *Heterogeneity.* In some ways, for Wirth, this is the most important (and distinctive) feature of cities. He argues that: 'The city has ... historically been the melting pot of races, peoples and cultures. It has not only tolerated but rewarded individual differences. It has brought together people from the ends of the earth because they are different and thus useful to one another, rather than because they are homogeneous and like-minded' (Wirth 1938: 50). The heterogeneity of the individuals making up the city's population and the social interaction between different personality types, said Wirth, resulted in the creation of a complicated caste and class structure reinforcing the effects of high population density. No single group had the full loyalty of individuals, since different groups were important for different segments of each individual's personality.

Wirth's arguments have been criticized for failing to capture some key features of urban life and exaggerating others. So, for example, in some academic and popular narratives 'community' has been associated with particular urban forms, often seen to be under challenge from development, redevelopment and 'soulless' town planning (see, e.g., among many others, the evocation of 'urban villages' in Gans 1962 and the depiction of community in Bethnal Green in Young and Wilmott 1957). Not only can these cases not be dismissed as pre-urban, but they are located precisely in the 'inner cities' whose experience Wirth was seeking to explain.

The rise of the **suburbs** may be a still more difficult phenomenon for Wirth's approach to incorporate (see Plate 9.5). Gans (1968) argues that they meet many of Wirth's broader criteria for inclusion as urban without exhibiting quite the same symptoms in their social relations. He claims (of the US suburbs of the 1960s in any case) that 'there is little anonymity, impersonality or privacy' of the sort identified by Wirth (Gans 1968: 40).

In the last decades of the twentieth century, the emergence of **edge cities** intensifies these challenges. Clark (2000) identifies a move away from monocentric urban forms (focused on a single central core) to more polycentric forms (in which a range of centres have a similar status). Garreau (1991) (again in the US context) has gone so far as to argue that he has identified new models of urban living in the form of edge cities, which represent quite a different 'way of

life' (see also Soja 1996 Chapter 8 on 'exopolis'). Garreau calls these places edge cities because, he says, 'they contain all the functions a city ever has, albeit in a spread out form that few have come to recognize for what it is. Edge, because they are a vigorous world of pioneers and immigrants, rising far from the old downtowns, where little save villages or farmland lay only thirty years before' (Garreau 1991: 4).

One way of dealing with such developments might simply be to deny that they are cities – or expressions of 'urbanism'. From such a perspective, following Wirth, they might even represent a move to a different form of social organization. Pile, for example, forcefully argues that suburbs are an attempt to eliminate or marginalize the tensions of urban life, to separate out different geographies and histories, and to de-intensify urban social interactions (Pile 1999a: 30; see also Pile 1999b for a thoughtful engagement with Wirth and an exploration of ways of defining cities). In other words, for him, suburbs are of little relevance

Plate 9.5　The spreading suburbs: Leavitown, USA. (Corbis UK/Bettmann)

Plates 9.6, 9.7 Making a megacity: Hong Kong and the Pearl River Delta. (Photofusion/Maggie Murray, and Geophotos/Dr Tony Waltham)

in the attempt to define cities, since their self-definition is as an alternative to the city. Pile (1999a) and Massey (1999) argue that while some of the features of urban life can be found outside cities it is the *intensity* with which they are experienced in cities that is significant.

There is a danger in taking quite so narrow an approach, however, since it may leave us unable to reflect the experience of the increasingly large number of people living in settlements that seem to be part of an emergent urban system. Even if Amin and Thrift (2002: 2) are right to suggest that 'The "cityness" of cities seems to matter', surely they are also right to go on to comment that 'it is debatable how far spatial propinquity remains a central feature of the sprawling and globally connected city'.

Even in Western Europe traditional urban forms are being eroded. De Ru (2004) identifies the emergence of an extended urban region that she calls Hollocore. She delineates a region that stretches between, and incorporates, Brussels, Amsterdam and the Ruhr Valley. She describes it as 'urban Europe's non-event', with a population of 32 million, yet without a single city of over 1 million (de Ru 2004: 336). One might question the extent to which the individual significance of some of these cities can be dismissed in this way, but her summary of the position is a powerful one: 'established cities lose residents while thinly populated areas gain. In the name of identity, city centres are stripped down to their historic pedestrian shopping streets, and appear more village-like than ever – frozen in a time that never was. Meanwhile the periphery fills up with a mix of business-, commerce-, leisure-, industry-, logistic-, villa-, office-, or brainparks, generic urban matter embedded in massive inversions of green. In the Hollocore, the city has become the void left in the wake of its own expansion' (de Ru 2004: 338–9). De Ru is contemptuous of those who seek to build a (more familiar) national metropolis within the wider region: 'The Dutch cherish the vision of a "Deltametropolis" (consisting of Amsterdam, Utrecht, Rotterdam and The Hague), the Belgians strive towards a "Flemish Diamond" (shaped by Antwerp, Gent, Brussels and Leuven), and the Germans dream of uniting the cities of the Ruhr Valley into a "Ruhrcity". Three hollow claims of a "unique metropolitan network" of "international stature" in a region where cities are increasingly indistinguishable' (de Ru 2004: 341).

Regional Case Study 9.2

The Pearl River Delta, China

'In 1995, this spatial system, still without a name, extended itself over 50000 km^2, with a total population of between 40 and 50 million, depending on where boundaries are defined. Its units, scattered in a predominantly rural landscape, were functionally connected on a daily basis, and communicated through a multimodal transportation system that included railways, freeways, country roads, hovercrafts, boats, and planes ... [it] is not made up of the physical conurbation of successive urban/suburban units with relative functional autonomy in each one of them. It is rapidly becoming an interdependent unit, economically, functionally, and socially.... But there is considerable spatial continuity within the area, with rural settlements, agricultural land, and undeveloped areas separating urban centers, and industrial factories being scattered all over the region' (Castells 1996: 407–9).

The architect Rem Koolhas argues that 'In the Pearl River Delta, we are confronted with a new urban system. It will never become a city in the recognisable sense of the word. Each part is both competitive with and has a relationship to each other part. Now these parts are being stitched together by infrastructures so that every part is connected, but not into a whole. We call this new model the City of Exacerbated Difference (COED). A city that does not imply the stability of a definitive configuration because each part is fixed, unstable and in a state of perpetual mutual adjustment defining themselves in relation to other parts' (quoted in Hanrou and Obrist 1999: 10–11).

If one set of criticisms of Wirth's approach relates to weaknesses it has in dealing with the complexities of urban life in the United States and Western Europe, then another relates to its lack of engagement with the massive new cities of the Third World. In some respects, of course, his criteria can be seen to fit particularly well with these cities – their populations are even larger than those of most US cities; their populations are dense and heterogeneous, too. The notion of permanence is, perhaps, rather more ambiguous, because of the extensive temporary and informal housing and fast-changing population, both in the core of cities and in extensive shanty towns or *favelas* on the outskirts. However, Castells' description of an emergent megacity around the Pearl River Delta in China – what he calls the Hong Kong-Shenzen-Canton-Pearl River Delta-Macau-Zhuhai metropolitan regional system – highlights some of the difficulties in uncritically pursuing Wirth's approach (Castells 1996: 407) (see Plates 9.6–9.7 and Case study 9.2).

Wirth's assumption that there are certain more or less automatic linkages between urban form and the lived experience of cities is difficult to sustain. Similar urban forms may be associated with rather different experiences of urban life, while apparently similar (urban) experiences may be associated with significantly different urban forms. Keith (2000: 424) counsels us against being misled by 'the mirage of a defining theoretical essence of the city'; there is a danger of losing the variety and excitement of urban life in the search for a conclusive definition. Nevertheless, it is possible to build on Wirth's arguments in ways that highlight how cities are actively defined and redefined by those who live in them and those who move to them and between them. This is probably particularly clear in the case of heterogeneity.

9.2 Some implications of heterogeneity

Cities bring together (juxtapose) people from different classes and different cultural backgrounds as well as combining a wide range of activity spaces. According to writers such as Jacobs, it is precisely this that gives them their strength as sources of innovation and dynamism. In her classic book, *The Death and Life of Great American Cities* (published in 1961), she argues that:

To understand cities, we have to deal outright with combinations or mixtures of uses, not separate uses, as the essential phenomena. . . .

Although it is hard to believe, while looking at dull grey areas or at housing projects or at civic centres, the fact is that big cities are natural generators of diversity and prolific incubators of new enterprises and ideas of all kinds. . . .

The diversity, of whatever kind, that is generated by cities rests on the fact that in cities so many people are so close together, and among them contain so many different tastes, skills, needs, supplies, and bees in their bonnets. (Jacobs 1961: 155–9)

Unlike those whom she accuses of seeking to impose a deadly uniformity on the city, Jacobs emphasizes the extent to which complexity and heterogeneity may create order out of apparent chaos. Like Wirth, she believes that urban form significantly influences the ways in which people live their lives. She positively celebrates those areas of cities that are characterized by mixed uses – commercial, retail and residential – mixed ages of buildings, and low-rise housing on streets of relatively small blocks. Because of the intricacy of interactions generated in such areas, she suggests that it should be possible to ensure that order is maintained under the watchful gaze of a series of residents and workers at different times, in a sidewalk symphony or 'an intricate ballet in which the individual dancers and ensembles all have distinctive parts which miraculously reinforce each other and compose an orderly whole' (Jacobs 1961). She fundamentally criticizes those planners (and architectural theorists) who seek to create more unified areas, which separate traffic and employment from residence and encourage standardization of dwelling units.

Jacobs's vision of the city is an inspiring one – and has indeed inspired much other writing on the city (see, e.g., Sennett 1990, Donald 1997). But it can also be criticized for itself being rather romantic, failing to recognize the ways in which the social life of cities generates inequality and division, power and powerlessness as well as diversity. Berman (1983: 324) notes the extent to which Jacobs's 'inventory of the people in the neighbourhood has the aura of . . . a Hollywood version of a World War Two bomber crew: every race, creed and colour working together to keep America free for you and me'. But then he goes on to highlight the absence of black and Hispanic people from her imagery, from her diverse neighbourhoods.

This is highly significant both in the context of developments which took place in the 1960s, but – more important – also because cities are characterized by substantial change and shifting population, particularly because of the ways in which they attract new populations through migration. As Berman notes:

In the course of the 1960s ... millions of black and Hispanic people would converge on America's cities – at precisely the moment when the jobs they sought ... were departing or disappearing. ... Many of them found themselves desperately poor, chronically unemployed, at once racial and economic outcasts, an enormous lumpenproletariat without prospects or hopes. In these conditions, it is no wonder that rage, despair and violence spread like plagues – and that hundreds of formerly stable urban neighbourhoods all over America disintegrated completely. ... [I]t was clear by the late 1960s that, amid the class disparities and racial polarities that skewered American city life, no urban neighbourhood anywhere, not even the liveliest and healthiest, could be free from crime, random violence, pervasive rage and fear.

(Berman 1983: 324–5)

Social heterogeneity and the juxtaposition (or physical proximity) of different groups also ensure that cities become arenas within which different groups and classes struggle for dominance, or search for forms of security in the face of challenge. Cities both bring different people together, requiring them to live together, but also provide ways of reinforcing and developing new forms of segregation and division. Urban spaces become sites across which conflicts take place and accommodations are negotiated, with particular places taking on symbolic meanings, which crystallize (for a time at least) the outcomes of those negotiations, sometimes as borders, sometimes as protected or open spaces. In their extreme forms, cities can become defined by what Davis (1998, Ch. 7) has called an 'ecology of fear', calling on imagery from the film *Blade Runner*, in which a range of protected enclaves is constructed by the middle classes to exclude

Plate 9.8 Making protected space in South London. (Photofusion/Maggie Murray)

those who may challenge their security (see Plate 9.8 and Case study 9.3). In other words, explicit attempts are made to create order by the spatial policing of heterogeneity (see also McLaughlin and Muncie 1999: 117–25).

Some of the consequences of the apparently insatiable search for 'defensible space' (Newman 1972) are rather paradoxical, however. Instead of creating greater feelings of security in suburban areas, or in areas with neighbourhood watch schemes, or even in shopping malls overseen by closed-circuit television, the creation of more protection seems to have been accompanied by increased concerns about crime and increased insecurity, potentially reinforcing 'a decline in communal trust and mutual support' (McLaughlin and Muncie 1999: 135). Davis (1998) concludes by offering an almost apocalyptic warning of the consequences of pursuing this route, arguing that it is likely to lead to explosions of violence, such as the Los Angeles riots, as well as lower-level but endemic expressions of disorder, such acts of arson.

In his powerful history of post-war Detroit, Sugrue (1996) explores some of the ways in which a city is constructed out of a complex series of interactions – battles and accommodations. Sugrue is critical of those who simply see matters as predetermined, instead emphasizing the active role of the city in constructing the lived experience of its residents:

> In the post-war city, blackness and whiteness assumed a spatial definition ... the completeness of racial segregation made ghettoization seem an inevitable, natural consequence of profound racial differences. The barriers that kept blacks confined to racially isolated, deteriorating inner-city neighbourhoods were largely invisible to white

Thematic Case Study 9.3

The creation of defended enclaves

Davis (1990) vividly describes the creation of protected enclaves in Los Angeles: 'The security-driven logic of urban enclavization finds its most popular expression in the frenetic efforts of Los Angeles' affluent neighbourhoods to insulate home values and lifestyles ... new luxury developments outside the city limits have often become fortress cities, complete with encompassing walls, restricted entry points with guard posts, overlapping private and public police services, and even privatised roadways. It is simply impossible for ordinary citizens to "invade" the cities of Hidden Hills, Bradbury, Rancho Mirage or Rolling Hills without an invitation from a resident. ... In the once wide-open tractlands of the San Fernando Valley, where there were virtually no walled-off communities a decade ago, the "trend" has assumed the frenzied dimensions of a residential arms race as ordinary suburbanites demand the kind of social insulation once enjoyed only by the rich' (Davis 1990: 244–6).

Even in its most extreme form – that is the literal creation of gated areas, accompanied by private policing – Los Angeles is not unique. Similar arrangements are emerging in most of the world's major cities, in luxury developments in London, as well as Detroit and Johannesburg (Robinson 1999). In some cities in less developed countries the contrast is particularly sharp, since the physical distance created by suburbanization is often absent, so that the enclaves may be right up against the shacks or slums of the dispossessed. In São Paulo, for example, Caldeira (1996) describes one development (Morumbi) that was constructed in the 1980s. The new housing offers 'the amenities of a club, [they] are always walled, have as one of their basic features the use of the most sophisticated security technology, with the continual presence of private guards.... [The] luxury contrasts with the views from the apartments' windows: the thousands of shacks of the *favelas* on the other side of the high walls which supply the domestic servants for the condominiums nearby ... for residents of the new enclosures, the inconvenience seems to be more than outweighed by the feeling of security they gain behind the walls being exclusively among their equals and far from what they consider to be the city's dangers' (quoted in Amin and Graham 1999: 39).

Detroiters. To the majority of untutored white observers, visible poverty, overcrowding, and deteriorating houses were signs of individual moral deficiencies, not manifestations of structural inequalities.

(Sugrue 1998: 9)

The social divisions that characterize cities are sometimes approached through notions like social polarization, particularly in discussions of global cities such as New York, London and Tokyo, which operate as key nodes in a global system of finance and economic management – of what Sassen (1991) calls 'global control' (see Chapter 14, p. 334, in which a second tier of global cities, the 'metaregional' capitals of Paris, Frankfurt, Sydney, Hong Kong, São Paulo and Mexico City, is also identified). Within these cities it is argued that there is an expanding gap between rich and poor, powerful and marginalized, as there is a growth in employment in financial and other specialist services, alongside a decline in manufacturing and traditional industry. At the same time as there is a growth in highly paid employment, there is a parallel growth in low-paid, low-skilled and casualized employment at the bottom end – servicing the high-end service sector and those who work in it.

There is certainly evidence that points in these directions – with some sections of the urban population linked into key networks of global economic power, while others are explicitly excluded, to be managed by or to provide services to the powerful. But there is also a danger of taking a mechanical view of these relationships, in which urban social relations may be read off from the position of cities within global systems of one sort or another. While it may be possible to highlight tendencies, or point to the contexts within which social interaction takes place, it is also important to recognize that the ever-changing resolutions of tensions and conflicts may point in rather different directions.

Storper explicitly criticizes the global cities approach for interpreting cities as 'machines' through which the logic of global capitalism works itself out. Cities should not be seen, he says, 'as a mechanical node in a bigger machine' (Storper 1997: 222). Instead he stresses the importance of viewing cities as complex and differentiated 'ensembles' made up of (informal as well as formal) relationships and conventions that exist between the agencies (people, organizations, interests) which underpin their economic operation. He prefers the metaphor of a 'crucible in which the ingredients,

once put in the pot together and cooked, often turn out very differently from what we can deduce from their discrete flavours' (Storper 1997: 255). Storper emphasizes the extent to which cities, their residents, social and economic actors, make themselves in a reflexive process of development and redevelopment, albeit 'constrained by the machine like forces of late modern capitalism' (Storper 1997: 256).

In this context, of course, heterogeneity and the juxtaposition of people with different backgrounds, interests, economic capacities and forms of social and financial capital, become a positive aspect of cities, rather than something from which its citizens need to be protected. The example of Bombay's bazaar in the 1960s suggests one way in which the reality of the city may work positively to erode and challenge divisions that sometimes seem inescapable (see Plate 9.9):

> In the 1960s, the bazaar swept across Bombay, sprawling along transport lines, on slopes, underutilized land, undefined or unpoliced pavements, and on any other interstitial space it could find and occupy. In the process, it blurred beyond recognition the physical segregation inherent in the colonial city structure that had survived until then. The bazaar completely altered the exclusivity of the two domains, the Indian town (in the north) and the Western city (to the south).
> (Mehrotra 1997: 40, quoted in Massey 1999: 128)

Amin and Thrift (2002: 157) neatly summarize what they see as the possibilities inherent in the modern city, which is 'so full of unexpected interactions and so continuously in movement that all kinds of small and large spatialities continue to provide resources for political invention as they generate new improvisations and force new forms of ingenuity'. Instead of starting from the fear of others implied by the search for protected spaces of one sort or another, they seek to identify ways in which urban citizens connect through a series of overlapping formal and informal networks (of leisure, work, friendship, commerce, interest community and so on). This is an urbanism of hope, rather than despair (see also Lefebvre 2003, which builds on a powerful critique of contemporary urbanization to present an almost utopian vision of what should be possible). It challenges those who interpret the urban experience as a battlefield in which different groups seek to protect themselves from others behind real or socially constructed walls, finding ways of inscribing power relations on the geography of the city. In a sense it highlights the continuing tension at

Plate 9.9   The bazaar, Mumbai (Bombay). (LinkIndia)

the core of 'cityness' – between the search for security, the struggle for order and the celebration of difference, the emphasis on opportunity.

9.3 Cities of connection and disconnection

The discussion of cities often focuses on internal processes – the divisions and connections within them. But it is also important to recognize that cities are the products of flows and interconnections across space which link them into wider networks of economic and political relationships. They are, as Massey (1999) argues, 'the intersections of multiple narratives' or multiple space-times. Allen (1999) explores the ways in which cities of power are defined through their positioning in global networks. So, for example, he argues that the importance of cities such as New York and Tokyo stems 'from the fact that the world of banking and finance seems to have little choice other than to go through the financial districts of those cities. They stand at the intersections ... of all that matters in global economic terms. In that sense, both

cities are locations for something far greater, far more significant in scale, than their actual size would have led us to believe.' They are 'the points at which the lines of authority and influence meet', drawing on a concentration of human resources and the local mix of skills and infrastructure (Allen 1999: 187). Summarizing the arguments of Castells, he notes that: 'global cities – or powerful cities – would not exist but for the connections which constitute them' (Allen 1999: 204). It is important to recognize that just as their positioning within these networks serves to define their power (through connection), so it may also confirm its absence (through their disconnection). So, to use an example from the UK, Liverpool was an important node in transatlantic trade during Britain's colonial era, but is now disconnected from major networks of trade, and this is reflected in its sharp decline even within the UK's urban hierarchy.

Castells has identified the growth of 'megacities' (Plate 9.10) as a key feature of the network society that he sees as defining the contemporary world. Although including the cities usually labelled 'global cities', this is a wider notion that also incorporates a wide range of cities of significant world regional importance. They have very large populations, but their key defining characteristic for Castells is that 'They are the nodes of the global economy, concentrating the directional, productive, and managerial upper functions all over the planet' (Castells 1996: 403). They include Shanghai, Mexico City, Buenos Aires, Seoul, Calcutta, Osaka, Jakarta and Moscow among many others. As well as being linked into global networks, they are at the heart of massive urban regions which pull in huge numbers of people – hence the term 'megacity'. If the megacities and some of their populations are connected into these networks, then – according to Castells – a further defining characteristic is that large sections of their population, and significant spaces within them, are disconnected and explicitly excluded. '[W]hat is most significant about megacities', says Castells (1996: 404), 'is that they are connected externally to global networks and to segments of their own countries, while internally disconnecting local populations that are either functionally unnecessary or socially disruptive.'

In that sense, the world identified by Castells is not only one characterized by what he calls 'a space of flows', but also one in which access to that space defines what is possible – those who are disconnected are effectively excluded from participation in the new global economy. Castells (1998: 70–165), himself, uses the term the 'Fourth World' to highlight the significance of this. This, he says, is 'made up of multiple black holes of social exclusion throughout the planet'. Some of these 'black holes' are familiar enough as examples of what might otherwise have been called the Third World or 'less developed countries' (e.g. much of sub-Saharan Africa, the poor rural areas of South America and Asia), but Castells is making a stronger point. The Fourth World, he argues, 'is also present in literally every country, and every city, in this new geography of social exclusion. It is formed of American inner-city ghettoes, Spanish enclaves of mass youth unemployment, French banlieues warehousing North Africans, Japanese Yoseba quarters and Asian mega-cities' shanty towns. And it is populated by millions of homeless, incarcerated, prostituted, criminalized, brutalized, stigmatized, sick, and illiterate persons' (Castells 1998: 164–5). In other words it is a fundamental aspect of contemporary city life.

The implications of this are explored by Amin and Graham (1999) who stress that 'physical proximity' increasingly coexists with 'relational difference'. In other words, the connected live alongside the disconnected – the *favelas* and shanty towns and informal living survive and develop alongside the gated homes of the rich. As we have seen, they use the example of São Paulo in Brazil, where they depict 'luxury enclaves rising above the roofs of the surrounding shanty town' (Amin and Graham 1999: 19). Similar examples could be drawn from cities across the globe, particularly – but not only – in the Third World.

Even cities that seem culturally and socially homogeneous, argue Amin and Graham, are made up of increasingly diverse and overlapping social worlds. Cities, they say, are 'places which bring together and superimpose diverse connections and disconnections' (Amin and Graham 1999: 9). In the case of Bangalore in India, for example, they note (drawing on the work of Madon 1997) that there has been a dramatic growth of software production linked into global networks and markets, which has generated a local concentration of skilled and relatively well-paid professionals in the city, as well as creating strong development pressure (see also Chapter 13, Case study 13.3). But this growth is accompanied by in-migration from rural areas that has led to the creation of shanty towns. These migrants 'struggle to find poor quality work, whilst facing the constant threat of harassment from city agencies and property developers who are keen to "modernize" Bangalore's fabric by developing shopping malls and corporate work spaces in the place of informal settlements' (Amin and Graham 1999: 29).

Amin and Graham use the apparently trivial example of high-class restaurants as a very effective way of illustrating the nature of interactions between groups within cities. Zukin (1995) presents a highly complex picture of the globalized social relations that are reflected in the local experience of New York restaurants. 'In New York restaurant cuisine', she notes, 'the local reterritorializes the global through the identification of neighbourhoods with particular national cuisines – e.g. in the forms of Little Italy and Chinatown – as restaurants produce an increasingly global product tailored to local tastes' (Zukin 1995: 156). More important, perhaps, the restaurants 'bring together global and local markets of both employees and clientele.... While it seems to be among the most "local" of social institutions, a restaurant is also a remarkable focus of **transnational** economic and cultural flows. As an employer, a restaurant owner negotiates new functional interdependencies that span local, regional, and global scales' (Zukin 1995: 158–9).

The heavy dependence on low-wage immigrant labour in restaurants is a reflection of the role cities

Plate 9.10   A gleaming new office building stands amidst the slum apartments in downtown Kowloon, Hong Kong, SAR. (Corbis/Kevin R. Morris)

play within global flows of migration, while the market for luxury restaurants is itself a reflection of cities' importance for global elites within worldwide networks of power. In these restaurants, the global elites (who, Castells argues, are linked to networks, rather than place) find themselves being served – in 'fleeting ties of local interdependence' – by those who are in rather a different position in the global pecking order – often migrant labour in casualized employment whose relationship to global networks is rather different but no less significant (Amin and Graham 1999: 14).

The emphasis in much discussion of cities and their role in a changing world often focuses on their economic role and on the role of elites. It is all too easy to forget the importance of the global networks that link the poor as well as the rich, the migrants as well as the global elites. Gilroy's discussion of the 'black Atlantic' is one useful corrective to this emphasis (Gilroy 1993). He carefully charts the emergence of a 'black Atlantic' political culture through music (soul, funk, reggae, hip hop and rap) which actively links black people in the cities of Kingston, New York and London. This is a global movement with its own networks and its own nodes. So, for example, Gilroy argues that hip hop emerged from the interaction of a

DJ style drawn from Kingston in Jamaica, with the emergent dance music styles of the Bronx in New York (alongside some borrowings from Hispanic styles, too) (Gilroy 1993: 103). In other words, one of the most successful forms of global contemporary popular music came together from the active interaction of people and music through urban networks created by migration (and diaspora), even if it ended up as part of more familiar networks of urban power through major media corporations, from Sony to MTV.

9.4 Conclusion

The search for the 'good' city has been a recurrent theme of debates about cities. As a result, cities have been the focus of a great deal of dystopian and utopian thinking, which has incorporated different understandings of the ways in which cities are constituted. Thinkers such as Howard (1902/1965) and Le Corbusier (1929/1987) have tended to emphasize the extent to which cities are unruly places that need to be managed; thinkers such as Jacobs (1961) and Sennett (1990) have stressed the importance of fluidity, uncertainty and mixing as defining the urban condition. They not only see the attempt to capture social relations in plans and urban design as hopeless but also argue that it serves to damage the vitality of those social relations. The challenge of urban governance is to capture both aspects of the urban experience, in ways that permit the expression of vitality, while allowing for forms of popular control of urban development.

With the help of Wirth, this chapter set out by trying to identify some key defining characteristics of cities. It has stressed the importance of recognizing the complex, contested and overlapping, internal and external, networks of relationships that help to define cities. It is finally important to remember that cities cannot simply be explained in terms of those networks (or in terms of the flows that characterize them). They are more than simply nodes in a system. On the contrary, it seems increasingly clear that the density of activities that is concentrated in cities remains of vital importance, not only in sustaining the networks but also for the people who live in them. They continue to be defined by the intensity of the social relationships within them, not only the excitement generated by the juxtaposition of people and groups with different cultures and understandings, but also the possibility of interaction within relatively large concentrations of people who share particular (sometimes specialist) skills, cultures and understandings. The lives of the world's major cities are simultaneously characterized by the experience of closeness alongside extreme processes of social differentiation.

Learning outcomes

Having read this chapter you should be able to:

- Develop a geographical approach to cities that emphasizes that they are socio-spatial entities. Cities are the products of living social relationships, which have clear expressions both internally (as people come together within cities) and in terms of their changing position within wider economic and social networks of connection. Cities as socio-spatial forms are not just outcomes of the ways in which social processes work themselves out, but themselves play a major part in shaping social processes.

- Understand that the world is becoming increasingly urbanized – most people now live in cities. But, since cities are not all the same, recognize that it is important to understand both what links them together (what makes it possible to talk of 'cities' as recognizable phenomena) and what helps to differentiate them.

- Identify and critically explore key defining characteristics of cities, building on features drawn from Wirth (for whom cities are defined as being large, densely populated, permanent and socially heterogeneous).

- Recognize the significance and implications of social heterogeneity and diversity for the ways in which people live their lives in cities, including the consequences of the existence of increasingly diverse and overlapping social worlds.

Further reading

Allen, J., Massey, D. and **Pryke, M.** (eds) (1999) *Unsettling Cities: Movement/Settlement*, Routledge, London. The chapters of this book focus on the open geography of cities, that is the interconnections between what happens within them and the cross-cutting networks which link them into wider social and economic relationships. Global economic change is linked to the experience of social groups and individuals living their lives within cities.

Amin, A. and **Thrift, N.** (2002) *Cities. Reimagining the Urban*, Polity Press, Cambridge. This is a brave attempt to open up new ways of thinking about the city – genuinely reimagining their potential and reflecting the full complexities of urban life. It celebrates the range of overlapping social relationships to be found in cities as the basis for a vision of democratic citizenship.

Bridge, G. and **Watson, S.** (eds) (2000) *A Companion to the City*, Basil Blackwell, Oxford. A multidisciplinary and stimulating collection of contemporary writings, which draws its examples from across the globe, and introduces a wide range of ideas about the city without seeking to impose any single approach on the reader.

Hall, P. (1998) *Cities in Civilization: Culture, Innovation, and Urban Order*, Weidenfeld & Nicolson, London. This book takes a historical view of cities. With the help of a series of fascinating case studies, from classical Athens to contemporary London, it presents an overwhelmingly positive vision of cities as the fundamental basis of civilization.

LeGates, R.T. and **Stout, F.** (eds) (1996) *The City Reader*, Routledge, London. A well-chosen collection of contemporary and classic readings on cities that provides a useful introduction to important debates about cities and city life.

Massey, D., Allen, J. and **Pile, S.** (eds) (1999) *City Worlds*, Routledge, London. Provides a relatively brief (185 page) introduction to and overview of geographical ways of thinking about cities. It highlights some of the tensions and ambiguities of urban life, exploring both the problems and potential of cities.

Pile, S., Brook, C. and **Mooney, G.** (1999) *Unruly Cities? Order/Disorder*, Routledge, London. Explores and questions the dominant image of cities as necessarily unruly and disordered. Sets out to identify both those aspects of urban life that are inherently fluid and diverse as well as those features that serve to regulate it – to make it manageable for both governed and governors. The book presents dystopian as well as utopian visions of cities, while also charting the practices of urban life.

And finally, just because everybody should read it at least once:

Davis, M. (1990) *City of Quartz: Excavating the Future in Los Angeles*, Verso, London. Despite its critics (see, e.g., Soja 1997: 27), this is an inspirational book, bringing together politics, society and economics in a very powerful analysis of contemporary Los Angeles and its development. It is a dystopia, but it is impossible to read it without being impressed by Davis's enthusiasm for the city whose problems he is dissecting. Most people who read it wish they could do what Davis does for Los Angeles for their own favourite city – unfortunately not all of us have quite the same style or detailed (and sympathetic) understanding to succeed.

For annotated, clickable weblinks and useful tutorials full of practical advice on how to improve your study skills, visit this book's website at www.pearsoned.co.uk/daniels

CHAPTER 10

Rural alternatives

IAN BOWLER

Topics covered

- Definitions of 'the rural'
- Implications of encroachments of urban on rural spaces
- Rural depopulation
- The concept of counterurbanization
- Service provision and rural development
- Planned rural development

A variety of socio-economic trends can be observed in rural spaces around the world and two main processes are implicated, namely globalization and urbanization. On the one hand, rural spaces are being drawn into competitive global relations through the increased international trading of their products and services and the restructuring of capital (McMichael 1996). On the other hand, rural spaces are being drawn into increasingly complex social and economic relations with national and regional urban systems through the interchange of capital, land, labour, raw materials and services. Chapters 12, 13, 14 and 15 look at various aspects of globalization; here we are concerned mainly with the impact of urbanization on rural spaces, but in developed and developing countries alike.

While urbanization is experienced throughout the world (see Chapter 9, p. 214), its impact varies significantly *between* rural spaces; consequently a number of 'rural alternatives' can be identified in terms of the uneven development trajectories of rural spaces. In addition, social groups and individuals *within* rural spaces also have varying experiences of urbanization and 'rural alternatives' can be considered in terms of varied 'life-styles' within the rural population (i.e. housing, employment, income, services and transport – Cloke 1996). This chapter explores these two meanings of 'rural alternatives' to reveal the varied human experience between and within rural spaces.

10.1 From rural functions to rural networks

10.1.1 Defining 'the rural'

A useful starting point for this chapter is a consideration of three constructs of 'rurality' as applied to rural spaces. At one level, the rural comprises those spaces left over once the urban has been accounted for; this *functionalist approach* assumes that rural space forms an analytical category containing non-urban, socio-economic structures and processes having causal or explanatory powers. Spotlight box 10.1 illustrates a recent attempt for the United Kingdom to identify 'indicators' that delineate the countryside in functionalist terms.

A second construct of the rural is based on the interpretation that, with the possible exception of greater rural distance costs, all socio-economic structures and processes are common to both rural and urban spaces (Hoggart 1990). This *critical political economy approach* links rural areas to the dynamics of the national and international political economy (e.g. through urbanization and globalization), with the causes of 'rural' structures and processes lying outside 'rural' areas. Rather, the rural comprises a distinctive spatial category within which the varying impacts of universal structures and processes can be examined.

Spotlight Box 10.1

Indicators for the countryside

Countryside indicators measure socio-economic impacts on rural spaces and identify targets for reducing those impacts. However, they tend to reflect the often unstated interests of those who create them, are dependent on measurable characteristics and are selectively inclusive/exclusive of what is measured. The countryside (i.e. rural) indicators in the study include:

* level of tranquillity (distance from sites of noise)
* darkness of the night skies
* sense of wilderness and isolation

* amount of derelict and vacant land and buildings
* sense of community
* level of rural services
* diversity of rural skills
* beauty of landscape
* frequency and availability of rural public transport services.

This list of indicators, whilst providing a sense of rural space, can also be interpreted as a range of socially constructed representations of 'the rural' (see text).

Source: Council for the Protection of Rural England (1995)

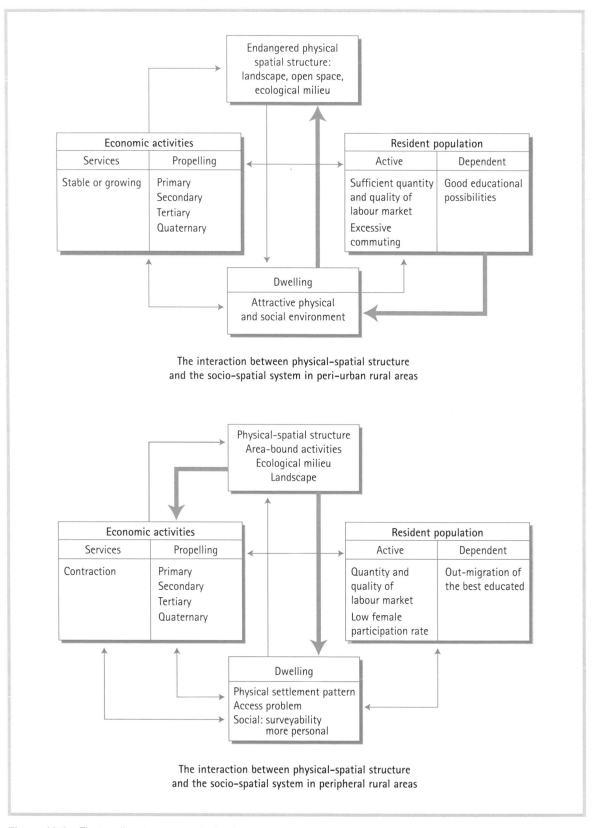

Figure 10.1 The 'rural' system conceptualized.
Source: After Veldman (1984)

This rural locality interpretation (Halfacree 1993) is considered to be most appropriate for the present discussion of 'rural alternatives'.

A third construct of the rural is based on a *social representation approach* (Mormont 1990). In this interpretation, the rural is constituted by lay, rather than academic, discourses of various social groups such as administrators, planners, service providers (e.g. tourism, health, electricity), farmers, recreationists and consumers. Here rurality becomes a social construct and, in Cloke's (1996: 435) words, 'reproduces the rural as an important intellectual category because human behaviours and decisions are influenced by the social constructs that indicate a place as rural.' This approach, which informs the present chapter, focuses attention on the divergent representations of rural space between the various lay discourses and the 'actors' who represent the rural (see Frouws 1998).

10.1.2 Constructing 'the rural' as a system

With rural space treated as a context for analysis, a conventional approach has been to examine its separate functions – such as food production, housing, transport, industry and service provision. Many researchers adopt this 'systematic' approach and commonly employ administrative boundaries to examine variations in human experience between rural spaces. More recently, however, given contemporary concerns over 'sustainability' and 'rural development', the various functions of a rural space have been conceptualized as comprising an open 'rural system'. Three components can be identified: space users, used space and space use. Figure 10.1, for example, shows Veldman's (1984) attempt to represent the interconnections between components of the physical environment, economic activities and the resident population within two types of rural space (peri-urban and peripheral). This approach recognizes the artificiality of examining, for example, economic activities separately from housing provision in rural spaces.

10.1.3 Constructing 'the rural' as a network

One problem of the rural systems approach, however, is conceptualizing space as a number of 'surfaces' with boundaries, each surface being constructed at a particular global, national or local scale. Just how these bounded surfaces at different scales are connected

together is a focus of much contemporary research in human geography (see Chapter 13, pp. 307–9). Connecting the local with the global is also significant for the present discussion, which examines the interaction between local rural spaces and a macro socio-economic process, in this case urbanization. However, if the rural is constructed as an actor-network (Murdoch 2000), the local can be connected to the national, and even the global, through the connectivity of human actors (individual people, firms, agencies, institutions, corporations or government departments) and non-human actors (scientific technologies, organizational forms, technical artefacts, money or knowledge in its material forms – texts and skills) (see www.comp.lancs.ac.uk/sociology/antres.html). With human and non-human actors having equal status in defining each other, they comprise formal and informal networks that connect through space to link the local with the global. More powerful actors enrol other actors into a network so as to 'act at a distance' in influencing local events. In relation to housing development in the countryside, for example, the regulations (documents as non-human actors) made by central government and civil servants (human actors), under development control policy, influence from a distance how rural land is used in a particular locality.

10.2 Encroachment of urban on rural spaces

From the discussion so far we find that the rural cannot be equated merely with the agricultural, since rural spaces fulfil many other functions. Indeed the term '*multifunctional countryside*' is increasingly used to describe rural spaces (Marsden *et al.* 1993; www.ruralfutures.org/reports/). So as to simplify the resulting complexity, three locational contexts are commonly recognized (OECD 1993), namely: (1) rural spaces within commuting distance of urban centres (i.e. peri-urban areas), (2) rural spaces at the peripheries of national economies, and (3) rural spaces intermediate between the previous two. This section looks in more detail at peri-urban areas and the consequences of two main forms of urban expansion: development at the edge of existing urban areas (i.e. the 'rurban fringe') and dispersed development (i.e. extended suburbanization).

Looking first at the transformation of rural spaces at the *rurban fringe* (Bryant and Johnson 1992), a continuum of state controls on land-use competition can be observed. These controls range from the direct state planning of former socialist societies, through the development control process in designated green belts around west European cities and the land zoning ordinances around many North American cities (Bunce 1998), to the relatively permissive policies around cities of many less developed countries. Whatever the cultural context, rurban fringes comprise a distinct 'rural alternative' both as a development trajectory and for the 'life-style' experiences of their inhabitants. Nevertheless, rurban fringes are heterogeneous: not only can 'inner' and 'outer' fringes be detected, but the dynamism of fringes varies significantly between cities and at different locations around individual cities. Studies by researchers such as Healey *et al.* (1988: 99–127) provide detailed empirical evidence on these generalizations.

10.2.1 Direct impacts of urban encroachment

Urban encroachment has direct and indirect consequences for 'rural alternatives' in rurban fringes.

Looking first at direct impacts, three main features can be identified: farmland loss, the land market and rural land use.

Farmland loss

Much of the land in rurban fringes can be classified as 'prime farmland' and its loss to urban development represents an irreversible reduction in the physical resource base for food production. In developed economies there is a long-standing debate over the scale and speed at which such 'foodland' is transferred to non-agricultural uses, leading in some cases to national enquiries, as in the US in the 1970s (National Agricultural Lands Study 1980) and in other cases to national debates, as in the UK in the 1990s (Council for the Protection of Rural England 1993). Land budget estimates of the short-term requirement for 'foodland' in developed economies tend to produce findings of a land 'surplus', although estimates rely heavily on continued gains in agricultural productivity through technological progress.

In developing countries, by comparison, the urban

Regional Case Study 10.1

The development of New Bombay, Maharashtra, India

During the early 1970s, the City and Industrial Development Corporation (CIDCO), on behalf of the Government of Maharashtra, purchased land for the development of New Bombay. The land and houses of 33 villages, and the land but not the houses of a further 62 villages (a total of 117,000 people), were acquired at agricultural land prices by voluntary sale or compulsory purchase. By the early 1990s, household surveys showed that the money from compensation and the new employment opportunities provided by CIDCO (e.g. rickshaw-driving, quarries, small shops/stalls, industry) had increased the social and economic differentiation among the villagers to the advantage of village leaders (*sarpanch*) and larger landowners (*patil*). Employment in agri-

culture had been almost completely eliminated, together with fishing and salt-making; only 33 per cent of households had been able to obtain urban-related employment; and unemployment had increased from 48 to above 60 per cent.

As a result, between 1983 and 1992 the proportion of households with a monthly income of less than 80 rupees increased from 9 to 29 per cent, while comparable figures for those with more than 280 rupees a month increased from 5 to 27 per cent. But the average level of material possessions of the villagers had increased (e.g. consumer durables such as a TV) and some households had used their compensation money to build homes of bricks and cement. As Shaw (1994) concludes: 'the gradual outward spread of the city is deadly in its impact on pre-existing agricultural systems, social structures and life styles.'

Source: Parasuraman (1995)

fringe is far more dynamic in terms of the transfer of farmland into urban-industrial developments (see Case study 10.1). Around the city of Aligarh in Uttar Pradesh, India, for example, Khan (1997) has reported an increase in the urban area of 149 per cent between 1951 and 1991, leading to the loss of 30 km² of agriculturally productive farmland. This scale of farmland loss has prompted a renewed interest in agricultural production within the rurban fringes of cities in developing countries as regards the maintenance of both farm families and local urban food systems (Gbadegesin 1991).

The land market

With the exception of compulsory purchase or appropriation by the national or local state, the transfer of land from rural to urban uses takes place mainly through the land market. Commonly there is a dual market between agricultural and non-agricultural (development) land in both developed and developing countries. The inflated sale value of land for urban development allows farmland owners a range of reinvestment opportunities either within agriculture at other locations or outside agriculture for alternative types of business investment, or retirement from economic activity. Clearly those with more land to sell are advantaged in relation to those with only small areas of land, or who are tenant farmers or farm workers with no land assets, with consequences for their subsequent life-styles.

Rural land use

Urbanization has direct consequences for related changes in land use in the rurban fringe, including fragmentation of farmland by transport routes and the provision of sites for waste disposal (e.g. landfill, incineration and the spreading of sewage sludge) and mineral extraction (e.g. construction materials such as sand, gravel and brick clay). The creation and operation of such sites have consequences for local land, air and water pollution and the sites are often contested by social groups in relation to their vested interests in maintaining livelihoods and residential quality of the rural locale. Research has shown how rural communities in countries as disparate as the US (Furuseth 1992) and India (Parasuraman 1995) mobilize to resist land use changes using a range of

actions, from participation in the planning process, thorough lobbying decision makers, to direct action which, at the extreme, has resulted in physical violence and the deaths of protesters. Similar processes of resistance have been observed at sites for transport, residential and industrial development where often the class interests of private capital are supported by national development priorities (Healey et al. 1988: 103).

10.2.2 Indirect impacts of urban encroachment

Turning now to the indirect rural life-style impacts of urban encroachment, two main elements can be identified: agricultural market opportunities and rural labour markets.

Agricultural market opportunities

There is a well-researched gradient of increasing agricultural intensification with decreasing distance to urban markets, a phenomenon observable in developed and developing countries alike. Contributory factors include reductions in transport costs in purchasing farm inputs and marketing farm products, and access to urban markets for perishable crops and livestock products such as fruit and vegetables and liquid milk. In addition products and services that exploit proximity to urban consumers characterize the rurban fringe; these include farm stalls/shops, direct marketing and farm diversification. The last form of development allows farm families to combine employment in agriculture with non-farm work on the farm, including agricultural contracting and the provision of recreational activities, such as horse riding, and farm accommodation (Ilbery et al. 1997).

The rural labour market

Urban encroachment also changes the structure of rural labour markets in rurban fringes, especially by raising the opportunity cost of working in agriculture. The highest rates of transfer of farm workers to full-time urban-industrial occupations occur in the rurban fringe; in addition income from agriculture can be combined with employment off the farm (termed 'other gainful activity' or OGA), such as work in factories, offices, schools, hospitals and transport. This phenomenon, also termed 'part-time' farming, is observable in the rurban fringes of developed, developing and former socialist countries alike (Cavazzani and Fuller 1982). The categories of farm diversification and OGA, when taken together, comprise 'pluriactivity' and this modification of the rural labour market is particularly prevalent in the rurban fringe. However, entry into a pluriactive life-style in developed and developing countries alike, and its various forms, is selective within the rural population to the advantage of individuals who are younger, have more advanced levels of educational attainment and occupy larger farms. In addition, women have a varied experience with pluriactivity (www.ofwn.org): for example, those with prior, non-farm work experience often develop and manage the

diversified enterprise on farms thereby enhancing their status in the farm family. Also women with higher educational attainment more frequently follow urban-based managerial and professional employment off the farm, while those with a lower level of educational attainment tend to take up poorly paid manual OGA within their locales.

10.3 Depopulation and counterurbanization

One common historical experience of most rural spaces has been the loss of population through rural-urban migration. The age (younger), gender (higher proportion of males) and educational (more advanced) selectivities of rural **depopulation** are well known, as are the causal 'push' and 'pull' processes. The former

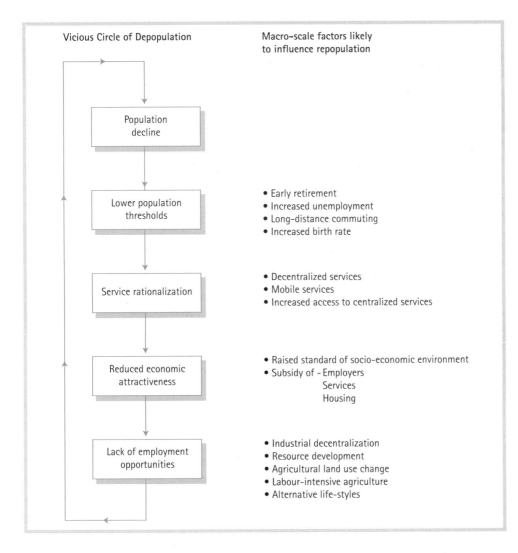

Figure 10.2

Factors influencing depopulation or repopulation.

Source: After Cloke (1985)

Plate 10.1   The effects of polder reclamation in the Kinderdijk area of the Netherlands. (Corbis/Paul Almas)

processes include relative or absolute poverty, the
restructuring of the farm sector, the decline of
resource-based industries (e.g. quarrying and textiles),
the paucity of non-farm employment opportunities,
falling levels of service provision, civil/military unrest
and natural catastrophes (e.g. drought and
earthquake). Amongst the 'pull' processes are the
growing employment opportunities of urban-
industrial locations, the enhanced social opportunities
of city life and the rising levels of service provision,
including access to food supply. That depopulation acts
as a 'vicious circle' in rural spaces is well documented
(Cloke 1985). The consequences include higher levels
of unemployment relieved by out-migration, an ageing
population, rising levels of poverty, erosion of the
extended family and multiple deprivation as regards
access to services such as health and education (Figure
10.2).

The greatest rates of rural depopulation occur
during the transition from an agricultural to an urban-
industrial economy. This was the experience of
countries such as the United Kingdom and Germany in
the eighteenth and nineteenth centuries (see Chapter 2,
pp. 49–60), whereas for many former socialist
countries and most developing countries it remains a

contemporary process (Shen 1995). But depopulation
can also be observed as a contemporary process in
some rural spaces within most developed countries: for
example, in Australia between 1976 and 1981, 46 per
cent of small inland settlements in New South Wales
lost more than 20 per cent of their population, while
12 per cent of such settlements experienced similar
levels of population increase (Sant and Simons 1993).
Thus contemporary depopulation is not characteristic
of particular types of rural space, although it is most
prevalent within developing countries and in the
peripheries of developed countries.

10.3.1 Evidence of counterurbanization

Beginning with the USA in the 1970s, evidence has
been emerging of population increases in rural spaces
in most wealthy, highly urbanized countries – a
process now termed **counterurbanization** (Fisher
2003). Resurgence of the rural population has been
based on net in-migration rather than natural rates
of increase (i.e. the excess of births over deaths) and
Figure 10.3 shows this distinction for départements
in France. Between 1975 and 1982, Paris and

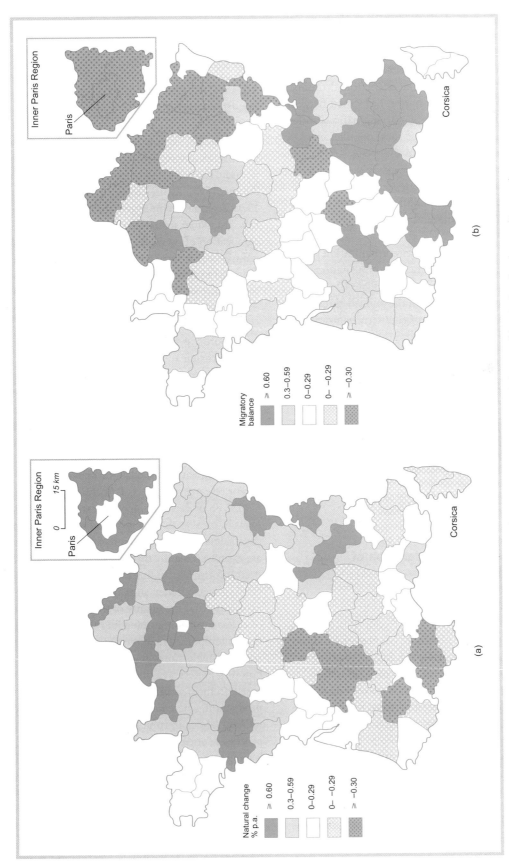

Figure 10.3 France: components of population change, 1975–1982, by department: (a) natural change; (b) net migration (per cent annum).
Source: After Ogden (1985)

industrialized areas of north-east France lost population through net out-migration, whereas peri-urban areas around Paris and much of rural France, especially in the south, experienced population gains through net in-migration. The strength and timing in the onset of counterurbanization has varied between countries while, in the 1980s, varying national rates in a widespread downturn of rural population increase were encountered. In the 1990s some countries, such as the US, showed evidence of a return to the historical experience of higher net in-migration rates for metropolitan areas, but in others, such as the UK, counterurbanization continued (Lewis 2000).

A precise definition of counterurbanization is problematic, not least because of varying national experiences with the phenomenon (Plate 10.1), but a consensus can be identified in the published research on the main characteristics of counterurbanization:

- higher population growth rates at progressively lower levels of the settlement hierarchy

- buoyant rates of population growth in peripheral rural areas

- population shifting from urban-industrial areas to locations more favoured in environmental terms

- a variety of causal processes

- the varied experience of rural spaces in the timing and magnitude of population increases.

10.3.2 The 'period' explanation of counterurbanization

A number of competing explanations of counterurbanization have been developed (Champion 1988). In a 'period' explanation, counterurbanization is interpreted as an historically contingent, short-term aberration in the long-term trend of urbanization. Summarizing, the 1960s and 1970s comprised an era of rising real incomes, increased personal mobility through car ownership, higher retirement rates, improved long-distance communications (e.g. rail, telephone, telematics) and rural infrastructure (e.g. roads, shops, housing). Taken together, these elements facilitated a relocation of the population into rural spaces during a 'mobility transition'. This explanation, however, is unable to account for the observable downturn in counterurbanization in the 1980s in some but not all countries, nor does it offer an insight into

the motives of the population relocating from urban to rural spaces.

10.3.3 The economic explanation of counterurbanization

An alternative explanation is offered by writers who interpret counterurbanization as one facet of the periodic restructuring of capital and labour (Sant and Simons 1993). Again summarizing, in the 1960s and 1970s regional sectoral specialization was replaced by a new functional spatial division of labour, including the ruralization of manufacturing and service-based industries (e.g. local government, hospitals, schools, tourism) in many developed countries (see Plate 10.2). In Italy, for example, between 1971 and 1981, 63 per cent of non-agricultural jobs created by private firms were situated in rural spaces (Lowe *et al.* 1998). In other words, as metropolitan areas lost employment in manufacturing and, to a lesser extent, service industries, so medium and small-sized settlements in rural spaces experienced gains in those sectors of employment (see Spotlight box 3.3). Population increased not only in the larger rural settlements, where employment growth was concentrated, but also in the surrounding villages as a result of increased commuting within local labour market areas.

For researchers such as Fothergill and Gudgin (1982), an understanding of new start-up or relocating firms in rural areas is to be found in the economic constraints of metropolitan locations. Such constraints are expressed mainly through the increasing space needs of firms, including the physical congestion of existing metropolitan sites, the high cost of land for expansion, and the quality of the work space in terms of access for loading/unloading and staff parking, and the contemporary requirement to locate production within a single-storey building. Fielding (1989), however, prefers a capital restructuring explanation based on the new hierarchical organization of production, where lower-cost, less unionized and militant labour in rural areas carries out routine labour processes (often in branch offices and factories), leaving long-established industrial areas with skill-based labour processes and metropolitan areas with specialized management functions and allied activities such as financial services. But not all rural spaces have been affected equally or at the same time, nor has all new employment been in routine labour processes or financed by external capitals. New communications corridors, as provided by motorways, science parks associated with universities, industries and services based on biotechnology and information technology, and interacting networks of small, locally financed firms in 'industrial districts' have all contributed to varying employment growth in different rural spaces.

10.3.4 The socio-environmental explanation of counterurbanization

While new employment has been a major component of counterurbanization, two socio-environmental motivations for urban–rural migration – Geyer (1996) summarizes them as 'environmentalism' – have also been recognized. The first of these is concerned with *retirement*, including the prior establishment of second homes for some migrants and return migration for others who, earlier in their life-cycle, left rural areas for urban employment. For example, Perry *et al.* (1986) found that, in the mid-1980s, 36 per cent and 25 per cent of the inflow of population into Finistère (France) and Cornwall

(UK) respectively was return migration. Coastal, mountain and other 'high quality' rural environments are favoured by retirees, for example the 'Gold Coast' of New South Wales in Australia (Walmsley *et al.* 1995). But whereas young adults are most often implicated in rural–urban migration, and families with young children in 'extended suburbanization', retirement migration tends to occur late in the life-cycle after children have left the parental home.

The second socio-environmental motivation in counterurbanization is access to a rural *life-style*, including the desire of members of the 'service class' (professional and managerial occupations) to acquire the material or 'positional' good of a rural residence. The search for a 'rural idyll' is also expressed in social terms: being part of the social life of a rural community. These are significant motivations in counterurbanization, as found amongst households in 'extended suburbanization' (Spencer 1995), the founders of new firms in rural areas (Keeble 1993), and the relatively small proportion of households that abandon urban-industrial employment and life-styles

Plate 10.2 New small rural business. Dene House Farm was previously a purely farming enterprise in the village of Longframlington, Northumberland, UK. The owners have diversified by providing leisure club facilities and holiday accommodation targeted at disabled groups. (Courtesy of the Alnwick Business Link Advice Centre)

for high environmental amenity living spaces. This last phenomenon, for example, has been reported by Jones *et al.* (1986) in relation to long-distance migrants from south-east England to northern Scotland.

10.3.5 Localized experiences of counterurbanization

Counterurbanization has contributed to the uneven development trajectories of rural spaces by providing a variety of localized experiences for population net in-migration. There is insufficient space to investigate the bases of these experiences in detail and a listing of the factors identified by Cloke (1985) must suffice:

- land market (price and ownership)

- physical environment (landscape amenity)

- settlement quality (heritage, conservation)

- housing market (stock and price)

- employment market (opportunities)

- social and community structure (services and prestige)

- accessibility (from urban centres).

As shown by Perry *et al.* (1986) for Finistère, for example, the operation of these factors produces a range of localized experiences of counterurbanization: varied combinations of employment (e.g. tourism, food processing, ferry services), retiree and alternative life-style migrants interact with situated characteristics such as the land and housing markets.

Individuals also experience a range of alternative life-style changes under counterurbanization. On the one hand, the operation of the housing market is socially selective to the advantage of those with high disposable incomes. Consequently 'service class' in-migrants gain from access to low density housing in smaller settlements set within rural environments. Moreover the 'newcomers' defend their new life-styles against further economic and residential development (termed 'nimbyism' – not in my backyard) and obtain further prestige and status by taking over the organization of local social institutions such as youth groups, interest societies and local government. On the other hand, 'newcomers' commonly exclude other social groups (defined by their lower incomes, ethnicity or family structure – Lowe *et al.* 1995) from local housing, as well as from professional and managerial posts in newly created employment and

from organizational positions in social institutions (see Plate 10.2). Counterurbanization, therefore, redefines the meaning of 'rural community' and brings about new forms of social exclusion for individuals in rural society, including many of the pre-existing inhabitants of rural spaces.

10.4 Service provision and rural development

In both developed and developing countries, access to services and employment form important but interrelated components of life-style, but social groups in rural spaces have a varied experience of such access. In **services**, the main difficulties lie in the incidence of low incomes as a constraint on access and the distances separating locations of provision from a dispersed and historically diminishing population. In employment, most traditional rural industries, such as agriculture and natural resource extraction, yield relatively low incomes and have been in decline through national and global competition, without significant replacement until recent years. Consequently the state tends to intervene through the planned provision of services and programmes of rural development (www.nal.usda.gov/ric/ruralres/aboutric.htm). In developing countries attention tends to be focused on services such as drinking water, health care (including family planning), education and employment; in developed countries, by comparison, more concern is expressed about transport, retailing, housing and care for the elderly. Nevertheless, individuals and social groups, particularly at the peripheries of national economies, experience life-styles that vary as regards access to services and employment.

10.4.1 The meaning of 'services' in a rural context

The term 'services' encompasses the social goods that comprise the fabric of non-production life, ranging from educational and health, through transport and utilities (e.g. water, electricity), to retail, religious, legal and financial (e.g. banks, post offices) services. Depending on the national or regional context, some services are provided entirely by public agencies, some entirely by the private sector in competitive markets, some through self-help organizations, and others through varied combinations of the three previous modes of delivery. While the private sector is driven by commercial imperatives, the public provision of services is underpinned by considerations of need, social justice, allocation and distribution. At issue are social and political judgments on (Humphreys 1988):

* the 'need' for various services, such as residential care for the elderly

* what is 'fair' and 'just' in society, for example as regards access to public transport

* the amount and quality of social (merit) goods to be provided, for example drinking water

* who will get how much of scarce resources and at what cost, for instance health care.

10.4.2 Locational access to services

Three main concepts underpin 'locational accessibility' to services, namely threshold, service area and order (Honey 1983). Threshold relates to the population size required to maintain the economic viability or economic efficiency of a service. Characteristically, the per capita cost of providing a service declines with the increasing number of people served (i.e. **economies of scale** – Powe and Whitby 1994), but begins to rise beyond an optimum size. Service area is configured as the division of space between service centres so as to equalize accessibility by the population to a particular

type or quality of service. Order concerns the cost/value or quality of a service, with the implication that (cheap) lower order services are used more frequently than (expensive) higher order services. For example, the (low order) surgery of a general practitioner can be compared successively with the (higher order) facilities of a 'cottage' hospital and metropolitan hospital (Case study 10.2). The three concepts interact in the provision of rural services: for instance, service areas vary in size according to population density and the order of the service, with provision commonly structured within a hierarchy of 'central places'.

Under rural depopulation, increasingly large service areas are required to reach threshold levels that maintain economies of scale, with a progressive reduction in the number of service outlets through either formal planning by public agencies or competition in the market. Most rural settlements show evidence of this rationalization process, perhaps as a redundant church, shuttered bank or disused shelter at a stopping place for a now withdrawn bus service.

It might be expected that counterurbanization would create a 'virtuous circle' of population growth and so reverse the decline of rural services (Figure 10.2). However, the threshold sizes and service areas of most services have tended to increase through time under the economic imperative of reducing per capita delivery costs (e.g. through the privatization or reduced funding of public services), while the new mobile population involved in 'extended suburbanization' has

---

Regional Case Study 10.2

Health care provision in Bangladesh

Health care is provided in a hierarchical structure of centres distributed throughout the country. By the mid-1990s 57 District Hospitals had been built, together with 351 rural Upazila Health Complexes (UHC), 4,500 Union Health and Family Welfare Centres (which contain a surgery), with support from visiting Health Assistants (UHC staff) and Family Welfare Assistants in village-based Health Posts. Even so, with continuing high birth and infant mortality

rates, and with life expectancy hovering around 52 years, problems remain. These include the 0.6 per cent of GDP allocated to health care by governments, the low health status ascribed to women, a shortage of qualified staff, the allocation of funds to buildings and salaries rather than medicines and equipment, corruption at the local level in the purchase and supply of medicines, and the preoccupation of salaried doctors with private patients at the expense of the public sector.

Source: Siddiquee (1996).

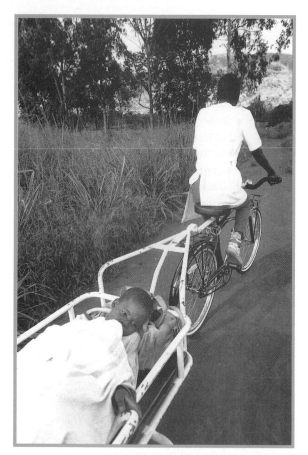

Plate 10.3 Getting across to services in developing countries – transfer to a health clinic. Inadequate transport and roads in many rural areas inhibits access even to the most essential services. (Panos Pictures/Penny Tweedie)

tended to obtain services in the urban centres of their daily commuting with damaging consequences for services such as the village shop. Indeed a general feature of recent years has been the increased centralization of services into urban centres to the disadvantage of communities in peripheral rural spaces.

10.4.3 Effective access to services

The normative assumptions on equality of access have to be suspended when the actual consumption of services is investigated. Income acts as the main constraint on access to most services and a succession of surveys has revealed a higher level of relative and absolute poverty in rural as compared with urban communities (e.g. Tickamyer and Duncan 1990). Within England, for example, Cloke (1996) estimated the incidence of relative rural poverty in a range from

47 per cent of households in Devon to 35 per cent in Northamptonshire. Within rural communities, the lowest incomes can be traced to four main life-style contexts, namely low-paid occupations (e.g. in agriculture, tourism, catering and self-employed small businesses), unemployment, old age and poor health/physical or mental disability. In each context, individuals and families experience multiple deprivation in terms of inequitable access to a range of services, such as housing, health care and education but also, critically, transport (McLaughlin 1986). Disadvantage in paying for the cost of public (bus, train) and private (car) transport limits personal spatial mobility and reinforces the difficulty of access to both services and employment. With 'deprivation' now a contested term within the discourse on service provision, increasing attention is being paid to social exclusion (Philip and Shucksmith 2003). This perspective emphasizes how the exercise of systematic economic and political power in society excludes some individuals from life-styles open to the majority. Such a perspective also brings 'neglected rural others' into focus (Philo 1992), such as single parents, ethnic minorities, New Age travellers, rural women, unqualified youth, the elderly and the disabled.

Social exclusion is aspatial and not confined to any particular rural context. Even so, the longer travel distances in 'remote' rural areas limit effective access to services compared with the shorter travel distances in more 'accessible' areas (see Plate 10.3). Here the significance of low income for access to personal transport is a defining issue, especially where public transport provision is itself problematic. Given the social exclusion experienced by those without personal transport, the state tends to intervene by either (1) increasing personal mobility by subsidizing public transport or self-help organizations (Banister and Norton 1988), (2) subsidizing services at locations from which they would otherwise be withdrawn on economic grounds, (3) concentrating the population into a limited number of larger settlements (e.g. key villages or growth centres) where services can be maintained at economic cost, or (4) facilitating the mobility of the services themselves (e.g. mobile shop, mobile health service) (OECD 1991).

10.4.4 Planned rural development

Huillet (1991) has observed that 'the appearance or disappearance ... [of services] ... is the best indicator of a region's economic health'. Unsurprisingly, therefore, state intervention in service provision often occurs within the context of programmes of rural development that include the diversification of employment opportunities. Such programmes are decreasingly sectoral (e.g. just rural electrification) but increasingly based on integrated development (e.g. jobs, housing, social facilities and transport taken together) with objectives beyond economic growth. While development programmes are commonly constructed for defined rural spaces – for instance, the Objective 1 (lagging behind) and 2 (problem) rural regions of the EU – Shucksmith *et al.* (1994: 213) have observed incisively that 'people not areas suffer rural disadvantage: … area-based approaches are more convenient because … they do not challenge the fundamental processes which structure disadvantage.'

Two modes of rural development are commonly identified, namely top-down and bottom-up. The former tends to be driven by statutory agencies placed by the state within a hierarchy of responsibility from the national, through the regional to the local. Development tends to be based on exogenous capital and decision making, for example state investment in modernizing rural infrastructure, such as electricity supply, and foreign inward investment by urban-industrial corporations seeking low-cost labour for the establishment of agricultural production for export or branch factories. This form of development has been experienced in developing and developed countries alike and is prone to instability through changes in state funding or global demand for a factory product during times of economic recession.

The failure of 'top down' rural development programmes (Schramm 1993), particularly as official aid programmes to developing countries, has been influential in shaping alternative approaches. 'Bottom-up' development, for example, is community-driven and based on endogenous development processes in which local people participate (Lowe *et al.* 1998). Such development can be interpreted, on the one hand, as the resistance of marginalized communities against the pressures for assimilation into wider social and economic structures or, on the other hand, as the rural expression of 'local economic development' that is usually associated with cities. Common characteristics of this form of development are the maximization of locally available resources, including human capital through training programmes, the retention of added value through the processing of raw materials in locales, and local networking to share experiences and collaboration through joint projects which can gain economies of scale and access to wider non-local markets. The development of local entrepreneurship, the production of high-value products or services for niche markets, and the regional association of a product (such as furniture) or a service (such as rural tourism) further characterize this form of rural development.

However, this top-down/bottom-up dichotomy is largely illusory for in practice the two modes of development tend to be combined: local and non-local actors and institutions are linked together in complex networks of collaboration (see Case study 10.3). Garmise *et al.* (1996) go so far as to claim that, taken together, networked institutions comprise

---

Regional Case Study 10.3

Local rural development in Kenya

In Kenya since 1983, Sub-Locational Development Committees (SLDCs) have operated within the government's District Focus of Rural Development Strategy. Administrative Districts have autonomy in setting their development priorities, while various development committees are responsible for the initial identification, prioritization and design of projects at District, Divisional, Locational and SLDC levels. Central government appraises and finances projects through a combination of funding from governmental and donor organizational sources. Membership of each SLDC comprises the assistant chief, the local head teacher, local civil servants and representatives of village residents chosen at public meetings. In Siaya District, for example, 16 SLDCs operate with 17 to 72 members, each holding 3 to 12 meetings a year with attendance rates varying from 50 to 90 per cent.

Source: Schilder and Beove (1996)

a new form of economic governance. The 1991 'Liaisons Entre Actions de Développement de l'Economie Rurale' (LEADER) programme of the European Union provides a good example of network-based rural development (Barke and Newton 1997; www.europa.eu.int/comm/leaderplus). Over 800 selected Local Action Groups (LAGs) in their 'territories' bring together local representatives of the community, entrepreneurs, service providers and government officers to work with officials from the Directorate General VI (Agriculture and Rural Development) of the European Commission. Together they produce and implement, with varying degrees of success, innovative, low-cost local development plans, partly financed from the Structural Funds of the EU.

10.5 Conclusion

This chapter has considered the theme of how people in rural spaces experience differing changes to their life-style as a result of the dynamic processes of urbanization and economic development. Life-style changes vary both between and within countries: experience of employment, access to services and quality of life vary between, for example, rurban fringes, the spaces of extended suburbanization, and spaces characterized by programmes of integrated rural development. The varied life-style experiences of social groups and individuals within rural spaces can be traced partly to constraints imposed by socio-economic structures, for instance as expressed through income levels, and partly to the characteristics of the types of rural space in which they are located. In this conclusion two brief observations are made on the

challenge of the new millennium as regards these themes.

The first observation concerns the changing functions of rural space. While many rural spaces will continue to serve a *production* function in economy and society, for example in developing countries in terms of producing food, minerals, timber and water, it is evident that other rural spaces, for instance in developed countries, are taking on multiple *consumption* functions as regards residence, recreation, tourism and environmental conservation (Marsden 1999). Estimates on the future use of rural land in developed countries predict a 'surplus' of farmland, with a reduced role for food production and its associated employment (Netherlands Scientific Council for Government Policy 1992). Conversely enhanced roles for urban land, environmental conservation and recreation are anticipated, together with their associated employment.

The second observation concerns the implications for rural spaces of the increasing emphasis being placed on sustainable development. Gaining political prominence through the 1992 Rio de Janeiro 'Earth Summit' and subsequent Agenda 21 national development programmes, sustainable development has environmental, social and economic dimensions, albeit to a large degree chaotic and contested. Sustainable development offers both constraints on and opportunities for life-style experiences of social groups in rural spaces. Sustainable development contextualizes the emerging struggle between central and local political and economic power in rural governance as evidenced in the tension between local, endogenous, community-driven rural development and its assimilation into the wider framework of capitalism.

Learning outcomes

Having read this chapter, you should understand:

- The problems of identifying and classifying rural spaces throughout the world.

- The varying development trajectories of different types of rural space.

- The varying life-style experiences of social groups and individuals within rural spaces.

- The impact of urbanization on all types of rural spaces through processes such as counterurbanization and rural development.

- The significance of the changing functions of rural spaces and sustainable development for human geography in rural spaces in the new millennium.

Further reading

Almas, R. and **Lawrence, G.** (eds) (2003) *Globalization, Localization and Sustainable Livelihoods*, Ashgate, Aldershot. Rural geographers and sociologists examine the impacts of global processes on the restructuring of local economies and communities. Case studies cover both developed (e.g. Norway, Australia, Finland) and developing countries (e.g. Brazil, Argentina).

Bernstein, H., Crow, B. and **Johnson, H.** (1992) *Rural Livelihoods: Crises and Responses*, Oxford University Press, Oxford. An Open University text that concentrates on the human experience of poverty, making a living, health and development in rural spaces within developing countries.

Ilbery, B. (ed.) (1998) *The Geography of Rural Change*, Longman, Harlow. A collection of essays by different authors on a wide range of contemporary changes in rural areas (land use, economy and society) but limited to the context of developed countries.

Marsden, T., Lowe, P. and **Whatmore, S.** (eds) (1990) *Rural Restructuring: Global Processes and Local Responses*, Wiley, London. A collection of research-oriented essays by a variety of authors drawn from different disciplines, which examine the theoretical basis of global restructuring processes but also the varying experiences of local spaces. Most of the discussion is for developed countries.

For annotated, clickable weblinks and useful tutorials full of practical advice on how to improve your study skills, visit this book's website at www.pearsoned.co.uk/daniels

CHAPTER 11

Social inequalities and spatial exclusions

PHIL HUBBARD

Topics covered

- Studying society and space
- Social exclusion, poverty and marginalization
- Stereotyping and the exclusionary urge
- Racial exclusions: real and imagined geographies of the 'ghetto'
- Gentrification as purification strategy

All societies are riddled by inequalities, some of which are more fundamental than others (for example, a person's skin colour, age or income typically plays a critical role in shaping their status in contemporary society, whereas their hair colour tends to be less significant). From a geographer's perspective, these social inequalities are of interest in so much that their imprint is observable in the spaces around us: in short, there is a definite, if complex, relationship between society and space, with social order bequeathing spatial order, and *vice versa*. This relationship constitutes the primary focus of *social geography*, a sub-branch of the discipline concerned with elucidating the spatial imprints of social relations and, conversely, demonstrating the centrality of space in social life.

This chapter explores some of the key concepts in social geography. It does this by focusing on the geographies of the least powerful and advantaged in society. As this chapter will demonstrate, one of the clearest ways in which social disadvantage is expressed is through its geographical manifestations, particularly in the existence of spaces described as being 'on the margins' – such as 'inner city slums', 'problem estates' or 'ghettos'. Such areas are a focus of simultaneous fascination and fear for most members of society (and many social geographers) because they represent spatial concentrations of those who fail to match mainstream ideas of how people should live and work. Even those who have never visited such spaces tend to be acutely aware of them, often regarding them as 'no-go' areas. As this chapter will demonstrate, the **stigma** surrounding these places compounds their poverty and isolation, as once a place is labelled as degraded and deprived, its residents find it hard to be accepted by mainstream society. Underlining the fact that space plays an *active* role in shaping people's standard of living, this chapter's focus on geographies of exclusion thus allows us to demonstrate why a sensitivity to place matters, and to illustrate some of the ways social geographers contribute to a tradition of socially relevant research.

11.1 Poverty and marginalization

Geographers' interest in social relations has often been driven by their desire to expose the injustices suffered by those living in **poverty**. In simple terms, poverty can be defined as the condition where individuals or households are unable to afford what might be perceived to be the normal necessities of life.

Inevitably, judgments of what constitutes an acceptable standard of living vary across time and space, meaning geographers are generally more interested in *relative* rather than *absolute* poverty. On a worldwide scale, for example, it is evident that indicators such as average income per capita provide only a very superficial insight into experiences of poverty in different nation-states (see Chapter 13, pp. 294–7). Consequently, recent attempts to measure poverty have generally defined people as being in poverty if they lack the financial resources needed to obtain the living conditions that are customary in the society to which they belong. For example, one study suggested 26 per cent of the British population were living in poverty in 1999, on the basis that their income would not secure at least three of the 35 items considered by at least 50 per cent of the British population as necessary for maintaining an acceptable standard of living (see Table 11.1). While some of these items may not have been considered a necessity in previous decades (and others remain luxuries that many in the global South can only dream of), comparison of these figures with previous studies suggests the incidence of poverty has actually risen sharply. In 1983, 14 per cent of British households lacked three or more necessities because they could not afford them, a figure that increased to 21 per cent in 1990 and over a quarter in 1999 (Pantazis *et al.*, 1999).

Roughly translated, such figures suggest that nearly ten million people in Britain cannot afford adequate housing, about eight million cannot afford one or more essential household goods, and four million are not properly fed by today's standards. While such deprivation is particularly pronounced among those who are out of work, many of those in the lower strata of the labour market are also living in poverty (the persistence of low-paid, part-time or precarious forms of employment being significant here, alongside the restructuring of welfare provision for those on low incomes). Coupled with the rapidly rising incomes characteristic of managerial and professional occupations (particularly in 'knowledge-rich' sectors such as law and finance), the consequence is an increasingly polarized society. For example, between 1979 and 1991, the average British income grew by 36 per cent, yet for the bottom tenth of the population it dropped by 14 per cent. While the rate of income polarization slowed in the 1990s – partly due to government policy – the gap between the 'haves' and 'have-nots' continued to grow (Dorling and Rees 2003). By 2001, the average income of the top fifth of households was around 18 times greater than for those

Table 11.1 Perception of adult necessities and how many people lack them (all figures show % of UK adult population)

	Percentage of respondents considering items necessary	Percentage of population unable to afford item
Beds and bedding for everyone	95	1
Heating to warm living areas of the home	94	1
Damp-free home	93	6
Visiting friends or family in hospital	92	3
Two meals a day	91	1
Medicines prescribed by doctor	90	1
Refrigerator	89	0.1
Fresh fruit and vegetables daily	86	4
Warm, waterproof coat	85	4
Replace or repair broken electrical goods	85	12
Visits to friends or family	84	2
Celebrations on special occasions (e.g. Christmas)	83	2
Money to keep home in a decent state of decoration	82	14
Visits to school, e.g. sports day	81	2
Attending weddings, funerals	80	3
Meat, fish or vegetarian equivalent every other day	79	3
Insurance of contents of dwelling	79	8
Hobby or leisure activity	78	7
Washing machine	76	1
Collect children from school	75	2
Telephone	71	1
Appropriate clothes for job interviews	69	4
Deep freezer/fridge freezer	68	2
Carpets in living rooms and bedrooms	67	3
Regular savings (of £10 per month)	66	25
Two pairs of all-weather shoes	64	5
Friends or family round for a meal	64	6
A small amount of money to spend on self weekly	59	13

Table 11.1 continued

	Percentage of respondents considering items necessary	Percentage of population unable to afford item
Television	56	1
Roast joint/vegetarian equivalent once a week	56	3
Presents for friends/family once a year	56	3
A holiday once a year (not with relatives)	55	18
Replace worn-out furniture	54	12
Dictionary	53	5
An outfit for social occasions	51	3

Source: Pantazis *et al.* (1999)

in the bottom fifth (£62,900 per year compared with £3,500) (see Figure 11.1).

While some nations do not exhibit the levels of social polarization apparent in the UK, it is important to note that levels of social inequality evident in many other nations are far in excess of this. However, in each and every case this social inequality is spatially expressed, whether as a division between rich and poor regions, an urban–rural divide or differences in wealth between cities. Yet some of the sharpest contrasts between wealth and poverty are *within* cities, and one of the major contributions geographers have made to debates surrounding poverty is to draw attention to the existence of areas of acute need (as

well as pronounced wealth) in urban areas. It has accordingly become something of a geographical cliché to juxtapose the assuredly affluent 'Westernized' city centres now typical of most of the world's most populous cities with the 'slum' dwellings that are often just a stone's throw away. For instance, profiling Jakarta (population 9 million), Cybriwsky and Ford (2001) contrast the city's Golden Triangle of prestigious residential districts (e.g. Cikini, Kuningan and Menteng) with the *rumah liar* ('wild houses') characteristic of its sprawling, chaotic *kampungs*. They thus conclude that Jakarta 'has extraordinary contrasts between the worlds of prosperity and poverty, and significant challenges ahead for continued

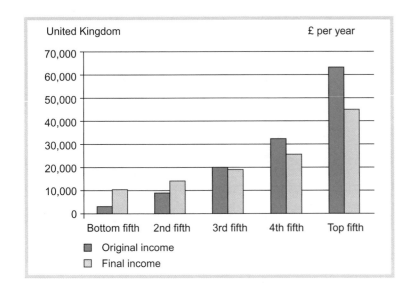

Figure 11.1
Average original and final income per household, 2002–2003.
Source: National Statistics (2004)

Table 11.2 Social characteristics of 12 'riot' estates and 20 'unpopular' estates from 1991 UK Census of Population

	British average (%)	20 'unpopular' housing estates (%)	12 'riot' housing estates (%)
Unemployment	10	34	31
Economic inactivity	36	44	45
Population under-16	19	31	31
Population under-24	33	46	48
Lone parents	4	18	15
Ethnic minority population	6	26	11
Pupils aged 15+ gaining 5+ GCSEs (grade A–E)	43	20	20

Source: A. Power and R. Tunstall, Dangerous disorder: Riots and violent disturbances in thirteen areas of Britain, 1991–92, Joseph Rowntree Foundation, 1997. Reproduced by permission of the Joseph Rowntree Foundation

development as a global metropolis' (Cybriwsky and Ford 2001: 209).

Yet such 'extraordinary' contrasts between landscapes of wealth and poverty are not just restricted to megacities or those cities living with the legacy of colonial rule. For example, most British cities are marked by contrasts between the affluent landscape of the central city and the 'pockets' of deprivation that surround it. Notable here are

Plate 11.1 The aftermath of joyriding: a housing estate in the UK. (Network Photographers/Michel Setboun)

Thematic Case Study 11.1

Geographies of institutional disinvestment

During the late 1990s, debates on poverty began to focus on the role institutional disinvestment played in exacerbating the problems encountered by those living in 'poor places'. One dimension of this was the withdrawal of financial services from poorer areas, with as many as 15 per cent of bank branches in the UK closed down in the early 1990s, the majority in low-income, inner city areas (Leyshon and Thrift 1994). Another was the closure of once-vibrant neighbourhood shops as retail chains concentrated on the development of large superstores catering to mobile and affluent consumers (Williams and Hubbard 2001). Given that many of these were located in out-of-town and edge-of-town locations, this left many neighbourhoods described as food 'deserts', where a range of affordable and varied food was available only to those who had private transport:

> Food deserts. . . are those areas of cities where cheap, nutritious food is virtually unobtainable.

Car-less residents, unable to reach out-of-town supermarkets, depend on the corner shop where prices are high, products are processed and fresh fruit and vegetables are poor or non-existent.
(*Independent*, 11 June 1997)

This has raised serious concerns about the health of those in poor places, with the lack of fresh fruit and vegetables and a reliance on fatty, fast foods implicated in the making of significant health inequalities. Confirming the intensity of food-access problems in some of the large local authority housing estates in British cities, Wrigley *et al.* (2003) noted significant improvements in diet following the construction of a new superstore in the Seacroft district of Leeds, an area that had previously suffered considerable retail disinvestment. This suggests that large-scale corporate intervention can significantly improve retail access and diet for people living in stigmatized neighbourhoods, albeit that small-scale, community-based initiatives might also provide a solution in some cases (Cummins and MacIntyre 1999).

those peripheral areas of local authority housing (built mainly in the 1950s and 1960s), which, in the 1980s and 1990s, came to be characterized by high rates of economic inactivity. Evidence from 20 of these 'unpopular' housing estates suggests they possess unemployment rates, on average, two and a half times higher than the surrounding urban area (Power and Tunstall, 1997), suggesting that they constitute pronounced spatial concentrations of poverty (Table 11.2).

The particular combination of problems encountered in such deprived estates created national headlines in the 1990s as confrontations between police and local youths (especially 'joy-riders' accused of stealing cars) escalated into fully-fledged riots (Plate 11.1). The spatial concentration of unemployment on these estates, especially among young men, has been hypothesized as the most important underlying cause of these disturbances. Figures for the 13 estates affected by serious disorder between 1991 and 1992 suggest that nearly 80 per cent of their occupants were either economically inactive or unemployed (Table 11.2). A

simultaneous withdrawal of economic infrastructure – including businesses, shops and banks (Case study 11.1) – meant that there was little money circulating locally. Neither did the mythical black market (an informal economy based on petty theft, drug dealing or cash-in-hand trading) offer any sort of escape for the majority of residents, given that black economies rarely flourish when there are few people who can afford to buy stolen goods, let alone legitimate ones (Green 1997).

Even so, it has been suggested that the spate of vehicle thefts occurring on such estates was not primarily economically motivated, as Campbell's (1993) vivid account of the Blackbird Leys estate, Oxford, suggests:

> The lads had stopped making cars, but that did not stop them stealing them. The collapse in manufacturing work for men was succeeded in Oxford by the rise in car theft, a crime that was emblematic of the Eighties and Nineties. . . . Car crime on the estate was about much more than trade, however. It was about a relationship between

young men, power, machinery, speed and transcendence. ... It was the spectacle of the displays that vexed and humiliated the police, uniting thieves, drivers and audiences in alliance against the authorities.

(Campbell 1993: 32–3)

The idea that car crime embodies a form of resistance suggests joy-riders were not motivated simply by greed or want, but by a more complex desire to reassert their masculinity by confronting authority. Such actions can, however, only be understood in the context of the economic transitions that led to significant changes in the relationship between masculinity, work, and domesticity on such estates. Here, Nayek (2003) suggests that the disappearance of a once-familiar **gender division of labour** organized around a male 'breadwinner' (and the expectation of women's domestic role) was the cause of much male anxiety in areas of high unemployment. With the shift to a service-based economy valuing 'feminized' attributes such as keyboard skills and communication proficiency over the robust 'masculine' qualities associated with the culture of manual labour, young men unwilling (or unable) to adapt to demands of service-based work found themselves increasingly cut off from the world of work. Without a regular source of income, or any real incentive to look for work, the traditional relationship between masculinity and spatiality broke down as these young men rarely left the estate for any particular reason. Yet, in the relatively secure space of a stolen car, Massey (1996) argues, masculine identities could be reasserted. Driving with speed – 'on the edge' – thus represented young men's attempt to reaffirm their masculinity, albeit in a manner that created conflict with other estate dwellers (as well as police).

While geographers have stressed that joy-riding and other uncivil actions are intimately connected to (gendered) processes of economic restructuring and institutional disinvestment, the media largely ignored such arguments in favour of a rhetoric depicting Britain's peripheral estates as breeding grounds of immorality, characterized by high rates of teenage pregnancy, widespread drug abuse, alcoholism and truancy:

The sense of decline and neglect in many of these areas is palpable: the built environment in many of these areas has now taken on all the classic, ominous characteristics (boarded-up windows, barbed wire surrounds) of the enclaves of high crime and violence associated with Los Angeles and its ghettos in the months leading up to the 1992

riots. Public space is often colonised by young men in baseball caps and cheap khaki.

(Taylor 1997: 6)

While there are as many as 2,000 local authority estates in Britain, accounting for around two million people in total (Goodwin, 1995), the term 'no-go' estate became a convenient way for the media and public alike to label a wide variety of social environments (only a small number of which experienced unrest). Such labelling exacerbated neighbourhood decline: those who could, moved out, leaving behind only the most vulnerable (Hastings and Dean 2003). Those left were consequently caught in a debilitating cycle of labelling and marginalization:

Labelled as failures, people accept and internalize this negative image. Outsiders – professionals, politicians, the media – reflect, reinforce and magnify that image. People expect to be treated badly and their image of themselves and the estate takes a further battering.

(Taylor 1998: 821)

Such arguments support Shields's (1991: 5) view that marginal places 'carry the image and stigma of their marginality which becomes indistinguishable from any empirical identity they might have had'. In this sense, it is clear that media stereotypes play a crucial role in creating and perpetuating social and spatial inequalities.

11.2 Social exclusion and stereotyping

Whereas poverty is visible and (perhaps) readily quantifiable, exclusion refers to processes that are sometimes less tangible. Broadly defined, the socially excluded are those unable to participate in activities taken for granted by the majority. Winchester and White (1988) suggest that this group includes the unemployed, the impoverished elderly, lone-parent families, ethnic minorities, refugees and asylum seekers, the disabled, illegal immigrants, the homeless, sexual minorities, prostitutes, criminals, drug-users and students. However, Winchester and White stress that exclusion – like poverty – is a relative concept, and that their list will not be applicable in all contexts. Furthermore, they stress that these categories are not mutually exclusive, with some individuals belonging to more than one excluded group. Even so, Winchester

Plate 11.2 Barriers to participation. A physically disabled demonstrator returns to her wheelchair, after crawling across the Plaza of the US Supreme Court. (Associated Press/Gerald Herbert)

and White identify a wide range of excluded populations, with the bracketing of students together with criminals and drug-users raising the puzzling question: what, if anything, unites these disparate groups? The answer is not necessarily that they are prone to poverty (although deprivation is a common experience for excluded groups). Rather, it is that their lifestyles lie outside the norms of society, making them unwelcome visitors in many of the spaces that are the loci of mainstream social life (Philo 2000).

In this sense, social exclusion has important non-economic underpinnings. Recognizing the multiple meanings of exclusion, many social geographers have incorporated a consideration of processes of sexism, disabilism, ageism, homophobia and racism into their studies of marginalisation. Collectively, such work has suggested that, while these exclusionary processes may be shaped by the economic imperatives of capitalism, they are analytically distinct. For example, Gleeson

(1999) suggests that people with a physical impairment or mobility problem found themselves excluded from spaces of capitalist production as a factory system became dominant in the nineteenth century, yet notes that they had been subject to exclusion from street spaces long before this (their 'imperfect' bodies provoking considerable anxiety among able-bodied citizens). In contemporary society, people with disabilities remain excluded from many workplaces, as well as those spaces of consumption and leisure designed around an able-bodied ideal (Plate 11.2). Here, the portrayal of the disabled as representing what Shakespeare (1994) describes as the 'imperfect physicality' of human existence is no doubt a major factor, with the able-bodied remaining anxious about those who are visibly different.

Hence, while social exclusion involves the physical exclusion of individuals from the social institutions, rituals and practices enjoyed by dominant groups, it

253

also has a symbolic dimension, being created in the realms of *representation*. Here, the term representation is taken to encompass the wide range of media – such as films, TV, Internet sites and newspapers – through which we come to understand the world and our place within it (Woodward 1997). Inevitably, such media are responsible for the creation of stereotypes that provide a partial and simplified view of the lifestyles of heterogeneous social groups. Although most people do not necessarily accept these stereotypes uncritically, they inevitably find themselves drawing on them in their everyday life. What is especially significant about these stereotypes is that they are ideological in nature, in the sense they are generally created by (and in the interests of) dominant social groups – typically white, able-bodied, heterosexual, middle-class males.

A vivid example of how such stereotyped images contribute to social and spatial exclusion is provided by the media reporting of HIV/AIDS from the mid-1980s onwards. In the ensuing moral panic that followed diagnosis of the virus and its mode of transmission, a sometimes hysterical media began to focus on the groups most readily identified as at risk – non-white ethnic minorities, intravenous drug users and gay men (Watney 1987). The presence of HIV infection in these groups was generally perceived to be no accident, but portrayed as a condition affecting those whose 'inner essence' diverged from that of normal society (and whose lifestyles were judged to be incompatible with the maintenance of 'family values'). Hence, sensationalist stories of sexual immorality, irresponsibility and wilful hedonism among those groups most affected by the virus were used to create social barriers between the 'healthy' and those regarded as sexually promiscuous, socially irresponsible and unclean.

Consequently, this symbolic marking of HIV-infected groups as deviant informed their social exclusion, encouraging widespread discrimination, prejudice and neglect. As Wilton (1996) has shown, this had clear spatial effects, with those infected with HIV exhibiting 'diminishing geographies' as their

Table 11.3   Relative acceptability of human service facilities, based on a US survey of 1,326 respondents

Facility type	Mean acceptability, 1 = low, 6 = high
School	4.75
Day care centre	4.69
Nursing home for elderly	4.65
Medical clinic for allergies	4.40
Hospital	4.32
Group home for mentally retarded	3.98
Alcohol rehabilitation centre	3.80
Homeless shelter	3.73
Drug treatment centre	3.61
Group home for mentally disabled	3.51
Group home for people with depression	3.47
Mental health outpatient facility	3.45
Independent apartment for mentally disabled	3.30
Group home for people living with AIDS	3.20

Source: Takahashi and Dear (1997: 83)

access to the workplace, the home and the street was subject to increasing constraint. In his analysis, Wilton stresses that these constraints were not a product of the physical onset of AIDS but of its social stigma. Symptoms of this stigma have included employers refusing to take on those with HIV, dentists refusing to treat them and, as Wilton relates, community groups opposing the construction of AIDS hospices in their neighbourhood.

Such neighbourhood opposition to community facilities and welfare services is so widespread in Western society – particularly in suburban landscapes – that few stop to question why people might object to such developments in their neighbourhood. Certainly, many NIMBY (Not In My Back Yard) campaigns are fought with reference to the detrimental environmental impacts of such developments (such as noise or air pollution). Such impacts are, to an extent, quantifiable; what is less measurable is the concern that homeowners have about the arrival of stigmatized populations. However, for Takahashi and Dear (1997), community opposition to facilities for those living with AIDS/HIV is indicative of the more general antipathy displayed toward Other populations. Their survey of homeowners in US cities reveals a 'continuum' of acceptance, where facilities for populations depicted as 'different' (such as homes for the elderly) are regarded more favourably than those for populations stereotyped as 'dangerous' or 'deviant' (such as those living with HIV) (Table 11.3). This frequently results in the concentration of facilities in inner city areas where home ownership rates are low and community opposition is least vocal (Case study 11.2).

Thematic Case Study 11.2

Community opposition to asylum seekers

In accordance with the United Nations' protocol on refugees, the UK is obliged to offer asylum to those in fear of persecution on the basis of their race, religion or culture. However, in the late 1990s, a mounting backlog of asylum applications, together with media stories of 'bogus' asylum seekers, combined to demonize this group in the public imagination. One strategy adopted by the government as it sought to diffuse concern about the 'wave' of asylum seekers 'flooding' Britain was the dispersal of asylum seekers away from the Channel ports where their visibility had been exploited by right-wing groups intent on exploiting racist fears of difference. As part of this process of planned dispersal, the Home Office proposed the construction of accommodation centres for asylum seekers in a number of rural locales. However, this policy often met with considerable local opposition. For example, in February 2002, residents of Newton in Nottinghamshire learnt that the Home Office was planning to convert nearby RAF Newton into an accommodation centre for asylum seekers. A series of protests organized by the Newton Action Group expressed vehement opposition to the proposal, with campaigners arguing that the development would cause increases in crime, vandalism and social disorder, and that property prices would fall. One resident claimed that parents would be 'terrified of letting their children play out', while another asked 'Can we be assured there are no child abusers, drug addicts or convicted rapists among them?' (see Plate 11.3). The fact that this is by no means an isolated example demonstrates how effective the media have been at exploiting public anxieties about the threat that asylum seekers pose to the sanctity and purity of the nation. As an extremely vulnerable group, asylum seekers have been less able than some to oppose these racist myths, and many have reported harrowing experiences of discrimination. For example, in the Sighthill area of Glasgow, the arrival of 1,200 Kurdish and Kosovan refugees in 2001 prompted a number of vicious assaults by local white youths who claimed the asylum seekers were being shown favouritism by the local council. Following a series of attacks, it took the murder of 22-year-old Kurdish asylum-seeker Firsat Dag to draw attention to the everyday racism fuelled by a media rhetoric that depicted asylum seekers as 'social security scroungers'.

Thematic Case study 11.2 continued

Plate 11.3   Protests against a planned asylum centre, Newton, Nottinghamshire 2002. (Nottingham Evening Post/Steve Footit)

This 'ghettoization' of facilities for those living with HIV, the homeless, asylum seekers and populations dependent on welfare demonstrates that the geographies of marginal groups are – to a lesser or greater extent – the product of dominant **imaginary geographies** casting minorities as 'folk devils' who need to be located elsewhere. As Sibley (1995: 49) contends, this elsewhere might be nowhere, as when the genocide of gypsies and Jews was undertaken by the Nazis, or it might simply be a space 'out of sight' of mainstream populations (such as the red-light districts that are the focus of sex work in many British cities – see Hubbard, 1999). In *Geographies of Exclusion*, Sibley (1995) offers a theoretically informed account of how these imaginary geographies fuel exclusionary practices. Drawing particularly on psychoanalytical ideas about the importance of maintaining self-identity (literally, maintaining the boundaries of the Self), he argues that the urge to exclude threatening Others from one's proximity is connected to ideas about the importance of bodily cleanliness, many of which may be inculcated in early infancy. Developing psychoanalytical ideas, Sibley details how fears of the Self being defiled are consequently projected (or mapped) onto those individuals and groups depicted as polluting or dirty. In turn, Sibley argues that individuals adopt a series of exclusionary strategies designed to retain 'psychic distance' between themselves and **abject** groups, building symbolic, psychological and physical boundaries between Self and Other. This engagement with psychoanalytical theory offers geographers an important insight into processes of marginalization, suggesting that the exclusion of Other groups can only be understood with reference to the deep-seated urge people have to *purify* their surroundings.

11.3 Racial exclusions

Foremost among the 'geographies of exclusion' explored by social geographers are those underpinned by racist processes of segregation. Unpacking the imaginary geographies that cast minorities to the margins, such research has exposed the way that media negativity identifies specific ethnic groups as threats to the social body, and hence as a group that needs to be excluded from mainstream space (see Smith 1989, Anderson 1991, Dunn 2001). In the urban West, such racist imaginaries have tended to result in the marginalization of those immigrant groups depicted as

a threat to white privilege. Sibley (1999) accordingly argues that the equation of whiteness with purity, order and cleanliness in northern European cultures has been responsible for the creation of negative stereotypes of (for example) Africans, Afro-Caribbeans, people from the Indian subcontinent and gypsies. He contends that each of these groups has consequently been allocated only a marginal role in national cultures because of their 'threatening' nature. In turn, such marginalization fuels spatial segregation, with the creation of ethnic minority enclaves in the poorest neighbourhoods of many European cities being a product of personal prejudice in the job and housing markets that is fuelled by negative media stereotypes of minority groups (Wacquant 1993). In other contexts, this segregation has taken more institutional forms: for example, under apartheid in South Africa, the Group Areas Act prevented black people from straying into residential areas declared 'white', bequeathing cities that are sharply divided on ethnic lines (see Case study 11.3).

While segregationist policies have long been illegal in the USA, whiteness remains profoundly encoded in the city–suburb distinction, with the suburbs seen to provide a refuge for the white middle-classes from the dirt, disorder, and, above all, criminality of the coloured inner city (Valentine 2001). Indeed, the imagined association between areas of predominantly black residence and the welfare-dependent 'ghetto' has become so pernicious in the US that it dominates white assumptions about Afro-American lifestyles, creating stereotypes that often bear little resemblance to black people's urban experiences. Events such as the Rodney King riots of 1992 (prompted by the acquittal of a white police officer for the filmed beating of a black Los Angeles resident) consolidated the reputation of the US ghetto as a crime-ridden environment; although the areas of unrest were typically those with the poorest income levels, high population densities and high school dropout rates, it was the 'racial character' of these areas that was highlighted in the subsequent reporting (Smith 1995).

This association between US minority populations and specific criminalized places has been cemented over the years in those media that have used 'ghetto' as a coded term for the imagined deviance of black people. As McCarthy *et al.* (1997) argue, the American middle-classes tend to know more about inner-city black people through long-distanced but familiar media images than through personal everyday interactions. Even 'new wave' black cinema such as

Regional Case Study 11.3

Ethnic division in apartheid South Africa

Apartheid-era South Africa is often cited as representing an extreme example of socio-spatial segregation, with a white minority population passing a series of acts in the mid-twentieth century designed to circumscribe the mobility of coloured, Indian and African people. Under apartheid laws, non-whites were denied access to major cities except if in possession of a pass that indicated they were gainfully employed in the city. As such, while the ruling white minority depicted non-whites as racially inferior, they recognized their importance as a source of unskilled and semi-skilled labour. Initially, non-white workers were housed in barracks or municipal hostels where they were subject to regimentation designed to curb the 'animal instincts' of natives and instill the importance of 'the proper habits of work and life' (Popke, 2001: 740). The understanding that non-whites possessed useful and productive bodies, yet were prone to indolence and incivility, was clearly connected to colonial representations that associated non-whiteness with savagery, animality and nature, characteristics that were repressed in 'civilized' Western cultures (Anderson 2000).

However, the continued growth of non-white 'shanty towns' around the periphery of major cities fuelled white fears of '*swaart gevaar*' (black danger), and led to national legislation designed to construct buffer zones between the white city and non-white townships. Both physically and symbolically on the margins of South African cities, the townships thus became abject landscapes, depicted by the white media as spaces of vice, disease and lawlessness. Repeated attempts at creating racially sanitized, planned townships were thus part of an explicit attempt to modernize South African society while maintaining the distance between poorer non-white groups and the affluent white city.

Rather than suggesting that the apartheid city represents a racist distortion of the processes of segregation evident elsewhere in the world, more recent commentaries on the apartheid era suggest it illustrates the racist conceits that underpinned twentieth century modernism (Robinson 1997, Popke 2001). As such, it is possible to identify planning and 'improvement' schemes as underpinned by racist imaginaries throughout the world (and not just in those nations with a colonial history). Indeed, despite the abolition of apartheid Group Area Acts, racist fears of difference continue to fuel socio-spatial exclusions in South Africa, with the growth in gated communities being merely one manifestation of this (Jurgens and Gnad 2002).

Menace II Society, Boyz 'n The Hood, Jungle Fever and *Straight Outta Brooklyn*, which has sought to show something of the reality of everyday life for Afro-Americans, has ultimately served to stereotype such places as landscapes of endemic violence and drug-dependence (Benton 1995). In this context, Spike Lee's 'realist' account of drug-dealing, *Clockers*, can be seen as a corroboration of dominant white myths that imagine drugs, guns and criminality to be part of everyday ghetto life. Such white fears and fantasies about the black inner city are acidly invoked in Wolfe's (1988) fictional account of the Wall Street trader, Sherman McCoy, who one night takes the wrong turning on his drive home only to be confronted with

a side of New York that he remains cosseted from in his 'yuppie' housing development:

> At the next corner, he turned – west, he figured – and followed that street a few blocks. There were more low buildings. They might have been garages, they might have been sheds. There were fences with spirals of razor wire on top. But the streets were deserted, which was okay, he told himself. Yet he could feel his heart beating with a nervous twang. Then he turned again. A narrow street with seven or eight storey apartment buildings; no sign of people, not a light in a window. The next block, the same. He turned again, and as he rounded the corner –

astonishing. Utterly empty, a vast open terrain...
here and there were traces of rubble and slag. The
earth looked like concrete, except it rolled down this
way, and up that... the hills and dales of the Bronx
... reduced to asphalt, concrete and cinders.

(Wolfe 1988: 65)

This depiction of the Bronx as a wasteland inhabited
by 'lots of dark faces' again reminds us of the ways in
which fears of difference (in this example, based on
skin colour) can provoke anxiety and feed the urge to
exclude. From the perspective of the white narrator in
Wolfe's book, the urban decay of the Bronx mirrors his
perceived view of the city's black population as Other:
they are street people rather than the 'air people' who
inhabit his social world (adopting Raban's, 1990,
memorable description of those who can afford to
distance themselves from marginal places).

For many Afro-Americans, ghetto life is similarly
typified by anxiety, but an anxiety originating from
problems of poverty (in 1990 the average white
household income was 1.72 times higher than that of
black households, with one-third of black families
living below the US poverty line). As the renowned
feminist and cultural theorist bell hooks relates, the
experience of living in the black ghetto is one that
consistently reinforces feelings of subordination,
marginality and inferiority:

To be in the margin is to be part of the whole but
outside the main body. As black Americans living in
a small Kentucky town, the railroad tracks were a
daily reminder of our marginality. Across those
tracks were paved streets, stores we could not enter
and restaurants we could not eat in, and people we
could not look directly in the face. Across these

Table 11.4 The five most segregated cities in the USA, measured by white/Afro-American Dissimilarity Index, 1980–2000

MSA/PMSA name	Year	Total population	Black or Afro-American population	Dissimilarity Index (D)
Detroit, MI	2000	4,441,551	1,045,652	0.846
	1990	4,266,490	942,794	0.874
	1980	4,387,613	892,347	0.874
Milwaukee-Waukesha, WI	2000	1,500,741	245,151	0.818
	1990	1,432,149	197,183	0.826
	1980	1,396,972	150,662	0.839
New York	2000	9,314,235	2,451,277	0.810
	1990	8,546,583	2,249,997	0.813
	1980	8,274,039	1,907,434	0.812
Newark, NJ	2000	2,032,989	473,829	0.801
	1990	1,915,600	424,037	0.825
	1980	1,963,131	409,244	0.827
Chicago, IL	2000	8,272,768	1,602,248	0.797
	1990	7,410,858	1,427,357	0.838
	1980	7,175,269	1,429,217	0.878

1 = complete segregation, 0 = complete integration

Source: US Census Bureau, www.census.gov/hhes

tracks was a world we could work in as maids, as janitors, as prostitutes, as long as it was in a service capacity. We could enter that world, but we could not live there. We always had to return to the margin, to cross the tracks, to shacks and abandoned houses on the edge of town.

(hooks 1984: 9)

Until the 1960s it was not uncommon to find specific ordinances and zoning laws controlling the development of black neighbourhoods. Although such laws have long since been overturned, the legacy of these remain in the racist practices of mortgage financiers, banks and estate agents, who may seek to maintain the blackness or whiteness of particular neighbourhoods in the interests of maintaining house prices (Short 1996).

It is thus unsurprising that the Otherness of the Afro-American population is mirrored in their continuing segregation, with black people remaining more spatially concentrated than any other US ethnic minority. However, Dissimilarity Indexes exploring how the Afro-American population is distributed relative to the white population suggest that their segregation has actually declined over the last twenty years (see Table 11.4). Yet the declines remain modest, and some metropolitan areas actually experienced an *increase* in residential segregation over the period 1980–2000 (the seven highest increases all in the Southern USA, including Columbus, GA, Goldsboro, NC, Athens, GA, and Danville, VA).

While the disappearance of work from US inner cities (and the associated out-migration of the black middle class) has been postulated as the major cause of concentrated black poverty in the USA, Mohan (2000) argues that such factors alone cannot explain the continuing segregation of Afro-Americans. Instead, he suggests it is necessary to consider the wholesale disinvestment that has accompanied the economic decline and stigmatization of Afro-American communities. Coining the phrase 'desertification' to describe the planned shrinkage of essential health, welfare and emergency services in the Bronx from the 1970s onwards, Wallace and Wallace (1995, 2000) have suggested that such disinvestment has been part of a deliberate attempt to cut back on public spending by withdrawing services from the very districts most in need. As they detail them, the long-term consequences have often been catastrophic (including increases in infant mortality rates, low birth weights, cirrhosis, TB, and AIDS-related deaths). At the same time, many

areas stigmatized as pockets of black poverty have become insurance no-go zones where small businesses find it impossible to get insurance because premiums are so high, and large percentages of the population are denied car insurance because of excessive costs. It is for such reasons that some commentators feel justified in speaking of the 'apartheid' of American cities (Wallace and Wallace 1998).

11.4 Gentrification: reclaiming the margins?

As already noted, all societies possess marginal places, whether these are spaces of poverty, sites of illicit activities or areas occupied by stigmatized populations (and perhaps all three). Often, the boundaries of such places have remained consistent over time, even if the individuals and groups they contain have changed (an example here might be the persistence of the inner city as a disordered urban zone). This is clearly connected to the locus of power in society, with powerful social groups able to physically and symbolically claim the centres, marginalizing the less powerful:

Centre/periphery distinctions tend frequently to be associated with endurance in time. Those who occupy centres establish themselves as having control over resources which allow them to maintain differentials between themselves and those in peripheral regions. The established may employ a variety of forms of social closure to sustain distance from others who are effectively treated as outsiders.

(Giddens 1991: 131)

In this chapter we have already examined numerous processes which tend to spatially isolate and exclude less powerful groups – e.g. discrimination in the housing market, institutional racism, NIMBY politics and so on. While such processes tend to carve up space in favour of powerful white elites, it is dangerous to suggest that poorer groups are unable to resist these processes. For example, some excluded groups may well resist forms of social closure by **transgressing** into the spaces of the powerful, challenging taken-for-granted expectations about where they should locate (Cresswell 1996). Such transgressions may trigger a **moral panic** (such as that which surrounded asylum seekers in the late 1990s) and encourage the introduction of new forms of social and spatial control. However, others may subtly change the social order (for instance, gay pride marches in Western cities

Plate 11.4 Castro, San Francisco: a centre for gay resistance? (Getty Images)

have often drawn attention to homophobia in society, encouraging the repeal of discriminatory legislation; while Muslims have pursued a variety of public actions designed to topple the Islamophobia that is rampant in the Western nations).

Accordingly, places on the margin can be seen as sites from which the relatively powerless can organize themselves into self-supporting cultures of resistance and cooperation. For example, so-called 'gay ghetto' areas in the US (e.g. Castro in San Francisco, or West Hollywood in Los Angeles) have been transformed from marginal spaces of persecution to relatively affluent centres of gay cultural life through political organization, creativity and activism, creating new gay identities in the process (Forest 1995) (Plate 11.4). Likewise, alternative economies thrive in many racialized areas (particularly in the cultural industries – food, music, fashion, arts and media), allowing ethnic entrepreneurs to bring different values and ideas to the attention of wider audiences, making them more mainstream in the process. This gradual **demarginalization** process may, over time, bring excluded populations into the mainstream.

On the other hand, it is worth noting that

mainstream society itself may attempt to claim back the margins, transforming landscapes of poverty and decay into prestigious (and *ordered*) space through selective investment and redevelopment. Such efforts are typically implemented under the rubric of state-led urban policy, with such investment in the physical and economic infrastructure being justified as attempts to improve the living conditions of the worst-off in society. These policies are underpinned by an increasingly complex diversity of agencies and partnerships, blurring the distinction between public and private sectors as property developers and policy-actors work in tandem to regenerate deprived areas. However, the net result of such improvement policy is often the onset of **gentrification**. A much-debated phenomenon in social geography, gentrification is the process by which poor neighbourhoods are transformed by an influx of capital in the form of affluent homebuyers and renters. In some instances, this involves individual gentrifiers buying existing properties and transforming them (often with governmental subsidy); in others, it involves the wholesale replacement of low-income housing with prestigious apartments. Either way, the net result is

Regional Case Study 11.4

Revanchist New York

As a leading writer on the uneven geographies of capitalism, Neil Smith has theorized gentrification as an attempt by developers to take advantage of the rent gap between actual and potential land value. Typically investing in areas ravaged by de-industrialization and the flight of wealthy residents, developers hope to make quick profits by attracting the wealthy back. Given that gentrifiers are prepared to pay a premium to participate in a putative 'urban renaissance', gentrification can be extremely lucrative for developers. For gentrification to succeed, however, it is necessary to remove 'Other' groups from spaces earmarked for gentrification, so that potential gentrifiers view them as safe neighbourhoods in which to live and invest.

It is here that Smith suggests the role of the local state is crucial. Smith (2002: 442) points to this when he details how squatters, the homeless, squeegee merchants and 'street people' were ruthlessly dealt with in New York following the election of Mayor Rudolph

Giuliani and appointment of Police Commissioner William Bratton. Espousing a rhetoric of Zero Tolerance for miscreants, these figures were pivotal in identifying the urban disadvantaged as a disorderly population. This urge to tame urban disorder was to trigger notorious police brutality against minorities, justified with reference to the need for improved quality of life, but actually intended, Smith argues, to make particular areas of the city safe for corporate gentrification (and the associated invasion of upper income groups). One notable example of this process was the wholesale removal of the homeless and their supporters from Tompkins Square Park at a time when developers were seeking to sell the Lower East Side as a space for 'family' residence (Plate 11.5). The seizure and subsequent 'purification' of the park (and surrounding areas) serves to underline how exclusionary urges may turn a genuinely public space into a space reserved for those who accord with mainstream ideas of living and working (see also Mitchell 1996).

Plate 11.5 The gentrification of New York. Ironically, the museum that preserves the Lower East Side's immigrant history is seeking to expand by evicting residents of a renovated tenement block. (Associated Press/Beth A. Keiser)

often one of displacement, with former residents frequently unable to afford to live in redeveloped gentrified areas.

The redevelopment of London Docklands in the 1980s provides ample evidence for this displacement thesis, with many local residents (some of them former dock employees) unable to afford the 'yuppie' housing developed under the auspices of the London Docklands Development Corporation. Similarly, Smith (N. Smith 1996) has graphically documented how the 'improvement' of New York's Lower East Side resulted in the systematic eviction of the homeless population from a number of shelters, parks and streets (see Case study 11.4). Smith sees this as symptomatic of the continuing fear amongst white middle- and upper-class citizens of Other populations, with the contemporary city characterized by increasingly vicious reactions against minorities including the unemployed, sex workers, the homeless and immigrants. Again, these tendencies are not merely isolated to the urban West, and it is easy to find instances of relatively powerless populations being displaced throughout the global South as shantytowns or informal settlements are bulldozed to make way for new prestige developments, shopping centres and highway developments. Though often resisted, the net result is the corporate gentrification of city centres worldwide, to the extent that the CBD of Manila now looks very much the same as the CBD of Sydney (Winchester *et al.* 2003). While such processes of corporate gentrification may ultimately attract mobile consumer capital, and tie Third World cities into a network of world cities, the consequence is of course the production of new landscapes of exclusion, with displaced residents and workers being consigned to the marginal spaces that exist beyond the gentrified core.

11.5 Conclusion

Social inequality is just one of many issues explored by social geographers. Nonetheless, it is one of the most important issues facing society in the twenty-first century. As such, geographers' research on residential segregation (e.g. Wacquant 1993), structural changes in urban retailing (e.g. Williams and Hubbard 2001), inequalities in health and well-being (e.g. Dorling and Shaw 2000), income polarization (e.g. Pinch 1994) and crime (e.g. Fyfe 2004) provides valuable insights into the particular combination of problems that beset our least well-off and most vulnerable populations. In many cases, such research also helps identify solutions to the problems faced by those living in marginal spaces. But work on geographies of social inequality does not just make a *practical* contribution to the policy-debates surrounding exclusion and polarization, it also makes a key contribution to *theoretical* debates about the role of social processes in constructing categories of identity and difference (such as class, gender, sexuality and race). In the final analysis, concepts such as exclusion and resistance are not solely of relevance to the issues of social inequality discussed in this chapter: they are fundamental to making sense of the diversity that characterises the contemporary world.

Learning outcomes

Having read this chapter, you should be able to:

- Understand that society and space are divided along axes of power that serve to segregate stigmatized 'Other' groups from mainstream society.

- Appreciate the complexity of the processes encouraging this segregation, including people's desire to distance themselves psychically and physically from populations they regard as threatening or polluting.

- Identify the key characteristics of spaces of exclusion, which may lack economic, social or political infrastructure, and be typified by high crime rates, health problems and a poor quality of life.

- Recognize the importance of stigmatization in creating a downward spiral of decline; once a place has obtained a reputation as a space of exclusion, it is unlikely to attract investment of a type that will benefit its inhabitants.

- Understand that efforts to reclaim spaces of exclusion in the name of improvement often instigate gentrification – a process that often triggers new forms of spatial purification and exclusion.

Further reading

Caldeira, T. (2001) *City of Walls: Crime, Segregation and Citizenship in São Paulo*, University of California Press, Berkeley. A fascinating account of the fear and anxiety that divides this Brazilian city which makes frequent comparisons with the forms of racism and hate evident in North American cities.

Cresswell, T. (1996) *In Place/Out of Place*, University of Minnesota Press, Minneapolis. The chapters of this lucid and lively book focus on the way that social orders are challenged by people acting 'out of place' (and how authority responds).

Hamnett, C. (2003) *Unequal City: London in the Global Arena,* Routledge, London. A detailed examination of the economic and social changes that have taken place in London over the last 40 years, noting significant spatial shifts in occupational structure and income. Though generalizing from a world city is dangerous, Hamnett's case study makes a number of important points about the geographical imprints of polarization.

Holloway, L. and **Hubbard, P.** (2001*) People and Place: the Extraordinary Geographies of Everyday Life*, Prentice Hall, London. Chapters 6, 7 and 8 of this textbook offer a fuller discussion of the issues discussed in this chapter.

Sibley, D. (1995) *Geographies of Exclusion: Society and Difference in the West*, Routledge, London. An important theoretical statement on geographies of exclusion, bringing geographical ideas into dialogue with psychoanalytical literatures. The book is illustrated throughout with examples ranging from the marginalization of gypsies to the exclusion of non-white voices from the spaces of the academy.

Smith, N. (1996) *The New Urban Frontier: Gentrification and the Revanchist City*, Routledge, London. Teases out many of the links between gentrification and the purification of space, with Smith's discussions of the 'regeneration' of New York having taken on added resonance in the post-September 11th era.

Valentine, G. (2001) *Social Geographies: Society and Space*, Prentice Hall, London. Issues of exclusion are addressed in this textbook through a series of thoroughly readable chapters focusing on different geographical scales (with those on 'The Street' and 'The City' addressing many of the themes explored in this chapter).

For annotated, clickable weblinks and useful tutorials full of practical advice on how to improve your study skills, visit this book's website at www.pearsoned.co.uk/daniels

CHAPTER 12

Geography, culture and global change

CHERYL McEWAN

Topics covered

- A definition of culture

- The 'cultural' and 'spatial turns' in the social sciences

- An evaluation of the extent of cultural globalization

- The impacts of globalization on local cultures

- An exploration of the concepts of multiculturalism and hybrid cultures

- Progressive ways for geographers to think about culture

Culture is a word on everybody's lips these days. Hardly a moment seems to pass when we do not hear on the radio or television, see in newspapers and magazines, or read in academic texts some account of the world of the cultural. Governments at all levels announce cultural policies and provide funding for cultural activities; intellectuals announce the need for cultural initiatives or bemoan the loss of traditional cultural values; famous cultural icons – musicians, artists, novelists – themselves culture makers, are increasingly sought out for their opinions on the state of the world.

(Kahn 1995: ix)

12.1 What is culture?

This chapter explores some of the challenges posed by and for culture in the twenty-first century. However, before considering these challenges, it is first necessary to define what is meant by culture. This itself is an extremely complex and difficult task. By the 1950s, for example, there were over 150 different academic definitions of culture. As Mike Crang (1998: 1) argues, despite sounding like the most airy of concepts, culture 'can only be approached as embedded in real-life situations, in temporally and spatially specific ways'. Cultures are part of everyday life. They are systems of shared meanings often based around such things as religion, language and ethnicity that can exist on a number of different spatial scales (local, regional, national, global, among communities, groups, or nations). They are embodied in the material and social world, and are dynamic rather than static, transforming through processes of cultural mixing or transculturation (discussed below).

Cultures are also socially determined and defined and, therefore, not divorced from power relations. Dominant groups in society attempt to impose their definitions of culture and these are challenged by other groups, or sub-cultures. The latter might include various types of youth cultures, gang cultures, and different ethnicities or sexualities, where identities are organized around different sets of practices and operate in different spaces from dominant cultures (Crang 1998, Skelton and Valentine 1998). Culture makes the world meaningful and significant. As Phil Crang (1997: 5) argues, we should think of culture as a process we are all involved in rather than as a thing we all possess (see Spotlight box 12.1).

12.1.1 The 'cultural turn'

As Kahn suggests, culture has generated a great deal of interest in recent years, for academics, policy-makers, and at the popular level. Geographers have turned their attention towards cultural explanations of global, national and local phenomena. They have explored issues such as the cultural embeddedness of economic processes (e.g. McDowell 1997), the relationship between cultures, identities and consumption (see Chapter 16), and cultural constructions of social relations of gender, ethnicity and class that shape people's lives (e.g. Radcliffe and Westwood 1996). However, the current popularity of culture is not simply a trend in academe, but is reflective of a broader cultural turn in (Western) society as a whole.

The world has changed fundamentally in the past two decades and these changes are deeply cultural in character. For example, enormous changes have occurred in 'advanced' economies since the early 1980s (the decline in manufacturing, the growth of services, the feminization of the workforce, increased flexibility

---

Spotlight Box 12.1

Cultures: a summary

- The systems of shared meanings which people who belong to the same community, group or nation use to help them interpret and make sense of the world, and to reproduce themselves.
- These systems of meanings include language, religion, ethnicity, custom and tradition, and ideas about 'place'.
- Cultures cannot be fixed, but shift and change historically through diffusion and transculturation.
- Culture is a process, rather than a thing, but is embodied in the material and social world.

- Cultures include those social practices which produce meaning, as well as those practices which are shaped by those shared meanings.
- Cultures give us a sense of 'who we are', 'where we belong' – a sense of our own identity and identity with others.
- Cultures are, therefore, one of the principal means by which identities are constructed.
- Cultures are not divorced from power relations.

Source: Adapted from Hall (1995)

– all characteristic of 'post-Fordism' as discussed in Chapter 3, pp. 79–81, and Chapter 14, pp. 325–35). However, as S. Hall (1996: 233) argues, if 'post-Fordism' exists, it is as much a description of cultural as of economic change. Modern consumption depends overwhelmingly on image (for example, the marketing of food and drink products and fashion clothing) (see Plate 12.1). Intensifying movements around the world of images, symbols, modes of thought and communications are unparalleled. As computer technology, video imagery and electronic music demonstrate, the material world of commodities and technologies is profoundly cultural. In addition, culture has become increasingly commodified; in other words culture is being translated into material goods that can be marketed and sold. Zukin (1995: 263) refers to the example of the annual Gay Games in New York to highlight how what was a political event has been turned into a cultural commodity. The Games draw the support of large corporate sponsors eager to capture the 'pink dollars' of a supposedly affluent gay population. The mainstream press covers the games as a cultural, rather than a sporting or a political event. As Zukin argues, culture now has great appeal – culture sells.

12.1.2 The 'spatial turn'

There has also been a 'spatial turn' in explanations in cultural and social theory. It is commonly accepted that the world is changing fast, and the rate of this change is probably greater than ever before. New technologies such as the Internet and satellite communications mean that the world is becoming more global and more interconnected. The increased speed of transport and communications, the increasing intersections between economies and cultures, the growth of international migration and the power of global financial markets are among the factors that have changed everyday lives in recent decades. There is no historical equivalent of the global reach and volume of 'cultural traffic' (Held *et al.* 1999) through contemporary telecommunications, broadcasting and transport technologies. The challenge for geographers is to find ways of understanding and interpreting these changes.

Culture can be said to operate at three spatial scales: local, national and global. Two main interpretations have dominated discussion. The first highlights the global aspects of change. At its simplest, this approach suggests that it is possible to identify processes of

Plate 12.1 Image marketing is central to popular consumption: the world of commodities is profoundly cultural. (Corbis/Benjamin Rondel)

cultural homogenization – the idea that everywhere is becoming the same – dominated by the USA and most easily recognized in terms such as 'Coca-Colonization', 'McDonaldization' and 'Hollywoodization' (Cochrane 1995: 250) (see Chapter 16, pp. 365–72). This *cultural globalization* involves the movement of people, objects and images around the world, through telecommunications, language, the media industries, radio and music, cinema, television and tourism. The second interpretation places emphasis on the local and the localization of people's everyday lives and experiences. Instead of homogenization, emphasis is placed on the diversity of culture, on the ways in which global icons such as Coca-Cola or McDonald's are reinterpreted locally so that they take on different meanings in different places. Geographers are responding to these changes by exploring the *interconnectedness* of global and local processes. For example, although the same event can be witnessed simultaneously around the world (e.g. an incident broadcast in a CNN news report, or an international sporting event), this event will be interpreted differently in different places. Furthermore, locality does not necessarily refer to the opposite of globality. For example, some environmentalists imagine the world as a locality, a 'global village'. Cultural theorists have a growing interest in how increasing globalization, especially of cultural production and consumption, affects people's sense of identity and place at both local and national levels (McDowell, 1994: 147). Thus a geographic or spatial perspective has become central to studies of culture more widely. These are some of the concerns that form the focus of this chapter. Subsequent sections explore in more detail ideas about a global culture, examine ways of rethinking local cultures, and explore progressive ways of thinking about cultures in the contemporary context.

12.2 Towards a global culture?

12.2.1 Imagining a global culture

Processes of cultural globalization have a very long history and are not peculiar to contemporary times. Through global patterns of trade and migrations, and through the spread of world religions and empires, people, objects and ideas have been circulating for centuries (see Chapters 1–3). However, contemporary

globalization is distinctive in extent, form, rapidity of change, intensity and impact. Today, the idea of a global culture is becoming as meaningful as the idea of national or local cultures. Proponents of the idea of an emerging global culture suggest that different places and cultural practices around the world are converging and becoming ever similar. As Shurmer-Smith and Hannam (1994: 76) argue, a global culture might be the product of two very different processes:

• The export or diffusion of supposedly 'superior' cultural traits (e.g. Western time-frames – the 24-hour day and the Gregorian calendar) and products (e.g. the motor car, television) from advanced countries, and their worldwide adoption ('Westernization', 'Americanization', 'modernization'). This is believed to create global cultural convergence – people around the world are becoming increasingly similar in terms of consumption, life-style, behaviour and aspirations (see Case study 12.1). It can be perceived positively (as 'modernization' or 'development') or negatively (as 'cultural imperialism', where 'we' assume that others in the world should aspire to be like 'us').

• The mixing, or hybridization of cultures through greater interconnections and time-space compression (the shrinking of the world through transport links and technological innovation), leading to a new universal cultural practice. This challenges the notion of unidirectional 'Westernization' and allows us to consider how Western cultures have influenced and are also being influenced by this mixing of cultures. Flows of music, food, ideas, beliefs and literature continue to percolate from around the world into the cultures of the West.

In reality, both these processes are flawed explanations for what is happening today. If a global culture exists, it is far from a product of unidirectional 'Westernization'. However, alternative ideas about cultures mixing to produce a universal global culture are also problematic. Cultures are mixing, but this does not necessarily mean we are all becoming the same.

12.2.2 Debunking global culture?

A different departure point for discussing global culture is that there is no such thing. Ideas about a global economy, a global politics, and a global culture imply some sort of worldwide commonality that does

Thematic Case Study 12.1

The globalization of culture: some examples

Music

Music lends itself to globalization because it is one of the few popular modes of cultural expression that is not dependent on written or spoken language for its primary impact (Held *et al.* 1999: 352). The relative ease of cultural diffusion has led to the spread of many genres and major artists all over the world. Madonna and Michael Jackson are associated with global products and images. The production, distribution and consumption of music has a particular geography. The global music industry is dominated by transnational corporations, with the US and UK dominating domestically generated popular music. 'World music' is now a significant part of the marketing strategies of these corporations, and exposes global audiences to local musical traditions from around the world. The recent popularity of Cuban salsa through its marketing in Western economies is one example. Migrations of people have also had cultural impacts on music, with increasingly 'hybridized' forms (see Case study 12.6).

Television

Until recently, television programmes tended to be produced primarily for domestic audiences within national boundaries, and could be subjected to rigorous governmental control. However, with the advent of cable, satellite and digital technologies, in addition to political and legal deregulation in many states, several television channels are now globally disseminated, and can to some extent circumvent national restrictions. The USA, France, Germany and the UK are the major exporters of television programmes, but Brazil, Mexico, Egypt, Hong Kong, Spain and Australia are increasing their output (Held *et al.* 1999: 360).

Sport

Sports are forms of cultural expression that are becoming increasingly globalized, as well as increasingly commodified. Football/soccer is the most obvious example. The World Cup was held in Japan/South Korea in 2002 in an effort to expand its global popularity. European leagues are increasingly internationalized, with some clubs fielding teams made up almost exclusively of overseas players. Clubs like Real Madrid and Manchester United are globally recognized icons due in part to global television coverage. Similar trends can be observed in US Major League Baseball. The New York Yankees are a global icon; many major league players hail from countries such as Cuba, the Dominican Republic, Puerto Rico and Costa Rica; the sport is becoming increasingly globalized through television coverage. Although football/soccer is increasingly globalized, it has different meanings around the world. In the UK, football was, until recently, a working-class sport. It remains an overwhelmingly male sport, despite the increasing numbers of women playing the game. The same can be said for Latin America, where very few women play the game because of social constructions of gender, which determine that it is 'unfeminine' for women to play football. In the USA, soccer is a sport played predominantly by white, middle-class people, but is also extremely popular among young white women and in Spanish-speaking communities. In South Africa, football is an overwhelmingly black sport, unlike cricket and rugby, which remain overwhelmingly white. Therefore, football/soccer is globalized, but it is also fundamentally localized.

Tourism

Tourism is one of the most obvious forms of globalization. The geography of tourism is skewed, since it is dominated by people of all classes from developed countries (North America, Western Europe and, increasingly, Japan and Australia). It can also be exploitative, particularly through the growth of international sex tourism and the dependency of some developing economies on the exploitation of women. However, it is a form of international cultural exchange that allows vast numbers of people to experience other cultures and places. It also locks specific places (tourist destinations) into wider international cultural patterns (Urry 1990).

not exist. While there clearly are global processes at work, the idea of a global culture is more problematic. Firstly, the image of rampant cultural imperialism by the West, and especially the US, is flawed since apparent cultural sameness is limited in scope, located only in the consumption of products and media images. The possibilities of this eroding centuries of local histories, languages, traditions and religions are far-fetched and people in different parts of the world respond to these images and products in different ways. Many millions of people do not have access even to a television. (In 2003, there were only 55 TV sets per 1,000 people in Africa, and 208 per 1,000 people in Asia, compared with 471 per 1,000 in Europe, and 754 per 1,000 in the US (*CIA World Factbook* 2003). There is no single global culture because of the unevenness of globalization.

Secondly, some theorists would argue that national cultures remain stronger than global cultures. This is borne out when we consider the many conflicts occurring throughout the world along the geopolitical fault-lines of national cultures (the ongoing conflicts between Pakistan and India are but one example of this). For the past 200 years, nation-states and national cultures have monopolized cultural power (state television is one example of a national institution influencing national cultures within national territorial boundaries). At the end of the twentieth century, this balance began to change, with international telecommunications and media corporations challenging the centrality and importance of national cultures. However, it could be argued that despite these changes a great deal of cultural life is still organized along national and territorial lines.

Thirdly, if culture is a system of shared meanings, then looking at the world today there are clearly many systems of shared meanings, and many different cultures. People in different places use different techniques and technologies to reproduce culture, such as oral histories, literature or television and cinema. These techniques have different patterns of dispersion, penetration and scale. Therefore, some cultures are more likely to become globalized than others – those reproduced through television and cinema have a greater range and speed of dispersion than those reproduced through oral histories. However, this does not mean that globalized cultures completely erode localized cultures; the ways in which these different cultures intersect is important. Those 'things' (products, symbols, corporate entities) that have become global signifiers are clearly globalized (they are

recognized the world over), but the ways in which people around the world make cultural responses to them are complex and multiple. Globalization of products and symbols does not necessarily equal Westernization. For example, Japanese consumer goods do not sell on the back of exporting Japanese culture but on a market strategy based around the concept of *dochaku* ('glocalism'). This involves a global strategy not of imposing standardized products but of tailoring Japanese consumer products to specific local markets. These goods are, therefore, both globalized and localized. This is now a popular strategy for multinationals in other parts of the world. Therefore, how intersections between cultures are played out at local levels are of significance, and this suggests that imagining a universal global culture is quite problematic.

12.2.3 Rethinking global culture

It might be more useful to think about a global culture in the restricted sense of what Featherstone (1995: 114) calls 'third cultures'. Instead of imagining a global culture that is erasing local and national cultures, we can think of local, national and global as three important spatial levels at which culture operates. Those aspects of culture that operate at the global level are 'third cultures'. As we have seen, the beginning of the new millennium is marked by the fact that national institutions are no longer in complete control of cultural globalization. 'Third cultures' are sets of practices, bodies of knowledge, conventions and life-styles that have developed in ways that have become increasingly global and independent of nation-states. Phenomena such as patterns of consumption, technological diffusion and media empires are part of these third cultures, and transnational and multinational corporations are the institutions that make them global. In this sense, global cultures exist but only as third cultures, outside national and local cultures, yet intersecting on both these scales in different ways around the world.

Acknowledging that 'our' global view might be very different from that of people elsewhere, living in very different contexts, is also important. (Indeed, the 'millennium' is culturally specific, and meaningless to millions of people around the world whose cultures are not centred on Christianity and Western time-scales.) It is clear that multiple global cultural networks exist, such as those connecting the overseas Chinese with their homeland, or those linking Islamic groups around the world. These networks disrupt any notion of a singular global culture. Power and inequality are bring into question the idea of global cultures. As Massey (1993) argues, a **power geometry** exists, which gives people with different access to power different notions of what global means. New institutions (like global media corporations) for the production, transmission and reception of cultural products are creating infrastructures supporting cultural globalization, including electronic infrastructure (radio, television, music, telecommunications), linguistic infrastructure (the spread of bi- and multi-lingualism, particularly the dominance of English), and corporate infrastructure (producers and distribution networks). As we have seen, these new institutions often operate at scales beyond the nation-state, and they are sites of power in the production of culture. The ownership, control and use of these institutions remain uneven across and within countries (Held *et al.* 1999: 370), thus creating a 'power geometry' that is centred overwhelmingly on the West. People have very different experiences of culture because of their different locations in the world and their relationship to these sites of power. The challenge for geographers is to map this power geometry, to identify sites of power, and to reveal the marginalization of some peoples around the world by cultural globalization. We might also suggest ways in which marginalized peoples might be empowered by engaging with, and perhaps transforming, the new institutions driving cultural globalization.

In summary, global processes are occurring, but they do not produce a universal global culture, nor are they distributed evenly around the world. Global cultural processes are not simply a result of a unidirectional 'Westernization', since culture flows transnationally. A number of different global cultures exist as 'third cultures' – in patterns of consumption, flows of knowledge, diffusion of technologies and media empires that operate beyond, but connecting with, the local and national scales.

12.3 Reinventing local cultures?

12.3.1 Locality and culture

In the past it has often been assumed that there is a simple relationship between local place and local culture. Places were thought of as having a distinct physical, economic and cultural character. Places, therefore, seemed unique, with their own traditions, local cultures, and so on that made them different from other places. It is clear, however, that processes of globalization are also posing serious challenges to the meaning of place. Places and cultures are being restructured. According to Massey and Jess (1995a: 1), 'on the one hand, previous coherences are being disrupted, old notions of the local place are being interrupted by new connections with a world beyond'. The appearance of 7-11 stores in the rainforests of northern Thailand is one example of how even the remotest of places are becoming increasingly internationalized, in this case through tourism. 'On the other hand', Massey and Jess continue, 'new claims to the – usually exclusive – character of places, and who belongs there, are being made.' A dramatic example of this is the revival of place-based ethnic identities in former Yugoslavia and the post-Soviet states, and the conflicts that have resulted. Therefore, modern life is characterized both by decentralization and globalization of culture and by the resurgence of place-bound traditions. Following this, the impact of the new global context on local cultures has two, possibly contradictory, outcomes.

12.3.2 Negative sense of culture

Where global processes are perceived to pose a threat to local culture there might be an attempt to return to some notion of the exclusivity of culture. At the extreme, this might take the form of exclusivist nationalism or even 'ethnic cleansing'. Reactions to the perceived threat to local cultures include nationalistic, ethnic and fundamentalist responses, which also entail a strong assertion of local cultures, such as reviving or inventing local traditions and ceremonies. These can create a level of local fragmentation, with a parochial, nostalgic, inward-looking sense of local attachment and cultural identity. In this sense, cultures are thought of as **bounded**, with very clear definitions of 'insiders' and 'outsiders' in the creation of a sense of belonging. As Sibley (1995) argues, this mode of thinking about cultures produces **geographies of exclusion**. One example can sometimes be found in the English countryside. Rural areas are often conceptualized as the preserve of English culture and identity. This idea is mobilized through the myth of the 'rural idyll', where rural cultures are thought of as timeless, unchanging

Plate 12.2 Exclusion in the countryside: travellers and other minority groups are often excluded from shops and public houses, marking their position as 'outsiders'. (Photofusion/Joanne O'Brien)

and unaffected by global processes. This cosy vision of a peaceful countryside excludes many people who live in rural areas but do not fit this stereotype. Gypsies, travellers, environmental protesters, hunt saboteurs, people with alternative life-styles and people from minority ethnic communities are deemed to be 'outsiders' and a threat to local cultures (Cloke and Little 1997). As Halfacree (1996) argues, the 'rural idyll' is a selective representation. It is exclusive in its class, race and status connotations, is profoundly conservative and demands conformity. It is based on a very inward-looking sense of place and culture (see Plate 12.2).

Conservative reactions to change can be thought of as a kind of cultural fundamentalism through which the process of cultural change is often bitterly contested. Gender plays an important role in this. Women are often considered as guardians of the borders of culture (Yuval-Davis 1997). They not only bear children for the collectivity, but also reproduce it

culturally. In closed cultures, the control of women's sexuality is seen as imperative to the maintenance of the purity of the cultural unit. It is important that women do not marry outside their cultural and ethnic group. Ethnicity and culture, therefore, are seen to be one and the same. In addition, symbols of gender play an important role in articulating difference between cultural groups. Women's distinctive ways of dressing and behaving very often come to symbolize the group's cultural identity and its boundaries. Women are often also the ones chosen to be the intergenerational transmitters of cultural traditions, customs, songs, cuisine and the 'mother' tongue, primarily through their role as mothers. This is especially true in minority situations where the school and the public sphere present different and dominant cultural models to that of the home.

Societies do not evolve smoothly from these closed, bounded perceptions of culture to more open, dynamic notions. The question of cultural power, identity and resistance also needs to be considered. For many groups cultural survival is seen to depend on a closed idea of culture, with strongly marked boundaries separating it from 'others'. An example of this can be seen in the continuing sectarianism in Northern Ireland, where two different cultures, religions and national identities collide with each other (see Case study 3.1). Another example of this is found in contemporary South Africa. Here, the transition from an apartheid state where racialized cultures were rigorously bounded and demarcated, to a post-apartheid state where an acceptance of cultural pluralism (the so-called 'Rainbow Nation') is key to the survival of the nation-state, has not run smoothly. On the one hand, the government attempted to create a non-exclusive, pluri-cultural society, with eleven official languages and constitutional guarantees for the protection of cultural difference. Television programmes are made and broadcast in all major languages, and the national anthem combines verses in English, Afrikaans, Xhosa and Zulu. On the other hand, some groups (e.g. some Afrikaaners, some Zulus) still claim a separate, culturally and ethnically defined homeland, and see this as the only way that their cultures can survive. Cultural fundamentalism is still a powerful force in post-apartheid South Africa. Elsewhere the mixing of cultures under the impact of globalization is often seen as threatening and as weakening the sense of cultural identity. This has created a revival of ethnicity that cuts across the political spectrum. Examples include the 'little England' reaction to European monetary union (characterized by a strong attachment to the symbolic importance of the pound), the rise of neo-fascism in Europe (characterized by anti-immigration and racism), and the strengthening of Islamic and Christian fundamentalism around the world. These phenomena are not all the same, but they do share a response to globalization that involves a closed, fixed, bounded and often place-specific definition of culture, and a strong resistance to changes heralded by cultural globalization.

12.3.3 Positive sense of culture

A more positive response to global processes would be to imagine cultures as fluid, ever-changing, unbounded, overlapping and outward-looking – akin to Massey's (1994: 151) 'progressive sense of place'. This involves people being more cosmopolitan (free of prejudice and tolerant of difference). Increasing interconnectedness means the boundaries of local cultures are seen to be more permeable, susceptible to change, and difficult to maintain than in the past. Rather than everywhere becoming the same, nation-states seem to be reconstituting their collective identities along pluralistic and multicultural lines, which take into account regional and ethnic differences and diversity. In Europe, this involves re-creation and invention of local, regional, and sub-state or new 'national' cultures (for example, the cultural renaissance of the Scots, Basques and Catalans, or the cultural assertiveness of minority ethnic communities in cities such as London, Paris and Berlin). Thus, what can be perceived as destruction of local cultures by globalization might in fact be the means of the creating new senses of locality. This still involves notions of local identity, but recognizes both the differences between cultures and their interconnectedness, taking account of the positive aspects of cultural mixing and increased cosmopolitanism. This new sense of local identity is based on notions of inclusion rather than exclusion.

This is not to say, of course, that all people within the same place will share the same culture and the same sense of locality. Within these more culturally pluralistic and cosmopolitan locales, different class fractions, ages, genders, ethnicities and religious groupings mingle together in the same sites, consuming the same television programmes and products, but in highly uneven ways. These groups often possess different senses of affiliation to places and localities, possess different cultural identities and belong to different cultural groupings (Featherstone

1995: 97). A progressive sense of culture does not foresee the locale as a 'melting-pot', where everything becomes the same, but rather recognizes the different experiences of people, and that increasing interconnections might create new, dynamic and exciting cultural forms. An understanding of this is crucial to the creation of a progressive notion of place and culture, which demands the recognition of cultures as fluid and open and interconnected, and an acceptance that older local cultures might decline as new local cultures emerge.

In summary, localities are important in maintaining cultural difference, but can also be sites of cultural mixing and transformation. Ideas about culture can be negative (bounded, fixed, inward-looking) or positive (progressive, dynamic, outward-looking). Bounded, fixed notions of culture can lead to localized resistance, racism, nationalism, and even 'ethnic cleansing'. Progressive ideas about culture involve the recognition of differences between cultures, the interconnectedness of cultures and their constant evolution.

12.4 Activity: thinking about 'own' and 'other' cultures

In the light of the above, this is a useful juncture at which to reflect upon how we think about our 'own' and 'other' cultures and how this might relate to notions of cultural superiority/inferiority. Consider first the report from the *Guardian* reproduced in Case study 12.2.

This recounts the remarks of a European political leader, which were quickly condemned around the world and subsequently retracted. In one sense, it reveals the sensitivities of global politics shortly after the shocking events in the US in September 2001. In another sense, it prompts us to think about how we see the world from the perspective of our own cultures and the implications of this. Berlusconi sees 'democracy', 'civilization', 'superiority' and 'human rights' as related to and informing each other but also as synonymous with 'European'/'Western' cultures. The explicit assertion is that these qualities do not exist in other (in this case, Islamic) cultures.

12.4.1. Some questions about cultures of democracy

Three questions are raised by this article. First, what is the assumed relationship between the West (specifically

Thematic Case Study 12.2

Berlusconi breaks ranks over Islam

Italy's prime minister, Silvio Berlusconi, yesterday went out of his way to stress what every other leader backing America's 'war on terrorism' has been desperate to deny – that the looming conflict is, at bottom, a clash of civilisations. On three occasions during a visit to Berlin, Mr Berlusconi enthusiastically … boasted of the 'supremacy' and 'superiority' of western civilisation and called on Europe to recognise its 'common Christian roots' … Standing beside the German chancellor … he declared that he and his host 'consider that the attacks on New York and Washington are attacks not only on the United States

but on our civilisation, of which we are proud bearers, conscious of the supremacy of our culture, of its discoveries and inventions, which have brought us democratic institutions, respect for the human, civil, religious and political rights of our citizens, openness to diversity and tolerance of everything.' Mr Berlusconi said: 'We should be conscious of … [our cultural] superiority …, which consists of a value system that has given people widespread prosperity in those countries that embrace it, and guarantees respect for human rights and religion.' Mr Berlusconi added: 'This respect certainly does not exist in the Islamic countries'.

Source: Hooper and Connolly (2001)

Europe and the United States) and democratic cultures? Second, what assumptions are being made about the superiority of European cultures and how they are exported around the world? Third, how might we question the supposed universalism of Western cultures and the assumption that it is appropriate to export these beyond the West?

First, let us think about relationships between place and the assumed origins of democratic cultures. A crucial point is that the West is assumed to have a special place in the history of democracy and its associated qualities (e.g. 'freedom', 'civilization', 'human rights'). Western cultures are depicted as the cradle of human rights, progress, enlightened thought and philosophical reflection on creating the conditions for the improvement of people's lives. Democracy, then, is seen as an intrinsic part of Western cultures. The second question concerns ideas about the supposed superiority of Western cultures. Democracy is thought of as produced solely within Western cultures. Berlusconi does not see it as a product of cross-cultural encounters or influenced by ideas from other parts of the world. This assumption is rooted in ideas about the supposed superiority of Western cultures. The third question concerns the supposed universalism of Western cultures. Western democracy is depicted as a common step forward for humanity as a whole and as encapsulating progressive and civilized cultures. This

assumption is based on the representation of non-Western cultures in negative terms and as less civilized. Western democracy is seen as the best way of improving people's lives and should be exported to other places assumed to be culturally inferior. How might we think differently about culture by looking at examples from other places?

12.4.2 Democratic cultures in other places

Now consider these three questions in terms of the two examples in Case studies 12.3 and 12.4. How do these examples challenge ideas of (a) democracy originating solely in Western cultures; (b) the superiority of Western cultures; and (c) the universal applicability of Western cultures?

These two examples do not suggest that non-Western cultures have 'better' cultures of democracy. Critics argue, for example, that *ubuntu* fails to take account of the current socio-political context of poor communities in South Africa and perpetuates gender inequalities; there are questions about the armed resistance of the Zapatistas within Mexico. However, they demonstrate that different cultural models exist and Western and indigenous cultures of democracy might inform each other in a respectful dialogue. In this sense, understandings of different cultures are

Regional Case Study 12.3

Ubuntu and democracy in South Africa

The relatively non-violent transition in South Africa from a totalitarian state to a multi-party democracy is not merely the result of the mimicry of Western democratic cultures, but also of the emergence of an ethos of solidarity and a commitment to peaceful coexistence amongst many South Africans in spite of their differences. *Ubuntu*, a cultural philosophy eroded for generations by the slave trade, Western missionary movements, colonialism and apartheid, is central to this. The South African government officially recognizes *ubuntu* as: 'The principle of caring for each other's well-being ... and a spirit of

mutual support ... Each individual's humanity is ideally expressed through his or her relationship with others and theirs in turn through recognition of the individual's humanity. *Ubuntu* means that people are people through other people. It also acknowledges both the rights and the responsibilities of every citizen in promoting individual and societal well-being' (*Government Gazette*, 2/2/1996). *Ubuntu* philosophy is one of generating agreement or building consensus. South African democracy is not simply majority rule; the Constitution and legislation explicitly protect the rights of minority groups. *Ubuntu* differs from Western cultures of individual rights by seeing democracy as built upon interdependent relationships.

Regional Case Study 12.4

Zapatistas: democracy and indigenous rights in Chiapas, Mexico

Struggles for democratization in Mexico have occurred since the early twentieth century. New philosophies began to emerge in the 1980s and 1990s with the Zapatista uprising in the impoverished Chiapas region. Globalization and free trade agreements are seen as threats to the livelihood of large sectors of the Mexican population, especially peasants and workers. The Zapatista uprising aims to democ-

ratize and decentralize Mexican society, secure autonomy and self-determination for indigenous communities and protect their cultural heritage. It aims to resist the negative effects of globalization and protect women's rights. The way the Zapatistas have conceptualized democracy does not flow from Western political philosophy but rather emanates from older, indigenous Maya social organization, based in cultures of reciprocity, communal values and respectful dialogue.

Source: Slater (2003)

enriched rather than based on Western-centred and Western-biased assertions of cultural superiority evident in the newspaper report. Expanding our geographies of reference and learning about examples from beyond our own cultures encourages the production of less partial/more informed visions and we can also begin to think about how cultures elsewhere have shaped and continue to shape our own.

12.5 Multi and hybrid cultures?

12.5.1 Hybridity

One of the major contemporary challenges concerns what we do with the concept of culture in the changing global scene, where nation-states in the West have been forced to tolerate greater diversity within their boundaries. Some people want to see national identity as homogeneous and assimilatory – in other words different cultures are subsumed into the dominant culture (the 'melting-pot' idea) (see Case study 12.5). Others call for the acceptance of ethnic pluralism and the preservation of minority ethnic cultures as a legitimate part of the national project. This is the politics of multiculturalism, which instead of thinking of different cultures as being absorbed by dominant cultures, relies more on a notion of a cultural mosaic, or a 'patchwork quilt' of cultures. Each culture is recognized as different and distinct, but these differences are understood and valued.

Multiculturalism might seem progressive, but it can sometimes reinforce difference because culture is seen as essentially connected to race, where racial difference is rooted in biological difference. (These ideas are no longer considered acceptable; anti-racists have demonstrated that 'race' is socially constructed and has little basis in biology (see Kobayashi and Peake 1994). We could just as easily have 'races' of blue-eyed and brown-eyed people.) Multiculturalism, therefore, still relies on a negative notion of bounded cultures. It might suggest tolerance, but often results in segregation and ghettoization. Similar arguments apply to attitudes towards homosexuality. The appearance of 'gay spaces' in cities such as San Francisco, London, Manchester or Cape Town might appear to signal greater tolerance and cultural diversity, but it points to segregation rather than integration. Indeed, in the US politically correct accommodation agencies now ask applicants to tick a box for 'gay' or 'straight' so that they can be housed suitably. As Reeves (1999) argues, 'Homophobia, overt and covert, is herding lesbians and gay men into ghettos – which are then celebrated as signs of a tolerant society.' It is OK to be gay in these segregated spaces; it is still not OK to be openly gay in other spaces of the city.

One way in which a more progressive idea of culture might be developed is through the concept of hybridity. Hybridity is a new way of thinking about the manifestations of culture such as ethnicity, gender and sexuality. It breaks down barriers, adhering to neither the 'melting-pot' nor 'mosaic' idea of cultural mixing, but rather seeing different cultures coming together and informing each other in different ways to produce something entirely new. This process has a long historical trajectory. Indeed, some would argue that cultures have always been hybrid forms – they have never existed in isolation from other cultures, and thus have always been subject to change and influences from elsewhere (Werbner 1997: 15).

One of the most obvious places we can observe hybridity is in popular mass culture, an immediate example being popular music. Innovations in music have always involved the fusion of different styles to create new sounds and rhythms. Rock-and-roll, rhythm-and-blues and Latin jazz are obvious examples; we might also think of contemporary forms of music that fuse rap and hip-hop with other styles. Authors such as Barthes (1972), Bourdieu (1984) and Bakhtin (1984) see popular hybridity as an exciting challenge to, or subversion of, dominant cultures and the exclusive life-styles of dominant elites. Such popular mixings and inversions, like the subversive elements of youth cultures (see Hebdige 1979, Skelton and Valentine 1998), are hybrid in the sense that they bring together and mix languages and practices from different and normally separated domains. They have the potential to disrupt dominant cultures by their 'out-of-placeness'.

In many ways hybridity is related to the notion of transculturation. Transculturation describes one of the key cultural processes that operate between hitherto sharply differentiated cultures and peoples who are forced (usually by the processes of imperialism or globalization, and primarily through migration) to interact. This interaction often takes place in profoundly asymmetrical ways in terms of relative power between different groups. Despite the illusion of boundedness, cultures evolve historically through borrowings, appropriations, exchanges and inventions. Cross-fertilization of cultures is endemic to all movements of people throughout history (see Case study 12.5). However, for those who aspire to bounded notions of culture and refuse this idea of perpetual hybridity, cultural mixing is felt to be threatening and a deliberate challenge to social order. In reality there are no fixed cultures in modern nation-states, but some people cling to ideas of pure or impure cultures. For others, however, hybridity remains the site of revitalization, resistance and fun (see Case study 12.6).

12.5.2 Diaspora

Related to the idea of hybridity is the notion of diaspora. This term was originally used to refer to the dispersal of Jewish peoples. However, it is now used in reference to the long-term settlement of peoples in

Thematic Case Study 12.5

Englishness: bounded or hybrid culture?

Nationalists around the world cling to a notion of bounded cultures that make them distinct from others. One effect of globalization has been resistance in the form of increased nationalism to what is perceived to be the erosion of national cultures. Increased mixing of cultures is seen to pose a threat to the survival of national cultures. Nationalists seek to preserve the symbols of nationhood, such as language, lifestyles and cultural forms in the face of what are perceived to be sweeping changes. However, cultures are not unchanging; a fundamental flaw in nationalist ideology is the adherence to a notion of static culture, and its reliance on a mythical history of the origin of the nation. For example, English nationalists define Englishness as distinct, which is used to justify anti-immigrationist ideas, anti-Europeanism and, in some cases, racism. But who are 'the English'? After the last ice-age many communities settled in the British Isles from all over Europe. They lived and fought with each other and in a short space of time produced a mixed group of people who eventually called themselves English. The British Isles have been subject to waves of invasion and settlement (e.g. Celts, Vikings, Romans, Anglo-Saxons, Normans). England and the British Isles have always been hybrid; peoples and cultures have mixed and evolved together. Some English nationalists avoid thinking about this point by arguing that the final invasion (by William the Conqueror in 1066) marks the origin of England and Englishness (Anderson 1983). This myth is also flawed. William the Conqueror spoke no English. Whom did he conquer? He conquered 'the English'. For many nationalists the founding father of England is French! It is also no small irony that one of the symbols of Englishness, the monarchy, changed its official name to Windsor in 1917 from Saxe-Coburg-Gotha, thus hiding its German origins. Even protectionist policies towards language are flawed; as with culture more generally, language is always hybridized and evolving. Most sentences in 'English' contain words that derive from German, French, Spanish, Latin, Nordic and Celtic languages. Similar myths of origin and notions of bounded cultures exist elsewhere in the world. In some places they emanate from deeply contradictory ideas (anti-immigrationist views in the USA and Australia, for example); in other places they have led to conflict (recent events in Bosnia and Kosovo, for example). The idea of nationhood, based on fixed, bounded and unchanging cultures, is an ideological creation that masks profound cultural divisions of gender, race, class and religion within a nation-state, and ignores the fact that, in reality, all cultures are hybrid and dynamic.

'foreign' places that follows their scattering or dispersal from their original homeland. It refers to a modern condition where a sense of belonging is not derived from attachment to territory, and where different peoples mix together through the processes of migration (forced or free). European imperialism and associated processes of globalization have set many of these migrations in motion. Diasporas are classic **contact zones** (spaces in which two cultures come together and influence each other) where transculturation or hybridization takes place. Diasporic identities are at once local and global, and based on transnational identifications encompassing both 'imagined' and 'encountered' communities (e.g. Irish-Americans belong to an imagined international community of people who have 'Irishness' in common, but whose identities are also informed by the communities in which they live in the US). In other words, diasporas are a direct challenge to the idea that there is a simple relationship between place and culture. They transgress the boundaries of the nation-state and provide alternative resources for constructing identity and fashioning culture.

The concept of diaspora space allows us to think of 'culture as a site of travel' (Clifford 1992), which seriously problematizes the idea of a person being a 'native' or an 'insider'. Diaspora space is the point at which boundaries of inclusion and exclusion, of

Thematic Case Study 12.6

Hybridity/Diaspora – some examples

British Bhangra

British Bhangra developed out of traditional forms of folk dances and music of the people of the Punjab. Traditional Bhangra arrived in Britain in the late 1960s with the migration of people from South Asia and East Africa. British Bhangra emerged in the mid-1980s among second-generation South Asian-British musicians, primarily in Birmingham and west London. This latter style combined the robust and energetic punctuated rhythms of double-sided drums (*dhol*), modern technology and urban sounds with Punjabi poetry, traditional Punjabi lyrics, acrobatic dance skills and celebrations of the Punjabi harvest festival (*bhaisakhi*). It played an important role in the proclaiming of a cultural identity for groups of youths defining themselves as South Asian-British. Although it appeals to the Punjab as a specific region of origin, Bhangra, with its fusion of instruments and musical styles from South Asia and urban Britain, has a large following among diverse groups of South Asian-British youths. Throughout the 1990s, Bhangra was characterized by ongoing processes of fusion with broader ranges of musical styles (ragga, reggae, soul, jazzfunk, rock, hip-hop, pop), both to reflect the lived experiences of South Asian-Britons, and as an attempt to 'cross over' into mainstream charts. It is the antecedent to newer and more fashionable forms of post-Bhangra and 'Asian Kool' which have been less marginalized by mainstream culture, and are part of an emergent and vibrant British-Asian urban youth culture.

Source: Adapted from Dudrah (2001)

Cuban Santería

Santería is an example of a syncretic (hybrid, mixed) religion. It is based on West African religions brought by slaves imported from what are now Nigeria and Benin to the Caribbean to work the sugar plantations. These religions were suppressed by the European plantation owners and in Cuba slaves were forced to convert to Catholicism. However, they were able to preserve some of their traditions by fusing together various West African beliefs and rituals and syncretizing these with elements from Catholic culture. One factor enabling this process was that many of the *orishas* (primary gods) shared many of the same characteristics of Catholic saints. This enabled slaves to appear to be practising Catholicism while practising their own religions. In Cuba, this religious tradition has evolved into what we know today as Santería, the Way of the Saints, whose traditions are transmitted orally from generation to generation. Despite suppression by Fidel Castro's Socialist Revolution since 1961, its influence is pervasive in Cuban life. Devotees are found in most households, Yoruba proverbs litter Cuban Spanish and high priests (*babalawos*) offer guidance based on ancient systems of lore. Santería was perhaps less persecuted by Cuba's communist regime since, being the religion of Cuba's poor and disenfranchised, it represented less of a threat than the Catholic Church, which could call on international help and was seen as the religion of the bourgeoisie. Today, with less religious persecution, Santería is experiencing a rise in popularity and is part of an emerging Cuban youth culture. Similar syncretic religions are found in Haiti, Puerto Rico and other Caribbean and Latin American countries.

Source: Adapted from Betts (2002)

belonging and otherness, of 'us' and 'them', are contested. As Brah (1996: 209) argues, diaspora space is 'inhabited' not only by those who have migrated and their descendants, but equally by those who are constructed and represented as indigenous or 'native':

In the diaspora space called 'England', for example, African-Caribbean, Irish, Asian, Jewish and other diasporas intersect among themselves as well as with the entity constructed as 'Englishness', thoroughly reinscribing it in the process.

Like notions of hybridity, the concept of diaspora is important since it allows for the recognition of new political and cultural formations that continually challenge the marginalizing impulses of dominant cultures.

12.5.3 Selling hybridity and the commodification of culture

In today's world, culture sells. Hybrid cultures, in particular, sell. Cities are now constructing themselves as cosmopolitan, and hybridity has become a form of 'boosterism' – where city authorities create marketing images to attract investment in the form of business and tourism. Hybrid culture is perceived as creating economic advantage. With increasing deindustrialization in Europe and North America, cultural strategies have become key to the survival of cities. Examples include the international marketing of cultural/religious festivals such as Mardi Gras in Sydney or New Orleans, or the importance of 'Chinatowns' and other 'ethnic' districts to tourism in cities throughout the developed world (see Plate 12.3). Ironically, hybridity is in danger of becoming just another marketable commodity. For example, treating the political work of some British-Asian bands as marketable hybridity trivializes black political activity and leaves problems of class exploitation and racial oppression unresolved. As Hutnyk (1997: 134) suggests, to focus on hybridity while ignoring (or as an excuse for ignoring) the conditions in which this phenomenon exists (the commodity system, global economic inequality, inequitable political relations), is problematic in that it maintains the status quo. Hybridity and difference sell, but in the meantime the market remains intact, power relations remain unequal, and marginalized peoples remain marginalized. Moreover, as culture is subsumed into capitalism, those marginalized peoples who might be capable of oppositional politics are also subsumed under the rubric of hybridity.

The notion of hybridity, therefore, can be problematic. In some Latin American countries, cultural elites and nation-states have appropriated the hybrid mestizo (mixed) identity, making it dominant.

Plate 12.3   Chinatown in San Francisco: 'ethnic' districts are promoted as major tourist attractions in many 'global' cities. (Corbis/Dave G. Houser)

This has been seen to be oppressive of 'Indian' populations, who have in turn been accused of ethnic essentialism (or emphasizing their racial difference) because of their desire to protect their cultures (Radcliffe and Westwood 1996). In the West, ideas of hybridity are currently popular with highly educated cultural elites, but ideas about culture, ethnicity and identity that develop in poverty-stricken underclass neighbourhoods are likely to be of a different nature (Friedman 1997: 83–4). Evidence of racial tensions in many North American and European cities points to the fact that class and local ghetto identities tend to prevail, with little room for the mixing pleaded for by cultural elites. The global, cultural hybrid, elite sphere is occupied by individuals who share a very different kind of experience of the world, connected to international politics, academia, the media and the arts. In the meantime, the world becomes more polarized in terms of wealth, and heads towards increasing balkanization where regional, national and ethnic identities are perceived as bounded, threatened, and in need of protection. As Bhabha (1994) reminds us, hybridity is an insufficient means through which to create new forms of collective identity that can overcome ethnic, racial, religious and class-based antagonisms – it sounds nice in theory, but does not necessarily exist outside the realms of the privileged.

As Hall (S. Hall 1996: 233) argues, however, we should not view the current fashionability of hybridity in a wholly negative light. Even as cultures are increasingly commodified, we should not forget the potential for the democratization of culture in this process, the increased recognition of difference and the diversification of the social worlds in which women and men now operate. This pluralization of social and cultural life expands the identities available to ordinary people (at least in the industrialized world) in their everyday working, social, familial and sexual lives. As Hall (*ibid.*: 234) argues, 'these opportunities need to be more, not less, widely available across the globe, and in ways not limited by private appropriation'. For Bhabha (1994: 9), it is the interconnections of different cultural spaces and the overlapping of different cultural forms that create vitality and hold out the possibility of a progressive notion of culture.

A challenge for geographers, then, is to think about the place and meaning of cultural hybridity in the context of growing global uncertainty, xenophobia (fear of foreigners) and racism. Why is cultural hybridity still experienced as an empowering, dangerous or transformative force? Why, on the one hand, is difference celebrated through a consumer market that offers a seemingly endless choice of identities, sub-cultures and styles yet, on the other hand, hybridity continues to threaten and shock? Conversely, why do borders, boundaries and 'pure' identities remain important, producing defensive and exclusionary actions and attitudes, and why are the latter so difficult to transcend? Is the sheer pace of change in cultural globalization producing these reactions?

To summarize, hybridity and diaspora are examples of more progressive ways of thinking about culture. It could be argued that all cultures are always already hybrid; they are never pure, have always evolved and changed through time and through contact with other cultures, and they continue to evolve. In today's world, hybridity is being commodified, which might make it less radical. However, despite this, it has the potential to democratize culture and to allow us to rethink culture in ways that are more tolerant of difference. Finally, cultural hybridity needs to be understood in the context of growing global uncertainty, xenophobia and racism.

12.6 Conclusion

In this chapter it is suggested that there are two apparently contradictory tendencies in thinking about cultures – the attempt to secure the purity of a culture by conceptualizing it as strong, fixed, bounded, permanent and homogeneous, and the hybridity of most cultures. We should therefore think of culture as a contested concept. A progressive way of thinking about culture is to reject the idea of boundedness and internal cohesion. In the modern world especially, culture is a meeting point where different influences, traditions and forces intersect. There is, therefore, a continual process of change in cultural practices and meanings. Globalization is undermining closed, fixed ideas of culture and leading to new ways of conceptualizing cultures (transculturation, contact zones, hybridity and diaspora). However, the fact that cultures are not fixed or homogeneous does not mean that we will stop thinking of them in this way. As Hall (S. Hall 1995: 188–9) argues, this is because some people need 'belongingness' and the security that closed conceptions of culture provide.

Despite this, recent years have witnessed a de-centring of culture, with nation-states increasingly superseded by transnational institutions that are producing cultural globalization and greater cultural diversity. There has also been a shift in the awareness of the cultural capital of the West, and an understanding of the cultural dominance of developed countries. At the same time, there are now more voices talking back, reflecting the cultural assertiveness of marginalized groups and making us aware of new levels of diversity. Thus,

> if there is a global culture it would be better to conceive of it not as a common culture, but as a field in which differences, power struggles and cultural prestige contests are played out. . . . Hence globalization makes us aware of the sheer volume, diversity and many-sidedness of culture.
>
> (Featherstone 1995: 14)

This points to a 'more positive evaluation by the West of otherness and differences' (*ibid.*: 89). For Massey and Jess (1995b: 134), globalization is not simply a threat to existing notions of culture, but a 'stimulus to a positive new response'. What does this mean for geography, and how should geographers rethink ideas about culture?

One possibility is to consider the impact on national and local cultures of cultural globalization. As Held *et al.* (1999: 373) argue, the most that can be claimed is that in the West there has been some degree of homogenization of mass consumption, particularly among the young, and that this is spreading among more affluent peoples in the developing world, especially in Latin America and east Asia. Thus, similar fashions, popular music, film and television are being consumed simultaneously in many different places. However, it is less clear that homogenization of other cultural practices, beliefs and identities is taking place. Products may be globalized, but it does not follow that they are consumed and translated in the same way the world over. For example, the broadcasters of the popular music channel, MTV Europe, were clearly aware that they could not simply replicate the tone and content of the US original; MTV Europe is marketed to a very different audience. National cultures in many places around the world do seem to have weakened. Moves towards devolution, regionalism and independence around the world are evidence of this (e.g. the Legga Lombardo in Italy, Basque, Catalan and Galician regionalism in Spain, the Bloc Québécois in Canada, and the many right-wing anti-statist groups in the US). However, it is not clear that cultural globalization is responsible for challenging and eroding national cultures. Furthermore, it seems that hybrid and transnational cultures, which might be important in rethinking local identities, have yet to make any significant impact on mainstream national cultures and national identities.

Culture is a process that operates at a number of different levels – local, national and global. Therefore, rather than conceptualizing a singular cultural landscape we should study the ways in which different cultural forms and processes come together in particular locations, creating both fragmentation and fusion. These different cultural forms (which Appadurai (1990) terms 'ethno-scapes' (cultural maps of ethnic identity), 'media-scapes' (representations of cultures in various media), 'ideo-scapes' (the range of ideas that people have about the world), 'techno-scapes' (the impact that technologies have in changing relationships over time and space), and 'finance-scapes' (the flows of money and capital at global and local levels)) might be the subjects of geographical inquiry. As Mike Crang (1998: 174) suggests, Appadurai's ideas are useful in that they produce different cultural maps that intersect in many different ways to produce unique cultural terrains. In this way, it is possible to analyze the complexities of cultural processes at different spatial scales. Even though most people remain physically, ideologically and spiritually attached to a local or a national culture and a local place, these various networks or cultural maps ensure that it is becoming increasingly impossible for people to live in places that are completely isolated and disconnected culturally from the wider world.

Finally, in reconceptualizing culture, we might also consider cultures of knowledge. People around the world have different cultures and systems of meaning. We cannot avoid reading the world from within our own cultures and interpreting it through our own systems of meaning. Understandings of global culture for the majority of the readers of this book are filtered through the logic of the West. Spivak (1985) argues that one of the key aspects of Western knowledge is 'worlding' – the power to describe the world. This has its origins in imperialism, with the increased abilities of Western imperial powers to write about and culturally represent what became known as the 'Third World'. This imperial 'worlding' means that areas beyond the West were, and still are, described and represented in Anglo-European terms. This is how culturally different parts of the world came to be perceived as 'barbarous' and 'uncivilized'. Western ideas and cultural forms are still considered superior.

'Development' means incorporating non-Western economies into global capitalism. Forms of government not based on Western forms of democracy are deemed invalid. Western cultures, therefore, have become hegemonic, or dominant. Similarly, one's own cultural positioning (on the basis of gender, ethnicity, class, location, sexuality, stage in life cycle, ability) also influences understandings of local cultures. The same processes operate at local levels; dominant cultures marginalize others on the basis of ethnicity, sexuality, gender, and so on. However, as we have seen, those dominant cultures also produce resistances that have the potential to create new ways of thinking about culture.

The challenge is to confront the limits of 'our' knowledge, to recognize other worlds, to acknowledge the legitimacy of other cultures, other identities and other ways of life. There is a need to consider 'cultural translation' (Bhabha 1994), which involves understanding the hybrid nature of culture, the influence of marginal cultures on dominant cultures, and that people in marginal cultural systems at local, national and international levels are also active in creating their own systems of meaning. They do not simply absorb ideas from, or become absorbed into, more dominant cultures. We might then develop cosmopolitanism in the twenty-first century that is global, sensitive to cultural difference, and dynamic.

Learning outcomes

Having read this chapter, you should understand that:

- Culture is complex; it is a process, rather than a thing, and is therefore subject to change over time.

- Cultures operate at the local, the national and the global levels.

- Global processes are occurring with relation to culture, but these are always filtered through localities – there is no singular global culture.

- Local cultures can be either negative (closed, bounded), or positive (progressive, hybridized).

- A progressive way of thinking about culture might be through new concepts such as hybridity and diaspora. These terms can be problematic (hybridity exists more in theory than it does in reality) but they hold potential for transforming how we think about culture in changing times.

- We must consider our own cultures of knowledge, and the ways in which these condition our ideas about global, national and local cultures.

Further reading

Anderson, K., Domosh, M., Pile, S. and **Thrift, N.** (eds) (2003) *Handbook of Cultural Geography*, Sage, London. This book offers an assessment of the key questions informing cultural geography and contains over 30 essays. It is an invaluable resource for students looking for an assessment of major issues and debates and the breadth, scope and vitality of contemporary cultural geography.

Cook, I., Crouch, D., Naylor, S. and **Ryan, J.** (eds) (2000) *Cultural Turns/Geographical Turns*, Prentice Hall, London. Aimed at advanced undergraduates and post-graduates, this book addresses the impact, significance and characteristics of the 'cultural turn' in contemporary geography and the intersections between space and culture.

Crang, M. (1998) *Cultural Geography*, Routledge, London. A useful introduction to the ways in which geographers think about culture, exploring the global and the local, cultures of production and consumption, and ideas about place, belonging and nationhood.

Massey, D. and **Jess, P.** (eds) (1995) *A Place in the World? Places, Cultures and Globalization*, Oxford University Press, Oxford. A comprehensive first-year text, which explores the conceptualization of place, and how this relates to identity, cultures and contestation.

Mitchell, D. (2000) *Cultural Geography. A Critical Introduction*, Blackwell, Oxford. Aimed at advanced undergraduates and post-graduates, this book considers the historical development of cultural geography, land-scape studies and cultural politics.

Shurmer-Smith, P. (ed.) (2002) *Doing Cultural Geography*, Sage, London. This is a good book for those students thinking of undertaking projects or writing dissertations in cultural geography. It explains the theory informing cultural geography and encourages students to engage directly with theory in practice.

Shurmer-Smith, P. and **Hannam, K.** (1994) *Worlds of Desire, Realms of Power: A Cultural Geography*, Arnold, London. An introduction to the 'new' cultural geography covering themes such as senses of place, and gender, ethnicity and class. Has a useful final chapter on potential undergraduate projects in cultural geography.

Skelton, T. and **Valentine, G.** (eds) (1998) *Cool Places: Geographies of Youth Cultures*, Routledge, London. Focuses specifically on youth cultures, exploring ideas about gender, ethnicity, class, sexuality, ability and so forth. A useful example of how geographers are rethinking ideas about culture at a slightly more advanced level.

Useful web-sites

www.culture.gov.uk The official web-site of the UK Government's Department for Culture, Media and Sport (DCMS). An animated site that includes information about the role of the DCMS, and government policy towards media and arts, heritage, libraries and museums, sport, the national lottery, and tourism. Also has links to other useful sites in each category.

www.alt.culture.com An encyclopaedia of modern culture spanning modern musical styles, film, cyberpunk and street fashion, extreme sports and political correctness. A useful site for those interested in global cultures of consumption.

www.un.org/womenwatch The official United Nations Internet Gateway on the Advancement and Empowerment of Women, and part of the global phenomenon of cyberfeminism, which might be considered an example of a 'third culture'. Has useful links to other sites advancing women's rights through new technologies.

www.asiansociety.co.uk A comprehensive site on British Asian culture, especially the Asian entertainments scene across the UK (Bhangra, TV, films and radio, etc.).

For an overview of the South African transformation, visit **www.nobel.se/peace/index.html** and its articles on President Mandela and Bishop Tutu. There are some interesting links to other sites on South African culture and history at **www.columbia.edu/cu/lweb/indiv/africa/cuvl/SAfrcult.html**, and **www.lonelyplanet.com/destinations/africa/south_africa/culture.htm** contains a useful introductory summary of the diversity of South African cultures.

For an introduction to Cuba's hybrid cultures see **http://folkcuba.com/** and **www.afrocubaweb.com/**

 For annotated, clickable weblinks and useful tutorials full of practical advice on how to improve your study skills, visit this book's website at www.pearsoned.co.uk/daniels

Production, exchange and consumption

EDITED BY PETER DANIELS

The idea of going 'global' is often associated with the economic aspects of human geography. This section of the book demonstrates how 'the economic' shapes many aspects of human and organizational behaviour and, thereby, the spatial patterns of growth and development. Many economic processes and their geographical outcome are increasingly a function of links and dependencies that extend far beyond the immediately local or national; they involve diverse interactions with places and people distributed around the world. Many of the symptoms of this 'globalization' are most obvious in the sphere of consumption (Chapter 16); Coca-Cola, McDonald's, Nike and Microsoft are brands that are almost universally known, if not accessible to all, worldwide. Such products and services are often slightly customized in line with local tastes, cultures or regulations (often reflected, for example, in the way that products are advertised in different markets), but essentially they offer a uniform experience or convey a common fashion statement wherever they are consumed in the world (see Chapter 16).

Consumption is both a consequence, and a driver, of the globalization of production. The shift from production systems largely focused on national markets to production facilities of the same company distributed around the globe has been accelerating. In part it reflects a search for new markets as national sales opportunities have become saturated; but it also reflects the demand created by a growing band of global consumers linked to common sources of information via the satellite, cable and Internet services that have metaphorically shrunk the world. During the first quarter of the twentieth century, Fordist methods of production helped both to fulfil and to create demand for high-volume, low-cost, standardized products at a time of steadily increasing incomes per capita, at least in the developed economies (Chapter 14). The more recent transition towards a post-Fordist production system with its emphasis on greater specialization, customization, and even more fragmented divisions of labour and horizontal rather than vertical integration of production has helped to continue the trend towards globalization while also reinstating the role of localities. The latter have been reasserting their role as the focus for production in conjunction with global-level influences. The massive production plants employing thousands of workers that were symbols of the Fordist era are being replaced by industrial districts made up of networks of firms, small and large, each contributing particular skills and knowledges relating to particular products. Together they form distinctive nodes of production such as Silicon Valley (California), Motor Sport Valley (UK), or the City of London.

This is not to say that large firms are no longer important in the globalization of economies. On the contrary, they are even more influential in shaping the geographies of production and consumption or the distribution of spaces that are strongly engaged with the global and those that are much less engaged (Chapter 13). Typical are the transnational corporations (TNCs) that often have more financial resources at their disposal than many national governments. They therefore exercise significant influence on decisions relating to the economy, investment in infrastructure, or regulation by nation-states and by international organizations such as the Organization for Economic Co-operation and Development (OECD) or the World Trade Organization (WTO).

Consumption and production are mediated or enabled by the circulation of finance capital (Chapter 15). National governments, companies large and small, or individuals require finance to allow them to execute investment strategies in the case of companies or purchasing decisions in the case of individual consumers. Advances in information technology and political shifts have greatly increased the global circulation of finance and encouraged innovation in the financial instruments traded daily on stock, commodity, currency and other exchanges in business centres around the world. Yet again, however, as with the other themes explored in this section of the book, the flows of finance capital and access to it are spatially uneven. This operates to the significant disadvantage of marginalized nations, regions and groups within society while reinforcing the advantage of those individuals, companies, and nation states that already have very good access to finance capital.

This section demonstrates how the geography of the economy incorporates global as well as local economic perspectives, reflecting changes over time and space. Some of the causes of the uneven economic development that results are highlighted. The analysis suggests that there are some important issues, questions, and challenges that economic geographers can usefully address as the twenty-first century unfolds.

CHAPTER 13

The geography of the economy

PETER DANIELS

Topics covered

- The changing nature of economic geography
- The rise of the global economy
- Is globalization inevitable?
- The significance of transnational corporations
- Role of localities in a globalizing economy
- New economies in the twenty-first century

The economy is everywhere. Wherever we happen to live or work our daily existence is influenced by economic relationships of one kind or another. Yet we tend to take for granted the system of activities and administration through which society uses the scarce resources at its disposal to provide for its material needs and to produce wealth. There is rarely a day that passes without references in the news media of most countries to the condition of the economy: the balance of exports over imports, the unemployment rate, consumer spending, announcements of factory closures, the creation of new jobs, or the prospects for the diverse activities that together make up a national economy. Indeed, in many countries separate sections

of the 'quality' daily newspapers or 24-hour satellite television channels are devoted exclusively to news and information about the economy and business (Plate 13.1).

The label 'national economy' is just one of many that tell us something about the scale or the functions involved. You may have come across economies labelled as global, local, market, command (or redistributive), informal, hidden, subsistence, peasant, Internet, digital, or space economy. The latter has particular resonance for geographers because it refers to the way in which **geographical space** is organized by the behaviour and actions of the activities that together make up the economy. Every economic activity

Plate 13.1
Economic and business information occupies an important place in daily newspapers and periodicals such as The Economist or Business Week. (Alan Thomas)

requires and occupies some unit of space. This may range from a small, one-room (say 15 m²) office used by a self-employed management consultant working from home, to the large sites (hundreds of hectares) occupied by motor vehicle production plants in Japan or the US, or the very extensive (hundreds of square kilometres) areas occupied by cattle ranches on the Argentine Pampas or sheep stations in the Australian Outback. Likewise, the families that provide labour for the production process require space in which to live. They are also a major source of demand for goods and services that are accessed via shopping complexes or the Internet (space is required for the warehouses from which goods are distributed). The character of different kinds of spaces is influenced by the availability of appropriate resources, including the means of production, so that some are better suited to certain activities than others.

13.1 Economic geography

Familiarity with the discipline of economics is helpful for studying the geography of the economy and some economists are interested in spatial analysis, which is the process of extracting or creating new information about a set of geographic features (e.g. regional or urban economists). Geographers who are interested in spatial structures with reference to anything from a peasant economy to the global economy comprise the major sub-discipline of economic geography which 'is concerned with the spatial organization and distribution of economic activity, the use of the world's resources, and the distribution and expansion of the world economy' (Stutz and de Souza 1998: 41). In order to measure and to explain spatial patterns it is necessary for economic geographers to understand the levels of, and reasons for, the interaction between the activities that make up an economy. For example, spatial interaction and/or economic dependence/interdependence takes the form of flows of passengers or freight by air, rail, road or sea within and between countries or the daily flows of money between international stock exchanges (see Chapter 15 for examples of some of the workings of global finance). Telecommunications networks enable flows of information and knowledge within companies as well as between them and their suppliers and clients. There are fast growing flows of e-commerce and other transactions linked to the expansion of the Internet and the World Wide Web. Such interactions change

over time and vary in spatial intensity and help to explain trends in local, regional, national or international economic growth, or structural changes in the composition and location of activities in different types of economy.

In its early guise economic geography was termed commercial geography: a largely factual compilation of the circumstances under which different commodities were produced and exchanged around the world. Those analysing the geography of commodities started as generalists, but gradually they began to specialize in the study of particular commodities, such as oil or coal, or manufacturing industries such as iron and steel. Although largely descriptive, some analysis of the factors affecting location shifted the emphasis towards explaining observed spatial variations in the pattern of commodity production and trade. The identity of economic geography was consolidated by the adoption of the principles of neo-classical economics (Case study 13.1) which is the study of the allocation of scarce resources amongst alternative ends, one of which is the maximization of social welfare. It was derived from the influential ideas of Engels and of Marx who promoted the idea that the economic basis of society determines the form of its social institutions.

Such economic determinism dominated economic geography for much of the 1970s and the 1980s. During the 1990s it was challenged by new economic geography. What precisely is 'new' is the subject of debate (see Bryson et al. 1999) but in essence it is about recognizing the importance of the 'cultural' when interpreting the 'economic' (see Chapter 16 for a discussion of the links between culture and the consumption of goods and services). It is argued that life-styles, beliefs, languages, ideas, imaginations and representations interact with the economic to produce culturalization of the economy rather than the economization of culture. Thus, goods and services incorporate cultural attributes in, for example, their design, marketing, packaging and potential benefits to users. Material possession of a good (car, camera, mobile phone) or consumption of a service (tourism, fast food, financial transaction) is only part of the experience (a real as well as an imagined event) of using or being seen to use or to consume. Consumption of goods and services involves beliefs about what it says about us as individuals or groups: social or job status, image or wealth. Economic geographers have previously recognized cultural factors when using terms such as socialist economy, Chinese

Thematic Case Study 13.1

Models of economic location

There has been much interest in economic geography in theories or models that help to explain the development of economic activities within a spatial context. *Neoclassical economic theory* was first used by Von Thunen in 1826 (Hall 1966) to model patterns of agricultural land use and later applied to industrial location (Weber 1929) and for explaining the distribution of services and of settlements (Christaller 1966). In each of these and numerous other cases the objective is to generalize about patterns of economic activity. However, such is the complexity of the real world that it is necessary to make some assumptions: decision-makers behave in a sensible (or rational) fashion; they possess complete and correct knowledge (often referred to as perfect knowledge); everyone is attempting to maximize profits; competition is unconstrained; economic activity takes place on a uniform land surface. The key assumption is that distance is the main influence on decision-making by households and businesses with the resulting spatial patterns of economic activity explained by examining the relationship between distance and transport costs. The outcome of this approach is an ideal or optimal pattern of land use or industrial location.

The main criticisms are the unrealistic nature of the assumptions and the oversight of many other factors that have an impact on the geography of economic activity such as:

- changes in the technology of production and consumption
- the variety of ways in which businesses are organized or make decisions on where to locate
- the role of social and cultural factors
- the effects of international and national political and regulatory environments.

The limitations of neoclassical models encouraged the development of alternatives. The *behavioural model* also attempts to arrive at generalizations but the focus is shifted to the role of the individual as the principal explanation for spatial patterns. The motives, opinions, preferences, and perceptions of the individuals making location decisions are incorporated and it is not expected that the outcomes are suboptimal (i.e. decision-makers do not have perfect information and are not assumed to maximize profits). Examples of the behavioural approach include studies of consumer behaviour (Potter 1979), office location (Edwards 1983), and location change by manufacturing industries (Hayter and Watts 1983). The limitations of the behavioural approach are that it is too descriptive and only highlights variations in behaviour rather than why, for example, an entrepreneur starts a certain business at a particular location. Too much attention is given to factors specific to the firm or households at the expense of more general processes operating in the wider environment.

The *structuralist approach* is a more holistic way to explain the location of economic activity and how it changes. It is based on the premise that behaviour is shaped or constrained by wider processes in the social, political and economic spheres. Notions of culture and of class rather than individual ideas determine the spatial structure of economic activity. The structuralist approach holds that firms work within a capitalist society in which production is essentially a social process structured by the relationship between capital and labour. It is not enough to explain patterns of economic activity by studying the various sectors such as manufacturing or services for their own sake; rather, it is necessary to examine the 'hidden' mechanisms or processes (e.g. social, political) that underpin economic patterns. These cannot be measured in the way typical of the neoclassical or behavioural approaches so that much of the structuralist approach involves developing theories about, for example, the way in which changes in economic conditions affect the requirements for production that in turn change the requirements of an economic activity at a given location and that might ultimately cause it to move elsewhere (see Massey 1981).

family production networks, or the various 'corporate cultures' encountered in transnational corporations (TNCs); but only recently have they considered the economic and the cultural to be intertwined rather than worthy only of separate study.

13.2 The economic problem

Each of us has a variety of needs and wants that, when they are combined, make up a society's demand for goods and services. This demand is infinite but the resources available are not, so the economic problem (Heilbronner 1972) is how to devise a system that combines the physical and human resources needed to produce and distribute the output of goods and services to attain given ends. There are numerous possible solutions but they can be distilled to just three coordinating mechanisms (Dicken and Lloyd 1990). The first is tradition, whereby the allocation of resources to production relies on a set of 'rules' based on past practice, such as the passing on of land from father to eldest son. This incorporates a social framework for resolving the economic problem.

The second coordinating mechanism is command. In a command economy a key objective is the redistribution of wealth which requires public ownership of the factors of production. A central political authority, elected or self-appointed, uses directives to set production targets for a defined period (say three or five years) for the sorts, and quantities, of goods and services to be produced. In order to achieve the targets and to distribute the output the centralized direction of national economic development is combined with control over the allocation of human and physical resources. Some countries such as Cuba and North Korea continue to use this approach to

solving the economic problem following its demise since the late 1980s in the former centrally planned (communist) economies of the Soviet Union and of Eastern Europe (e.g. East Germany, Poland and Czechoslovakia).

China is one of the world's major emerging economies at start of the twenty-first century which is undergoing a process of reform to a market economy. This is the third approach to solving the economic problem. Market economies rely on decentralized decisions by consumers and by firms on, for example, the quantities of goods to produce, what prices to charge, and how and when exchange transactions take place. The workings of the modern market system are complex and difficult to generalize. Witness the diverse interpretations by professional economists on the significance for an economy of the latest unemployment figures, the impact of a change in bank interest rates, or of a steep rise in the price of a barrel of crude oil. Market economies are often equated with capitalist economies; in such systems most of the resources and the means of production are controlled by a relatively small proportion of individuals and firms who are seeking to improve their economic well-being through competition (see Spotlight box 13.1). Assuming that there is more than one producer of a given good or service, each is seeking to create the most favourable exchange terms (such as lower price, better quality, superior design, or efficient after-sales support) in order to retain or to enhance their share of the market. Success depends in part on the wage-labour market that transforms the labour provided by individual workers into goods or services. Because workers do not own the means of production or the goods/services that they produce they are vulnerable to changes in the requirement for their labour. This may result in slower wage growth, pressures to increase

Spotlight Box 13.1

Perfect competition

The intensity of competition among firms encourages efficiency and helps to keep prices low. The ultimate expression of this is *perfect competition*. In these purely theoretical circumstances the actions of any one individual buyer or seller have a very limited

impact on market prices. This is because everybody has access to all the information they need, all products are the same, and firms earn only the base minimum profit. If firms earn excess profits (more than the base minimum) other firms will enter the market. This will continue to happen until profits are driven down so that only normal profits are made.

Spotlight Box 13.2

Transnational corporations

The definition of a TNC is that it:

- controls establishments/economic activities in at least one other country apart from its home country;
- possesses an ability as a result of its size or the ownership of particular knowledge or skills to move its operations and its resources quickly between international locations, i.e. it is relatively footloose;
- can exploit or take advantage of differences between countries, regions or cities around the globe in factor (land, labour, capital) and non-factor (information, knowledge, regulation) endowments;

- owns and controls overseas activities, although this is not a pre-requisite for TNC status since there are many other ways, such as franchising, licensing or joint ventures, of achieving a presence in markets outside the home country.

You may well encounter a related term: multinational corporation (MNC). This suggests a firm that has premises or production plants in several countries. TNC is therefore a much more all-embracing term than MNC; there are many more TNCs than there are MNCs and they provide the basis for a more realistic assessment of the scale of international investment.

productivity, less favourable working practices, the imposition of short-term rolling contracts or, in the worst case, loss of their jobs.

Although theoretically founded on the principle of freedom of action, in practice market economies are the object of regulation by national governments and international institutions. The market is not necessarily the most effective and equitable way to allocate the rewards of production. Society needs to control the behaviour of those who own the means of production as well as providing mechanisms for sharing the benefits as fairly as possible. The geography of the economy has been increasingly influenced by the actions of national governments or groupings of nation-states (or trade blocs) such as the European Union (EU) and the North American Free Trade Area (NAFTA). International bodies such as the World Trade Organization (WTO), the United Nations Conference on Trade and Development (UNCTAD), the International Monetary Fund (IMF) and the World Bank also exert considerable influence over economic development.

The geography of the economy has also been significantly shaped by the activities of very large transnational corporations (TNCs) with markets and operating facilities that extend well beyond their home countries (see Spotlight box 13.2). The incidence of

TNCs varies by industry sector and by country. The majority are small relative to the limited number of very large global TNCs but they generate significant flows of capital, knowledge, information, expertise, products, raw materials and components amongst their own establishments and between countries. There are few parts of the world that are not in some way affected by the activities of TNCs and for some countries it makes the difference between inclusion or exclusion from the economic mainstream.

13.3 The economic system

We will return to some of these issues later in this chapter since they are the source of some of the most interesting questions for economic geographers to address during the twenty-first century. Before doing so it is useful to briefly elaborate the notion of an **economic system**. What is it and how is it structured? This is important because economic geographers are keen to understand the processes that produce spatial patterns of economic activity. A system is a set of interrelated objects that are connected together as a working unit or a unified whole. In this case the objects are, amongst others, homes, shops, offices, warehouses, manufacturing plants, and oilfields that

Plate 13.2 Large-scale distribution facilities at strategic locations with easy access to motorway networks serving national or global markets are a vital part of modern production. (Corbis/Walter Hodges)

are connected by the movement of labour, information, raw materials, finance, e-mail messages, knowledge, and freight necessary to sustain their particular role in the economy.

Conventionally this system is divided into **economic sectors** and there are four of these:

1. Activities engaged in the exploitation of natural resources, such as agriculture, fishing, mining and oil extraction, make up the **primary** sector. Much of the output from the primary sector has limited use and value until it has been transformed in some way to become part of usable goods.

2. Whether the transformation occurs near the source of the primary commodity or at a location some distance away (therefore requiring transfer between point x and point y) it requires a **secondary** sector or manufacturing. The products created by manufacturing processes may be immediately suitable for final use by consumers or may be components of other final products.

3. Either way, they will have to be distributed to the places and markets where they can be assembled,

consumed, or purchased (Plate 13.2). This is the role of the **tertiary** sector, which includes wholesale and retail trade, transportation, entertainment and personal services. Improvements in transport and telecommunications and their increasing integration with the rapid advances in computing technology since the early 1980s have transformed the operation and reach of secondary and tertiary sector firms. International markets are now more accessible and consumers' expectations of products and of service have been much enhanced. The growth of international purchasing by firms and individuals via the Internet is symbolic of these changes (Plate 13.3).

4. Goods and services that are carefully designed and refined to meet the specialist requirements or needs of sub-groups of individuals or firms or specific types of national or regional market have also been made possible by the **quaternary** sector. This includes banking, finance, business and professional services, the media, insurance and administration. These are the activities that assemble, transmit and process the information and knowledge that is

37,000. Clothing manufacturing employed 140,000 in 2002, down by more than 55 per cent since the mid-1970s. The decline of the coalmining industry is even more dramatic; just 10,500 were employed by the industry in 1999 compared with more than 190,000 just 15 years earlier (1984). The primary sector, already a very small part of the economy in most developed countries, is also becoming even smaller as more productive methods of farming are introduced and crop yields are enhanced by improved crop disease resistance and better fertilizers. As a counter to these trends, the tertiary and quaternary sectors have expanded and diversified as information and telecommunications technology (ICT) has shifted production towards goods and services whose functionality is determined, more than anything else, by the knowledge and expertise incorporated in their design and specifications. These sectors are growing most in the newly industrialized countries that are 'catching up' with the developed countries where service industries already account for more than 65 per cent of total GDP (Figure 13.2).

These structural and functional readjustments to the economic system are closely linked to

Plate 13.3 Customers can now purchase a wide range of goods and services across international borders from the comfort of their own homes using Internet sites such as Amazon.com. (Alan Thomas)

required by the other three sectors to enable them to adjust effectively and efficiently to the changing geographic, economic, and social parameters of doing business in the twenty-first century.

13.4 A dynamic economy

It should be apparent from this very brief outline of the economic problem and the economic system that relationships between the sectors, their internal composition and the balance between them are constantly changing. This takes place at all levels, from the rural locality to the largest metropolis, from peripheral to core regions, from developing to developed economies. This dynamism is nothing new but during the closing decades of the twentieth century it was dominated by changes which will set the agenda well into the twenty-first century (such as more flexible production systems, see Chapter 14).

Since the 1970s the share of employment in the secondary sector has been declining steadily, even in the developing economies and post-socialist economies that have invested heavily in manufacturing as a means of achieving economic growth (Figure 13.1).

Take the case of the United Kingdom where, for example, the steel industry employed 165,000 in 1980; by 2002 the figure had decreased to approximately

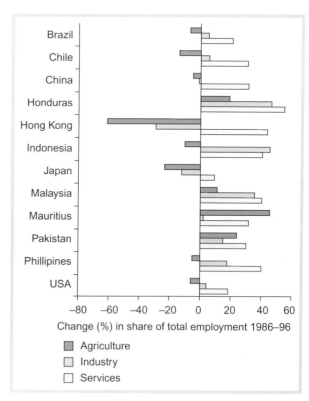

Figure 13.1 Changing structure of employment by sector, selected countries 1991-2001.

Figure 13.2 Service industries share of total GDP, by country, 2001.

internationalization of economic activity, or the spread of economic activities beyond national boundaries. This is not a new phenomenon, but it has been taking place more quickly and has involved an ever-widening range of economic activities since the 1980s. Internationalization should not be confused with globalization, which is a more contemporary phenomenon that involves integration across national boundaries of markets, finance, technologies and nation-states at a level that has not been witnessed in the past. It has enabled nation-states, TNCs, as well as individuals, to extend their reach (markets, travel) around the globe faster, further, deeper and at lower cost than could ever have been imagined, even ten years ago (Plate 13.4). As Dicken (1992: 1) has put it: 'These two themes are intimately connected in the sense that the pervasive internationalization, and growing globalization, of economic life ensure that changes originating in one part of the world are rapidly diffused to others.'

In principle at least, globalization implies convergence of economic, social, and cultural values in the localities, regions and nations that come together as the 'world economy'. Economic diversity is replaced by economic uniformity. However, as the other contributions to this section show, we should beware of such sweeping assumptions. 'Globalization' is convenient shorthand, but it should not lead us to overlook the significant variations in the economic health or the levels of participation in the global economic system of different parts of the world. This is clearly apparent on a map showing countries drawn at a scale proportional to their share of world gross domestic product (GDP) (Figure 13.3). Almost a quarter of the world's total GDP is accounted for by the United States. Almost half of the world's population is in low-income countries (Table 13.1) but they produced only 3.4 per cent of world gross national income (GNI) in 2001, equivalent to just US$430 per capita. The high-income countries have one sixth of the world's population but almost 81 per cent of total GNI in 2001, at a rate per capita that is 62 times greater than for low-income countries such as Ethiopia or Nicaragua where severe poverty (more than 50 per cent of the population living below US$1 a day) is often a major problem. 'Every country may not feel part of the global[ization] system, but every country is being globalized and shaped by it' (Friedman 1999).

Plate 13.4   Globalization has been accompanied by increasing geographical concentration of corporate control as symbolized by the density of office development and the skyline of Manhattan, New York City. (T. Paul Daniels, Bromsgrove)

Figure 13.3   A proportional map of the world, share of GDP by country, 2002.
Source: Data compiled by CIA (2003)

Table 13.1 GNI-based and GDP-based measures for the relative performance of low, middle and high income countries, 2001 (all data in US dollars)

Country groups	Population (millions)	GNI[1] ($ billions)	GNI per capita[2] ($)	Agriculture[5] value added[4] (% of GDP)[3]	Industry[6] value added (% of GDP)	Services[7] value added (% of GDP)
Low income[8]	2,460	1,069	430	24.7	31.7	43.6
Middle income[9]	2,640	4,957	1,860	9.5	35.5	55.0
High income[10]	950	25,372	26,510	1.9	28.6	69.5
World	6,050	31,400	5,120			

Source: Extracted from data in World Bank (2003) *World Development Indicators* database.

Notes:
1 GNI (formerly GNP) is the sum of value added by all resident producers plus any product taxes (less subsidies) not included in the valuation of output plus net receipts of primary income (compensation of employees and property income) from abroad. Data are in current US dollars.
2 GNI per capita (formerly GNP per capita) is the gross national income, converted to US dollars using the World Bank Atlas method, divided by the midyear population.
3 GDP (in current US$) is the sum of gross value added by all resident producers in the economy plus any product taxes and minus any subsidies not included in the value of the products.
4 Value added is the net output of a sector after adding up all outputs and subtracting intermediate inputs.
5 Agriculture corresponds to ISIC divisions 1–5 and includes forestry, hunting, and fishing, as well as cultivation of crops and livestock production.
6 Industry comprises value added in mining, manufacturing (also reported as a separate subgroup), construction, electricity, water, and gas.
7 Services include value added in wholesale and retail trade (including hotels and restaurants), transport, and government, financial, professional, and personal services such as education, health care, and real estate services.
8 Low-income economies are those in which 2001 GNI per capita was $745 or less.
9 Middle-income economies are those in which 2001 GNI per capita was between $745 and $9,205 (90 countries).
10 High-income economies are those in which 2001 GNI per capita was $9,206 or more (52 countries).

13.5 The rise of a global economy

The transition to a global economy has a number of consequences (Amin and Thrift, 1994). First, *there has been a marked increase in the power of finance over production*. Finance capital now takes many forms and moves almost seamlessly and with great speed, especially between the world's stock, currency, commodity, and futures exchanges in 'global cities' such as Tokyo, New York, London and Hong Kong (see Chapter 15). Electronic trading has ensured volatile and fast-changing financial markets that can transform the economic prospects of companies and, more importantly, of national or regional economies overnight.

Take the case of Asia, which by 1997 had reached its peak as an engine for worldwide economic growth. This centred on the region's consumption of vast quantities of raw materials used by its flourishing consumer goods conglomerates (especially those of Japan, South Korea and Taiwan) that exported merchandise such as electronic products to every corner of the world. But in early December 1997 global speculators took the view that Thailand's economy was much weaker than the published figures (or speculator sentiment) warranted. This caused a free fall in the value of the nation's currency (baht) and the government of Thailand announced that it was closing almost all of the country's leading finance houses, mostly private banks, because they had almost instantly become bankrupt. With the notable exception of Singapore, this triggered a dramatic loss of confidence in the economies of the other emerging markets in the region. The finance capital, often in the form of foreign direct investment (FDI) that had flowed into the region on the back of its success, was rapidly diverted elsewhere.

The economic contagion did not stop there. Prices for commodities such as crude oil, aluminium and copper began to collapse as demand fell, triggering impacts on the economies that supplied them. Russia, which since the early 1990s has been trying to refashion its economy in the image of a market rather than a centrally planned system, was one of those hit hardest. With many of its pre-market economy factories not up to the task of producing internationally competitive goods there was a heavy dependence on the income from sales of crude oil, and foreign loans to fund economic readjustment. Foreign banks and corporations were attracted by unrealistically high interest rates on financial instruments such as government bonds (as much as 70 per cent by early 1998) that they were persuaded to purchase in the belief that the IMF would help Russia if there was a risk of default, thus protecting their loans. In the event, the collapse of crude oil prices triggered by the Asian crisis had a massive impact on Russian government revenues, making it difficult to pay interest on government bonds. When Russia found it necessary to devalue its currency (the rouble) in August 1998 it was not long before it defaulted on its government bonds.

Although the Russian economy is small by world standards, events there had further consequences elsewhere. With some of the world's largest banks reporting losses on their Russian investments of several hundred million dollars, they needed to compensate by selling assets in countries that were more financially sound. The Brazilian economy had been performing quite well during the mid-1990s but investors lost confidence during 1998 and, although the government raised interest rates to 40 per cent in order to retain capital, there was a flight of investment to safer havens such as the United States or some of the European markets (United Kingdom, France or Germany).

A second consequence of the rise of a global economy is the pivotal role of *knowledge*. Since knowledge, its creation and exchange, is embodied in people rather than machines, economic welfare relies more on 'producing' an educated and skilled workforce. There is disagreement over whether this is linked to growth of GNI or to more efficient production. Third, *technology has become transnationalized*, especially amongst knowledge-intensive economic activities such as financial services or telecommunications. This does not necessarily mean better access to the factors of production; the increased complexity of the opportunities so created means that only those producers and institutions with the resources to manage technology can really take advantage.

Fourth, globalization of technological change, the mobility of finance capital, and transportation and communication has been accompanied by the emergence of *global oligopolies*, i.e. there are few sellers of a product or a service (Case study 13.2). The large share of the world market for computer operating systems and software held by the Microsoft Corporation is a good example. There is now a sense in which corporate and economic survival requires 'going global'. Fifth, the power of individual nations to regulate their own economic development or to exercise a strong influence on the outcome, for example, of trade agreements has been diluted by *the rise of transnational institutions* that coordinate, steer and even regulate aspects of the world economy (see examples on p. 292). Sixth, *the 'wiring' of the world* has stimulated more extensive cultural flows and has 'deterritorialized signs, meanings and identities' (Amin and Thrift 1994: 4). Last but not least, *these symptoms of globalization present themselves as new (global) economic geographies*. These have been characterized in various ways, including the centralization of economic power in global cities (Sassen 1991), as a 'borderless' global economy (Ohmae 1990), as a 'space of flows' (Castells 1989), or as a new global division of labour (Dicken 1992).

13.6 Is globalization of the economy inevitable?

Globalization of the economy is well established but will it become all-pervasive? Not necessarily. Perhaps we have become dazzled by the rhetoric of globalization. Two examples will suffice. First, labour markets have largely continued to function within national boundaries. Of course many millions of people (see Chapter 4 for a fuller discussion of international migration) move temporarily or permanently to jobs in countries other than their own. But the numbers involved do not stand comparison with the massive migration of labour from Europe to North America or to Australia and New Zealand in the late nineteenth century and the early part of the twentieth century. In Asia, Latin America or Europe, labour markets are essentially national for, amongst others, linguistic, cultural and educational reasons. The

Thematic Case Study 13.2

An oligopoly: Nestlé

Nestlé is a Swiss-based company founded in 1866 that is now the top-ranking food and beverage producer in the world with sales of over US$64 billion in 2002. It employs 230,000 and has factories and operations in almost every country in the world. Nestlé has achieved this position by acquiring other related companies and brands; in 2001 it purchased Ralston Purina, making it the leading pet food manufacturer in the world; it purchased the leading French bottled water producer (Perrier) in the late 1990s to become number one for that product. Nestlé is the leading producer of instant coffee (Nescafé was the first instant coffee on the market), and a major manufacturer of sweets and chocolates, and has increased its dominance of the ice-cream sector by merging the Nestlé Ice Cream Company with the leading US company (Dreyers) in 2003. In addition to food products, Nestlé is involved in cosmetics (it has a large stake in L'Oréal), nutritional supplements, and eye care. Its long term strategy is to continue expanding worldwide, especially into emerging markets such as China, India, Latin America, and Russia where it has recently invested US$120 million in a new coffee-processing plant.

Other oligopolies include Coca-Cola, Pepsico, Cadbury Schweppes, Pearson, Interbrew, and Gillette. For these leading oligopolies the key objective is to protect or to acquire world-class brands. There are currently some 40 consumer brands worldwide that sell more than one billion dollars annually (Coke, for example, sells over US$15 billion). The companies with these exceptional products are growing and expanding internationally at more than 10 per cent per annum.

European Union (EU) has actively tried to dismantle the administrative and legal barriers to cross-border movements of labour and the ability to practise certain professions but this has not, for example, triggered a massive shift of labour from the less prosperous (southern) regions to the more prosperous (northern) regions.

Second, globalizing markets should become more uniform with respect to prices and product specifications because, for example, of the liberalization of trade rules or the pressure of competition on high-cost producers to lower their prices. Yet, price differentials for the same products (DVD recorders, digital camcorders, personal computers, cars, washing machines, books, etc.) have remained as wide, or even wider. By levying value-added taxes (VAT) on goods or services, partially complying with the universal aspects of trade agreements such as equal treatment for all firms or labour whether from the home country or from outside, or applying import quotas, national governments exert considerable influence. Companies may also be culpable where they rely on inefficient distribution networks, use less cost-effective transport, or charge premiums for minor product modifications in order to meet environmental or other regulations in particular markets. The possibilities are numerous so that there is potential for, rather than real, product price convergence.

Thus, a truly integrated world economy may be some way off but it will certainly be a central part of the development agenda for the early decades of the twenty-first century. ICT will continue to set the pace, changing the way in which information and knowledge are transferred and accessed, or modifying the nature of work and the skills required. This may change the balance of demand for labour between richer and poorer economies and between and within each of the major economic sectors. Satellite and other surface-based telecommunications media have also made it much more difficult for governments to regulate their economies (using trade barriers as well as internal controls). It is now more difficult to levy taxes on transactions of all kinds (financial, currency, retail, entertainment, information, etc.) that can take place electronically without the impediment of national boundaries. This adds to the potential for further displacement of jobs and growth opportunities for less

Plate 13.5　The world headquarters of Microsoft Corporation, Redmond Washington, USA. (Corbis/Matthew Mcvay)

developed economies and to increased uncertainty as TNCs can relatively easily move production from one country to another (Spotlight box 13.3).

Further integration of economies will also depend upon national and international political priorities and the extent to which individual nations are prepared to give up some of their sovereignty. Greater economic integration in Asia, for example, is dependent on a reduction in cultural differences/values across the region. The new pan-European currency (the euro) was launched as an electronic currency on 1 January 1999 and became legal tender (replacing the old national currencies) in 12 of the 15 EU member countries on 1 January 2002. Only the UK, Sweden, and Denmark remain outside the 'eurozone'. The new currency has only recently begun to demonstrate that it can be a viable alternative to established international currencies such as the US dollar, the Japanese yen and the UK pound sterling.

13.7 Trade and comparative advantage

One important reason for creating the euro is to reduce the cost of cross-border trade and investment amongst the participating countries. Such trade is one of the most obvious manifestations of the globalization of the world economy. Since 1950 the value of the world's total GDP has increased by a factor of approximately nine, but world trade has increased more than eighteenfold. More recently, trade in services has also been growing rapidly and now accounts for more than 23 per cent of world exports (by value) (see Table 13.2). World exports as a percentage of GDP have grown from around 7 per cent 50 years ago to more than 16 per cent today. In addition to the more open climate for trade created by organizations such as the World Trade Organization (WTO), expansion has taken place because trade is generally considered good for economic development.

The export side of trade is especially important since it helps to protect existing jobs as well as to create new ones. Imports are viewed more cautiously by most

Spotlight Box 13.3

Comparative advantage

Comparative advantage explains the tendency for countries (or regions/localities within countries) to specialize in certain goods and/or services even if they have the ability to fulfil their needs from domestic production. As long as countries or regions specialize in those products or services in which they have comparative advantage, they will gain from trade. Advantages can stem from spatial variations in, for example, mineral or land resource endowments, from variations in the educational levels of the labour force, in access to markets, or differences in levels of technology. For comparative advantage to work effectively it is necessary to assume a system of free trade, hence the significance of the trade liberalization that has been high on the agenda of many countries since the 1980s. There is a nagging concern, however, that comparative advantage is not reflected in actual patterns of world trade. We would expect the biggest flows to be between the countries with the largest cost differences. It seems that consumer tastes and geographical proximity are actually more important than cost differences. This explains the fact that the vast majority of trade is between countries with relatively small cost differences, often involves similar rather than different goods (such as cars, electrical goods of all kinds, certain kinds of business services such as management consulting), and occurs on a 'nearest neighbour' basis. Well over half of EU goods and services trade takes place between the Member States while Canada and Mexico are the major trading partners of the US.

Table 13.2 Trends in merchandise and commercial service exports, 1980–2000

Economy	Sector	Exports (% of world total)			Annual change (%)	
		1980	1990	2000	1980–90	1990–2000
Developed economies[1]	Merchandise	62.3	72.4	64.1	7	5
	Services	77.1	79.2	73.6	8	6
Developing economies[2]	Merchandise	29.7	23.0	28.9	3	9
	Services	19.5	18.2	23.0	7	9
Economies in transition[3]	Merchandise	7.7	3.1	4.3	–	–
	Services	–	–	3.4	–	–
Total world exports (value, US$bn)	Merchandise	2,028	3,395	6,252	5	6
	Services	364	783	1,475	8	7

Source: WTO (2003), extracted from data compiled by Trade in Services Section, Economic Research and Statistics Division.

Notes:
1 Developed economies: North America, Western Europe, Australia, Japan and New Zealand.
2 Developing economies: Latin America, Africa, Middle East and Developing Asia (i.e. Asia excluding Australia, Japan and New Zealand).
3 Economies in transition: Central and Eastern Europe, the Baltic States and the CIS.

governments because they have the potential to cause job displacement. However, according to trade theorists as far back as Mill (1821) and Ricardo (1817) the benefits of international trade actually arise from the goods received (imports) rather than those exported. This arises from the principle of comparative advantage (Spotlight box 13.3). Porter (1990) has suggested the concept of competitive advantage (the advantage that a country or a business has over its competitors because of the quality or superiority of its products or services which will persuade other countries or customers to buy from it rather than from competitors). This is actually a more useful way of explaining how individual nations participate in the world economy. It is broader than comparative advantage in that it incorporates differences in values, cultures, histories and institutions as well as variations in endowments of the factors of production.

History also plays a part. The City of London proved to be an innovative location for financial and insurance activities during the late eighteenth century. Originally stimulated by Britain's extensive trade with its colonies and subsequently diversifying as the industrial revolution really got under way during the nineteenth century, London's finance houses had the established prestige, experience and knowledge that later entrants such as New York or Paris had yet to accumulate.

The combination of absolute, comparative and historical advantage has endowed some countries and TNCs with significant 'clout' within the globalized economy. The home countries of TNCs tend to be the developed economies of Europe and North America but the lower wage costs or longer working hours in the developing countries or the newly emerging economies of Eastern Europe now make them an attractive alternative, especially in the context of an ongoing international division of labour (Coffey 1996, see also Chapter 14 for an illustration of division of labour). Manufacturing and service industries' production is increasingly subdivided into a series of discrete processes undertaken by separate workers or groups, increasing the scope for companies to seek locations offering lower wage costs and less restrictive regulations. But factors such as lower productivity, higher training costs and less efficient transportation and telecommunications infrastructure can often outweigh the wage cost advantages of the developing economies. This is not to say that if the circumstances are right, footloose production will not be relocated. A good example is the recent emergence of the Indian

sub-continent as a favoured location for outsourcing software development, call centres, technical support and related work from the US and the UK (Case study 13.3). Other examples include the relocation by US firms of ticket processing and other high volume, labour-intensive, routine functions to Jamaica and insurance claims processing to Ireland.

13.8 The rise of transnational corporations

The example of India will be repeated elsewhere (such as the Pearl River Delta region, southern China). The success of this region and others like it depends partly on the actions of TNCs. If we group the top ten TNCs by their assets and according to whether their home countries are developed, developing or transition economies, the scale difference between global household names such as General Electric, Toyota Motor, or BP as measured by employment, and their opposite numbers, is immense (see Table 13.3). The leading TNCs from the developed countries are primarily in the automotive, petroleum exploration, and telecommunications sectors. The top ten from the developing countries include two primary sector TNCs while those based in the transition economies are primarily in transportation. However, size of TNC does not determine the extent of their transnationality (as measured by the transnationality index, see Table 13.3). While Vodafone has by far the highest index for the 30 corporations shown in Table 13.3, Latvian Shipping is actually more transnational even though it is one-tenth the size (by employment).

The pros and cons of global economic integration are also seen by some as a function of the behaviour of TNCs. Depending on the viewpoint, they are seen as the saviours of otherwise disadvantaged economies because of the technology or the job opportunities that they provide or, conversely, they are portrayed as ruthless exploiters of low-cost labour and inadequately protected resources in environmentally sensitive areas. Either way they cannot be ignored because, whatever measures are used to illustrate globalization, TNCs invariably figure prominently (Daniels 1999). They are a major source of foreign direct investment (FDI) and sales outside their home countries by affiliates are growing faster than exports and were worth about US$7 trillion in 1995 (Spotlight box 13.4). This is not insignificant, since it exceeded the total value of all world exports of goods and services. TNCs are also

Regional Case Study 13.3

Outsourcing to a developing economy: India

Even though India is a poor country, generally lacking basic infrastructure, its education system produces a large number of highly trained engineers, computer scientists and mathematicians who, in addition to their native dialects, speak the global *lingua franca*, English. Time zone location means that its workers are at their desks when those of Europe, the US, or Japan are heading home or already in bed. Providing that they can access their markets/clients Indian companies are well placed to supply knowledge-based services to the global market. Since the mid-1990s, Bangalore (southern India) has emerged as a location for hundreds of computer software companies, including Infosys (first launched on the Bombay stock exchange in 1992 and on the Nasdaq stock exchange (New York) in 1999 where it was valued at US$2.7 billion), WIPRO, and Tata Consultancy Services. Using two telecommunications satellites, the Bangalore firms provide online update, repair and computer code writing services to multinational clients such as British Airways, Boeing Corporation, Goldman Sachs and Levi Strauss. Suburban districts of several Indian cities, such as Noida and Gurgaon in Delhi, are attracting significant numbers of call centre and back office processing jobs relocated by major international financial institutions such as HSBC, Prudential, Citicorp, and GECapital. This business process outsourcing is transforming the economic landscape as new office buildings, glittering new shopping centres, and multiplex cinemas appear alongside much less ostentatious traditional housing and retailing. The former are symbols of the inward investment by TNCs, the increasing spending power of some Indian workers, and the way in which economic globalization is shaping change, economic and social, in places several thousand miles apart. It is estimated that some 200,000 jobs will be outsourced from the UK to India by 2010. However, such is the demand that wage inflation at 20–25 per cent per year is threatening to erode India's cost advantages. This may ultimately lead to outsourcing that bypasses India for China or the Philippines, or even to outsourcing by Indian-owned firms to other lower-cost locations.

major investors in buildings, equipment and people at locations outside their home country (FDI) although relative to all investment, especially by the developed economies, the proportion is relatively small (6–7 per cent) (Daniels 1999).

How can we explain the inexorable expansion of TNCs? Given that most companies are driven by capitalist principles, they need to achieve profit growth for stakeholders and to enable reinvestment in their plant, infrastructure, and human resources. Saturation of their home markets reduces the opportunities to become larger or to produce more, so that the scope for making savings is through *economies of scale*, i.e. reducing the cost per unit of output as firm or plant size increases. Increasing market size does not necessarily involve the location of new production facilities in overseas markets since it is now often cost-effective to transport finished goods, especially high value-added products, to geographically dispersed locations. Transport developments such as containerization, air freighters, and faster customs clearance and transfer facilities at seaports and airports have made it possible for TNCs to outsource the manufacture of component parts to global suppliers while continuing to assemble the finished product at a small number of key locations. Examples include car assembly, aircraft construction, computer chips, laptop computers, and a wide range of other high value to weight technology-intensive products. This is part of a recent trend towards the integration of international production systems over ever larger geographic scales.

At the beginning of the twenty-first century the location decisions of TNCs and their partners in international production systems are dictated by three drivers (UNCTAD 2002a). *Cost differentials*, the first of these drivers, are long established; production costs will always be evaluated relative to the efficiency and productivity of a location. The second driver involves *asset seeking*, especially skills and knowledge, on the global scale using ICTs to utilize these assets amongst a

Table 13.3 The top ten TNCs (ranked by foreign assets and transnationality index) in developed, developing, and emerging economies, 2000

T index[1] (%)	Corporation	Country	Industry[2]	Rank (by foreign assets)	Employment	
			DEVELOPED COUNTRIES		**Total**	**Foreign (%)**
81.4	Vodafone	UK	Telecommunications	1	29,465	81.5
76.7	BP	UK	Petroleum expl./ref./distr.[3]	7	107,200	82.4
67.7	ExxonMobil	US	Petroleum/expl./ref./distr.	3	97,900	65.4
59.7	Vivendi Universal	France	Diversified	4	327,380	64.2
57.5	Royal Dutch/Shell	UK/ Netherlands	Petroleum/expl./ref./distr.	6	95,365	60.0
57.4	Fiat	Italy	Motor vehicles	10	223,953	50.1
53.8	Telefonica	Spain	Telecommunications	9	148,707	47.9
40.3	General Electric	US	Electrical and electronic equipment	2	313,000	46.3
35.1	Toyota Motor	Japan	Motor vehicles	8	210,709	–
31.2	General Motors	US	Motor vehicles	5	386,000	42.8
			DEVELOPING COUNTRIES			
78.6	Neptune Orient Lines	Singapore	Transport and storage	9	8,734	78.3
54.8	Cemex	Mexico	Non-metallic mineral products	2	25,884	59.7
50.3	Hutchison Whampoa	Hong Kong/China	Diversified	1	49,570	54.8
42.7	LG Electronics	Republic of Korea	Electrical & electronic equipment	3	46,912	42.8
35.8	Petróleos de Venezuela	Venezuela	Petroleum expl./ref./distr.	4	46,920	11.6
34.9	Samsung Electronics	Republic of Korea	Electrical & electronic equipment	8	60,977	27.8
29.5	Petronas	Malaysia	Petroleum expl./ref./distr.	5	23,450	16.2
28.9	Companhia Vale Do Rio Doce	Brazil	Mining and quarrying	10	17,364	35.6
18.5	Samsung Corporation	Republic of Korea	Diversified/trade	7	4,740	3.7
15.8	New World Development	Hong Kong/China	Diversified	6	23,530	3.4

Table 13.3 continued

T index[1] (%)	Corporation	Country	Industry[2]	Rank (by foreign assets)	Employment	
		EMERGING ECONOMIES			Total	Foreign (%)
87.3	Latvian Shipping	Latvia	Transport	3	1,748	64.3
63.2	Atlantska Plovidba	Croatia	Transport	10	509	–
59.4	Primorsk Shipping	Russian Federation	Transport	4	2,777	47.1
53.7	Novoship	Russian Federation	Transport	2	7,406	1.2
46.9	Gorenje Group	Slovenia	Domestic appliances	6	6,691	8.8
39.7	Pilva Group	Croatia	Pharmaceuticals	9	7,857	33.7
38.8	Far Eastern Shipping	Russian Federation	Transport	7	8,873	3.0
34.7	Lukoil Oil	Russian Federation	Petroleum and natural gas	1	130,000	15.4
31.6	Pedrovka Group	Croatia	Food and beverages	8	6,827	7.6
4.3	Hrvatska Elektroprívreda	Croatia	Energy	5	15,877	–

Source: Extracted from UNCTAD (2002a), based on UNCTAD/Erasmus University database and an UNCTAD survey of the top TNCs in Central and Eastern Europe. The United Nations is the author of the original material.

Notes:
1 The *transnationality index* (T index) is the average of three ratios: foreign assets to total assets, foreign sales to total sales and foreign employment to total employment.
2 Using US Standard Industrial Classification as used by the US Securities and Exchange Commission.
3 Exploration/refining/distribution.

wide range of locations. The third driver that has become a key influence of TNC location decision-making is *clustering*. Although geographic concentration of the related activities of production is long established, cluster formation has now become a global as well as a local or regional process. TNCs have recognized that co-location with competitors, knowledge providers such as business and professional services, other service providers, and suppliers offers significant advantages. Access to these sources of tacit knowledge is vital for competitive advantage and TNCs must trade these off against the opportunities that arise from their ability to be footloose.

Not all economies of scale are material; they can also arise from opportunities for global advertising and marketing in support of a global brand or image of the kind associated with some soft drinks, fashionable clothing, or cars. The development of large companies

can also be explained as the outcome of *vertical integration*: the efficiency gains, improved quality control, and protection of proprietary information achieved by acquiring the forward (customer) and/or the backward (supplier) linkages in the production chain. In some cases these acquisitions will be outside the home country of a firm, thus transforming it into a TNC. The 'follow the leader' effect on the rise of TNCs should not be overlooked. Corporations that undertook production at plants outside the US during the 1920s and 1930s, such as IBM, Kodak and Ford, were generally successful and it was logical for competitors to follow suit. More recently, service activities, such as accountancy, legal, financial and advertising firms, that provide TNCs with specialist knowledge and advice on critical aspects of their overseas expansion and trading have followed their clients to similar locations. Business and professional

Foreign direct investment

For statistical purposes, *foreign direct investment* (FDI) is defined as the acquisition by an individual or enterprise resident in one country of assets located in another. There are numerous difficulties in identifying such flows and collecting reliable information about them. Especially important is the fact that many take place as transactions internal to TNCs and are therefore difficult to enumerate. If financial transactions result in a material interest in a foreign enterprise (usually defined as a shareholding of at least 10 per cent) they are included as part of FDI.

World FDI inflows exceeded US$1,271 billion in 2000 with developed countries accounting for almost 80 per cent of the total. Flows had almost doubled since 1998 but then declined to an estimated US$760 billion in 2001. Thus, while FDI is clearly contributing to the increasing internationalization and integration of economies with firms producing more and more outside their home country, it is also highly volatile. A slowdown of the world economy in 2000–2001 caused a decline a corporate cross-border mergers and acquisitions (M&As) which form a large part of FDI. The value of worldwide M&As (domestic and foreign) during the first eight months of 2001 was half the value reported during the same period in 2000. The tragic events of September 2001 in New York further accentuated the decrease of FDI flows. It is also the case that FDI between developed countries is more important than FDI into developing countries. However, some of the biggest inflows in recent years, although still a relatively small proportion of the total, have been recorded by the newly emerging economies of Central and Eastern Europe.

FDI inflows, by region, 1998–2001 (billions of US$)

Region	1998	1999	2000	2001[1]
World	693	1,075	1,271	760
Developed countries	483	830	1,005	510
Developing countries[2]	188	222	240	225
Africa[3]	8	9	8	10
Latin America and the Caribbean	83	110	86	80
Asia and the Pacific	96	100	144	125
South, East and South-East Asia	86	96	137	120
Central and Eastern Europe	21	23	25	25
Including the countries in the former Yugoslavia	22	25	27	27

Source: UNCTAD, FDI/TNC database (UNCTAD 2001). The United Nations is the author of the original material.

Notes:

1 Preliminary estimates, September 2001, on the basis of 51 countries that accounted for more than 90 per cent of FDI inflows in 2000.

2 Including the countries in the former Yugoslavia.

3 If South Africa is included, the figures are 8 in 1998, 10 in 1999, 9 in 2000 and 11 in 2001.

service firms, for example, have created the level of support and expertise that their clients already enjoy at their home country headquarters.

FDI by TNCs is unevenly distributed. Although the number and size of developing country TNCs is increasing, most of the FDI inflows are generated by developed country TNCs and most of their investment (more than 60 per cent in 1998) goes to other developed economies. There are signs that the balance is shifting towards the developing economies and newly emerging markets such as China and Eastern Europe that have significant consumer potential. For those countries able to attract it, FDI helps to secure access to capital, technologies and organizational expertise. The net effect is modernization of infrastructure, to increase industrial capability and to improve the quality and breadth of much-needed financial, business and professional services. China, a vast potential market comprising more than one-fifth of the world's population, was closed to FDI until very recently but is now one of the leading recipients. Prior to the collapse of their economies almost two-thirds of the investment flows to developing countries during the late 1990s went to Asia (excluding Japan). Some of the states of Latin America such as Brazil and Chile were also attracting significant FDI, along with Eastern Europe before the knock-on effects of the Asian financial crisis occurred (see p. 298). During 2002 and 2003 investor confidence in these regions recovered but much of the African continent continues to be bypassed for any form of FDI, even though many of its countries are resource-rich. With the exception of South Africa, consumer purchasing power is low and average incomes per capita are not anticipated to grow in a way that will raise the consumption of major consumer goods to a level that justifies the attention of most TNCs.

To summarize, globalization of the economy has strengthened the role of market forces while, as a result of advances in telecommunications and transportation technology, easing some of the constraints imposed by geography. As the real cost of international telecommunications has declined steadily over the last fifteen years, global financial integration has strengthened. Globalization has also diversified the opportunities for new kinds of international trade, especially in services, and enabled easier transfer of the technology and innovation that encourages economic development and participation. The importance of non-government organizations, TNCs, and regional trading blocs such as the EU and NAFTA for shaping the economic geography of globalization has been greatly enhanced and will continue to increase.

13.9 Role of localities in a globalizing economy

While the geography of the economy is increasingly accommodating a global dimension, this is not at the expense of individual localities. As Porter (1990: 158) has noted, 'it is the combination of national and intensely local conditions that fosters competitive advantage.' It has been suggested that 'territorial policies ... should now focus more on mobilizing local resources to build competitive strength and create jobs. This change of emphasis can add impetus to the restructuring of national economies by reinforcing the capacity for self-generating change' (Johnston 1998). Localities are taken to mean subdivisions within nations such as cities or regions or places that often have a particular economic identity because of the kinds of activities that take place there. Some localities even smaller and less complicated than cities, such as rural districts that are the foci for producing very particular kinds of wine (France), dairy products (Denmark) or woollen goods (Wales), for example, are also included. Many are identified with named territorial units such as administrative areas, while others are industrial districts or agglomerations whose identities are derived from specific economic activities, such as Motorsport Valley (Pinch and Henry 1999) in southern England or Silicon Valley, California (see Chapter 14 for more details of Motorsport Valley). While these are often highly specialized and self-contained localities, often with systems of governance and regulation that fit their particular needs, they are inextricably linked with the wider national and global economic system.

These 'new industrial spaces' thrive on the dynamism, innovation and 'untraded interdependencies' (knowledge and information that circulates through the transfer of key workers between firms or via social and other networks) that are made possible by agglomeration. Even TNCs, that are now often portrayed as being 'placeless' because their operations are so extensive and relatively mobile, started from businesses that were initially nurtured in a particular locality with its own economic and other characteristics, and there is evidence that the locally shaped attributes of firms are carried through into their organization and transformation into TNCs (Dunning 1981).

On the other side of the coin is the question of how, once they have become globalized, TNCs leave an imprint on the localities where they have manufacturing plants, offices, research and development laboratories, distribution facilities or retail floorspace. Part of the answer is determined by the way in which a TNC operates. Those that rely heavily on in-house production and exchange of goods and services will have a less marked impact on a locality when compared with those that outsource their requirements for components or for services ranging from security and cleaning to marketing or personnel recruitment. Local orientation will also vary by type of corporate function with headquarters having stronger connections to suppliers and services than, say, a manufacturing plant. The way in which corporate operations are managed will also be influential. Some TNCs grant a good deal of autonomy to their subsidiary or branch establishments for their operational and strategic planning. Others exert highly centralized control and direction from their world headquarters or from a small number of regional headquarters covering, for example, Western Europe or the Asia Pacific.

It is actually very difficult to generalize about the embeddedness or otherwise of TNCs in localities; even within one organization this may vary according to the style and social or community networks accessible to individual managers or teams (see Spotlight box 13.5). This reflects the importance of the most visible aspect of corporate local embeddedness: the links with suppliers of goods and services. As the demand for flexible production and 'just-in-time' supply of inputs such as car components has escalated so has the involvement of TNCs with their suppliers become much closer. This is partly driven by the need to maintain the very exacting quality control standards required for the production of sophisticated products for ever more demanding clients and customers. It is also driven by the need to share the technological expertise, which many small firms cannot expect to be able to acquire very easily if they are to fulfil the expectations of the TNCs' clients. This may extend to TNC staff working at the premises of supplier firms or being involved with the training of their staff. The 'localness' of supplier firms becomes quite important in these circumstances. On the other hand there are numerous examples of TNCs bringing together personnel from around the globe at one training establishment or at a research and development facility in the home country to transfer the necessary information, knowledge and training to their preferred supplier firms. Whatever the model used, such arrangements make it more difficult for local

Spotlight Box 13.5

Embeddedness of economic activity

The advantages of local agglomeration for some economic activities have been eroded as more of them, especially TNCs, have a presence that is simultaneously local and global. This hybrid identity is sometimes referred to as *glocalization*. There has therefore been a shift in the literature on industrial agglomeration towards the idea that an institutional atmosphere is important for maintaining identities and competitiveness. This sociology of institutions contributes to the idea of *embeddedness* (Grabher 1993). Economic action is now seen as social action and is affected by a network of relations between 'actors' (firms, institutions, individuals) that may be located within an agglomeration but can also be located in different agglomerations. Instead of trying to understand and explain the continuation of existing industrial agglomerations or the emergence of new ones by using economic considerations such as vertical disintegration of changes in product life-cycles, attention has turned to social and cultural explanations. Levels of inter-firm collaboration, social consensus, the nature and intensity of institutional support and networks for local business and the circumstances encouraging skill formation, innovation and the circulation of ideas are the kinds of factors that are considered important (see Storper 1993, Amin and Thrift 1994).

Plate 13.6 A street vendor walks past an old house marked with the Chinese character for demolish'. Parts of China, such as Shanghai or Guangzhou are experiencing rapid economic transformation which is expected to continue well into the 21st century. (Associated Press/Joanne O'Brien)

national coordination because some localities within a nation, especially cities, are usually better positioned to act as independent competitors for globally mobile investment or employment. Localization has boosted urbanization, not least the concentration of economic activities, while globalization, the rise of the service economy and increasing international competition offer opportunities as well as threats to urban areas (Sassen 1994, European Commission 1999). There are fewer barriers to separate local markets as trade liberalization is limiting the prospects for national intervention in flows of capital and trade. Metropolitan areas are major sources for national economic growth in developed and developing economies. They are the ultimate expression of the benefits of localization such as scale and agglomeration economies, access to diverse and lower cost information, and an environment conducive to innovation. They are also more easily promoted as locations for investment and good factor productivity. The strength of economic growth in urban centres can also be used to stimulate vibrant rural industrial sectors such as Guangzhou, north of Hong Kong, or Zhejiang and southern Jiangsu adjacent to Shanghai, all in China (World Bank 1999) (Plate 13.6).

13.10 New 'economies' in the twenty-first century?

There are a number of new types of economy that are rapidly making their mark at the beginning of the twenty-first century. These are the Internet, e-commerce or digital economies whose presence was already apparent at the end of the 1990s but which are now rapidly growing and diversifying. It has been estimated that there were 655 million Internet users worldwide (approximately 11 per cent of total population) at the end of 2002, up from 100 million in 1997. Developing countries accounted for almost one third of new Internet users in 2001. The great bulk of commerce and international trade online, including the Internet, consists of the sale of services such as telecommunications, information technology, publishing and the media, travel and tourism, retailing, transportation and professional services like consulting, architecture and engineering. Although consumer sales gain much of the publicity, much of the growth of electronic commerce is taking place between and within business. Global ecommerce sales were estimated to be worth US$ 2.3 billion in 2003,

economies that are not already part of established supplier–TNC networks to become participants. As some global–local relations deepen, others remain as shallow as or shallower than ever before.

Localization as a counterpoint to globalization of the economy is also being sustained by national governments that are decentralizing resources and decision-making to local and regional governments (World Bank 1999). Spending by these subnational institutions was 10 per cent of the national total for Mexico in the mid-1980s, rising to more than 30 per cent in the mid-1990s; equivalent figures for South Africa are 21 per cent and 48 per cent, and for China, 45 per cent and 58 per cent. This allows localities to exercise greater autonomy over the institutions and instruments that they put in place, using the insights gained from detailed knowledge of local networks to capitalize on economic advantages or to create an economic and social environment that will attract new investment. Such decentralization still requires some

rising to almost US$13 billion, and according to the most optimistic forecasts, e-commerce will comprise about 18 per cent of worldwide retail and business-to-business transactions in 2006. The potential for growth in direct cross-border sales to consumers is enormous as Internet access, user confidence, payment systems and Internet/Web security continue to improve.

Apart from numerous legal, regulatory, security and other challenges that it presents, the global digital economy is likely to trigger changes in the geography of the economy. Just as the industrial revolution brought about significant economic and social changes (both positive and negative) in the shape of new jobs, new industries and new industrial regions, so will the digital economy stimulate its own revolution. Perhaps the Internet really will remove the friction of distance on economic interaction once and for all but only if a truly global infrastructure is put in place. One of the first challenges is to 'plug-in' national economies to the Internet. To be excluded will be to widen the gap that already exists between those countries 'inside' and those 'outside' the world economy. A networked readiness index (Figure 13.4) reveals wide differences in the geographical distribution (number, density and processing power, for example) of the computers,

telecommunications networks and software required to participate in the global economy. Internet connectivity is improving in Africa but it is starting from a very low baseline. In 2002 only 1 in 440 Africans had access if the countries with most users (such as Kenya and South Africa) are excluded from the calculation (UNCTAD 2002b). China's population of Internet users is already one of the largest in the world but transforming this potential into e-commerce will take some time because inefficient transport networks are a serious obstacle for business-to-business development.

A second challenge posed by the digital economy is the availability of human resources. More jobs will be created by the digital economy than will be lost but they are generally in higher-skilled and better-paid occupations. All economies face the big challenge of retraining existing workforces and training future workers to fill the new digital economy jobs. History suggests that the response to this challenge will be uneven, with comparative advantage enabling some economies and some localities to respond more quickly and effectively than others. Perhaps less certain is how this will modify existing spatial patterns of economic development. A digital economy will, for example, enable a wide range of purchasing and banking

▓	>5
▓	4–4.9
░	3–3.9
▫	<3

Figure 13.4   The Networked Readiness Index, 82 countries, 2002–2003.
Source: World Economic Forum (2003)

transactions to be undertaken from home via the telephone, personal computers, televisions, or on the move using mobile phones that provide access to Internet services. An international door-to-door delivery time of 24 hours or much less is now commonplace, depending on the type of good or service or the location of the supplier. Moreover, cross-border transactions and payments are as easy as those made to the local supermarket or bookstore. With shopping sites on the Internet presenting themselves as 'virtual malls' and encouraging users to place their selections in 'baskets' or 'carts', they are attempting to replicate offline shops. More than 40 per cent of Internet users shopped online in the US and in Western Europe in 2003, and by 2008 it has been estimated that 40 per cent of online buyers will purchase travel and 58 per cent will purchase books (http://cyberatlas.internet.com/markets/retailing) This, combined with trends such as increased working from home or teleworking, whether self-employed or as an employee of a TNC, points to the potential for significant changes.

The challenge for economic geographers will be to understand how the digital economy impacts on the structure and pattern of economic activities at different scales, on the structure of localities and their relationship with the global economy, or on the relative importance of nations and TNCs in regulating or shaping the outcome.

Learning outcomes

Having read this chapter, you should be aware of:

- The nature and changing concerns of economic geography during the last quarter of the twentieth century.

- The importance of economic restructuring and the shift from manufacturing to services in the modern economy.

- The significance of the globalization process in the structure and spatial development of the world economy.

- The major role performed by transnational corporations in trade, foreign direct investment and other indices of globalization.

- The challenges posed to economic geographers by the rapid growth of the digital economy and its potential for changing established patterns of economic activity.

Further reading

Bryson, J., Henry, N., Keeble, D. and Martin, R. (1999) *The Economic Geography Reader: Producing and Consuming Global Capitalism*, John Wiley, Chichester. Charts the revival and expansion of economic geography during the 1990s using selected papers that explore the evolving economic geography of the advanced capitalist economies. Covers globalization, new spaces of production and consumption, restructuring of welfare and new landscapes of work.

Dicken, P. (2003) *Global Shift: Reshaping the Global Economic Map in the 21st Century*, 4th edition, Sage, London. A broad-ranging overview of the evolution of internationalization and the globalization of economic activities. Focuses on the complex interaction between transnational corporations, nation-states and the rapid developments in information and communications technology, drawing on fields outside geography such as political science and economics.

Healey, M. J. and Ilbery, B. W. (1990) *Location and Change: Perspectives on Economic Geography*, Oxford University Press, Oxford. Provides a good broad understanding of the evolution of the economic landscape with particular reference to key economic sectors in developed market economies and to a lesser extent developing and planned economies. Includes overviews of trends and theoretical developments in economic geography prior to the onset of globalization.

Knox, P., Agnew, J. and McCarthy, L. (2003) *The Geography of the World Economy: An Introduction to Economic Geography*, 4th edition, Hodder Arnold, London. A synthesis of the factors shaping the development of contemporary economic patterns in market-oriented and centrally planned economies. Strong emphasis on interdependence of economic development at different spatial scales, from local through national to international.

Useful web-sites

www.ilo.org International Labour Organization. Documents and statistics on a wide variety of employment and labour market issues for individual countries and on a comparative basis for countries around the world.

www.un.org United Nations. Reflects the wide-ranging responsibilities of the UN but includes a useful section on global economic and social development, including an annual report on trends and issues in the world economy. Links to other web-sites.

www.wto.org World Trade Organization. Offers a wide range of documentation and statistical material about global trends in trade in goods and services, regional aspects of trade, electronic commerce and background research and analysis on all aspects of world trade. Links to related web-sites.

www.europa.eu.int/comm/index_en.htm The European Commission. Readers of this chapter will find the links to the various Directorates-General (DGs) sites within the Commission (Industry, Transport, etc.) useful starting points for monitoring economic development patterns, problems and policies across the EU.

www.oecd.org Organization for Economic Co-operation and Development. Represents mainly the developed economies and provides free documents, summaries of OECD economic surveys and statistics.

www.unctad.org United Nations Conference on Trade and Development. Focuses in particular on the interests of less developed countries in relation to international trade and foreign direct investment. Publications, statistics and links to related web-sites.

 For annotated, clickable weblinks and useful tutorials full of practical advice on how to improve your study skills, visit this book's website at www.pearsoned.co.uk/daniels

CHAPTER 14

The global production system: from Fordism to post-Fordism

JOHN BRYSON AND NICK HENRY

Topics covered

- Global production system
- The division of labour
- Spatial division of labour
- Fordism
- Post-Fordism
- The global triad
- New industrial spaces

As a world we have never produced or consumed as much as we do today or enjoyed as high a quality of life. Other chapters deal with the environmental implications and inequalities of such life-styles which the world is now, finally, beginning to face up to. This chapter is about the very large, and complex, capitalist economic system producing the goods and services we consume in our everyday lives. We begin by exploring the meaning of a *global* production system and argue that we have, in fact, had different forms of 'global' production system for centuries. The continued development of this system over time (and space) is attributable to economic specialization driven by the continuous extension of the division of labour. This key concept is the subject of the second major section of the chapter. The remaining sections examine Fordism, the global system of production which dominated the second half of the twentieth century, its subsequent economic failure (crisis), and the post-Fordist global system which has now taken its place in the twenty-first century.

Post-Fordism is a global system of production like never before. When journalists, politicians and TV reporters mention economic globalization, they are referring to the unprecedented speed, scope and extent of the post-Fordist global production system. The final section outlines the characteristics of this production system and its distinctive global economic geography.

14.1 A global production system?

The story of economic geography is one of continued geographical expansion ('globalization'); in this sense, globalization is not just a twentieth- and twenty-first-century phenomenon. There has been a 'global' production system for centuries, although it is difficult to ascribe a date to the origins of this system.

Massey (1995a), for example, writes about Chang Ch'ien, ambassador to the Emperor of the Chinese Han Dynasty, sent in 138 BC in search of allies. He made it to Bactria (today part of what is now Afghanistan), which was, in fact, an old outpost of Greek influence previously conquered by Alexander the Great. Other historical empires to name might be Carthage, Muscovy or the Roman Empire. Each of these empires included trade with 'far-flung' and 'distant' places: 'global production systems' of their time. The Roman economy stretched well beyond the borders of its empire, beyond the line of the Rhine and the Danube, and eastwards to the Red Sea and the

Indian Ocean. It has been recorded, for example, that the Roman Empire had a trade deficit of 100 million sesterces a year with the 'Far East' and that ancient Roman coins are still being dug up in India today (Braudel 1984).

Early 'global' production systems were complex and volatile. The difference between the Roman Empire's world (globe) and that of the twenty-first century is one of scale, scope and extent. Today, individuals, goods and services can travel vast distances in very short periods of time and it is this that has made the terms 'global' and 'globalization' very much late twentieth-century creations. Economic globalization is the process by which the economic relationships between places are constantly being reworked and shrunk (Allen and Massey 1995) by developments in information communications technologies (ICT) (Bryson *et al.* 2004) as well as production systems. As a result the economic globe is continually changing shape; it is also becoming smaller as the ties that bind places to the global production system become ever tighter. It is the scale, scope and extent of the current 'global' production system, and processes of globalization, which allows differentiation from earlier forms of production system and geographical expansion.

It is geographical expansion and the development of new technologies (from ships to ICT) throughout history that has seen countries, products and people become enmeshed in the 'global' capitalist production system. This is a history of how increasing numbers of people, companies, regions and countries gradually gained specialist economic expertise. The issue is how to explain this process.

14.2 The division of labour

The key to explaining this process is found in two places: in Josiah Wedgwood's eighteenth-century porcelain factory in Stoke-on-Trent (Staffordshire, England, also known as 'the Potteries') and in the work of Adam Smith, the economist.

14.2.1 Wedgwood and 'making machines of men'

In 1759 Josiah Wedgwood established a porcelain factory in Stoke-on-Trent. At the beginning of the eighteenth century most pottery was produced for the

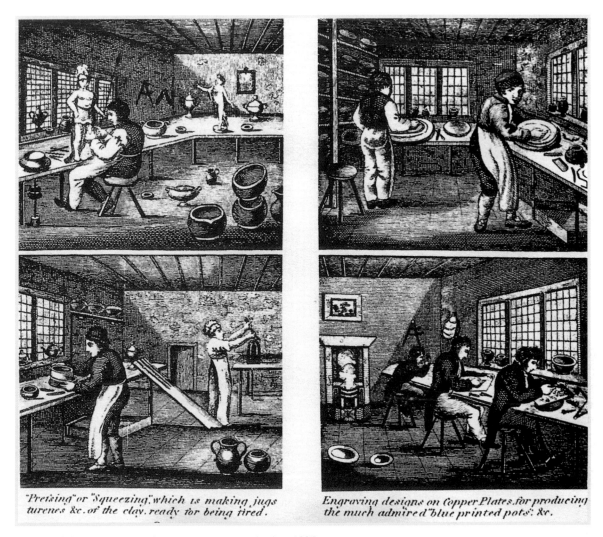

"Pressing" or "Squeezing", which is making jugs turenes &c. of the clay, ready for being fired.

Engraving designs on Copper Plates for producing the much admired "blue printed pots" &c.

Figure 14.1 The division of labour in pottery production, 1827.
Source: Forty (1995: 35)

local market. By 1750, some potters had developed expertise in particular types of pottery and had begun to export to other regions of England. When Wedgwood established his workshop most potters sold their products by sending completed pots direct to market or to a merchant. In 1774, Wedgwood decided to alter this practice by producing a catalogue. Catalogue-selling only works when factories can produce goods that are uniform in quality as well as detail. The problem for Wedgwood was that potters liked to show their skill and creativity rather than produce standardized pots. To prevent product variation, he introduced enamel-printed transfers that were mass-produced and applied to the pots before firing. The hand painters could no longer stray from the designs printed in the catalogues.

Individual potters were still able to introduce variation into the production process; the shape of the pots could differ, if not the decoration. To overcome this problem, in the 1750s Wedgwood divided the work required to manufacture a pot into seven different tasks (Forty 1995: 32) (see Figure 14.1). Each employee undertook a single task, and because no employee was responsible for a single pot, no individual was able to alter the design. This was Wedgwood's attempt to convert his employees into 'such machines of men as cannot err' (Wedgwood, cited in Finer and Savage 1965: 82). At each stage of the production process, however, employees were still able to make small variations to the design. Wedgwood kept dividing the production process into more and more specialist tasks. This increased his control of the process, as well as permitting him to replace skilled labour with less skilled, cheaper labour. Josiah Wedgwood's effort to standardize the production of pots by splitting the production process into smaller and smaller tasks is a process known as 'the division of labour'.

14.2.2 Adam Smith and the division of labour

In 1776 Adam Smith published *The Wealth of Nations*, in which he established the foundations of economic theory. For Smith:

> the greatest improvement in the productive powers of labour, and the greatest part of the skill, dexterity, and judgement with which it is anywhere directed, or applied, seem to have been the effects of the division of labour.

> (Smith 1977: 109)

To illustrate the importance of the division of labour, Smith explored pin-making, suggesting that an individual without knowledge of the production process would be unable to make even one pin in a day. The production process of pin-making can be divided into eighteen distinct operations: from the drawing out of the wire, to making the pin head. If one person performs all of these tasks then they might make 20 pins in a day. If, however, individuals specialize in particular tasks, by introducing a division

of labour, Smith shows how ten people could make 48,000 pins in a day.

Smith reasoned that the increase in the quantity of work that results from the division of labour is the product of three factors. First, the dexterity of employees improves and this contributes to an increase in speed. Second, time is lost when an employee has to change from one type of work to another; tools have to be found, the working area altered or the worker must move to a different part of the shopfloor. Time is also lost as the worker adjusts to a new task. Third, machines permitting cheaper and faster production could be introduced to replace workers; a division of labour precedes the replacement of a task by a machine. It is through a division of labour that distinct parts of the production process become visible, and as they become visible machines can be developed to replace labour.

14.3 The division of labour: spatial and gendered

14.3.1 Spatial divisions of labour

Adam Smith's division of labour explores specialization within factories. His theory, however, made no explicit reference to geography. The obvious extension of the division of labour was to incorporate geography into

Plate 14.1 An early factory. Matthew Boulton's Soho Factory, Birmingham, 1774 This factory manufactured buckles, buttons and jewellery. (Birmingham City Council Department of Leisure and Community Services)

Regional Case Study 14.1

ICT and the establishment of data processing factories in the Caribbean and call centres in India

Developments in ICT have made the transfer of service jobs to low-cost locations possible (Bryson *et al.* 2004). What is occurring is an intriguing on-going international division of labour, but with a difference. The difference is that branch plants in developing or less developed countries used only to be associated with the assembly of products designed by and for the developed world: now, many such branch plants process data and/or interact with customers located thousands of miles away.

Data processing factories, for example, in the Caribbean are linked via cable and satellite with service workers in Ireland, the Dominican Republic, Jamaica, Mauritius and the United States. To Freeman this development 'signals an intensification of transnational production and consumption – of labour, capital, goods, services and styles' (2000: 1). The Barbados Data Processing Centre, for example, employs 100 women who enter data from 300,000 ticket stubs for one airline's 2,000 daily flights, while on another floor women enter data from medical claims sent for processing by one of America's largest insurance companies (Freeman 2000: 1). This company employs nearly 1,000 people, mostly women, who work eight-hour shifts and who

are constantly monitored for productivity and accuracy.

Call centre operations are labour intensive, labour accounting for over 65 per cent of their running costs. In the 1990s, in the United Kingdom, American Express and British Airways (BA) transferred their customer services divisions from the UK to Delhi and then Bombay. BA was attracted by India's large pool of English-speaking graduates who could be employed on starting salaries of £1,500–£2,500. India currently dominates the market for the provision of English-speaking call centres. Every year India produces two million graduates, mostly taught through English and desperate for well-paid employment in a country where the national average wage is £300. In the UK, call centres are staffed by students and temporary workers and are considered to be twenty-first century sweatshops with high staff turnover rates and low salaries (£12,000–15,000). In India, they are desirable places regarded as 'hip and funky places to work, somewhere to hang out with like-minded, outward-looking young people' (Spillius 2003: 44). Six years ago, the call centre industry did not exist in India, but by 2003, 1,500 Indian call centre providers employed 102,000 young people in the 'remote services' industry; the industry is growing by 70 per cent a year (Spillius 2003: 41). Over the next five to ten years over 200,000 British call centre jobs will be exported to South Africa, Malaysia, the Philippines and China, but India is currently becoming the world's back office.

the process and to produce what has been termed a spatial division of labour (Massey 1995b, Knox and Agnew 1998). Once a production process had been subdivided into its component parts it is a comparatively simple step to develop factories (Plate 14.1) dedicated to a particular part of the production process, such as a steel mill, paper or packaging factory. This spatial division of labour began with manufacturers establishing specialist units that were contiguously located, but soon developed into a more dispersed spatial division of labour. Different parts of the production process were located close to either

sources of raw materials, cheap labour or the market place. Very quickly, some regions and countries came to specialize in particular types of economic activity. For example, an (international) division of labour soon developed between the industrial countries (core) producing manufactured goods and the non-industrialized countries (semi-/periphery) supplying raw materials and agricultural goods as well as a market for manufactured products.

This spatial division of labour can occur *within* firms. One location (a major advanced city such as London) is used for the headquarters (HQ), another for research and development (R&D), and yet another (a peripheral region, South Wales, or country, China) for the manufacturing branch plant. Moreover, this process is not confined to manufacturing companies (Case study 14.1) (Bryson *et al.* 2004).

14.3.2 Gender divisions of labour

The division of labour is also gendered (and racialized) producing a gender division of labour. In most industrialized countries women are about as likely to be in paid work as men. However, only rarely do women work alongside men, performing the same tasks for the same remuneration. Most jobs are clearly identified as either 'men's work' or 'women's work'. 'Female' jobs usually pay less than 'male' jobs, and are frequently seen as less prestigious (Bryson *et al.* 2004: 117–19). A gender division of labour occurs as a result of perceived, but rarely actual, differences between women and men.

In the labour market certain jobs (tasks) are labelled on the basis of a set of feminine and masculine characteristics. For example, women are employed on assembly lines in the semiconductor industry as it is considered that, in comparison to men, they are more manually dextrous – they have 'nimble fingers' (what of 'male' professions such as chefs, pianists, electricians?) – and are able to endure long hours of boredom (Elson and Pearson 1981, Wilkins 1983). Another important perceived feminine characteristic is related to motherhood, and the suitability of women for the caring professions such as nursing and teaching (why can't a man be an *au pair*?). Finally, female jobs frequently involve the body as an ornamental object. One of the best examples of this process is in the case of the *female* flight attendant who is required 'to deploy "skills" and abilities which they are deemed to possess simply by virtue of their sexual difference from men' (Tyler and Abbott 1998: 434). Tyler and Abbott sat in on a number of selection days for female flight attendants. They found that applicants were rejected for the following reasons: 'the applicant was too old, the applicant's skin was blemished, the applicant's hair was too short, too messy or too severe, the applicant's nails were too short or bitten, the applicant's posture was poor or the applicant's legs were too chubby ... another applicant ... was rejected on the basis that she was slightly "pear shaped"' (Tyler and Abbott 1998: 441–2). The world of work would be a different world indeed if 'not being pear-shaped' was a selection criterion applied to men!

The gender division of labour also operates spatially: from pools of labour in the 'third' world to clusters of women living in suburban areas or rural regions. Branch plants, call centres and back offices are located close to supplies of (often cheap) female workers (see Case study 14.1). One of the best examples of a spatial gender division of labour occurred in the 1960s and 70s in the United Kingdom with the relocation of government offices to South Wales. The civil service was aware that South Wales possessed a large potential reserve of unemployed women. Offices that were relocated there included the Companies Registration Office (Cardiff) and the Driver and Vehicle Licensing Agency (Swansea). Both offices require low-skilled, poorly paid, malleable office workers to undertake routine administrative functions such as checking that forms have been filled in correctly, data entry and processing and data retrieval (Winckler 1987). The best location may not necessarily be the cheapest but the location with the relatively best-educated or least-unionized workforce.

14.4 Fordism

The growth of capitalist production and today's global economy is the story of the continued expansion, both geographical and functional, of the division of labour. The nineteenth century was a period in which successful family firms that were located in the core economies grew by developing new markets and by the acquisition of less successful companies. It was a period also in which family capitalism, or companies mainly run and controlled by families, started to be replaced by forms of organized capitalism. This involved the separation of ownership from control, as management increasingly became a professional occupation. The professionalization of management is part of the continual escalation in the complexity of the division of labour; from the general manager to specialists in finance, human resource management, public relations and so on. This period is also associated with escalations in the division of labour, the international division of labour and the gender division of labour.

One of the key figures in this process was Henry Ford, the American automobile manufacturer. Ford pioneered the development of mass production based on assembly-line techniques and the introduction of scientific management into the workplace. He also encouraged mass consumption through advertising, higher wages and the supply of relatively inexpensive products. Mass production cannot occur without mass marketing to encourage mass consumption. Both these business techniques were developed and refined in the United States before being exported to other countries. There are two important locations to consider in this story: the cities of Chicago and Detroit.

14.4.1 Chicago and the meat-packing industry

The history of the Midwest of the United States is the history of the city of Chicago (Plate 14.2). This history has two sides: a history of Chicago's relationships with the Great West, its hinterland (Cronon 1991), and a history of the development of Chicago and its industries (Miller 1997), especially the meat-processing industry. Meat-processing is an interesting industry as it is a product of an early division of labour: the separation of the production of food from its final

consumption. Chicago became the centre of the United States' meat-processing industry as a consequence of two related processes.

First, in the 1880s, meat-packing became the first assembly-line industry as it proved to be impossible to mechanize the production process. An assembly line is organized around a division of labour, with the production process divided into a number of parts with each part located adjacent to each other so that tasks are undertaken in a set sequence. Rather than using technology, the division of labour was used to reduce production times and cost. The first assembly line, or 'sort of dis-assembly line' (Lacey 1986: 104), was divided into thirteen processes; live animals went in at one end and ten minutes later reappeared as dressed meat products. In 1893 a workforce of 25,000 men, women, boys and girls processed nearly 14 million animals creating products valued at US$200 million (Miller 1997b: 199).

Second, as the populations of American cities increased it became impossible for each city to rely on

Plate 14.2   The city of Chicago in the 1930s. (Popperfoto)

local meat supplies. Alongside this change, cattle production expanded dramatically on the unfenced plains of the Midwest (Cronon 1991: 221). Cattle were sent to Chicago in slatted railway cars, and shipped from there to be slaughtered in the cities on the East Coast. In 1878, the owner of a Chicago slaughterhouse hired an engineer to design a refrigerated railway car that could be used to ship processed meat rather than live animals. The result was dramatic. The trade in transport of live animals was undermined and Chicago developed into the centre of the United States beef-butchering industry. With the refrigerated car, Chicago 'moved from a center for the transfer of live cattle to a center for slaughtering and packing cattle' (Miller 1997b: 208). A transportation innovation resulted in a new spatial division of labour, with Chicago developing a specialism in the production and supply of meat products.

14.4.2 Detroit, Henry Ford and Fordism

Most economic geographies of Fordism begin with Henry Ford's development of the Model T motor car, the first successful mass-produced car. Yet the introduction, in 1913, of the assembly line in Ford's Detroit plant relied on the knowledge, experience and expertise developed in Chicago's meat-packing industry. This knowledge along with the ideas of Adam Smith and F.W. Taylor, the founder of scientific management and Taylorism (Bryson 2000), were developed and refined to such an extent that each job in the Ford factory was carefully timed and subdivided into a number of smaller operations of equal duration. An employee's job was reduced to a single task that usually lasted considerably less than a minute. Ford workers when asked what they did in the factory would, for example, say 'My work was to put on bolt no. 46' (Ling 1990: 145).

Fordism, the production system named after Henry Ford, was a combination of:

1. The extended division of labour.

2. Improvements in the manufacture of parts so that they became increasingly standardized.

3. The realization that machines should not be grouped together by type on the shop floor (e.g. all presses together) but arranged in the sequence required to manufacture a product.

4. Linking each part of the production process

together by a constantly moving conveyor belt: the assembly line.

The secret of Ford's success lay in standardization, efficiency and marketing. In the first decade of the last century most car manufacturers sold a small number of cars but at prices substantially above production costs. Ford reversed this logic by maximizing sales of low-priced cars, each of which carried a small profit. As more cars were sold, production costs declined as economies of scale were achieved and Ford was further able to reduce the cost of the Model T. These cost reductions stimulated additional sales as customers found that they could now afford to buy a Ford, but only a Ford. Ford's other secret was that he only manufactured one car – the Model T. This provided him with a cost advantage, as he did not have to manufacture a wide range of different parts. The Ford Model T was originally built like any other motor car; it sat on a cradle while workers swarmed around it. The demand for the car was to change this. In 1909 18,664 cars were sold, but this had increased to 34,528 by 1910, and to 78,440 by 1912 (Lacey 1986: 105). Yet the Model T was still expensive as the price reductions enabled by mass production were still to be introduced. In 1908 a new Model T cost US$825, which compared with the average teacher's salary of US$850 (Lacey 1986: 94). By 1917 the price had declined to just US$360 and as low as US$290 in 1934. The demand for the Model T was generated by advanced publicity and the fact that the car contained some genuine innovations, for example it was the first car with an engine protected against the elements by steel casings.

For Henry Ford the success of the Model T was a real problem. Production could not keep up with demand. The solution was the development of the first assembly line in an automotive factory and an increasing division of labour. Production went up, but at the expense of the quality of life of Ford's employees. Staff turnover in the Ford factory in 1913 was running at a staggering 370 per cent (Ling 1990: 147). In the same year, Ford surprised his competitors by raising pay from US$4 to US$5 a day and simultaneously reducing working hours from nine to eight hours a day. This was an attempt to retain staff, but at the same time the existing two nine-hour shifts were replaced with three eight-hour around the clock shifts. More cars were produced, yet each worker was paid more (Lacey 1986: 117). The five-dollar day was widely considered by newspapers and rival car

Plate 14.3   Ford Model T assembly at Highland Park, 1914. (Topham Picturepoint)

companies to be a serious error of judgment on Ford's part. However, his critics were unaware of the savings in labour costs created by the new continuously moving assembly lines: in 1912 it took 728 worker-minutes to assemble a car in comparison to only 93 minutes a year later (Lacey 1986: 120).

14.4.3 The rise of the branch plant

The Ford Motor Company's final innovation was the development of branch plants. The Model T left the main Ford plant in Detroit as little more than a chassis, motor and body. Important accessories like headlights were installed by branch agencies established and run by Ford. Branch agencies also undertook repairs at a time when most cars had to be returned to the factory for repair, but they also meant that Ford had the facilities to sell cars direct to the consumer. It was a short step from a branch agency to a full assembly

branch plant and what prompted this step was transportation costs.

The Model T was transported from Detroit by fitting as many vehicles as possible into a rail freight car. Railway companies charged for this service on the basis of a minimum weight of 10,000 pounds per 11-metre freight car. Such a car held four fully assembled Model Ts but these only weighed 4,800 pounds. Ford was being charged for 5,200 pounds that was not being transported and, on top of this, railway rates for carrying completed cars were 10 per cent higher than normal freight rates as cars were considered to be fragile and difficult to manoeuvre. The solution to this problem was the establishment of assembly plants located close to the main markets (a spatial division of labour within the firm). Freight cars packed with enough parts to assemble 26 cars were sent from the main Detroit plant. The development of assembly plants halved Ford's transportation costs, but the savings were not passed on to the consumer.

Assembly plants also embedded the Ford Motor Company into local communities, encouraging local newspapers to adopt the company. Free publicity and the impression that Ford was a local company resulted in additional sales of the Model T.

14.5 The geography of Fordism

Fordism has its own distinctive geography. From the 1920s the heartland of US Fordism developed between Boston and Chicago and from Cincinnati to Milwaukee (Figure 14.2). In the United Kingdom the West Midlands consolidated its position as the workshop of the world (Bryson *et al.* 1996) whilst, in Germany, manufacturing production consolidated itself in the Ruhr. These regions became the centres of mass production activities and of reasonably well-paid but monotonous jobs. This does not mean that other regions did not become industrialized, but that the scale and intensity of manufacturing production in these areas was different in scale and scope.

14.5.1 The new international division of labour

The obvious development from branch plants located within the confines of a nation-state is the establishment of branch plants located in overseas countries. An international division of labour (a form of spatial division of labour) occurs when the process of production is no longer confined primarily to national economies. This is not a new process, as regions of the world have specialized in particular segments of the production process for a long time. Examples include wool production in England during the fourteenth century and the export of cocoa beans from West Africa from the nineteenth century onwards to support the rapidly growing European chocolate industry.

As mass consumption developed alongside improvements in mass communications (Williams 1998) ever-larger companies developed with the capacity to mass-produce affordable goods. It was a period of consolidation of businesses, of mergers and acquisitions, and of the development of monopoly capitalism. By 1957 America's top 135 corporations

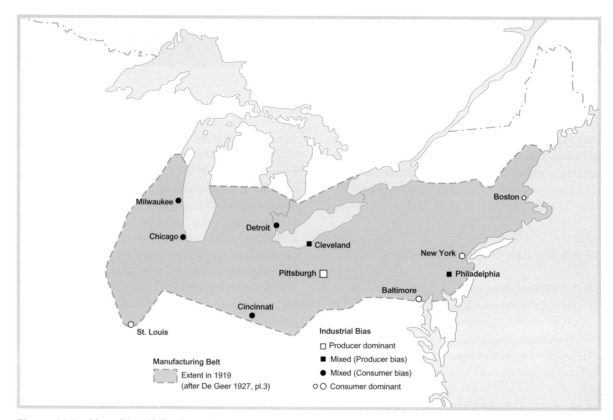

Figure 14.2   Map of the US Fordist region: the American manufacturing belt in 1920
Source: Knox *et al.* (1998)

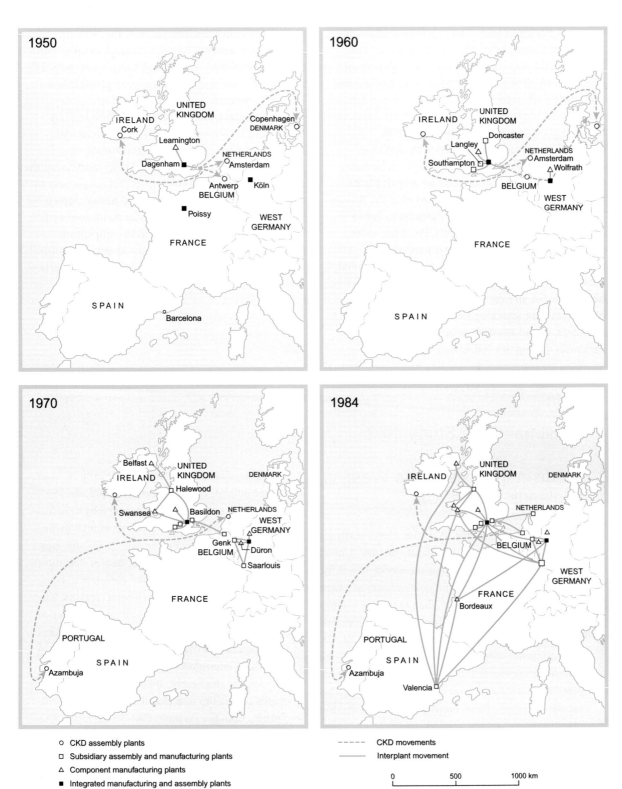

Figure 14.3 Ford in Western Europe, 1950–1984.

Source: Chapman and Walker (1987)

1950

IRELAND
Cork
Leamington
Dagenham
Amsterdam
NETHERLANDS
Antwerp
BELGIUM
Köln
Poissy
WEST
GERMANY
FRANCE
UNITED
KINGDOM
Copenhagen
DENMARK
SPAIN
Barcelona

1960

IRELAND
UNITED
KINGDOM
Doncaster
Langley
Southampton
NETHERLANDS
Amsterdam
Wolfrath
BELGIUM
WEST
GERMANY
FRANCE
SPAIN

1970

Belfast
IRELAND
UNITED
KINGDOM
Halewood
DENMARK
Swansea
Basildon
NETHERLANDS
WEST
GERMANY
Genk
Düron
BELGIUM
Saarlouis
FRANCE
PORTUGAL
Azambuja
SPAIN

1984

IRELAND
UNITED
KINGDOM
DENMARK
NETHERLANDS
BELGIUM
WEST
GERMANY
FRANCE
Bordeaux
PORTUGAL
SPAIN
Azambuja
Valencia

○ CKD assembly plants
□ Subsidiary assembly and manufacturing plants
△ Component manufacturing plants
■ Integrated manufacturing and assembly plants

---- CKD movements
—— Interplant movement

0   500   1000 km

owned 45 per cent of the country's industrial assets (Baran and Sweezy 1966: 45). The period after the Second World War was one of economic growth, of nearly full employment, of prosperity, of growing car ownership and, in the words of Harold Macmillan, in the United Kingdom it was a time when 'you've never had it so good'. It was also a period in which multinational companies developed through the creation of branch plants (Dicken 1998). American companies like Ford had established branch plants in Europe as early as 1910, but until just after the Second World War such plants were very similar to Ford's American assembly branch plants. From the 1950s large companies reorganized their production processes and began to develop a New International Division of Labour (NIDL) with parts being supplied from countries all over the world. Effectively companies were searching for the cheapest locations to manufacture and assemble components. The key industries driving this NIDL were the car and electronics industries. The result of the NIDL in the car industry was a highly complex network of cross-border flows of completed vehicles and component parts. In 1967, Ford consolidated its European plants into a European division and began to develop cars especially for the European marketplace (see Figure 14.3). A European division of labour was introduced with engines assembled in Germany, transmissions in France and electrics in the United Kingdom. Today, the onset of multinational companies that act globally to exploit the international division of labour (remember the example of call centres in India mentioned earlier, see Case study 14.1) has resulted in competition between nation-states and regions to attract companies that are seeking the best location to undertake parts of their production process.

Regional Case Study 14.2

Fordism in crisis in the West Midlands region, UK

After the Second World War the West Midlands became the industrial heartland of the United Kingdom and the workshop of the world. It was the centre of the United Kingdom's car industry at a time when Britain made more cars than any other country except the United States. Indeed, the West Midlands conurbation accounted for over 8 per cent of UK manufacturing output. Between 1951 and 1961, the region's employment increased by 14 per cent compared to a national average of only 8.5 per cent. This decade of employment growth was driven by five manufacturing sectors: vehicles, metal goods, metal manufacture, mechanical engineering and electrical engineering. In 1961, just under 65 per cent of the labour force was employed in manufacturing compared to a figure of 39 per cent for the United Kingdom. Between 1951 and 1966 the region gained 361,000 jobs. The West Midlands was at the heart of the Fordist boom in the United Kingdom.

The crisis of Fordism emerged in the early 1970s in the West Midlands economy. Between 1971 and 1993, just over half a million manufacturing jobs were lost in the region, a decline of 50 per cent in total manufacturing employment. By the late 1980s the West Midlands had shifted from being one of the most prosperous regions in the national economy to being one of the most depressed 'peripheral' regions. The West Midlands crisis was the result of a number of related factors (see Bryson et al. 1996), but at the heart of the problem was overseas competition and cheap imports that undermined the profitability of the area's companies. By the early 1980s many large West Midlands companies had closed or severely rationalized production. For example, GKN in the early 1970s employed 17,000, but by 1980 this had fallen to only 5,000. The industrial decline of the West Midlands left a legacy of industrial dereliction, empty obsolete factories, and high long-term unemployment. The industrial heartland of the UK economy had become an industrial wasteland.

14.6 The crisis of Fordism

The post-war years of economic growth and expansion peaked in the mid-1960s. The 'crisis' hit different companies and countries at different times. In America the heyday of Fordism can be identified with the debut party of Henry Ford II's daughter, Charlotte. This took place on 21 December 1959 and cost US$250,000. At that time very few people in the US had heard of Toyota as American car manufacturers dominated the market for cars in the United States. Fordism was a sellers' market, in which manufacturers dictated what could be purchased; the wishes and the whims of consumers were secondary. During the heyday of Fordism the quality of products declined, but their cost increased. All of this was to end in the 1960s when European, followed by Japanese, car companies began to export cheap, high-quality cars to the United States.

The crisis of Fordism emerged in the early 1970s drawing to a close the post-war industrial boom. Economic growth in the industrialized countries fell, unemployment increased and household disposable incomes fell. Unemployment in the advanced economies averaged around 4.5 per cent in the 1960s, but grew to 10 per cent by the mid-1980s (Knox and Agnew 1998: 209). The crisis was caused by the combination of a series of economic pressures including the inflexibility of Fordist industrial production, the high cost of production in the heavily unionized advanced economies and increasing overseas competition. At an international scale, industrial production was again reorganized away from high-cost to low-cost locations and a further spatial division of labour evolved. One of the most pronounced consequences of this crisis was the decline in the industrial manufacturing base of the advanced economies. This decline was most evident in former core industrial zones such as the US Midwest and the West Midlands region in the UK (Case study 14.2). **De-industrialization** affected all aspects of the economies of these regions as spiralling multiplier effects undermined the client base of small subcontracting firms, and rising long-term unemployment and industrial dereliction discouraged new inward investment (Massey 1988).

With hindsight, it is now generally recognized that the crisis of Fordism marked a crossroads in the historical development of capitalism. Just as the industrial revolution or the onset of Fordism marked distinctive phases of capitalism, so it is now believed that with the crisis of Fordism came an epoch-making transformation in the organization of capitalism. The 'post-Fordist' debate concerns the nature and direction of such epoch-making change (Amin 1994). This transformation of the organization of production – for example, of technology, firms, labour, markets (consumption patterns) – has brought with it a further transformation in the geography of the 'global' production system.

14.7 Post-Fordism

There is still a strong debate about the contours of post-Fordism. How do we characterize and explain its defining features? One full-blooded description of post-Fordism is that given by Hall (1988, quoted in Amin 1994: 4):

A shift to the new 'information technologies'; more flexible, decentralized forms of labour process and work organization; decline of the old manufacturing base and the growth of the 'sunrise', computer-based industries; the hiving off or contracting out of functions and services; a greater emphasis on choice and product differentiation, on marketing, packaging and design, on the 'targetting' of consumers by lifestyle, taste and culture rather than by categories of social class; a decline in the proportion of skilled, male, manual working class, the rise of the service and white-collar classes and the 'feminization' of the work force; an economy dominated by multinationals, with their new international division of labour and their greater autonomy from nation state control; and the 'globalization' of the new financial markets, linked by the communications revolution.

While there is general agreement with this description, there is greater disagreement as to why these changes have come about: disagreement on how to explain these new economic realities through the use of theory. A wide selection of theories have been put forward – including 'post-industrialism', 'the fifth Kondratiev', 'neo-Fordism', 'flexible accumulation', 'disorganized capitalism' and 'informational capitalism' – to account for these major changes in the economy and society.

Four major processes driving economic change are commonly identified:

- *Technological change:* the rise of information technology; flexible machines in flexible manufacturing systems.

- *Labour practice change:* the use of flexible, multi-skilled core workers combined with numerically flexible peripheral workers (e.g. temporary, contract, part-time, zero hour).

- *(Combining to produce) industrial organization change:* the growth of the flexible firm; 'lean and mean' multinationals and small and medium-sized enterprises, concentrating on their 'core' specialisms as part of networks of collaborating firms (both small and large).

- *Market demand/consumption change:* increasingly differentiated (customized, segmented, niche, lifestyle) demands for goods and services.

Furthermore, these processes are taking place within the context of 'globalization'. In the contemporary economy, the term is used to denote at least two distinct processes. First, it depicts the quantitative shift of increasing integration of the world economy as more and more parts of the globe are drawn into the networks of capitalist production and markets. Simple examples would be the rise of the NICs (Newly Industrializing Countries) such as Korea, Singapore, Malaysia and Thailand to name but a few (Thompson 1999), the growing presence of China as it liberalizes

its economy towards capitalist production (Ogilvy and Schwartz 2000) and, of course, the growing, but highly uneven, insertion of the post-socialist states into the global economy following the fall of the Iron Curtain (Stenning and Bradshaw 1999). It is in this sense that the global production systems of today are 'more global' than those of the Roman Empire or even those depicted in Fordism's new international division of labour.

Second, however, is a qualitative dimension that has driven the use of the term 'globalization' to describe the current economic era. In this sense, the term global economy is used to stress how production and investment crosses the globe with very little regard for national boundaries (Allen 1995b, Dicken 1998). Many markets are global, not national (for example, the EU or Latin America), and the major players in such an economy talk of global products – Fiat's Tipo 178 world car, for example, or Coca-Cola or Microsoft software. They question the need to have a 'home-base' in a nation-state as against a headquarters located, as with all the other parts of the production process, in the best location to meet its particular requirements (such as access to skilled labour and finance capital).

Table 14.1 Alternative production systems in post-Fordism

Characteristic	Flexible specialization	Fordist mass production	Japanese flexible production
Technology	Simple, flexible tools/machinery; unstandardized components	Complex, rigid single-purpose machinery; standardized components; difficult to switch products	Highly flexible (modular) methods of production; relatively easy to switch products
Labour force	Mostly highly skilled workers	Narrowly skilled professionals 'conceptualize' the product; semi/unskilled workers 'execute' production in simple, repetitive, highly controlled sequences	Multi-skilled, flexible workers, with some responsibilities, operate in teams and switch between tasks
Supplier relationships	Very close contact between customer and supplier; suppliers in physical proximity	Arms-length supplier relationships; stocks held at assembly plant as buffer against disruption of supply	Very close supplier relationships in a tiered system; 'just-in-time' delivery of stocks requires 'close' supplier network
Production volume and product variety	Low volume and wide (customized) variety	Very high volume of standardized products with minor 'tweaks'	Very high volume; total partially attained through production of range of differentiated products

Source: Adapted from Dicken (1998: Table 5.2)

'*Tripp Trapp*' and the Norwegian furniture Industry

Scandinavia's contribution to post-war design is highly visible in most furniture retailers. The concept of Scandinavian design, as an amalgamation of design traditions associated with Sweden, Denmark, Finland, Norway and Iceland, is an artificial construction created out of a strategic alliance developed between these countries during the 1950s to encourage trade and to project an image of a distinctive Scandinavian design culture. This alliance was formed around an exhibition that toured the USA and Canada from 1954 to 1957 whose primary purpose was to promote Nordic craft-based modern designs. In 1997–1998, Sweden, Finland, Norway and Denmark combined to develop another travelling exhibition of Nordic design that was first exhibited in St Petersburg.

Norway is traditionally associated with the export of raw materials (oil, gas, timber, fish) but over the last twenty years has been developing a reputation as a design nation. Norway's furniture industry is now an active consumer of design services and has become known for the production of novelty products. One of Norway's leading furniture designers is Peter Opsvik, the designer of the best-selling children's *Tripp Trapp* chair that has sold around three million units. There are 469 furniture and furnishing companies based in Norway employing over 9,000 or around 3 per cent of Norway's total manufacturing employment. Until quite recently this industry was orientated towards the domestic market, but has begun to engage in design-led exports. It is concentrated in the county of Møre og Romsdal (western Norway) which accounted for 41 per cent of the industry's employment in 1990 (Rusten 1997: 179).

The majority of Norwegian companies are small family-owned and managed firms; in 1994, 78 per cent employed fewer than 20 and only 3 per cent employed more than 100 (Rusten 1997: 177). The furniture production system is designed to produce a range of standardized as well as customized design-rich products. The same furniture design can be produced in a variety of materials (wood, fabric, leather) and qualities (inexpensive mass-produced or expensive handmade or finished). To achieve economies of scale most firms subcontract parts of the production system, but retain the main manufacturing or assembly elements. However, some firms have subcontracted or outsourced the manufacturing process retaining only research and development, design and coordination as in-house functions. Most of the subcontractors are based in Norway, but foreign-based subcontractors will become more important. Of note is that there are many different ways of producing products and many different ways of combining the expertise, skills and knowledge that exist inside and outside a firm.

Specialist firms supplying furniture parts (hardware – hinges, wheels, handles, fabric, etc.), processes and services are co-located with client firms in parts of Møre og Romsdal. These firms form a specialist furniture cluster that enables clients and subcontractors to respond rapidly to changing market conditions and fashions. Some furniture companies prefer to use local suppliers, recognizing that this contributes to local economic development. Others search for the cheapest or highest quality producers. The existence of a local specialist cluster of furniture producers and subcontractors/suppliers provides the region with a competitive advantage. Most small manufacturing firms would be unable to function without subcontracting elements of the production process; their size makes it impossible for them to internalize all essential business and production functions. The majority of Norwegian furniture firms are situated in dense networks of relationships with local suppliers based on trust or on legal contracts with suppliers located outside the local cluster. The availability of local suppliers enables small firms to develop flexible production systems that are designed to produce small volumes of customized goods.

Websites:

http://odin.dep.no/nhd/engelsk

www.norway.org

www.ncad.ie

14.7.1 Post-Fordist production system(s)

If the Ford Motor Company – with its assembly-line mass production in branch plants around the world to meet the demand of a mass market – is taken as the exemplar of the Fordist production system, then today's post-Fordist system is best understood through a number of 'typical' examples. There have been a number of alternative responses to the challenges of the new production era. Currently, the post-Fordist production system comprises a *mix* of different ways of organizing production at a number of spatial scales. Table 14.1 contrasts two of the major alternatives to Fordist mass production.

Examples of all *three* forms of production (flexible specialization, mass production and flexible production), and, indeed, further hybrid mixtures of them, can be identified in the post-Fordist global production system. What is common to all, however, is the response to the processes of change identified above: all represent the struggle for competitiveness in a global economy through some form of flexible production. For example, at the heart of flexible specialization is a skilled (craft) worker using flexible machinery to produce small volumes of customized goods. This has led to the resurgence of small, independent, entrepreneurial firms which develop as part of dense networks of interrelated, specialist firms (see Case study 14.3). In contrast, Japanese flexible production implies flexible specialization but for the production of extremely high volumes of products and on a global scale (Fujita and Hill 1995) (see Case study 14.4). Meanwhile, the Fordist mass production system has been shaken to its roots (see, for example, Case study 14.2). Adaptation has seen some companies survive through further global extension in the search for lower costs of production and higher economies of scale through increased production for new markets (sometimes called global Fordism).

The response to change has produced three characteristics of this new post-Fordist global production system.

14.7.2 The big are getting even bigger

The move to a global economy has triggered an ever greater number of multinational firms (see also Chapter 13). In 1969, the number of multinationals – companies that produce goods or market their services in more than one country – in the top 14 industrial countries of the world was estimated at 7,000. By the early 1990s, the figure had risen to about 24,000. In 2001, there were about 65,000 multinationals, with about 850,000 foreign affiliates employing around 54 million. Foreign affiliates now account for one-tenth of world GDP and one-third of world exports (UNCTAD 2003: 1). Multinationals are also becoming more multinational; they are involved in more countries and in greater numbers of more varied overseas production relationships.

Multinational production (production by companies overseas) now outweighs exports as the principal route to meeting overseas demand and it is estimated that two-thirds of world trade is accounted for by multinationals. Furthermore, up to a third of world trade is actually cross-border, but *intra-firm*, trade. That is, different parts of the same company, different parts of the corporate spatial division of labour, sending goods and services between their sister factories and offices, and crossing national borders when doing so (see Figure 14.3). Moreover, despite the growth in total numbers of multinational firms, the *share* of the global economy accounted for by the top 100 multinationals has continued to increase. In 2000, the largest 100 firms accounted for 14 per cent of total world sales by multinationals and employed over 14 million workers worldwide (UNCTAD 2003: 89). Sales of the world's top 500 companies nearly tripled between 1990 and 2001, while world GDP in current prices increased by only 1.5 times over the same period. It is frequently suggested that some of the largest multinationals are bigger than the economies of some of the smaller nation states. Comparing the sales value of multinationals with country GDP reveals that of the 50 largest 'economies' in the world in 1999, 14 were multinationals and 36 were countries. Nevertheless, this methodology is flawed as the total value of a firm's sales does not equate with GDP. Converting sales into a measure of added value similar to GDP reveals that the largest multinational company is ExxonMobil, which ranked 45th in the combined global ranking of counties by GDP (UNCTAD 2003: 90).

14.7.3 And Small may be Beautiful once again

On the other hand, some of the most dynamic centres of economic growth under post-Fordism have become so due to the rapid growth of large numbers of new,

Thematic Case Study 14.4

Global Toyotaism

The post-Second World War rise of Japanese manufacturing multinationals to global power is one of the stories of twentieth-century capitalism. Flexible specialization is used in the main to refer to network systems of production among small producers. Global Toyotaism refers to how, in Japan, flexible specialization is more often found in collaborative relations among large parent companies and smaller supplier firms (see Table 14.1; Figure 14.4, *keiretsu* diagram). Economies of scale are gained through the flexible mass production of a variety of products in a system. Craft flexibility and the most advanced information technology are combined – automation with a human mind – in a distinctive organizational system that has produced higher assured quality, flexibility and conti-

nuity in production. On the shopfloor, work teams, job rotation and learning-by-doing take precedence in a system which integrates conception and execution throughout the production process.

This integration is continued throughout a supply chain based on JIT (just-in-time) principles. The parent firm, suppliers and subcontractors continuously consult to assure flexible delivery of high-quality, error-free parts to the assembly line to be fitted just-in-time. This system fosters spatial agglomeration in Toyota-style industrial districts. As the Japanese giants have globalized production, they have 'transplanted' similar-style districts comprising a parent plant combined with suppliers (see Section 14.8.3 on new industrial spaces).

www.toyota.com

small and medium-sized enterprises (SMEs) (Birch 1987, Storey 1997). In particular, a connection has been made between the production system of flexible specialization and the dynamic opportunities it has afforded for small firms (see Case study 14.3). This contrasts starkly with the Fordist era when the cutting edge of capitalist organization was seen as (i) the very large firm and the very large plant and (ii) the need to be large because of vertical integration: doing for yourself (bringing 'in-house') as much of the production process as possible (if in different places).

14.7.4 'No one can do it all any more': networks of production

The increasing pace of technological change, the need for 'global presence', and the drive for flexibility of production and products (speed in response to rapidly changing market demand) has created a situation where no one firm, no matter how large, can do it all on its own. The result is the rise of networks of production (production *dis*integration) such that the

global economy is labyrinthine in its interrelationships between firms, large or small and global or local (see Dicken 1998, Ch. 7). These production relationships take a myriad of forms (compare Case study 14.5 with Case study 14.3). Examples include strategic alliances, (international) subcontracting, franchising, subsidiaries, joint ventures, R&D networks and sponsored spin-offs to name just a few. In 1996, the world's 41 largest car producers shared 244 strategic alliances between them, with Peugeot of France involved in 22 agreements with other car companies (Knox and Agnew 1998: 208).

Some of the most well known forms of networks include the Japanese *keiretsu* and South Korean *chaebol* (see Figure 14.4). Whilst very different in form from the production network of Benetton or Nike, all these examples are successful in producing flexibility and global presence through production networks and indirect labour. Gone are the days of Ford's River Rouge plant in Detroit, which at its height used 70,000 workers to transform the raw materials going in at one end into over one million black Model T Fords exiting at the other.

Thematic Case Study 14.5

Nike

We grow by investing our money in design, development, marketing and sales, and then contract with other companies to manufacture our products.

Nike, the US sports footwear, clothing and equipment company, is one of the leading examples of network production. In contrast to Toyota, exploiting the international division of labour through the exhaustive use of international subcontracting has driven Nike's strategy from its very beginnings. Co-founder Philip Knight's founding theory (in 1964) was to produce high-quality running shoes designed in the US, manufactured in Asia, and then sold in America at lower prices than the then popular West German-made running shoes. Today this US$10-billion company, headquartered in Beavertown, Oregon, has a global brand based on high quality design and marketing, employs around 22,000 staff, and manufactures in more than 900 contract factories around the world. In total, suppliers, shippers, retailers and service providers who operate with Nike employ almost one million people on six continents. Throughout its short history, Nike has subcontracted virtually its entire production to a changing array of factories, predominantly in South-East Asia. It has shifted in unison with the changing international division of labour in the search for lower production costs. Whilst the company has exclusive partners for high-end products, its main product base of volume training shoes is produced by volume subcontractors which the company is able 'to turn on and off like a tap' in response to poor performance or changing market demand.

www.nike.com

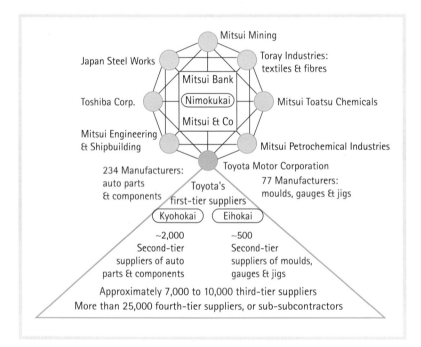

Figure 14.4

Toyota: a Japanese keiretsu.

Source: Harryson (1998)

14.8 Post-Fordism: a new global production system

The transition from Fordism to post-Fordism has dramatically redrawn the economic geography of global production. It has produced also one of the paradoxes of modern times. On the one hand, the global economy – and the largest firms that comprise it – has never been as large, or reached as far, as it does today. Few places and people in the world are not touched by this global spatial division of labour, let alone by the consumption dynamic which creates the demand for this global production system. Yet, on the other hand, the onset of post-Fordist global production has triggered a *reconcentration* of the division of labour. A major characteristic of today's global economy is the 'resurgence of the region' (Sabel 1989) or the reagglomeration of production ('clusters') (Scott 1988; see also Chapter 11). Particular places – cities, towns, regions – are becoming famous as *the place* to produce particular products. Probably the most well-known example is Silicon Valley, the massive agglomeration of semiconductor and computer producers in California. Other examples would be the three financial centres – London, New York and Tokyo – that dominate the world's financial and business service industries. These New Industrial Spaces (NIS) are just one of three distinctive characteristics of the global geography of post-Fordist production: global reach *and* global concentration, an increasingly complex international (global) division of labour, and local geographical concentration (agglomeration).

14.8.1 The global triad

The concept of the global triad encapsulates both the global nature of the world economy and, even at this spatial scale, elements of its geographical concentration. The term describes how the world economy is now essentially organized around a tripolar, *macroregional* structure whose three pillars are North America, Europe and east and south-east Asia (see Figure 14.5). By 1994, these three regions' share of total world manufacturing output had increased to 87 per cent and they generated 80 per cent of world merchandise exports. The triad also accounted for 90 per cent of the world stock of foreign direct investment (FDI) (Dicken 1998: 60–1). Of the world's largest 100 multinationals in 2000, 88 were headquartered in the US, EU or Japan. Switzerland, South Korea and Canada accounted for six of the remaining companies (UNCTAD 2003).

Nevertheless, increased concentration within the triad hides, for example, the exceptional economic growth of what have become known as the Asian Tiger NICs (see South Korea above), and the opening-up of the former Soviet Union and east central Europe to capitalist production. Meanwhile, Africa has seen its tiny share of world trade and investment decline over the preceding decades. Overall, whilst the tentacles of

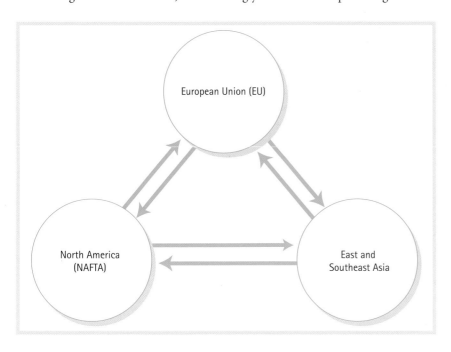

Figure 14.5
The global triad

the global production system continue to spread this is overridden by an increased concentration at the global level within the three macroregions of the triad.

14.8.2 The newer international division of labour

Under Fordism, the new international division of labour (NIDL) was taken to describe the increasing foreign direct investment (often through branch plants) by advanced economy (US, European) multinationals in other advanced economies and, eventually, in the less developed countries (for example, Taiwan and South Korea). Indeed, this increasing trend became part of the crisis of Fordism (deindustrialization) for the advanced economies as home-country plants were closed down and production shifted to new plants in the global periphery (see Case study 14.3).

With the transition to post-Fordism, it has been argued that the global economy is being shaped by a '*newer* international division of labour' (Coffey 1996). There are several processes at work shaping this global division of labour. First, whilst foreign direct investment remains important, firms from the advanced economies are using other, less direct, ways to exploit the cost advantages of the less developed countries. For example, and drawing on the network approach to production, 'manufacturing' firms in the advanced economies perform less and less of their own physical production. Rather than building manufacturing branch plants in LDCs, they concentrate on the higher value-added elements of design and marketing and engage in international subcontracting with local companies in the LDCs (Bryson *et al.* 2004: 50–75). Nike is a prime example of this strategy (see Case study 14.5). Secondly, an increasing share of the activities being relocated is involved in the provision of services. One example is the growth of the Indian computer software industry over the last two decades (Lakha 1994; see also Chapter 11). In 1988, the top five software exporters in India were either fully foreign (US) owned companies or partnerships with overseas (US) companies. Another example is that of 'back offices' in the Caribbean (Case study 14.1) and in Ireland (Allen 1995b: 56). The latter highlights also the third aspect of the newer international division of labour: namely, that an increasing share of outward FDI by firms in advanced economies is destined for other advanced economies. It

is this trend which has contributed to the increased dominance of the global triad mentioned above. However, and fourth, within this triad some of the LDCs which were among the original destinations for FDI by core nations have begun to undertake FDI themselves. The Asian Tigers are a prime example of this. Moreover, there are two distinctive patterns to this FDI. First, as the original NICs have seen their economies develop, so too have their labour costs. One result has been their replication of the search for low-cost labour through investment in their less developed neighbours such as Indonesia and Bangladesh. Second, there has been the growing pattern of 'reverse' FDI (Lall 1983). Whilst Japan's investments in the US and Western Europe – often directed at the deindustrialized regional economies such as South Wales (Cooke 1995) and the US Rust Belt (Florida 1996) – are well established, more recent sources of FDI include the Korean *chaebols* such as Samsung and LG and Malaysian multinationals such as Proton.

14.8.3 The new industrial spaces

The term new industrial spaces (Scott 1988) is used to describe the set of dominant economic regions and spaces which have grown in tandem with the transition to post-Fordist flexible production. It is these economic nodes which encapsulate the shift in the geography of production – in particular, the process of reagglomeration of production – with the move to post-Fordism. They take a number of different, overlapping, forms:

- *New industrial districts:* Dense agglomerations of, normally, small and medium-sized firms specializing in the high-quality production of a particular good or service. The archetypal example is the 'Third Italy' – home of Benetton – with individual towns and districts renowned for their specialist production of goods such as textiles (Como, Prato), ceramics (Sassuolo), shoes (Ancona) and special machines (Parma, Bologna, Modena). Other examples include Jutland in Denmark (furniture, shipbuilding), Småland in Sweden (metalworking) and Baden Württemberg in Germany (machine tools, textile machinery).

- *High technology (sunbelt) areas:* A diverse range of technopoles, corridors and innovative complexes, often in previously unindustrialized areas, which have grown through their capture of the 'sunrise'

Plate 14.4 An aerial photograph of Silicon Valley. (Air Flight Service Mapping Photographers)

high technology industries of information technology, electronics, pharmaceuticals, R & D, etc. (Castells and Hall 1994). The most famous example is Silicon Valley in California, birthplace of the integrated circuit and the personal computer (Plate 14.4). The experience of Santa Clara County (as it was known then), a mainly agricultural region, is the growth story of the era. In 1950, it employed around 800 manufacturing workers. Today, Silicon Valley employs nearly 400,000 high technology workers alone. Companies born in the region include Hewlett Packard, Intel, Apple and Sun Microsystems. Other, more modest, examples include Britain's Western Crescent/M4 Corridor,

Boston's Route 128, the Ile de France Sud, and Japan's Technopolis programme.

- *Flexible production enclaves within old industrial regions:* This varied classification attempts to capture the variety of economic development that has taken place in previously industrialized regions. Often rocked by the decline of previously core 'sunset' industries (shipbuilding, coal, steel, heavy manufacturing, etc.), some alternative economic growth has been achieved through firms and sectors applying flexible production techniques. This might occur through in-situ restructuring of firms both large and small or the attraction of new, often foreign, investment. Examples of the former would include the resurgence, largely based on small firms, of the textile and film industries of Los Angeles, or the post-production industries of Soho, London. Examples of the latter include the large concentration of overseas investment from Japan and Korea in South Wales or the resurgence of the US Midwest through Japanese 'transplant' investment (see Case study 14.4). This highlights the local and regional impacts of the 'reverse' FDI discussed above.

· *Global cities* (Sassen 1994, 2001; Bryson *et al.* 2004): These are the pinnacle of the global urban hierarchy as cities once again become key sites of production with the move to service industries in the advanced economies (see Chapter 13). Cities such as London, Tokyo and New York are the command and control centres of the global economy. This means more than just being the headquarter sites for global companies. In addition, they are the industrial districts of the financial and business services and the centres of the corporate services that provide the expertise for the management of global production. These centres include the 'servicing' industries – the cleaners, security guards, couriers and pizza delivery personnel – who service the corporate elite. They are sites also of the elite cultural and retail industries; the galleries, theatres

and fashion houses. A second tier of global cities includes, for example, the 'metaregional' capitals of Paris, Frankfurt, Sydney, Hong Kong, São Paulo and Mexico City.

14.8.4 The global mosaic of post-Fordist production

Even this simplified outline of the post-Fordist geography of production offers, then, a complex, multiscale picture of economic growth. The economy is global like never before and at this scale it is no longer a simple case of allocating countries to the core, semi-periphery and periphery. Yet, continued global reach has gone hand-in-hand with increased geographical concentration (localization). This is no more starkly shown than in Britain's Motorsport Valley (see Case study 14.6) in which some of the largest global companies in the world have beaten a path to semi-rural Oxfordshire – with an array of comparatively tiny firms – to produce their flagship motor-sport products (racing cars). The ever-changing international division of labour is producing increasingly complex, and interdependent, relationships between, and across, metaregions, nations, cities and localities, between firms large and small, and between economies old and new.

Regional Case Study 14.6

The 'global agglomeration' of Motorsport Valley

Within an hour's drive, two hours at the most, it is possible to get anything in motorsport done, have anything designed, anything built.

Over 75 per cent of the single-seat racing cars used in more than 80 countries across the world are UK-built. Of the ten Formula 1 teams, six are wholly UK-based and a seventh, Minardi, is split between the UK and Italy. All the chassis and engines for the US-based Champ Car World Series are produced in Motorsport Valley and the UK is the base for the majority of the World Rally Championship teams including giants such as Mitsubishi, Ford and Subaru.

The motorsport industry employs in excess of 40,000 people, mostly in several hundreds of small and medium-sized firms clustered in a region known as Motorsport Valley, the heartland of which is in a 50-mile radius centred around the Oxfordshire and Northamptonshire area. It is a truly global industry with finance, sponsorship, drivers, engineers, parts and expertise coming to the UK from all over the world. Motorsport Valley is now recognized as *the place* in the world to produce your motorsport product – no matter how large the company and the global choice of locations on offer.

www.motorsportvalley.com

The 'Motorsport Valley' name and mark is registered by the MIA.

What is undeniable, however, is the substantial reorganization of global production with the move to post-Fordism and the new, uneven, geography of winners and losers that this entails. For the 'heavy industry' communities and Fordist regions and cities of old, for example, this has often meant devastating change as new power-house regions and nations such as California and South Korea have displaced them on the economic map. For other regions, such as sub-Saharan Africa, their position has barely changed, their economic marginality as evident under post-Fordism as it ever was. For yet others, such as the post-socialist states, the onward march of global capitalism has thus far implied an uncomfortable ride on the roller-coaster path of economic development.

14.9 Conclusion

The history of economic development is one of increased economic specialization moving hand-in-hand with geographical expansion; the combined and uneven extension of the (spatial) division of labour. In different historical periods of production, whether, for example, the Roman Empire, family capitalism, Fordism or post-Fordism, those involved may well have felt themselves to be part of a 'global' production system. From the standpoint of today, the post-Fordist system of flexible production is 'more global' than anything that has gone before. Similarly, globalization is taking place at a scale and intensity greater than previously experienced.

The paradox of this process is not 'the end of geography' but, in contrast, an even greater sensitivity to the geographical choices on offer within the global production system. Indeed, with further global expansion has come greater geographical concentration – at a number of spatial scales – as certain places have become almost the only place to produce particular goods. One thing is certain: today's global production system is more complex and interdependent than ever. It is the role of the economic geographer to both characterize and understand this complexity.

Learning outcomes

Having read this chapter, you should understand:

- Throughout history there have been 'global production systems'. Nevertheless, there is something distinctive about today's post-Fordist global economy.

- Economic production is a story of economic specialization and the continual reworking of the division of labour.

- A spatial division of labour occurs with the geographical separation of production tasks, for example, writing, printing, binding and distributing a book.

Such a division may be local (different buildings), regional (the UK's North–South divide) or global (the world's North–South divide).

Different economic geographies can be explained by distinctive patterns of spatial divisions of labour.

Fordism and post-Fordism each have their own characteristic economic geography, their own map of a global production system.

The emerging economic geography of the twenty-first century is an interlinked mosaic comprising continued global expansion of capitalism hand-in-hand with local geographical concentration.

Further reading

Allen, J. and **Hamnett, C.** (eds) (1995) *A Shrinking World? Global Unevenness and Inequality*, Oxford University Press, Oxford. An accessible edited collection that provides an understanding of the processes that are shrinking the world and the impact of these processes on a range of global economic objects and activities.

Bryson, J.R., Daniels, P.W. and **Warf, B.** (2004) *Service Worlds: People, Organizations, Technologies*, Routledge, London. The most recent account of the different divisions of labour amongst service activities and occupations as well as the role of ICT in altering the geographies of production systems.

Bryson, J., Henry, N., Keeble, D. and **Martin, R.** (eds) (1999) *The Economic Geography Reader: Producing and Consuming Global Capitalism*, Wiley, Chichester. A more advanced collection of edited readings of the 'great and good' in economic geography. For students wishing to develop their understanding, introductions and references are particularly useful.

Dicken, P. (2003) *Global Shift: Transforming the World Economy*, 4th edition, Paul Chapman, London. The definitive textbook on the global economy, its evolution, structure and geography. The third edition is as useful as the fourth.

Knox, P. and **Agnew, J.** (1998) *The Geography of the World Economy*, 3rd edition, Edward Arnold, London. A comprehensive and concise 'encyclopaedia' of the history and geography of the world economy. Heavily influenced by a 'world systems theory' approach to understanding economic geography.

Massey, D. (1995) *Spatial Divisions of Labour: Social Structures and the Geography of Production*, 2nd edition, Macmillan, London. This extremely important book in economic geography theorizes, and exemplifies, the concept of spatial divisions of labour. Based on the experience of the UK economy, this is one of the definitive accounts of why economic geography matters.

Peck, J.A. and **Yeung, H.W.** (eds) (2003) *Remaking the Global Economy*, Sage, London. This collection provides a useful overview of recent debates on the global economy.

For annotated, clickable weblinks and useful tutorials full of practical advice on how to improve your study skills, visit this book's website at www.pearsoned.co.uk/daniels

The global financial system: worlds of monies

Topics covered

- Money in economic geography

- The social, cultural and political significance of money

- Geographies of monies

- Global monies, local monies and monetary networks

- Money, space, and power: geographies of inclusion and exclusion

Money is simultaneously everything and nothing, everywhere but nowhere in particular.

(David Harvey 1985b: 167)

This chapter explores some of the worlds of money we inhabit, or rather the economic, social, cultural and political networks through which we establish value and manage exchange. Our attitudes to money, how we obtain it and use it, are not simply an economic phenomenon. Geographers and other social scientists are now grappling with the complex social, cultural, political and geographical aspects of money.

As the quote from Harvey suggests, however, money can be a difficult subject. Novelists, film-makers and social theorists alike have grappled with its seemingly contradictory characteristics. Money can be everything and nothing, everywhere and nowhere; money can be

The functions of money

- Money is a *unit of account*: it is the base of economic accounting systems.
- Money is a *measure of value*: it is the commodity against which the values of other commodities can be calibrated.
- Money is a *store of value*, in that you can sell a good for a certain amount of money and then use that money to buy something else at a later date. Money thus allows for the separation of the sale and purchase of commodities over space and time *if* it is a reliable store of value, if money is not devalued over time and space. Preserving this ability of money to store value is one of the reasons why governments are so concerned about the phenomenon of inflation.
- Money is a *medium of exchange and circulation*.
- Money is a *means of payment and a standard for deferred payments*.

These different functions of money often come into conflict with each other. As a store of value, governments, firms and individuals alike want monies to be relatively stable and maintain their value, and hence purchasing power, between different times and spaces. As a medium of circulation, however, it is important for money to be extended as credit.

Banks create credit by extending loans to individuals, firms and governments. In periods of rapid expansion, there is temptation for banks and other lending institutions to create large volumes of credit. Sudden loss of confidence in the quality of credit, as happened in the Wall Street Crash of 1929 (see Plate 15.1) or, more recently, in South-East Asia and Russia, can trigger financial panics and devaluation as depositors, investors and financial institutions flee from credit moneys and seek out safer havens in which their moneys will not be devalued (see Figure 15.1 on page 355).

Plate 15.1 The Wall Street Crash 1929: the stock market collapse of October 1929 brought the 1920s boom to an end and led to widespread bank failures, bankruptcies and drastic reduction in the availability of credit. (Corbis UK/Bettman)

the root of all evil and a source of independence and freedom and so the examples go on. Even simple questions like 'what is money?' present some difficulties. Davies (1994: 29) suggests that money is 'anything that is widely used for making payments and accounting for debts and credits'. Historically, a wide range of items, including beads, shells, whales' teeth, cattle, salt, skins, tobacco, beer, gold and silver, to name only a few, have been used as money (Davies 1994). Such variation suggests that our question should be altered. Rather than asking what money is, we should perhaps ask what functions does money perform? This question takes us to the heart of some of the important economic functions that money performs and to the essence of money as a source of social power.

Money performs several functions (Spotlight box 15.1). We are used to thinking of money 'oiling the wheels' of commerce and also being a measure of worth or value. Yet these two facets of money can be contradictory. As a store of value, it is desirable for money to be a stable, fixed representation of value. As a medium of circulation and as a form of capital – money that is thrown into circulation with the intent of generating more money – it is desirable to have money freely available as credit, divorced from its moorings in the 'real economy'. As Davies (1994: 29) argues, then, the history (and geography) of the evolution of money reveals 'unceasing conflict' between borrowers, keen to expand the *quantity* of money in circulation, and lenders, keen to limit the supply of money and, at all costs, preserve the *quality* of monies in circulation. Managing this conflict, or more accurately, reacting to financial crises resulting from it, has been and still is the mission of various regulatory authorities.

15.1 Money in economic geography

Many accounts of money and 'the global financial system' are written by economists or finance specialists, who pay little attention to the geographical anatomy, workings and implications of financial systems. Is money something that we should think about geographically? As the Harvey quote at the start of the chapter suggests, money seems to be 'everywhere but nowhere in particular' and many contemporary debates about globalization stress the increasingly integrated, global machinations of the financial system in which 'virtual' monies can be moved around the globe at the touch of a button. Indeed, O'Brien (1992)

argued that, as far as international finance is concerned, we have reached 'the end of geography', that location is of declining significance for firms and consumers in these global times. One of the purposes of this chapter is persuade you of the existence and importance – economically, culturally and politically – of different geographies of money and financial institutions that make up what tends to get called 'the global financial system'

Having said that, economic geographers have, until quite recently, paid relatively little attention to issues concerning money and finance (Leyshon 1995a) and many economic geography textbooks still devote little, if any, attention to the subject. One rationalization for this state of affairs is simply that money and finance have, historically, appeared well down the list of priorities for economic geographers. As Martin (1999: 3) argues, 'since its inception, economic geography has been pre-occupied with the industrial landscape.' Manufacturing industries were viewed as the 'drivers' of the economy and received more attention than industries concerned with circulation or consumption.

A second explanation for the relative neglect of matters financial is that '[m]ost geographers are interested in tangible things, and money ... is very intangible' (Laulajainen 1998: 2). This argument takes us back to Harvey and the difficulty of nailing down the different, contradictory qualities of money. Moreover, the organization of monies into different financial markets can appear dauntingly complex. The 'global financial system' can sometimes seem like something 'out there', outside us, a fast-moving, volatile 'thing', running out of control, increasingly uncoupled from the 'real' economy and operating beyond the regulatory reach of any one nation-state (Langley 2003). This global sense of money and finance can be difficult to grasp. We may have a sense that this international financial system is important, but sometimes it is difficult to understand just how its machinations affect us. For example, how did the south-east Asia financial crisis of the late 1990s change the lives of people in Europe, Africa, the United States or elsewhere in the world?

For all the difficulties, real and imagined, involved in understanding the workings of international, national and local finance, money and finance are now attracting a great deal of interest from geographers, other academics, governments and the popular press. This is partly explained by a number of recent financial and political crises, including those in Pacific and south-east Asia, Russia and Brazil, and growing unease about levels of household indebtedness in the UK, US and elsewhere. These crises have fuelled a growing sense of unease, in popular and academic writings, over the volatility and instability in global financial markets and a concern that regulatory bodies and nation-states are increasingly unable to control financial flows. Also, as Leyshon (1995a) argues, money and finance have assumed a growing social and cultural prominence since the 1980s. This is related to the growth of financial services employment, the ascendance of Thatcherism and Reaganism and the growing media and popular attention devoted to the rhythms and practices of financial services in London, New York and elsewhere.

The growing body of work on money and finance has grappled with a number of challenges including:

- the economic, social and cultural aspects of money

- the links between our everyday experiences of money and the workings of 'the global financial system'

- the changing geographies of money in these increasingly global times

- the creation of progressive, non-exclusionary forms of financial institutions in the contemporary world.

In dealing with these issues, geographers are not talking about 'the global financial system' as though it were a machine, but rather seeking to understand 'global finance' as a social, cultural and political phenomenon. Revealing the *social construction* of 'global finance' is important because the workings of global finance make a real difference to the well being of individuals and the choices/predicaments facing governments. The workings of the financial system shape who has access to credit, how, where and how much they pay for it and whether governments can afford welfare, education, healthcare and other services.

In the subsequent sections of this chapter, some of the above issues are explored in more detail. In the next section, it is argued that in addition to its important economic functions, it is important to recognize some of the different social and cultural aspects of money and how it is bound up with individual and group identities. The focus is broadened in the following section which considers the geographies of money that connect individuals and groups in distant places and at different times. By way of conclusion, the final section considers how some groups and regions are included or excluded from particular geographies of money, before exploring some of the ways in which we can think about progressive, inclusive monetary geographies.

15.2 Everyday worlds of money

Although we may be used to thinking of money as primarily an economic phenomenon, geographers and other social scientists are now recognizing that money is saturated with cultural, social and political significance (Dodd 1994, Tickell 2003). In this section, we examine some of the social and cultural aspects of money in different societies. We focus initially on money and power, before moving on to consider money and identity, money as a transformer or leveller and money as a mediator of our experience of social relations.

15.2.1 Money and power

Many of the social and cultural aspects of money are related to its role as 'the very incarnation of social power' (Harvey 1982: 245). Again, however, we confront the contradictory qualities of money. As a means of expressing social power money has both desirable and less desirable qualities. Leyshon and Thrift (1997) talk of two discourses about money, one of suspicion and one of liberation.

To understand the discourse of suspicion, we can turn to the work of Georg Simmel. For Simmel (1978: 277), money is 'the most perfect representation' of the tendency to reduce quality to quantity. As a store of value and unit of account money transforms social relations – qualitatively different commodities and experiences – into an abstract quantity, namely their price. This was what Marx (1973: 205) meant when he talked of money destroying communities and, in so doing, becoming 'the real community'. In contemporary capitalism, money has become *the* mediator and regulator of economic relations between individuals, a measure of wealth and a means of expressing social power. All manner of social, political and cultural issues – from healthcare, housing, education, leisure and sports, the environment, working conditions, and so forth – are debated not in terms of what they are worth, but in terms of *how much* they are worth, in terms of what can be afforded. Money 'affects our very ideals of what is good and beautiful and true' (Mitchell 1937: 371). Money, in this sense, has corroded the importance of other meanings and measures of 'value'.

Yet, although money can be treated with suspicion because it reduces qualities to quantities and introduces 'impersonality and colourlessness' into society (Simmel 1991: 19), there is also a discourse of liberation that accompanies money. First, money provides a degree of individuality, freedom, security and independence for those that hold it. This is especially true in societies like Britain where laws regarding private property buttress ideologies of liberty, equality and freedom. The idea that money is empowering and liberating is central to the desire to possess money, to work for it, to accept it as payment and to save and invest it. Second, as Simmel (1991: 20) argues, money 'creates an extremely strong bond among members of an economic circle'. Simmel argues that people have far more connections to other people in modern, monetized societies than was the case in feudal society. In the **division of labour** in capitalist societies (see Chapter 14, this volume), for example, money links people together in offices, factories, homes and shopping malls in different parts of the world. Money may in some senses be corrosive of social bonds, but it also creates a community that can forge new connections and hold people together.

15.2.2 Money and identities

Thus far, the discussion about money and power has been rather abstract. One of the ways in which we can bring this discussion closer to home is to think about the role of money in the construction of identity. If we think of identity not as something that is fixed or innate but as a 'capacity to keep a particular narrative going' (Giddens 1990: 54), then money can be an extremely important part of telling and maintaining stories about ourselves. In the remainder of this section, we consider four brief examples.

First, money is very obviously tied up with how individuals identify their class positions, what type of accommodation they live in, what type of schools they attend, and whether they can afford to stay on and pursue higher education and so forth. Being one of Britain's 'very rich' is associated with social and cultural networking practices which include particular (very expensive) rituals of conspicuous consumption, including balls and dinners and attendance at annual sporting events like Ascot and Henley. Such activities maintain relatively tight-knit networks, often linked by

marriage. Within such moneyed networks, children can be sent to exclusive schools to mix with children from other wealthy families and be endowed with the social, if not financial, capital necessary to smooth their journey along their chosen future path.

Second, money stirs passions about nationhood, sovereignty and *national identities*. Examples abound. A glance through the *Oxford Dictionary of Current English* (1985: 737) suggests that 'sterling work' is derived from sterling, 'of or in British pound, of coin, precious metal, genuine, of standard value or purity (of person or qualities etc.), of solid worth, genuine, reliable'. Another example relates to 'arguably the most radical political-economic and cultural transformation of space in Western Europe since the French Revolution' (Swyngedouw 2000: 71), namely the birth of the euro (€). Each euro note has an image of an arch and on the reverse a bridge, yet none of these are real bridges in (or between) European states. Instead, they are all simulated bridges, symbolizing connection, interchange, communication; they hint of the economic, political and cultural project of European integration (Pollard and Sidaway 2002).

Third, just as money can be a powerful symbol of class or national identity, it is also integral to the performance of *gender identities*. By way of example, we can think of how central the role of 'the male breadwinner' is to the construction of many masculinities. Research into changing expectations of the role of fathers in Rochdale in the UK found that men, women and children felt that being a 'provider' was the main role for fathers, even in circumstances where women in the same household were working and, in some cases, earning more than the men (Brindle 1999). Feelings of guilt and of not contributing to the family unit were attributed to men who were unemployed or working in a low-paid job. Films like *The Full Monty* and *Roger and Me* illustrate only too well that loss of employment in the car, steel and mining industries is not simply an economic catastrophe for the men involved. For those whose masculinity is rooted in 'industrial work' and the status of a 'breadwinner', the loss of regular employment and earnings can entail wrenching changes in social status, self-esteem and relations with partners and children.

Fourth and finally, another form of identity which is very directly linked to money, and in many societies assuming increasing importance, is the *virtual financial identity* individuals (and indeed countries) are assigned by credit-scoring companies that vet applications for credit. Those who seek credit in the form of credit cards, mortgages and other loans are the subject of 'credit checks' that assess a wide range of factors concerning employment status, incomes, expenditures, outstanding debts, repayment histories and so forth. In the data banks of major credit-scoring companies, the credit scores that make up our virtual financial identities are constructed and then shared with the financial institutions we approach for credit.

Credit-scoring companies are not the only groups attempting to construct virtual identities. Many organizations are now in the business of either constructing virtual identities of their customers (for example supermarkets in Britain use 'loyalty cards' as a means of tracking buying habits) or else buying customer information from third-party providers. The aim is to use customer information to inform decisions about targeting marketing initiatives towards those socio-demographic groups which will generate the greatest profits for the lowest marketing related expenses.

15.2.3 Money talks . . .

Money, as an incarnation of social power, can create new links between groups of people. As such, money can be a force for making people more tolerant of difference. For example, the importance of Chinese investment in Birmingham, London, Manchester and other British cities is now being recognized and they are seeking to embrace ethnic and racial diversity as part of their economic development strategies (Henry *et al.* 2002). In many British cities, 'Chinese Quarters' and other distinctive neighbourhoods associated with minority ethnicities are taking shape. Similarly, the 'pink pound' – the earning, investing and purchasing power of gay consumers – is now being courted by financial services providers, film and music media, travel firms, airlines, health and fitness firms, restaurants and bars. Surveys in Britain, the United States and Australia characterize gay consumers as groups that are relatively affluent, white, active,

Plate 15.2 Work and identity: mining is not simply a means of earning a living; it is also a source of status and camaraderie for the men involved. (© Sheila Gray)

financially literate, and likely to travel overseas and eat out regularly (Anon. 1999). Although this stereotype is not representative of gay couples on low incomes or with children, the economic significance of pink pounds and dollars – already acknowledged in cities like London, San Francisco, Toronto and Sydney – is being recognized in a growing number of places.

With its ability to forge new associations between individuals and groups, money is also, potentially, a leveller. Money can disrupt traditional cleavages along the lines of class, gender, ethnicity and sexuality. Pop stars, actors, footballers and other high earners can join the ranks of the financial elite regardless of their class backgrounds, their ethnicity, where they went to school and so forth. The realization that 'money talks', that it can be transformative, a bridge to a different life (McLuhan 1964) is integral to the appeal of the forms of gambling like lotteries. Lotteries and casino gambling are now very big business in Europe, North and South America, Australia, Hong Kong and, more recently, China. Global lottery revenues rose 50 per cent (to US$100 billion) between 1990 and 1995 (Emerson 1995) and in 2000, the worldwide lottery industry sold $140 billion of tickets, with $100 billion

of this from North America and Europe (www.lafleurs.com 2003). The growth of the Internet has aided the development of on-line gaming and lottery playing; in the USA, for example, there were over 133,000 online lottery retailers in 2000 (www.lotteryinsider.com 2003)

15.2.4 Money as a mediator of social relations

The idea of money being a language that can connect and liberate different individuals and groups suggests that money can, in all sorts of ways, mediate our experience of social relations. By way of example, we can develop our previous discussion of gender identities and think about how money, its organization and uses are structured in ways that are gendered. Viviana Zelizer (1989), for example, suggests that how monies are exchanged within households is usually subject to a very different, and usually gendered, set of rules than those that govern the use of money in the market place. Zelizer (1989: 351) argues that social structures

Plate 15.3   Money changing lives? The possibility of a big win, however remote, helps to fuel various forms of gambling.
(Associated Press/Luke Palmisano)

systematically constrain and shape (a) the *uses* of money, earmarking, for instance, certain monies for specified purposes; (b) the *users* of money, designating different people to handle specified monies; (c) the *allocation* system of each particular money; (d) the *control* of different monies; and (e) the *sources* of money, linking sources to specified uses.

Research in Britain has shown that men and women tend to be responsible for different forms of expenditure. Women are primarily responsible for expenditures on clothing, school-related expenses, food and charity donations and men responsible for expenditure on trips and holidays, repairs, redecorating, car maintenance and so forth (Pahl 1989).

Gendered assumptions about earnings and uses of money are also writ large in financial services industries. Financial spaces like the City of London and New York are renowned as elitist, masculinist environments (Lewis 1990, McDowell 1997) and many financial products target 'a male breadwinner' and are designed around gendered assumptions and stereotypes about work histories and sensitivity to risk. Pension products, for example, are designed to privilege those clients (most usually men) who work full-time and invest premiums regularly, and without interruption, over a 40-year period (Knights and Odih 1995). Similarly, Ravenhall (1999), citing a survey by *Money Magazine*, argues that male financial advisers treat men and women differently, with women being steered towards so-called 'widows and orphans investments', i.e. relatively conservative products with lower risks, and potential returns, than other investments. In Bangladesh, by contrast, it is often women who are the clients of banks, who receive credit and organize themselves and others into village groups to oversee the repayment of debt. Later in this chapter we will examine the establishment of Grameen Bank in Bangladesh. Its founder, Mohammad Yunus, established the bank believing that conventional banks were anti-poor and anti-women and insisted on collateral before making loans. Ninety-four per cent of Grameen Bank's clients are women. Women are regarded as better borrowers because 'they had most to lose, were more careful, had longer vision and were more concerned with the family than men' (Caulkin 1998: 2). In different ways in different societies, then, money can mediate the experience of, in this example, gender relations. Households and financial institutions

can operate to reinforce or challenge prevailing assumptions and stereotypes concerning women and money.

In summary, money has cultural, political, social and not just economic significance and this is reflected in the negative and positive discourses about the power of money. Money, as an 'incarnation of social power' is integral to the construction and performance of different identities. It is a 'leveller', a language for linking different individuals and groups, and it mediates our experiences of social relations.

15.3 Geographies of monies

How can we understand the links between some of our everyday experiences of money, described in Section 15.2, and the workings of so-called 'global finance'? This section broadens the focus to consider the geographies of money that connect individuals and groups in distant places and at different times. In these global times, some would argue that geography is relatively less important. Below, we examine some of the arguments concerning the globalization of money and finance, before outlining some of the problems with this view. The section concludes by introducing the concept of *monetary networks* as a way of thinking about money and its 'global' and 'local' qualities.

15.3.1 Global monies?

Many geographers, particularly those working in a **political economy** tradition, have argued that capitalism seems to be speeding up and spreading out (Harvey 1989, Massey 1994). Capitalism is seen to be spreading out in that more people in different countries and regions are becoming bound up with the logics of contemporary capitalism. It is also speeding up in that the pace of life seems to be increasing in different parts of the globe. In the second section, we noted the ability of money to transform qualitatively different commodities and experiences into an abstract quantity, a price. This has led many theorists to argue that money is a vehicle for the *homogenization of space*, i.e. a vehicle for making different spaces more similar. While there are ongoing debates about the extent to which capitalism is generating a *global culture* (see Chapter 6), it is difficult to argue with the contention that financial markets in different countries have, since the mid-1970s, become more interconnected (Harvey

1989, Martin 1994). Financial space has become more homogeneous.

When we talk about the *globalization of finance*, however, we are talking about more than just the growth of international financial transactions, or the growing presence of multinational companies in domestic financial markets. Globalization implies

> a strong degree of *integration* between the different national and multinational parts ... the emergence of truly transnational banks and financial companies ... that integrate their activities and transactions across different national markets. And above all, it [globalization] refers to the increasing freedom of movement, transfer and tradability of monies and finance capital across the globe, in effect integrating national markets into a new supranational system.
>
> (Martin 1994: 256)

The emergence of this *supranational* system coincides with the increasing difficulties faced by *nationally* based regulatory authorities like the Bank of England or the US Federal Reserve. The 'discourse of suspicion' we mentioned in Section 15.2 rears its head in many popular representations of 'global finance':

> [T]he world is now ruled by a global financial casino staffed by faceless bankers and hedge fund speculators who operate with a herd mentality in the shadowy world of global finance. Each day they move more than two trillion dollars around the world in search of quick profits and safe havens, sending exchange rates and stock markets into wild gyrations wholly unrelated to any underlying economic reality. With abandon they make and break national economies, buy and sell corporations and hold the most powerful politicians hostage to their interests.
>
> (Korten 1998: 4)

Whether or not you agree with Korten's view, it is true that the degree of integration of the financial system profoundly shapes and connects the lives of people thousands of miles away from each other. In this sense, Leyshon (1995b) talks of money being able to 'shrink' space and time, bringing some (but not other) parts of the globe relatively closer together through the working of financial markets. A good example of this is considered in Case study 15.1, which describes how the fates of property developers in Bangkok (Thailand) ultimately affected the fortunes of a bartender, Graham Jones, in Whitley Bay (UK). The fates of individuals

and companies in Britain and Thailand are connected through the workings of foreign exchange markets that, in turn, link rises and falls in national currencies to the competitiveness of different nation-states.

This example highlights only some of the interconnections between changes in the value of the Thai baht and how these affected a range of businesses across Asia and Europe. We could extend the example by thinking through how redundancies in the north-east of Britain would also have been felt by those not in paid work. For those in households dependent on the wages of someone made redundant there would be further belt-tightening or the search for alternative sources of income. For the unemployed, more redundancies in the area might mean more competition for new jobs that are created and so forth.

15.3.2 Forces for globalization

How has this globalization of money come about? There are a range of factors to consider here. First, through the 1980s and 1990s, different governments and international institutions, like the International Monetary Fund (IMF), have pursued neoliberal, 'free market' policies and encouraged the deregulation of financial markets (by eliminating exchange and capital controls) and the liberalization of flows of capital across national borders.

Second, there have been important advances in telecommunications and computing technologies, advances that can be summarized under three headings: computers, chips and satellites (Strange 1999). Computers have transformed payments systems. For many hundreds of years, payments and transfers of money were completed in cash – coins and notes – and written down in ledgers. Coins and notes were superseded by cheques which in turn have now been superseded by electronic monies that can be moved around the world at the speed of light. Semiconductor chips in computers mean that consumers in many countries can use credit and debit cards to pay for goods. In some countries, computer chips are now embedded in plastic 'smart' cards to allow the use of 'digital money', or, more accurately, electronic representations of currencies, to pay for groceries, telephone use and so forth. Finally, there are the systems of communication, using earth-orbiting satellites, that are integral to the operation of computing systems, e-mail, the Internet and other

Regional Case Study 15.1

From Bangkok to Whitley Bay ...

1 Devaluation in Thailand, July 1997

In 1997, several property companies collapsed in Thailand; property prices and the stock market started to fall. Currency speculators, already nervous about slowing growth in the region, started to sell the baht (Thai currency) as they expected the currency to be devalued. In July, the baht was devalued. As a result, Thai exports became cheaper and, to stay competitive, Indonesia, Malaysia, South Korea and the Philippines allowed their currencies to fall sharply. Some Korean firms who needed foreign currency to pay off loans started dumping cheap microchips, forcing microchip prices down from US$10 a unit to US$1.50.

2 Factory closure in North Tyneside, UK, 31 July 1997

Rapidly falling semiconductor prices meant losses of £350 million for German electronics company Siemens. Siemens semiconductors were produced at plants in Tyneside (UK), Taiwan, Germany, France and the US. Managers in Munich announced plans to close their £1.1 billion plant in North Tyneside that employed 1,100 people and had opened only in 1994.

3 The cleaning company

The Siemens factory provided 10 per cent of business in the North East for Mitie, a cleaning company. One-third of the workforce of 90 people were facing redundancy and their boss had his salary bonus cut 10 per cent, in line with the loss of work from Siemens.

4 The hotel

Siemens and its contractors, like Mitie, were the single biggest sources of business for the Stakis Hotel. The hotel responded to the closure by switching their market focus. As the flow of German executives and their UK contractors slowed and then ceased, the hotel sought to attract more families.

5 The taxi firm

Foxhunters, a taxi firm, had a contract with Siemens that generated more than a dozen runs a day, usually to the Stakis Hotel, the airport or the university. Since August, Siemens business had dried up and 85 drivers were chasing work for 50. Despite working longer hours, drivers' takings were down by between £100 and £200 a week. Drivers started to economize by bringing in their own lunches and by cutting down their trips to the local pub.

6 The local pub

Takings at Cameron's had fallen by £600 a week since August. The landlord, who blamed the Siemens shutdown for the reduced trade, cut his opening hours. Attempts to drum up more trade by price reductions had little effect. The landlord was hoping for more business around Christmas. He and his partner had less money for spending on their leisure activities, which included visiting places like Whitley Bay.

7 Graham Jones, bartender in Whitley Bay, October 1998

Graham was fired from his job at a pub in Whitley Bay; there was not enough business to occupy two bartenders in the public bar.

Source: Adapted from Carroll (1998)

forms of communication that are used by banks, governments and other financial players. These innovations have made 24-hour trading possible as stock markets in different countries are linked by computer.

The growth of the Internet through the 1990s has been closely linked to the rise of discourses about a 'new economy' (see Aglietta and Breton 2001, Thrift 2001). The Internet has not only altered how business can be done in financial markets – think, for example, about the rise of online banking in North America and parts of Europe – but also changed the cast of characters involved in undertaking that business. New communications and software technologies have spawned the proliferation of financial web-sites, information and intermediaries able to provide financial information to firms and consumers. Online providers, without the costs of maintaining the bricks and mortar associated with a high street presence, provide insurance, banking, pension and other financial products to firms and consumers with access to Internet technologies.

Third, and closely related to developments in technology, there have been innovations in tradable financial products that have made it easier and faster to move money around the globe. One example of such an innovation is that of derivatives (Tickell 2003). Derivatives are contracts that specify rights/obligations based on (and hence 'derived' from) the performance of some other currency, commodity or service. In the 1970s, financial derivatives were created to allow financial managers to deal with risk and they have become increasingly sophisticated and extended to more markets. Since 1997, for example, energy suppliers, transport agencies, construction companies, wine bar owners and other firms exposed to weather risk (the possibility that weather could have an adverse impact on their profits and cashflow) can now purchase weather derivatives contracts to protect themselves against this risk (Oldfield et al. 2004). With this extension, however, is increased speculative trading of derivative contracts, some high profile losses (for example, Barings Bank in London) and growing concern for regulatory authorities (Strange 1999, Tickell 2003).

Innovations in computing and software technologies and in financial instruments are fundamental to arguments about the growing *financialization* of the economy; this is an argument that asserts that the speculative accumulation of capital has become an end in itself in contemporary capitalism, that the financial system has come to feed on itself, so to speak, rather than supporting firms and industries and other elements of the 'real' economy (Strange 1999). Round the clock trading of financial instruments on exchanges in different places and time zones opens up opportunities for speculation and arbitrage, the ability to profit from small differences in price when the same financial product is being traded on more than one market. One indicator of how growth in the international financial system is outpacing the growth of the 'real' economy is provided by foreign exchange trading data. Data from the Bank for International Settlements reveals that the daily turnover of foreign exchange trading in 1973 was $10–$20 billion, roughly twice the amount necessitated by world trade. By 1998, daily trade in foreign exchange averaged $1.5 trillion, roughly *seventy* times that necessitated by world trade (Langley 2003). For individual consumers with access to the technology who want to get involved, there are now on-line financial bookmakers encouraging clients to enjoy spread betting (and possibly tax-free profits) on price movements of stocks, stock indices, currencies, interest rates, commodities and even house prices.

These, then, are some of the ways in which money has become more global since the 1970s. But what motivates such changes? What motivates the implementation of 'free market' policies, the development of new technologies and financial instruments like derivatives? For those working in a *political economy tradition*, like David Harvey (1989), the motivation for these changes is the search for profit. Capitalism is fundamentally about the accumulation of surplus. Competition drives capitalists to seek out new markets, new products and to reduce the *turnover time of capital* (see Spotlight box 15.2). In different parts of the globe, money is the language through which the imperatives of capitalism are being communicated.

So, there is a very strong economic rationale for the globalization of money. And it is difficult to argue with the contention that money has become more globalized since the 1970s, that it has, increasingly, connected people in distant places and homogenized financial space. For Richard O'Brien (1992), the growth of the international financial system is tantamount to 'the end of geography', the notion that geography is becoming relatively less important as money, in its different forms, is able to overcome the friction of distance and link distant places together. Others argue that this view is too simple, that

Spotlight Box 15.2

The circulation of money as capital

In capitalist societies, the circulation of money as capital is as follows. From left to right in the equation, money (M) is invested by producers to purchase commodities (C), namely labour power (LP) and the means of production (MP), say pieces of wood and wood cutting machinery. Labour power and the means of production are combined in production (P) to make more commodities (C′), in this example, let us say chairs, which are then sold for more money (M′) than was originally invested (M).

$$M \rightarrow C \rightarrow \begin{Bmatrix} LP \\ MP \end{Bmatrix} \rightarrow P \rightarrow C' \rightarrow M' \rightarrow \ldots$$

The purpose of production in capitalist societies is to produce profit and accumulate capital. The *turnover time of capital* is the amount of time it takes for money to complete this circuit. The shorter the turnover time, the more often money can be lent out and the more profit can be made. Producers therefore have a very strong incentive to, where possible, reduce the turnover time of capital; time *is* money.

geography remains critical to our understanding of global finance. In the next part of this section, we will consider these views in more detail.

15.3.3 Debunking 'global' monies

A rather different departure point for talking about the 'globalization of finance' is to argue that the financial system is really not all that global, that it is more like a *web of connections* between different financial systems, some of which are bound together more tightly than others. This argument develops from noting a number of problems with the notion of a 'global financial system' and global monies.

The first problem stems from the simple observation that 'global finance' is largely the province of North America, Europe and parts of Asia, most notably Japan, Hong Kong and Singapore; it is an idea centred on the experiences of the West (see Table 15.1). To continue with one of the examples mentioned in the previous section, consider the geography of Internet banking. In 2003, it was estimated that there were just over 102 million users of on-line banking. More than 47 million of these users lived in Western Europe and another 40 million lived in the USA and Japan. Central and Latin America and Africa provided just 5.1 million users, less than 5% of the total (www.epaynews.com 2003). Many parts of Africa, Asia and Latin America are not only not included in 'the global financial system', they are being

actively excluded as banks refuse to lend money until outstanding debts have been repaid. There is what Massey (1993) describes as a *power geometry* at work when some commentators describe the financial system as 'global'. For some financial workers in New York, London and Tokyo, international finance may be regarded as 'global' in the sense that all countries and regions of the globe deemed creditworthy and capable of producing profits have been included; sub-Saharan Africa ceases to exist in such a conception of 'the global'.

Second, and resulting from this geographical concentration of the management of 'global finance' in North America, Europe and parts of Asia, some currencies circulate more widely and have a greater spatial reach than others. The Japanese yen, the euro and most especially the US dollar are very useful in international markets because they are accepted as forms of payment, unlike, for example, Indian rupees. Historically, the country that occupies a dominant economic and political position has underwritten the soundness of the international financial system and had its currency accepted internationally as the currency in which commodity prices are quoted and in which payments are made. Before the Second World War, Britain and the pound sterling fulfilled this role; after the Bretton Woods conference in 1944, the US dollar became the key international currency. More recently, as the economic dominance of the US has declined, the yen and the euro have become relatively more important in international markets. Countries

Table 15.1 Total value of share trading 2002 expressed in US$

Exchange	Total value of share trading (000 of US$)	Region % of world total
North America		**41.8%**
Amex	642,181.0	
Bermuda	413,743.5	
Chicago	532,040.0	
Mexico	32,285.6	
NASDAQ	7,254,594.3	
NYSE	10,311,155.79	
Toronto	408,164.9	
South America		**0.23%**
Buenos Aires (Argentina)	1,277.4	
Lima (Peru)	1,186.6	
Santiago (Chile)	3,011.2	
São Paulo (Brazil)	46,300.2	
Europe		**38.64%**
Athens (Greece)	23,461.5	
Budapest (Hungary)	5,908.1	
Deutsche Bourse (Germany)	1,212,301.6	
Euronext (Amsterdam, Brussels, Lisbon, Paris)	1,988,358.6	
Helsinki (Finland)	178,202.4	
Irish	33,270.3	
Ljubljana (Slovenia)	1,526.5	
Copenhagen (Denmark)	53,262.2	
Istanbul (Turkey)	69,936.7	
Italian exchange	634,634.6	
London (UK)	4,001,339.9	
Luxembourg	495.5	
Malta	47.9	
Oslo (Norway)	56,126.8	

Table 15.1   continued

Exchange	Total value of share trading (000 of US$)	Region % of world total
Spanish Exchanges (BME)	653,220.8	
Swiss Exchange	599,749.1	
Stockholm (Sweden)	279,943.5	
Vienna (Austria)	6,108.95	
Warsaw (Poland)	7,811.27	
Middle East		**0.07%**
Tehran (Iran)	2,071.2	
Tel Aviv (Israel)	12,676.2	
Africa		**0.35%**
JSE South Africa	78,391.8	
Asia/Pacific		**18.91%**
Australian	295,399.4	
Colombo (Sri Lanka)	318.5	
Hong Kong	194,003.6	
Jakarta (Indonesia)	13,049.8	
Korea	596,631.6	
Kuala Lumpur (Malaysia)	32,923.2	
Mumbai (India)	68,538.6	
National SE India	128,534.6	
New Zealand	88,783	
Osaka (Japan)	124,017.4	
Philippine	3,092.7	
Shanghai (China)	211,643.9	
Shenzhen SE (China)	140,660.6	
Singapore	63,047.8	
Taiwan	633,632.3	
Thailand	41,288.9	
Tokyo (Japan)	1,564,243.9	

Source: Adapted from *World Federation of Exchanges* at: www.iasplus.com/stats/stats.htm (accessed 1 December 2003).

that do not use the US dollar, the yen or the euro domestically have to convert their currency to, say, US dollars in order to facilitate international trade.

So, some currencies, like the US dollar, are truly international while some others are national. There are also over 2,000 *local currencies* in operation around the globe, currencies that facilitate exchange only within very specific spatial and social contexts. Local Exchange Trading Systems (LETS), for example, are associations whose members list their services needed/offered in a directory. Members of LETS then trade with each other in a local unit of currency, for example, 'bobbins' in Manchester and 'solents' in Southampton (Williams 1996). Since their establishment in Canada in 1983, LETS schemes have spread to Europe, Australia, New Zealand and North America, allowing members not only access to credit, but also the chance to engage in productive activity to earn such credit (Williams 1996). Some local currency schemes are devised in times of hardship to allow local people to trade goods and services when they are unemployed and have little money to spend. Other schemes are motivated by ecological concerns, the desire for community development and social cohesion, and the desire to construct alternative local economic geographies as a form of resistance against the global gaze of capitalism (Lee 2000).

Third, as Martin (1999: 6) argues, for all the talk of 'global finance', there remain different *geographical circuits of money* that form the 'wiring' of an economy, along which 'currents' of wealth, consumption and power are conveyed. Martin (1999) identifies four geographies of money: locational, institutional, regulatory and public. The *locational geography* refers to the location of different financial institutions and markets. Financial institutions and specialized functions (for example, stock markets and foreign exchange markets) tend to be agglomerated in large urban centres, with London, New York and Tokyo sitting at the top of the global hierarchy. Different countries also have different *institutional geographies*, in that they organize their financial institutions and markets in distinct ways. In Britain, for instance, we are used to thinking of the 'big four' retail banks – NatWest, Lloyds-TSB, HSBC and Barclays – and relatively new banks like the Halifax, Woolwich and others that have converted from building societies. In the USA, by contrast, there were over 9,000 Federally insured commercial banks in 1997 (Pollard 1999). Meanwhile, in parts of Asia and west and south Africa,

banks and other community based financial institutions like Rotating Savings and Credit Associations (ROSCAs) (see Spotlight box 15.3) are important in organizing and funding economic activity. ROSCAs take different forms in different regions and vary with the class, gender and ethnicity of their members (Ardener and Burman 1995).

These varied institutional geographies are products, in turn, of contrasting *regulatory geographies*. There are a wide range of supranational, national and regional regulatory spaces that govern the workings of financial institutions and markets. To return to the example of how the Asian financial crisis cost Graham Jones his job in Whitley Bay, a key player in how the crisis unfolded was the International Monetary Fund (IMF). The Chief Executive of Siemens, Heinrich von Pierer, argued that Korea was able to sustain a 'suicidal pricing strategy' for its memory chips (a pricing strategy that led to losses of £350 million for Siemens) because it was receiving financial aid from the IMF (Milner 1998). To give another example, as Martin (1999: 9) observes,

> [M]oney has had a habit of seeking out geographical discontinuities and gaps in these regulatory spaces, escaping to places where the movement of financial assets is less constrained, where official scrutiny into financial dealing and affairs is minimal.

Offshore financial centres like the Bahamas, Cayman Islands, Jersey and Guernsey are attractive because of their low tax rates and minimal regulation. In 1990, an estimated 50 per cent of the world's money stock either resided in or passed through a tax haven (Laulajainen 1998).

Finally, when Martin (1999) discusses *public geographies*, he refers to the role of states in distributing monies across regions in the form of goods and services, infrastructure, health, education and so forth, and in transferring monies in the form of various social and welfare programmes.

> What becomes clear from this discussion is that the world remains made up of a patchwork of different financial spaces or systems, and although there has been considerable integration of financial practices and processes across such systems in recent years, we are a long way from the seamless global financial space described by some commentators.
>
> (Leyshon 1996: 62–3)

Spotlight Box 15.3

Rotating Savings and Credit Associations (ROSCAs)

A ROSCA is 'an association formed upon a core of participants who agree to make regular contributions to a fund which is given, in whole or in part, to each contributor in turn' (Ardener 1995: 1). ROSCAs are known by different names, depending on their form and scale, the social classes of their members and their location. In South India they are known as *kuris, chitties* or *chit funds*, in Cameroon as *njangis* or *tontines*.

ROSCAs were well developed in China, India, Vietnam and parts of West Africa by the end of the nineteenth century. Variants of ROSCAs also existed in Scotland and parts of northern England. Members, usually ranging in number from a few to several hundred, make regular contributions, in cash or in kind, to a fund. The fund, or part of it, is then given to each member in turn. The order in which the fund is given out is determined in a variety of ways; lottery, age, kinship seniority, or by rules established by the organizer.

In addition to encouraging regular savings and providing small-scale capital and credit for their members, many ROSCAs have very strong moral and social dimensions. Trust, social solidarity and responsibility are emphasized and members have a strong interest in ensuring that no member defaults on their regular payment. Some ROSCAs are women-only and their potential to empower women, by giving them greater control over income and credit, has attracted the attention of anthropologists, sociologists and feminists (see Ardener and Burman 1995).

15.4 From 'global monies' to monetary networks

The phrase 'global finance' therefore seems to be something of a misnomer. While it is true that different monies and markets have become more interconnected since the 1970s, what we have been describing as 'the global financial system' seems to comprise some elements that are international in their reach, juxtaposed with other local and national currencies, regulations and institutional geographies. One way of juggling with this complexity is to think about the idea of *monetary networks.*

Recall that in the introduction to this chapter 'the global financial system' was described as a social, cultural and political *network of relations* in which individuals have different capacities to participate. Martin (1999: 11), for example, describes financial markets as 'structured networks of social relations, interactions and dependencies – they are communities of actors and agents with shared interests, values and rules of behaviour, trust, co-operation and competition.'

When we talk about 'structured networks' what do we mean? Dodd (1994: xxiv) argues that monetary networks are, above all, *networks of information* that have five abstract properties (Spotlight box 15.4). These networks are formed by the production, interpretation and circulation of information between actors through time and space. Thus we can think of foreign exchange traders in London, Tokyo, New York and elsewhere constituting a network of relations through which foreign exchange trading is organized, regulated, understood and carried out.

How do networks actually work? How are all these different forms of information, that are monetary networks, produced and then moved around? Nigel Thrift (1996, 1998) has argued that the forms of information that constitute monetary networks are always ingrained in *practices*. To give an example of what he means, we can return to the workings of global telecommunications networks that we discussed earlier. Telecommunication networks rely on the workings of thousands of workers and machines to install, maintain, repair and operate them. Similarly, the Internet consists of thousands of miles of fibre-optic cables, most usually laid alongside existing transport infrastructure such as railway tracks, and computing hardware and software linked together,

Spotlight Box 15.4

Properties of monetary networks

- Networks contain a *standardized accounting system* into which each monetary form within the network is divisible. Thus any goods, priced in terms of this standardized accounting system, are exchangeable.
- Networks rely on *information regarding expectations of the future*; 'money is accepted as payment almost solely on the assumption that it can be reused later on' (Dodd 1994: xxiv).
- Networks rely on *information regarding their spatial extent*; institutional frameworks will limit the territories in which specific monetary forms may be used.
- Networks are based on *legal information*, usually in the form of rules that govern forms of contractual relations of network members.
- Networks presuppose *knowledges of the behaviour and expectations of others*. In order to trust money, those holding it need to anticipate, and trust, that it can be used and reused within the network.

Source: Dodd (1994: xxiv)

managed and sustained by thousands of programmers, engineers, users and machines (Bray and Kilian 1999). What we think of as the 'global financial system' is really an assemblage of consumers, workers, computers, telephones, office buildings, bits of paper, financial reports and so forth. There is simply no escaping the *materiality* (i.e. the bits of cable, bank branches, computers, telephones, desks, trading floors and so forth) and *practices* (sitting at trading desks using telephones/computers, paying for goods with particular coins, credit cards and so forth) that constitute these networks.

When we talk of 'the global financial system', then, we are really talking about combinations of human beings, technologies and documents which come together to sustain, in particular spaces and times, the flows of information that constitute different monetary networks. Note the use of the plural. Rather than thinking of one 'global financial system' we can think instead of there being a *multitude of monetary networks that overlap and intersect*. This is an important point, because different monetary networks, as we have seen, have very different geographies. Some monetary networks, like those of the foreign exchange markets, have the capacity to connect people and machines in London, New York and Tokyo at the touch of a button. Other monetary networks, like LETS schemes in Manchester, are much more compact and connect people over much shorter distances. Rather than thinking about networks as being either 'global' or 'local', Bruno Latour (1993: 122) argues that we should

think of networks as being 'more or less long and more or less connected'.

Monetary networks differ not only in the *length of connections* they are capable of sustaining, but also in the *speed of connections* they are capable of sustaining. We have talked already of the computing networks that facilitate foreign exchange trading and allow banks to move monies around the world in, usually, less than 25 seconds (Warf 1999). One of the attractions of ROSCAs (Spotlight box 15.3) for men and women in Ghana and elsewhere is the speed at which news of hardship or an emergency can be spread, and the order of rotation of the fund changed, to help out a member in trouble. ROSCAs may not have quite the speed of connection of foreign exchange trading (assuming, of course, that all the computers are working), but they can react with a speed rarely matched by local banks, which are often perceived as distant, impersonal and slow (Ardener 1995).

In summary, financial markets have become increasingly globalized since the 1970s, although truly 'global' markets are concentrated in a select number of major financial centres. There remain distinct regional, national and supranational geographies of money, such that the 'global financial system' can be conceptualized as a web of intersecting networks. If we conceive of 'the global financial system' as an intersecting web of networks – some fast, some slow, some long, some short – then we can envisage a whole host of networks that, to differing degrees, connect different geographies of employment, technology, regulation, government policies, leisure trends and so forth. Such a conceptualization provides a way of thinking about how, over a period of fifteen months, the fates of property developers in Thailand became intertwined with foreign exchange markets, semiconductor companies, cleaning companies, hotels, taxi ranks, bars and the loss of a job for Graham Jones in Whitley Bay. This approach also reinforces the point that when we talk about 'global finance', we are not talking about a machine operating 'out there', but a socially, economically, culturally and politically constructed network of relations with which we engage, in some way, shape or form, on a daily basis.

15.5 Conclusion: money, space and power

From the preceding discussion, it should be clear that many monetary networks are very uneven and exclusive. Different monetary networks are not open to all who would like to participate and there are some very distinctive *geographies of financial inclusion and exclusion* that have already been mentioned. The benefits of some of the most extensive, fastest monetary networks are, for the most part, extended only to those areas where individuals, companies and governments can produce a profit for the major lenders in North America, Japan and Europe. Access to such networks is carefully screened; countries, firms, banks and public agencies, like individuals, are credit-rated. On the basis of their credit rating, decisions are made by major financial institutions on whether to offer credit or buy debt and, if so, for how long, and at what rate of interest.

So, who does this screening? What are the geographies of this credit scoring? The main international agencies are Moody's Investment Services, Standard and Poor and, in insurance markets, A.M. Best, all US firms (Laulajainen 1998). The Securities and Exchange Commission (SEC) in the US recognized only one non-US rating agency, Ibca, in the UK, until 1994 (Laulajainen 1998). Not surprisingly, only the major capitalist economies are assigned high ratings when it comes to government-issued debt in the form of bonds (Figure 15.1). Standard and Poor's so-called 'Investment grade' scales rate types of debt on a scale from 'AAA' ('Superior') through 'A' ('Good') to 'BBB'. Below this are the 'Speculative' grades which range from 'BB' ('Questionable') through 'D' ('Default').

This is not to say that only the major capitalist economies are able to secure credit. In the late 1970s and early 1980s, total lending to Africa, Asia and most especially Latin America increased sharply. Yet rising interest rates, falling commodity prices and a rapid appreciation of the value of the US dollar left debtor countries increasingly unable to cope with repayments. Affairs came to a head in August 1982, when Mexico suspended its debt repayments, soon to be followed by Brazil, Argentina, Peru, Venezuela, Ecuador and more than twenty other countries by the end of 1983 (Porter and Sheppard 1998). The response of banks to the debt crisis was to drastically reduce their lending to these countries, to effectively exclude them from international bank lending networks.

More recently, investors in North America and Western Europe have been encouraged by fund managers and financial media to invest in so-called 'emerging markets', which include parts of the Russian Federation and former Soviet Republics and other parts of Asia, Latin America and Africa. Again, the lure for investors is the prospect, real or imagined, of higher rates of return than those that can be secured in what are represented to be the more conservative, less volatile, less 'adventurous' confines of Western European and North American equity markets (see Sidaway and Pryke 2000). Again, the US based credit ratings agencies are key players in the construction of these financial markets. Research in emerging economies demonstrates that the effects of a downgrade in, say a country's bond rating, are not confined only to the price of its bonds; a downgrade can also reduce a country's stock market returns and 'spillover' to neighbouring states as international investors anticipate further downgrades in the region (Kaminsky and Schmukler 2002).

The concentration of wealth and power in the major capitalist economies, most especially the US, is such that they hold sway, economically and politically, over the networks through which international lending is conducted and its terms negotiated. Even supranational agencies like the World Bank and the IMF, designed and established after the Second World War to avoid biases in simple bilateral lending, rely on funds from major capitalist countries who exert considerable influence over where the funds go and how they are spent. If South Africa and Russia and other post-socialist economies want to be included in the monetary networks of the World Bank and the IMF, then instituting reform packages based around market-led growth, a reduction in government

Figure 15.1 Sovereign risk of bonds denominated in a foreign currency, May 1996.

Source: Adapted from Laulajainen (1998: 56)

intervention in the economy and a general mimicking of the political and economic organization of major capitalist economies is the price of admission. Such structural adjustment packages have spread widely through east and west Africa, Latin America and Asia since 1982 (Porter and Sheppard 1998). For those that do not toe the line, life can be made very difficult. Money, as we saw in Section 15.2, is a very powerful tool of social and political control.

Such geographies of inclusion and exclusion operate within countries too as banks, private investors and other lenders look for prosperous growth regions and secure rates of return. In the UK, for example, the distribution of venture capital investments has traditionally favoured firms located in the prosperous south-east of England (Mason and Harrison 2002). In Los Angeles, in patterns repeated in many other US cities, major retail banks have closed bank branches in low-income neighbourhoods like South Central Los Angeles and opened new branches in prosperous areas like Beverly Hills (Pollard 1996). The financial institutions that tend to move into South Central are so-called 'alternative' financial institutions like pawnbrokers and cheque cashing establishments, which often charge very high fees for basic banking services.

Financial exclusion, whatever its spatial extent, tends to reinforce existing patterns of wealth, poverty and inequality. Those individuals and nations that already have wealth are deemed the most creditworthy and are offered the cheapest forms of credit to expand businesses, pay off debts, buy new goods and services and so forth. And, as discussed in Section 15.3.4, those consumers that have the financial, cultural and technical capital necessary to access to the internet will be the beneficiaries of new and sometimes cheaper financial products available on-line. For low-income groups and countries still reeling from crippling debt repayment schedules, credit is difficult to find and usually very expensive. For those groups suffering varying degrees of financial exclusion, what alternatives are available?

15.5.1 Some responses and alternatives to financial exclusion?

Although some alternative monetary networks, like those of cheque-cashing shops, may be exploitative and expensive to use, there are plenty of examples of other monetary networks that are designed to be inexpensive and inclusive. One very famous example is the Grameen Bank, formed in 1983, which now employs over 14,000 people in 35,000 villages in Bangladesh (Case study 15.2). The successes of the Grameen Bank and other micro-credit institutions in Indonesia, Bolivia, Kenya and elsewhere have attracted a great deal of attention and spawned innumerable imitations and different varieties of micro-credit schemes in Africa, Asia, Latin America and elsewhere. Indeed, the prospect of linking access to credit for the poorest with job creation has made micro-finance schemes increasingly popular in Britain, the US and Western Europe, and not just in those regions of the UK and the US that are experiencing financial exclusion.

It is also important to note that some 'alternative' financial networks may be preferred by individuals and households who do not want to use financial institutions like traditional retail banks and instead want to participate in networks that encourage broader social, environmental and other assessments of 'value'. The growing popularity of various micro-credit schemes, like the LETS and ROSCA schemes we discussed earlier, is based not only on their economic successes, but also on their status as 'alternative' and, potentially, progressive schemes. One of the interesting features of micro-credit schemes, like the Grameen Bank, is their often strong emphasis on creating social bonds and a sense of shared responsibility between members; these are schemes in which members recognize other measures of 'value' besides monetary ones. Another feature of the many experiments with microfinance is that it is women who make up the bulk of the membership of many programmes. As Goetz and Gupta (1996) point out, however, access to credit is not necessarily synonymous with control over the money and, more generally, empowerment for women in environments where they frequently face extreme constraints on their lives. As we mentioned in Section 15.2, how monies are exchanged within households is usually subject to a very different set of rules than those that govern the use of money in the market place (Zelizer 1989).

To conclude, these examples of monetary networks illustrate only too well that money, how it is obtained, how it is used, how it is managed and by whom and what significance it has in society, is an economic, social, political, cultural and *geographical* phenomenon. Although many accounts of 'the global financial system' pay little regard to the geographies of money and finance, this chapter has argued that small

Regional Case Study 15.2

Grameen Bank

The basic principles of Grameen Bank lending are as follows:

- the bank lends only to the poorest landless villagers in Bangladesh
- the bank lends, overwhelmingly, to women; 94 per cent of its clients are women
- loans are extended without borrowers needing collateral
- the borrower, not the bank, decides what the loan will be used for
- borrowers pay sufficient interest to keep the bank self-sustaining.

The Grameen Bank has lent out over US$4 billion since its establishment in 1983. The average loan is for about US$100. Rather than insisting on collateral or security, women wanting a loan join a village group, attend weekly meetings and assume responsibility for the loans of other group members. Thus, there is strong peer pressure for individual women to maintain payments, otherwise the group is unable to secure further loans. 97 per cent of Grameen Bank loans are repaid.

The success of Grameen Bank has spawned imitations in 58 countries in many parts of Africa, Asia and Latin and North America. Grameen has now branched out into fisheries, mobile telephones, energy and the Internet.

Sources: Bornstein (1996), Caulkin (1998), www.grameen.com/bank/GBGlance.htm

---

regional banks, foreign exchange markets, ROSCAs, and a host of other monetary networks – some relatively slow, some relatively fast, some longer than others – make up the 'global financial system' that is in many ways linking more people and diverse financial institutions and practices together in distant places. These monetary networks intersect most densely through institutions and practices in North America, Japan and Western Europe . This geopolitical reality profoundly shapes patterns of uneven development and structures which regions, industrial sectors, governments and individuals are deemed credit worthy and on what terms. 'Global finance' is not something 'out there', but a socially constructed, geographically rooted network of relations with which we engage, in some way, shape or form, on a daily basis.

Learning outcomes

After reading this chapter, you should be aware that:

- Money has geographical, economic, cultural, social and political significance and mediates our experiences of social relations.

- The 'global financial system' is a socially constructed web of intersecting networks operating at different speeds and over different distances.

- There are distinct regional, national and supranational networks of money.

- There are many geographies of financial inclusion and exclusion.

- There are a range of financial networks which may be more inclusive and politically progressive than others.

Further reading

Budd, L. and **Whimster, S.** (eds) (1992) *Global Finance and Urban Living*, Routledge, London. This selection of essays considers how the international financial system affects the lives of people in London.

Leyshon, A. and **Thrift, N.** (1997) *Money Space: Geographies of Monetary Transformation*, Routledge, London. This is a compilation of Leyshon and Thrift's essays on money.

Martin, R.L. (ed.) (1999) *Money and the Space Economy*, Wiley, Chichester. A wide-ranging collection of essays covering geographies of banking, financial centres, money and the local economy, and money and the state.

McDowell, L. (1997) *Capital Culture: Gender at Work in the City*, Blackwell, Oxford. This is a more specialist book that examines the masculine cultures and practices of parts of the City of London.

Sassen, S. (2001). *The Global City: New York, London, Tokyo*. Princeton University Press, Princeton, NJ. This book looks at the growth and internationalization of the financial system and its effects on the economic base and social structure of its three 'command centres', London, New York and Tokyo.

Useful web-sites

www.ft.com The *Financial Times* web-site contains useful links to financial and other firm, sector and country data, as well as breaking news stories.

http://cnnfn.cnn.com/ The US-based CNN financial web-site, like the *Financial Times* site, contains numerous links to business, financial and market news, by country, sector and firm.

www.worldbank.org/ The World Bank web-site is home to a host of information on World Bank publications, research, data and related organizations.

www.bis.org/statistics/bankstats.htm This site, maintained by the Bank for International Settlement, provides sources of banking data and statistics.

www.imf.org/ The home page of the International Monetary Fund. This site provides a wealth of information about the IMF and its mission in addition to country information.

www.grameen-info.org/ The web-site of the Bangladesh-based Grameen Bank. It has information on the mission of Grameen Bank and how it works, together with material on other microcredit schemes.

http://adams.patriot.net/~bernkopf/ A site dedicated to central banking issues, that includes links to Central Banks and their research departments, Ministries of Finance, Bankers' Associations and Institutes, and multilaterial financial institutions.

www.creativeinvest.com The site of Creative Investment Research, a US-based firm specializing in 'socially responsible' investment and providing many links on women and minority-owned financial institutions.

 For annotated, clickable weblinks and useful tutorials full of practical advice on how to improve your study skills, visit this book's website at www.pearsoned.co.uk/daniels

CHAPTER 16

Consumption and its geographies

PHILIP CRANG

Topics covered

- Why consumption matters
- Global geographies of consumption
- Local geographies of consumption
- Consumption, knowledge, and geographies of (dis)connection

16.1 Consumption matters

Watching television, eating a burger, wearing designer-label clothing, going clubbing, drinking a Starbucks cappuccino, shopping, skateboarding. All of these are examples of consumption. All will get a mention in this chapter. At first sight they seem trivial, mundane, unimportant things. They may seem unworthy of academic scrutiny, especially when set alongside more obviously important issues like global corporate geographies, geographies of poverty, or new forms of global monetary regulation. Not only that but these instances of everyday consumption may also not seem terribly *geographical*. In this chapter I argue that such snap judgements would be wrong. In fact I will suggest precisely the opposite: that these everyday practices of consumption have great significance, economically and culturally; and that they are fundamentally geographical, implicated in productions of the 'global', 'local' and non-academic or 'lay' geographical knowledges. In sum, I argue that in significant part we both produce and understand our geographies through our everyday, mundane activities as consumers.

There are a number of reasons why we are not

Plate 16.1 Man relaxes in the company of Ronald McDonald and Starbucks in Shanghai, China. This chapter argues that consumption is implicated in both global cultural flows and local fashionings of urban space. (Corbis/Reuters/Claro Cortes IV)

predisposed to think of consumption as important in this way. Technically, consumption can be defined as the utilization of the products of human labour. On that basis, it seems necessarily to occur after and follow from those activities of production. Let's hypothesise that I eat a burger in McDonald's. I couldn't do this if burgers weren't being made there, if McDonald's hadn't developed its transnational corporate structure so there is a McDonald's outlet where I live, if I didn't work to earn money to spend, and if complicated financial systems weren't established to enable such commercial transactions as me buying a burger to take place. But we must not mistake the chronology of the life of an individual burger and its consumption – burger made first, burger eaten afterwards; money obtained first, spent afterwards – with issues of import or causality. Yes, consumption depends on production. But, obviously, production also depends on consumption. The nature of these relations are complex and open to both change and dispute, but in outline there is now widespread agreement that consumption is neither the inevitable, final stage of a uni-directional economic production chain nor simply a by-product of monetary circulation.

Specifically, in the form of 'retail capital' consumption in fact plays an increasingly central role within many economies. It accounts for a growing proportion of paid employment and often exercises great powers over manufacturing through the organizational power of retailers over their suppliers (for work in this vein see Wrigley and Lowe 1996). Moreover, in parts of the economy one can also see how economic production has been reconfigured away from just material manufacture and towards the fashioning of 'signs and spaces' (Lash and Urry 1994) devoted to marketing and sales. Nike would be exemplary. According to Robert Goldman and Stephen Papson (1998) Nike don't so much make sportswear as make and sell the brand Nike, constituted through the sign and typography of the Nike name, the logo of the 'swoosh', and philosophies such as 'just do it'. The sportswear is the vehicle for that brand. So, Nike sub-contract manufacture. They invest far more in advertising and promotion (around 10% of annual revenues) than in capital expenditure (around 3–5%) (*ibid.*: 13). And that capital expenditure is largely devoted to distribution infrastructure, administrative and design infrastructure, and display or retail infrastructure (such as 'Niketown' retail outlets) rather than material production. Nike's business and its corporate geographies are constructed to be market led

and market leading. They are consumption focused. This is a wider trend, with discourses of consumption – of thinking and acting in ways that are consumer focused – coming to increasing prominence in both commercial and public organizations (see du Gay 1996).

More generally, every day on news bulletins all over the world 'consumer confidence', or its lack, is cited as the cause of the 'productive' sectors of an economy booming or contracting. Consumers, or more often 'the consumer' or 'the market', are constantly being invoked as the reasons why actions elsewhere in the economy must be undertaken. Bananas must be a certain size because 'the consumer' wants them to be; hence Caribbean growers are passed over by UK food retailers in preference to those in South America. Costs must be cut in banking as 'the consumer' is unwilling to choose institutions that do not offer the best deals; so branches close and jobs are lost. Supermarkets switch food suppliers, or at least threaten to and thereby exert their power, claiming their concern with matching the quality standards that 'the consumer' demands. In the words of Danny Miller, this figure of 'the consumer' has become a sort of 'global dictator': 'Today, real power lies unequivocally with us, that is the waged consumers of the First World' (1995: 10).

This also suggests that any squeamishness about studying consumption simply because so many in the world are not affluent consumers is profoundly misplaced. For Miller 'The acknowledgement of consumption need not detract from the critique of inequality and exploitation, but this critique is foundering precisely because the enormous consequences and attractions of consumption are left out of the analysis' (*ibid.*: 21). A fundamental challenge for the twenty-first century is to fashion economic forms that fulfil those attractions of consumption, but deal too with its consequences. But what about these attractions of consumption? Here we come to its social and cultural importance. For consumption is rarely, in fact almost never, simply a matter of meeting material needs. Rather, consumption plays a central role in our wider social and cultural lives and those wider contours of society and culture inflect our consumption practices. Consumption is centrally important to economies and their geographies, but it is not purely economic.

Just what social and cultural processes are implicated in our consumption practices, then? To simplify somewhat, one can identify two main sets of ideas on this. On the one hand there are those who emphasize how consumption is socially and culturally produced, how we as individual consumers are, to put it crudely, made to consume by being immersed in a wider consumer culture. Here, for example, the philosopher Herbert Marcuse wrote about how in capitalist societies there is a social production of 'false needs' (Marcuse 1964), a sense of wanting and

requiring far more than is materially necessary. More specifically, numerous studies have debated the role of advertising in stimulating consumption, creating desires for products that would otherwise not exist (for a classic study in this vein see Packard 1977, where the role of adverts on our subconscious is emphasized; for more recent and subtle interventions on advertising

Plate 16.2
'BMW-Alpina: the car with flair for people with flair'. Advertising, commodity auras and the production of false needs: you may want to own a BMW convertible, but do you really need to? (Vintage Ad Gallery)

and its role in mediating production and consumption see Nixon 1997 and Jackson and Taylor 1996). In this portrait, then, it is the producers who dominate consumer culture. Indeed consumer culture is a way of ensuring that mass production is accompanied by the appropriate levels of mass consumption. All in all, consumers are portrayed as rather sad and pacified figures:

> irrational slave[s] to trivial, materialistic desires who can be manipulated into childish mass conformity by calculating mass producers. This consumer is a cultural dupe or dope, the mug seduced by advertising, the fashion victim . . ., yuppies who would sell their birthright for a mass of designer labels. Ostensibly exercising free choice,

this consumer actually offends against all the aspirations of modern Western citizens to be free, rational, autonomous and self-defining.

> (Slater 1997: 33)

On the other hand, it is possible to see consumer culture as less passive, as less controlled, as less subordinate. Here the emphasis is not on consumers being made to consume, but on consumers *using* things for their own ends. In this perspective, consumption is less the death of the commodity, the end of an economic chain that began with the commodity's production, and more its resurrection. In this spirit, some stress the ways in which (many) consumers creatively re-work the products being sold to them, giving them new meanings in the process, and

Plate 16.3 Originally designed to be a stylish mode of travel for Italian women, the Italian motor scooter became a crucial component of the British mod subculture. The meaning of things we consume are not simply controlled by those who design them, but are remade through their consumption. (Corbis/Hulton-Deutsch Collection)

using them as launch pads for their own symbolic and practical creativity (see for example Willis 1990). A classic example would be the Italian motor scooter (see Hebdige 1987). Having been originally aimed in the post-war years at a market of young, urban women in Italy, in the UK the Lambretta and Vespa scooters became a central identifier of the 'Mod' subculture. Initially attractive to Mods for its Italianicity and design aesthetics, by the late 1960s the scooter was increasingly symbolically recast and materially customized to act as a subcultural icon opposed to the British, brute masculinity of 'Rockers' and their motorbikes. More generally, there have been a range of approaches that have emphasized how consumption involves the use of commodities within at least partially autonomous cultural fields: within existing status structures and patterns of class distinction, so that consumption is regulated by socially structured and inherited tastes (Bourdieu 1984); within the private dream worlds of consumers, so that consumption is about fulfilling our fantasies and desires (Campbell 1987); or within domestic relations of love and care, so that consumption is about concretizing our relationships to our nearest and dearest (cooking and sharing a meal, buying treats and gifts, helping someone choose their clothes) (Miller 1998).

In all these different approaches, then, to understand consumption one has to understand the social and cultural relations within which it is incorporated. One cannot simply read off our worlds of consumption from the corporate strategies of producers. Furthermore, what are identified here are the more positive social possibilities of consumption, the feelings of agency and self-autonomy it can engender. It could be argued, for example, that in comparison to many people's worlds of work, worlds of consumption can seem comparatively fertile spaces for creativity, self-fulfilment and, indeed, liberation (Miller 1995). It is through consumption that we can make the world meaningful to us, make our worlds.

Of particular interest to geographers is the geographical character of the worlds produced through consumption. In the remainder of this chapter, we examine this geographical character through three interrelated issues: the production of the global through consumption; the production of the local through consumption; and the geographical knowledges produced through consumption. First then, our focus will be on the character of the global geographies consumption produces. Some argue that

consumption has increasingly corroded cultures and their meaningful geographies of place and locality, ripping apart the historical sedimentations that mark places and people as different from each other, and smoothing them over into more homogenized markets that better suit the supra-local operations of capital. So, are our worlds of consumption increasingly all the same? Does contemporary consumption promote a global space characterized by the end of geography? Reviewing examples from East Asia, Russia, Israel, the Congo and London, I will suggest not, arguing that consumption forges different, localized experiences of the global and incorporates and produces cultural and geographical difference rather than eliminating it.

Secondly, the discussion then shifts scales to think about the production of the local through consumption more directly. Not only do we make the world global in diverse forms through consumption, we also make local environments through consumption. Taking the examples of shopping centres, nightclubs and skateboarding in urban public space, we will be looking at the roles played by such places in the very process of consumption itself, asking just what it is that is going on in them, and what this tells us about the character of consumer cultures more generally. My argument will be that in part such local places represent strategies of commercial interests designed to promote and regulate consumption. However, they also involve people making and re-making places into forms and through practices that accord to a different logic than that of commerce and economic rationality.

Finally, I will turn directly to the connections between local and global geographies of consumption. Consumer worlds are always simultaneously local and global. For example, the food I eat, the television I watch, the music I listen to in my very local, domestic spaces are sourced from all over the world. The everyday acts of consumption I undertake in my kitchen and living room are dependent upon huge networks of provision. The last part of this chapter therefore reflects on these networks and what I and other consumers know about them. In part through a critical consideration of ideas of a (geo)ethical consumption, it considers how consumption enacts connections between human beings and places that are also, at the same, often disconnected from each other in terms of knowledgable contact. One of the characteristics of our world of consumption is its connecting of the local and the global, the immediately present to the absent, and this raises difficult geographical questions, both intellectually and personally. These, it is suggested, are appropriately important issues for consideration by human geography at the start of the twenty-first century.

16.2 Global geographies of consumption

16.2.1 Consumption, cultural imperialism and the end of geography

Consumption is at the heart of some of the most significant intellectual debates of our age, being seen as central to nothing less than the cultural fate of the world. These debates centre on the extent to which modern consumption both produces and reflects a profound erosion of cultural traditions and differences, homogenizing the world into increasingly similar landscapes, people and social systems as it spreads from its heartlands in North America and Europe. Dick Peet, for example, argues that global capitalism

and its consumer culture comprises 'a powerful culture which overwhelms local and regional experience . . ., breaking down the old geography of society and culture' (Peet 1989: 156). Sack concurs that through consumption there is a : 'trend toward a global economy and culture, which seems to require that places all over the world contain similar or functionally related activities and that geographical differences or variations that interfere with these interactions be reduced' (Sack 1992: 96).

More prosaically, it is hard to miss the frequent commentaries on the worldwide spread of products and brands such as McDonald's or Coca-Cola, to the extent that they have entered academic language – in the form of theses of the 'McDonaldization of society' (Ritzer 2000) or of 'Coca-colonization' – as bywords for a consumerist homogenization (Plate 16.4). Travellers' tales too are replete with mournful stories of how one can trek to the farthest ends of the earth, only to find people doing exactly what the traveller could have seen at home; in the title of Iyer's account from the 'not-so-far East', having 'video nights in Kathmandu' (1989). Some of these issues of cultural homogenization are discussed in Chapter 12. This

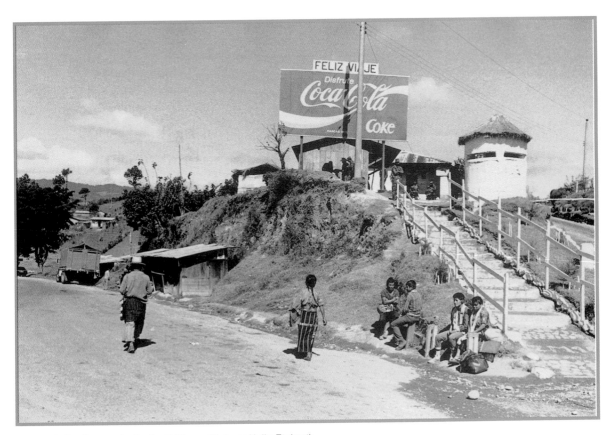

Plate 16.4   Coca-colonization? (Panos Pictures/Julio Etchart)

section will consider them in more detail, focusing upon the issue of consumption.

For many, what modern consumption enacts is a form of 'cultural imperialism' (see Tomlinson 1991 for an excellent review). This thesis takes a number of forms. For some it is evidenced by the 'dumping' of First World, and especially American, consumer products onto Third World markets: 'authentic, traditional and local culture in many parts of the world is being battered out of existence by the indiscriminate dumping of large quantities of slick, commercial and media products, mainly from the United States' (Tunstall 1977: 57). What concerns critics here is both the dominance of First World and American multinational companies (Mattelart 1979) and the predominance of products that promote First World and American values (Dorfman and Mattelart 1975) (for a general review of these emphases see Fejes 1981). Also raised are concerns about the consequent loss of indigenous, 'authentic' local cultures, so that the 'real' culture of an area and a people are lost.

For others, the issue is not just the conquest of some national and regional cultures by others, but the ways in which consumerism destroys all culture as we have traditionally understood it. The suggestion here,

then, is that what consumption represents is the contemporary dominance of a very peculiar cultural form – consumer culture – which in many respects is barely cultural at all. Consumption, some suggest, is all about the political economy of market shares, niches and profit, and as such fundamentally opposed to the intrinsic values of cultural meaning and communication. For others, such as George Ritzer in his recent account of what he calls 'the globalization of nothing' (2004), consumption thereby tends to produce empty forms and spaces, devoid of ties to particular times and places (Spotlight box 16.1). What matters here is partly the loss of cultural diversity – the making of a duller world – but perhaps more importantly the loss of cultural alternatives. Uniformity is doubly problematic as it entails conformity. This results in a form of culture that is inadequate and unfulfilling for many of those it is imposed upon:

A cultural system which would be adequate for the poorest people in that system would mean a set of instrumental, symbolic and social relations that would help them to survive in meeting such fundamental needs as food, clothing, housing,

Spotlight Box 16.1

Consumerism as the end of culture?

'Culture' was one of the crucial terms through which anti-liberal forces counted the cost of modernity. *Consumer culture*, in this perspective, is merely an ersatz, artificial, mass-manufactured and pretty poor substitute for the world we have lost in post-traditional society. In fact, it is the antithesis and enemy of culture. In it individual choice and desire triumph over abiding social values and obligations; the whims of the present take precedence over the truth embodied in history, tradition and continuity; needs, values and goods are manufactured and calculated in relation to profit rather than arising organically

from authentic individual or communal life. Above all, consumerism represents the triumph of economic value over all other kinds and sources of social worth. Everything can be bought and sold. Everything has its price. 'Consumer culture', therefore, is a contradiction in terms for much of modern western thought.

(Slater 1997: 63)

The social world, particularly in the realm of consumption, is increasingly characterized by nothing. In this case, 'nothing' refers to a social form that is generally centrally conceived, controlled and comparatively devoid of distinctive substantive content ... we are witnessing the globalization of nothing.

(Ritzer 2004: 3, 2)

medical treatment and education. Such needs are not met if they are identified with the consumption of Kentucky Fried Chicken, Coca-Cola, Aspro, or Peter-Stuyvesant cigarettes.

(Hamelink 1983: 15)

What is questioned here, then, are the values that underpin consumer culture, and the ability of consumption to satisfy the 'real' as opposed to the 'false' needs of consumers. Clearly these are fundamental issues. However, the accounts related above are partial and rest upon some problematic assumptions: that the authenticity of a culture is dependent upon its resistance to external forces and artefacts; that non-Western consumers are duped or coerced into having consumption tastes that are inappropriate to their real needs, just as the mass of Western consumers have been too (on this basis it seems that only the critics of consumer culture know what it is any of us 'really' need); and that consumers passively consume Western goods, and make of them exactly the same the world over. We shall now turn to some of these criticisms of the cultural imperialism approach to consumption, through the lens of work that emphasizes processes not of 'homogenization' but of 'indigenization'.

16.2.2 Consumption and cultural indigenization

Let us start with someone who may seem an unlikely proponent of this idea of indigenization. Marshall McLuhan wrote one of the most famous social science books of the post-war period, as he sought to understand the character of a contemporary society so saturated by the mass media (McLuhan 1964). His argument about the increasing linkages between different places, and phrases expressing this such as the notion of the 'global village', have passed into common parlance, echoed in numerous advertising campaigns promoting the unifying qualities of new communication technologies. However, his argument always went beyond suggesting that this increasing interconnection between people and places – or in the much cited language of the Marxist geographer Harvey (1989), this 'time-space compression' of the world – meant simple homogenization. He elaborates on this in a later paper:

The question arises, then, whether the same figure,

say Coca-Cola, can be considered as 'uniform' when it is set in interplay with totally divergent *grounds* from China to Peru? . . . The 'meaning' of anything . . . is the way in which it relates to the user . . . [For example] it might be suggested that the use of American stereotypes in the United Kingdom yields a widely different experience and 'meaning' from the use of the same stereotypes or *figures* in their American *ground*. The interplay between the *figure* and the *ground* will produce a wide range of different effects for the users.

(McLuhan 1975: 44; emphases in original)

McLuhan argues, then, that consumer culture does not float above the world in some abstract space of capital or modernity. It lands and takes root in particular places and times. In so doing it not only impacts on its surroundings; it is also shaped by its locations.

Especially within anthropology a number of studies have developed this theme by exploring how globally distributed facets of consumer culture – whether particular goods, styles, or cultural orientations – are locally incorporated. Take, for example, one of the exemplars of homogenizing American consumer culture: McDonald's. On an average day over 47 million customers are served at one of 31,000 McDonald's restaurants in more than 19 countries. Over one-third of these restaurants are in the United States, but based on 2003 figures the United Kingdom has over 1,000 outlets, Brazil over 500, China nearly 200, Thailand nearly 50. Given their date, and notwithstanding more recent commercial problems, these figures are almost certainly substantial underestimates; a recurrent promotional corporate statistic is that a new McDonald's restaurant opens somewhere in the world every three hours. Not only this, of course, but McDonald's are famed for their uniformity; the same decor, the same basic menu (with very small variations, such as McSpaghetti in the Philippines or the Maharajah Mac in India), and the same service style the world over.

And yet, McDonald's may not be just the force for cultural homogenization that this suggests. This was certainly Watson's (1997a) conclusion, based on studies of the cultural geographies of McDonald's in East Asia (Case study 16.1). These studies found that McDonald's has been localized, indigenized, and incorporated into traditional cultural forms and practices, even as it has played a part in the on-going dynamism of those traditions. And exactly how this

Regional Case Study 16.1

Cultural geographies of McDonald's in East Asia

In the introduction to a collection of findings on the cultural geographies of McDonald's in East Asia, Watson poses the following questions:

> Does the spread of fast food undermine the integrity of indigenous cuisines? Are food chains helping to create a homogeneous, global culture better suited to the needs of a capitalist world order? ... But isn't another scenario possible? Have people in East Asia conspired to change McDonald's, modifying this seemingly monolithic institution to fit local conditions? ... [Perhaps] the interaction process works both ways.
>
> (Watson 1997a: 5–6)

His conclusion is that there is indeed a two-way interaction:

> McDonald's has effected small but influential changes in East Asian dietary patterns. Until the introduction of McDonald's, for example, Japanese consumers rarely, if ever, ate with their hands; ... this is now an acceptable mode of dining ... [On the other hand] East Asian consumers have quietly, and in some cases stubbornly, transformed their neighborhood McDonald's into local institutions. In the United States fast food may indeed imply fast consumption, but this is certainly not the case ... [in] Beijing, Seoul, and Taipei ..., [where] McDonald's restaurants are treated as leisure centers, where people can retreat from the stresses of urban life.
>
> (*ibid.*: 6–7)

has happened varies across East Asia. In Beijing, McDonald's has lost its American role as a place of fast and cheap food. Instead, Yan suggests that it has become a middle-class consumption place, somewhere for a special family outing, somewhere 'customers linger ... for hours, relaxing, chatting, reading, enjoying the music' (1997: 72). McDonald's here is seen as American, but Americana means something stylish, exotic and foreign, and as such actually results in the meanings and experiences of McDonald's in Beijing being very un-American (Plate 16.5). In contrast, in Japan, whilst there is a similar leisurely use of McDonald's, it is not a place of exotic social prestige, but a youth hangout, a place where someone in a business suit would be out of place (Ohnuki-Tierney 1997). In Hong Kong, McDonald's was likewise marketed to the youth market, and adopted as such initially, but here its American origins are of little contemporary consequence (Watson 1997b). Whilst initially patronized in the 1970s by adolescents seeking to escape products associated both with China and with Hong Kong's perceived provincialism, by the 1990s McDonald's has become routinely local, just another mundane part of the Hong Kong landscape: '[t]oday, McDonald's restaurants in Hong Kong are

packed – wall to wall – with people of all ages, few of whom are seeking an American cultural experience. The chain has become a local institution in the sense that it has blended into the urban landscape ... McDonald's is not perceived as an exotic or alien institution ...' (Watson 1997b: 87, 107).

Hence the meanings and practices of consuming McDonald's – often cast as an icon of global homogenization – vary from place to place, from culture to culture (for a parallel account on an American global TV product, the series *Dallas*, see Case study 16.2). Melissa Caldwell's work (2004) on McDonald's and consumerism in Moscow develops the significance of these variations. In Russia too McDonald's has been indigenized or 'domesticated', and not just at the level of everyday practice. Counterintuitively, McDonald's – which as outsiders we might read as an American export to Russia – has come to be understood by consumers in Moscow as an 'authentically Russian product', to the extent that eating there is seen as responding to 'nationalist-oriented consumer campaigns' (*ibid.*: 5). Whilst initially perceived and promoted as an exotic American import, more recently McDonald's has become Russian, explicitly marketed as 'Nash Makdonalds'

Plate 16.5   McDonald's in Beijing: American fast food becomes exotic, stylish Americana. (Popperfoto/Reuters)

(which Caldwell translates as 'our McDonald's'). Since the mid-1990s Russian consumer culture more generally has evidenced a patriotic enthusiasm for goods understood as 'Nash'. This Nash ideology entangles notions of national space with those of everyday familiarity and trust. McDonald's has thus been transformed from American exotica to both mundane familiarity and thence to imagined Russianness.

Parallel accounts of domestication have been produced through research on a range of other global products: for example, in the context of a study of modern Trinidad, Danny Miller writes about how both the American television soap opera *The Young and the Restless* and 'Coke' are authentically Trinidadian (Miller 1992, 1997a). Particularly telling is Jonathan Friedman's study of the subcultural research of European designer fashions in the Congo (Case study 16.3). Here, not only does *haute couture* come to mean something different to what it means in Paris or Milan, it comes to mean it through relations of consumption, identity and personhood that are different. Rather than one consumer culture spreading globally, this suggests

many consumer cultures, their forms and meanings being locally specific. At a broader level this marks a conceptual shift away from seeing the modern world as a singular universal entity, and towards recognizing the many 'specific', 'prismatic', 'comparative' or 'global' modernities being forged through consumption practices worldwide (see for example Breckenridge 1995 on South Asia; Featherstone *et al.* 1995 generally; and Miller 1994 on Trinidad).

Thus, rather than necessarily destroying local, indigenous cultures, the consumption of non-local products may be irrelevant, indeed even helpful to, the on-going production of authentic local cultures. Miller argues that in a world of globalized consumption 'authenticity has increasingly to be judged a posteriori not a priori, according to the local consequences not local origins' (Miller 1992: 181). It matters less where things come from than what they are used for. Indeed, it may be that the non-local origins of products may be positively beneficial, giving 'the periphery access to a wider cultural inventory' and providing new raw materials for local cultural and economic entrepreneurs (Hannerz 1992: 241; for a specific case,

Regional Case Study 16.2

Media products and cultural indigenization: the case of Dallas

For proponents of the cultural imperialism thesis, television programmes are primary instruments for the imposition of Western, American, consumerist values on their global audiences. *Dallas* was the most watched programme in the world in the early 1980s. Set, unsurprisingly, in Dallas, Texas, it was a weekly soap opera following the fortunes, relationships and fabulous squabbles of an extended oil baron family, the Ewings. Among many other places, *Dallas* was very popular in Israel where two media researchers, Liebes and Katz, were investigating what was made of it by a variety of Israeli audiences. What they found was that, far from the programme simply imposing a world view on its audience, different groups of people understood the programme in very different ways and enjoyed it for very different reasons.

> Theorists of cultural imperialism assume that ... [meaning] is prepackaged in Los Angeles, shipped out to the global village, and unwrapped in innocent minds. We wanted to see for ourselves. ... In the environs of Jerusalem, we watched the weekly episode in the homes of Arabs, veteran settlers from Morocco, recent arrivals from Russia and second-generation Israelis in kubbutzim – *Dallas* fans all .... Each cultural group found its own way to 'negotiate' with the program – different types of reading .... We found only very few innocent minds, and a variety of 'villages'.
>
> (Liebes and Katz 1993: xi)

The recent immigrants from the Soviet Union denied watching it (even when they did) or dismissed it as American, capitalist propaganda. They saw it in ideological terms. Middle-class kibbutzniks had a more psychological take on the series, concentrating on the personalities of the protagonists and enjoying the task of understanding their motivations. The interviewees from North African and Arabic backgrounds emphasized the convolutions of the plot, framing the programme as an epic family saga. Different people with different backgrounds, and different 'readings' of the same show.

the economy of *Salaula* or Western 'second-hand clothes' in Zambia, see Tranberg-Hansen 1994). Moreover, when it does matter where things come from, when their origins are part of what they are valued for, then those origins are understood in locally specific, imaginative ways. America, for example, is mythologized the world over, but in a range of different forms (see, for example, Webster 1988).

16.2.3 Consuming difference

This last point – on the mythologies or 'imaginative geographies' (Driver 1999) attached to the things and places we consume – leads on to a further set of qualifications of the cultural imperialist thesis. These concern the assumption that the goal of those promoting consumer culture is simply to eradicate cultural and geographical differences, viewing them as unhelpful encumbrances to the smooth operation of the world of consumption. In fact, one can see cultural and geographical difference being actively embraced, even produced, within consumer culture. Global arrays of cultural and geographical difference are produced and sold within consumer culture.

Exemplary here, it is often suggested, are the themed environments of consumption spaces such as theme parks and shopping centres. In Disneyland and Disney World, for example, the amusements and other consumption opportunities are not just randomly arrayed, but organized into distinct lands (for example Frontierland themed on the American West, or Adventureland themed on landscapes of European colonialism and imperialism) (see Gottdiener 1982, Wilson 1992, Bryman 1995). This echoes much earlier consumer festivals and fairs such as the Midway attractions at the 1893 Chicago exhibition (Rubin Hudson 1979) or Luna Park at Coney Island, New York

Regional Case Study 16.3

Western goods and cultural indigenization: the case of 'Les Sapeurs'

'Les Sapeurs' are a group of consumers in what was the People's Republic of the Congo (see Friedman 1994). They are men of the politically marginalized Bakongo ethnicity, drawn especially from the poor in large cities such as Brazzaville and Pointe Noire, who aim to move through a graded and spatialized consumption process of 'Western' fashions. Beginning with the conspicuous wearing of imported ready-to-wear items, Sapeurs look to progress to a stay in Paris ('*l'aventure*') accumulating *haute couture* from French and Italian design houses whilst usually living in extreme poverty, before returning to Brazzaville's 'Sape clubs' to perform competitively in 'la danse des griffes', a catwalk display of one's dressed self that is augmented by the sewing of a host of designer labels to jacket lapels. All of which, especially at a first hearing, might be rather too easily dismissed as an extreme case of fashion victimhood, or as a bizarre oddity, thrown up by the maelstrom of globalized consumer culture as inflected in the Congolese context. However, for Friedman, far from being a cultural oddity, la sape is symptomatic of much more general local–global geographies of consumption, in which globally distributed commodities and materials – here Western European designer fashions – are encompassed within local cultural logics of identity practice. In the Congo, understandings of identity are marked by a conception of a cosmic 'life force', which can be tapped into through consumption, and which, especially for the socially marginalized, is located elsewhere, external to the local political and cultural hierarchies. As such, the conspicuous consumption of designer labels becomes something very different from what one might first make of it. It is not any sort of dandyism, or an abandonment of serious cultural projects. Rather it is 'a life-and-death struggle for physical and social survival' (Friedman 1994: 106) that accords to the existing, indigenous cultural logics of the Bakongo.

(Kasson 1978), and is also a pattern to be found in many large shopping centres, most famously the huge West Edmonton Mall in Canada (Crawford 1992; see also Goss 1999a). Indeed Gottdiener (1997) goes so far as to posit that consumer culture has produced a general 'theming of America', so that theme parks become models for what is happening across the urban landscapes outside them. Sack (1992: 98) also points to the production of 'pseudo-places' in consumer culture. Harvey (1989) sees the same general trend towards highly packaged, and indeed simulated, experiences of cultural and geographical difference, such that microcosms of the whole world are encountered in particular consumption spaces, from the television screen, to the High Street, to the shopping centre and theme park.

Rather than eradicating cultural and geographical difference – and thus producing the end of geography – the world of consumption (re)produces geographies, framing certain local places of consumption as global centres. What then becomes the issue is not the end of geography, but what sort of new geographies are taking its place, and the kinds of understandings of people and place being circulated through consumer culture.

For instance, these kinds of questions have framed a number of pieces of research on the global arrays of food now being sold and consumed in the United Kingdom, as well as elsewhere (see, for example, Cook and Crang 1996; Cook *et al.* 1999, 2000a). This explicit globalization of food provision is captured in an appreciative review of the diversity of restaurants in London, penned by the listings magazine, *Time Out*:

> The world on a plate. From Afghani ashak to Zimbabwean zaza, London offers an unrivalled selection of foreign flavours and cuisines. Give your tongue a holiday and treat yourself to the best meals in the world – all without setting foot outside our fair capital.
>
> (*Time Out*, 16 August 1995)

A central concern in research thinking about such developments has been just what sorts of geographical understandings and knowledges are being produced through this facet of consumer culture; what conceptualizations of people, place and culture are being mobilized. Take, for example, the following anecdote from the tabloid British newspaper, the *Daily Star*. Under the heading 'korma out for a meal', this item reported on the popularity of 'Indian' restaurants and take-aways amongst British consumers ('curry lovers have sparked an eating out boom in Britain'). At the same time, however, its notion of Britishness was pointedly narrow as, in the very next sentence, it reported that 'just 26 per cent [of those dining out] chose British restaurants' (*Daily Star*, 5 December 1995). Indian restaurants are a popular British pastime, but not British themselves, it seems. The logic here is all too clear. Indianness is something (non-Indian, white) British people may enjoy consuming, but that does not make it, or the British-Asian (in fact primarily British-Bangladeshi or British-Pakistani, not British-Indian) restaurateurs and waiting staff who embody it, culturally British. Difference is appreciated and enjoyed, at least when it comes to food, but it also distances those with whom it is associated. Here, an ethnicized geographical difference is incorporated into a very white British way of seeing the world, in which everything non-white becomes a resource to 'liven up the dull dish that is mainstream white culture' (hooks 1992: 21).

Of course, Britain's culinary multiculturalism, and its fashionings of 'Indian' food, are much more complicated than this anecdote allows: 'Indian' food is constructed and consumed in many forms in the UK – from the 'exotic', to the 'authentic', to the 'mundanely British' curry house, to the modern, contemporary 'Asian Kool' up-market restaurant – each enacting different consumer tastes, market strategies and 'multicultural imaginaries' (Cook *et al.* 1999). More generally, the politics of cross-cultural consumption are not fixed and are open to intervention (generally see Jackson 1999; for examples about British-Asian fashion see Dwyer and Crang 2002 and Dwyer and Jackson 2003). But what is common is how through consumption ideas of the global and the local, the foreign and the domestic, the there and the here, of difference, are not eradicated but actively produced (see also Lury 2000).

16.3 Local geographies of consumption

16.3.1 Shopping and place

I have been arguing, then, that consumption is not productive of global homogeneity but instead involves various interrelations between the global and the local (Crang and Jackson 2000). We now turn to the local geographies of consumption more directly through attention to the sites where consumption takes and makes place. We start in a shopping centre. Shopping centres (or malls in North America) are routinely cast as the cathedrals of our consumerist age, the symbols of all we value and hold dear. But what goes on inside a shopping centre, and what does this tell us about the kinds of places and 'public' space being produced in the world of consumption?

One response would be that in shopping centres we are manipulated into behaving in certain ways, and especially into buying things, by the power of the shopping centre as a place. Shopping centres are in part designed according to a 'merchandise plan' (Maitland 1985: 8) that tries to maximize the exposure of consumers to goods. In older shopping centres and North American malls this plan tended to be understood in quite mechanical terms. Shopping centres were seen as functional 'machines for shopping'. 'Generator' stores are used to pull us to the shopping centre in the first place, and once we are there 'magnet' stores every 200 metres or so make sure we don't just pop in and out, but explore every part of

Plate 16.6 Eating into Britishness: are these Indian curry houses 'British'? (Photography courtesy of G.P. Dowling)

Regional Case Study 16.4

The shopping mall experience: Woodfield Mall, Schaumburg, Illinois

Leaving the comfort of her air-conditioned car, the shopper hurries across the sun-broiled or rain-swept tarmac and thankfully passes through doors into an air-conditioned side mall. Drawn down this relatively dark, single-level tunnel, lined with secondary units, by the pool of light ahead, she reaches one of the squares ahead located at the end of each arm of the main mall system. She discovers this to be a two-storey space. . . . On one side of the cavern is a Magnet department store, but before deciding to enter it she turns to see that a second tunnel, much taller than the first with natural light spilling though holes punched in its roof, leads off into the heart of the building. . . . This route now culminates in the central space, Grand Court. . . . Its floor has been quarried and shaped to form an amphitheatre and pools. . . . Emphasis is placed upon the creation of a plausible internal world, full of visual interest and variety, in which one is tempted not simply to pass through, but to stay.

(Maitland 1985: 26, 28)

it. Here, then, 'the public mall [is defined] as essentially the passive outcome of a merchandising plan, a channel for the manipulation of pedestrian flows' (Maitland 1985: 10). There are well-known tricks to make us behave as the shopping centre management and its shops want us to: escalators arranged so we have to walk past shop fronts or merchandise when going between floors; hard seats so we don't linger too long without getting up and seeing some more potential purchases; no water fountains so we have to buy expensive drinks (see Goss 1993b).

In more recent shopping centres, however, the manipulation may be more subtle. Increasingly, retail planners and developers see their job as providing spectacular places that people will want to spend time in. This is because research suggests that the longer people stay, the more on average they spend, and because as competition between centres hots up each one needs more than just the usual shops to attract us in the first place – everything from fountains to funfairs (see for example Goss 1999b). Here, then, the manipulation is less physical and more emotional. The hope is to excite, inspire, relax and please (Case study 16.4). Of course, these are promotional intentions rather than shoppers' experiences. As Chaney (1990) puts it in his review of the MetroCentre in Gateshead, north-east England, the designer's utopian impulses to produce a place of perfection often result in rather underwhelming, 'subtopian' forms. Nonetheless, there is much here to be taken seriously. There is the attempt

to produce a fabricated space in which the individual consumer can be made to feel like consuming (see Goss 1999a). There is the emphasis on creating an internal, closed-off, privately owned but partially public environment, divorced from the harsh exterior world, not only climatically but also socially, through the operation of security systems that ensure the absence of anyone who might threaten this consumer paradise or disrupt the pleasure of the shopper. Thus, these privately owned and managed spaces rework the older public spaces of the street or market:

They reclaim, for the middle-class imagination, 'The Street' – an idealized social space free, by virtue of private property, planning and strict control, from the inconvenience of the weather and the danger and pollution of the automobile, but most important from the terror of crime associated with today's urban environment.

(Goss 1993b: 24)

Such shopping centres recognize that there is much more to going shopping than buying things (Glennie and Thrift 1992). Shopping is in part about experiencing an urban space, seeing and being seen by other shoppers. It is a social activity. But they seek to manage this social experience. The street gets re-cast as a purified space of leisurely consumption, cleansed of nuisances that inhabit the 'real' streets, like 'street people'.

However, this fashioning of the mall can be disturbed. Retail spaces can be reclaimed by those who

Plate 16.7 The new Bullring shopping centre, Birmingham, UK was opened on September 4th 2003. The round tower that was part of the 'old' Bullring has been retained. (Photographer Ravi Deepres, courtesy of Bullring)

use them. Leading the charge are a mixed assortment of senior citizens and adolescents. Take the case of Bert, 78 years old, and a regular visitor to a shopping mall in New England (US):

> 'I come here at quarter to eleven and I leave at twenty minutes past one . . . I do this five days a week, except I missed one day this year. . . . And I have my lunch here at noontime . . . . I move every half-hour. I go from here down to where the clock is, I go down in front of Woolworth's, then I come back again and go up there and take my bus by the front of JC Penney.'
>
> (Quoted in Lewis 1990: 126)

For Bert, the mall is not a place to shop, but somewhere sheltered to go outside his house, a chance to meet up with friends. For the mall management there is the option of evicting these low-spending visitors, but ejecting old people forcibly onto the pavement is hardly good public relations. Instead, and as part of wider moves to see malls as the new civic centres, Bert's mall has actually established its own walking club for the elderly, with over 300 members,

for whom the doors open at 6 a.m., rather than 9 a.m. when shops open.

Adolescents also use the mall as a meeting place, somewhere to hang out. They also spend very little. 'Suburban kids come to malls to look around, meet and make friends, and hang out – because there is nowhere else to go' (Lewis 1990: 130). Or as Ed and Tammy, self-confessed 'mall rat' and 'mall bunny' respectively, put it:

> '[Gesturing around himself] I met all these people here. I've met lots of other people, too. One place where you can always find someone. . . .'
>
> (Ed; quoted in Lewis 1990: 130)

> 'I used to come here every Saturday from 11am to 9.30pm and just walk around with my friends, like Gina here, and just walk around and check out the guys.'
>
> (Tammy; quoted in Lewis 1990: 130)

Unlike the elderly, the mall is less indulgent of these teenagers. There are constant skirmishes with the security staff over how long ten people can share one

Thematic Case Study 16.5

Skateboarding and the consumption of urban space

Skateboarding responds meaningfully to the city ... saying and living the city on its own terms.

(Borden 2001: 208)

A curb is an obstacle until you grind across it. A wall is but ledge until you drop off it. A cement bank is a useless slab of concrete until you shred it.

('In the streets today', *Thrasher*, vol. 1, no. 1, January 1981, p.16; quoted by Borden 2001: 192)

Skaters create their own fun on the periphery of mass culture. Sewers, streets, malls, curbs and a million other concrete constructions have been put to new uses.

('Lowboy, Skate and Destroy', *Thrasher*, vol. 2, no. 11, December 1982, p.25; quoted by Borden 2001: 191)

Skateboarding shows that pre-existing uses of space are not the only possible ones, that architecture can instead be productive of things, and consumed by activities, which are not explicitly commodified.

(Borden 2001: 247)

Coke at the food court, and some mall rats and bunnies do get banned, getting new hair cuts or clothes in attempts to get back in.

So what are we to make of our trip to the mall? What does it tell us about the local geographies of consumption? Well, it certainly tells us that place matters to consumption. But how? Here the answer is more ambivalent. In part what we see is a pseudo-public space symbolizing and enacting claims for consumption to be all there is to civic life, making citizens consumers in a very concrete way. Yet one also sees the small-scale, trivial resistances that people enact, their refusal to be just consumers in the sense of purchasers, to use this space for their own ends. And that too has a wider symbolism, signalling the possibility that consumption is never just about satisfying those selling to us, but also about fulfilling our own needs for social affiliation and participation. Two more examples extend this argument.

16.3.2 The geographies of skateboarding and dancing

As geographers are increasingly recognizing, shopping centres are not the only sort of consumption space to explore (Crewe and Gregson 1998). Indeed the kinds of semi-illicit place consumption undertaken by the elderly and the adolescent in malls can be extended to a wider account of the consumption of urban space. Writing from the discipline of architecture, Iain Borden (2001) pursues this possibility in a study of skateboarding, space and the city. Paying particular attention to 'street skating', he argues that skateboarders re-make urban spaces designed for other purposes: sometimes explicitly symbolic monumental space (e.g. town halls, national theatres, historical monuments, tourist attractions); but more often everyday spaces of neglect, the left-overs of rational, economically focused planning (e.g. mini-roundabouts, the spaces under bridges and urban highways, mini-malls). Thus spaces of public, official symbolism and of arid architectural functionality become the stages for energetic practices of place consumption. Handrails become tools for 'ollie nose grinds'; cement banks opportunities for 'shredding'. Skaters consume the city with their boards, recomposing it in the process. Skateboarding enacts a way of using the city, of consuming the city, that precisely emphasises use over commercial or other values (see Case study 16.5).

Skateboarding is the consumption of urban space through skilled performances of embodied competencies. A parallel case, though often taking place in more clearly demarcated commercial arenas, would be practices of conviviality and togetherness.

Clubs (in the sense of nightclubs) are exemplary. In London alone there are over 500 club nights on offer every week. In Britain the industry has a turnover of over £2 billion per annum, and it is estimated that 42 per cent of the population now visit a club once a year, 43 per cent of 15–24-year-olds once a month (Mintel 1996).

Ben Malbon (1997, 1999) argues that clubs are spaces of play. This should really not need saying. It is patently obvious that people go to clubs to have fun. It is also patently obvious that this fun is achieved through a variety of means. Through enjoyment of the music; through an enjoyment of being part of a crowd of people with whom, temporarily, one can feel at one; through the feelings of competency and bodily expressiveness that dancing can give you; through the sharing of a special place and time with friends; through behaving in ways one cannot in other social situations, for example at work. Perversely, this fun needs to be taken seriously. We need to understand the playfulness of many consumption practices. We need to analyse the complicated micro-spatial practices through which play is enacted; the ways in which one gets to feel, and enjoy being, part of the clubbing crowd; the ways in which one dances so as to lose oneself in the music. We need to grasp the importance of place to playful consumption; the ways in which one has to be in the right place, with the right music and the right crowd, to have these feelings of exhilaration, communality, and vitality; the way that 'the emphasis is placed on the near and the affectual: that which unites one to a place, a place that is experienced among others' (Maffesoli 1996: 128). And we need to appreciate the wider significance of such playful consumption for our productions of local experience, such that consumer culture is not just made up of atomistic, avaristic individual consumers, but also facilitates temporary, some would say 'neo-tribal' gatherings, through which people come to feel part of something bigger than themselves (see Maffesoli 1996: 72–103).

Clubbing is, then, a place-specific practice of consumption. Whereas in the shopping centre, senses of communal belonging tend to be forged by those occupying the space for purposes beyond those for which it was commercially designed – such as the elderly and the adolescent – in clubs that communal belonging is the product that people are paying to consume. In many ways, clubs are therefore rather unusual places of consumption, characterized by an exceptional emotional and bodily intensity. However,

they also point to wider qualities of our local places of consumption: how we use them for experiences of 'public' life and senses of social belonging; and how those experiences of public life and social belonging are produced through practices and places of consumption.

Skateboarding and clubbing represent, then, how consumption is about far more than the purchase and end or 'death' of a manufactured commodity. Consumption is about the (re)making of the spaces of everyday life. Commodities are the props and tools that we use to manufacture these local geographies.

16.4 Making connections

So far this chapter has explored some of the debates over the impacts of modern consumption on both global geographies (of cultural difference) and local geographies (of public space and life). Especially with regard to the latter, it has been deliberately optimistic. By way of conclusion, though, I want to think about an emergent concern with reconnecting the local contexts of everyday consumption with the global networks of commodity flows and production that make them possible. This reconnection has emerged as a fundamental intellectual, political and personal challenge in the field of consumption studies (for a wonderful, short evocation of some of the issues involved see Cook et al. 2000b). Miller, for example, emphasizes how the 'study of consumption should ... be increasingly articulated with, and not become an opposition to, the study of the mechanisms by which goods are produced and distributed' (1995: 17). Sack arrives at a similar conclusion on the basis of his determination to develop a moral framework for analysing consumption. Consumers are, he says, profoundly ignorant of the geographies of provision that bring goods to them: 'A shop that sells Colombian coffee does not reveal the social structure that produces the coffee, the economic impact of coffee production on the Colombian economy, or the way coffee growing affects the Colombian environment.' (Sack 1992: 200). Such ignorance, he argues, makes it impossible for consumers to behave with a sense of responsibility to those they are in fact connected to through the webs of commodity flows. In this, modern consumption is deeply amoral. Crucially, then, to think about reconnecting practices and places of consumption and production demands that we somehow articulate play and morality, pleasure and exploitation, creativity and responsibility.

But what sorts of articulations can one make between worlds of consumption and production? One perspective is to see consumers as blinded by a veil that is draped over the things we consume, through a process of what Marx called 'commodity fetishism'. In this portrait the true history or biography of a product, and its basis in the work of others, is obscured to consumers. Instead, one is confronted by false advertising imagery that seeks to associate the product with other fantasy worlds. The task then becomes to puncture these fantasies with some sharp shafts of reality. Hartwick points to the startling incongruities and inequalities that such a puncturing can bring to light:

> Michael Jordan, basketball player extraordinaire, receives US$20 million a year to endorse Nike; Philip Knight, cofounder and chief executive officer of Nike, is worth US$5.4 billion; Indonesian workers are paid US$2.40 a day and Vietnamese workers are paid US$10.00 a week to make the sneakers. . . . The Walt Disney Company is publicly criticized for its treatment of Haitian workers making Pocahontas shirts for 28 cents an hour; Disney's chief executive officer makes US$78,000 a day.

> (Hartwick 1998: 423)

Smith's account of Starbucks Coffee shops makes similar points. He highlights how unfairly traded coffee sees only 25 per cent of the retail price returning to the producing country, with even less going to those who actually cultivate the crop (the smallholders, waged workers and seasonal pickers). He contrasts the stylish coffee-shop culture Starbucks epitomizes with the ways in which that coffee is sourced. For the poorly paid women who pick the coffee beans, their working day begins somewhere between three and four in the morning. Getting back home at six in the evening, they still have evening meals to prepare, water to fetch from the river, a family to feed and clear up after. Few live in houses with electricity or running water. Along with malnutrition, a major health hazard are the pesticides used on the coffee plants (M.D. Smith 1996: 512–13).

So one way to reconnect worlds of consumption and production is to acquaint consumers with the harsh realities of commodity production. It is hard to criticize such an approach. It is a very simple message, but it is always worth getting it across. However, it does have its problems. Rather than articulating worlds of consumption and production together, it can tend to dismiss worlds of consumption, and their positive

possibilities, altogether. Consumption just becomes inherently bad. That is neither true, as hopefully the discussions earlier in this chapter have demonstrated, nor terribly attractive in terms of what it suggests one should do. Particularly if one's own worlds of production, as a worker, are thoroughly unrewarding – admittedly not usually the case for academic critics of consumption, but true for many others – then giving up all the pleasures of consumption is pretty hard to stomach. Certainly the last two decades in Western politics suggest that failing to recognize the genuine senses of autonomy and creativity that consumption can facilitate, is a sure way to alienate those very same workers in whose name one wishes to re-centre the world of production.

There are, though, more subtle variations on this theme. So-called 'ethical consumerism', in particular, is a fascinating phenomenon (see also Freidberg 2003). Here, consumption is not disavowed, but it is channelled through concerns with the wider provisioning systems, and their human and non-human participants, of which the consumer is a part. Green consumerism, fair-trade, vegetarianism and other animal welfare concerns are the main strands. Whilst containing some elements of self-denial, and most definitely of hard work, empirical research on ethical consumers in London challenges any sense of ethical consumption as joyless moral prescription (Bedford 1999). Instead it emphasizes the positive pleasures gained through such consumer movements: the feelings of achievement, the senses of empowerment. However, in its full-blown forms, ethical consumerism is not only hard work and expensive, but requires the adoption of consumption life-styles that are very difficult to maintain unless one is immersed in wider friendship and family networks of like-minded people. Consumption is, after all, thoroughly social. Moreover, the levels of knowledge required to understand the incredibly complex biographies of products are more amenable to academic study than they are to someone who has something else to do. Even the most committed ethical consumers therefore rely on shorthand rationales, judgements and knowledges such as 'Nestlé is bad, Brooke Bond is good'; 'meat is murder, milk is allowed'; and so on.

What this suggests is that rather than simply replacing a superficial and false advertising fetish with the real history of commodities, it is more realistic to see ethical forms of consumption as re-working the fetish or advertising form, and re-working the pleasures of consumption rather than disavowing them in a blanket condemnation of consumerism. Thus, for example, if we consider the huge growth in the market for organic produce in Britain and other Western countries at present what we see are products that are branded as in some way ethical, in the UK with a Soil Association approved designation and logo. What is more, one of the attractions of organic food is that it can easily be seen as benefiting not only the environment but also those to whom it is being fed (in terms of 'healthiness'). The 'organic' in organic food becomes understood as more generally 'good', drawing on vague, positive associations of the 'natural': good for the environment, good for my family, good for me, good to eat. Of course, not all ethics have such an easy semantic slippage. In particular, the Fair-Trade logo has struggled to make the same market impact. Connecting care for one's nearest and dearest – a very strong existing consumer ethic – to caring for the distant strangers involved in commodity production and distribution has proved trickier.

More broadly, what such an analysis suggests is that consumption is not only a way in which we make geographical entities such as the global and local; it is also an arena for the production, circulation and reception of geographical knowledges. It is tempting, especially perhaps for geography academics and students, to respond by looking to export the factual knowledge of our subject to this everyday arena, to correct the ignorances that we all operate under as consumers. But that, I have been suggesting, is both too easy – offering a very simplistic account of both geographical knowledge and the politics of consumption – and impossibly difficult – given the complex histories and geographies of the multitude of things we consume. The geographies of consumption offer no such easy resolution. We consumers are both global dictators and local freedom fighters. We all have to continue to think about and develop practically the connections between these two roles.

Learning outcomes

After reading this chapter, you should have:

- An understanding of the importance of consumption to modern economies and cultures.

- An understanding of the global geographical impacts of modern consumption.

- A sensitivity to the continuing importance of cultural and geographical difference within contemporary consumer cultures.

- An understanding of the character of public space fashioned within the worlds of consumption.

- An understanding of connections that exist between worlds of consumption and production.

- An ability to think about the geographies of your own everyday consumption practices.

Further reading

Bell, D. and **Valentine, G.** (1997) *Consuming Geographies: We Are Where We Eat*, Routledge, London. A lively exploration of the geographies at stake in food consumption, at every scale from the body to the home, the city, the nation and the world.

Cook, I. and **Crang, P.** (1996) The world on a plate: culinary culture, displacement and geographical knowledges, *Journal of Material Culture,* **1**(2), 131–53. Modesty prevents me from saying how wonderful this article is, but I can say that it offers an account of the global array of foods found in the UK that emphasizes the global, local and knowledgeable geographies of consumption.

Goss, J. (1993) The magic of the mall: form and function in the retail built environment, *Annals of the Association of American Geographers*, **83**, 18–47. A brilliant, if somewhat sceptical, analysis of shopping malls and the consumer worlds they embody.

Hartwick, E. (1998) Geographies of consumption: a commodity-chain approach, *Environment and Planning D, Society and Space*, **16**, 423–37. A pithy comparison of the advertising imagery used to sell gold and the ways in which that gold is produced. Argues for a knowledgable and ethical connection of consumers and producers. Simplistic but powerful.

Miller, D. (1992) The young and the restless in Trinidad: a case study of the local and the global in media consumption, in Silverstone, R. and Hirsch, E. (eds) *Consuming Technologies: Media and Information in Domestic Spaces*, Routledge, London, 163–82. A typically insightful and exuberant analysis from perhaps the best contemporary writer on consumption. Here he analyses the consumption of an American soap opera in Trinidad. His conclusion is that paradoxically this soap, despite its American origins, is authentically Trinidadian.

For annotated, clickable weblinks and useful tutorials full of practical advice on how to improve your study skills, visit this book's website at www.pearsoned.co.uk/daniels

Geopolitics, territory, states and citizenship

EDITED BY JAMES SIDAWAY

Late in 1998 General Augusto Pinochet, one-time President of Chile (1973–90), paid a visit to London for private meetings and medical treatment. Whilst in hospital he was arrested by British police. The arrest took place on behalf of the Spanish courts who had charged Pinochet with crimes against Spanish citizens in Chile who were murdered by the brutal military regime that Pinochet once headed. Pinochet directed the Chilean armed forces in their violent overthrow of the democratically elected Chilean government on 11 September 1973, and he continued to preside over Chile for more than 15 years afterwards. His detention in London reminded the world of Pinochet's politics and the crimes committed by his right-wing regime. It also raised complex questions about sovereignty. Could a former head of state be prosecuted in one country for alleged crimes committed in another? Did this mean that Spain and the UK had the right to violate Chilean 'sovereignty'? Or is 'sovereignty' and the right of the 'nation-state' an excuse for the otherwise inexcusable? Which law should apply? Spanish, Chilean, English, European, International? What about crimes committed abroad or in their own countries by the Spanish and British armed forces in the past? What geographical *scale* of justice should apply? For his supporters in Chile, Pinochet had 'saved' the country from Soviet/Cuban-led communism. He had after all (with decisive support from the USA) overthrown a leftist government, and many of those imprisoned and murdered by his regime were seen as 'communists'. Many other Chileans celebrated Pinochet's arrest. Those who felt themselves or their families and friends to have been victims of Pinochet's regime saw the possibility of some justice. The former British Prime Minister Margaret Thatcher reminded whoever cared to listen to her that her friend General Pinochet had assisted Britain during its 1982 war with Argentina to recapture the Falkland Islands (another struggle over sovereignty and territory). In the end, Pinochet was released and returned to retirement in Chile. However, the Chilean story and career of General Pinochet always had *global* dimensions. These incorporate a variety of domains, scales and modes of political analysis, imagination and action; a variety of 'political geographies'.

This section will enrich understandings of what can be meant by 'political geographies' and their myriad intersections. By the time they finish the section, readers will have an understanding both of some of the ideas that Pinochet drew upon to justify and codify his rule and of powers (and sometimes the fragility) of states, nations, and sovereignty. We aim to whet appetites for the complex topics of political geographies, to encourage critical thinking about often controversial, contested and complex issues and topics.

Those who have reached this section after reading the rest of the chapters (congratulations!) should already be aware that many elements of geography touch on political issues. Such 'politics' may not be confined to those things that are usually designated by the term: governments, elections (or the lack of them), political parties and so on. Rather, readers will have noticed that debates about culture, economic change, history, changing gender relations and many other things described in this text are 'political'. They are political in the sense of being about power, albeit at all kinds of levels and in lots of different ways. So, in an important way, *all* geographies are political geographies. Though focused on geopolitics, territory, nation-states, citizenship and collective actions, the chapters that follow aim to bring that home. Firstly, Chapter 17 examines the diversity of geopolitical traditions, showing how the meanings of the term 'geopolitics' have varied and continue to shift and how they articulate local and global visions of geography and power. Chapter 18 examines 'territoriality'; the ways that human individuals and human social groups claim or are assigned to particular areas. Whilst this is perhaps most clearly expressed in the form of nations and states (the focus of Chapter 19), Chapter 18 stresses that territoriality operates in myriad ways; for example, in the manner in which areas within many cities become associated with particular ethnicities or, more widely, in the ways that some spaces are seen as public and others private. Readers will therefore (hopefully) find many connections with earlier sections in of the book here. Chapter 19 introduces the historical and geographical variability of nations and states. It stresses that neither the nation nor the state is to be taken at face value. Or in other words, it examines how nation-states are complex symbolic systems that crucially depend upon particular visions and associations of territory, place and space. In the same vein, Chapter 20 will extend understandings of the contested and *uneven* formats of citizenship and collective action in an interlinked, but conflict-ridden world.

CHAPTER 17

Geopolitical traditions

JAMES SIDAWAY

Topics covered

- Origins and history of geopolitics
- Diversity and dissemination of geopolitical discourses
- Changing 'World Orders'

[US National Security advisors] Robert McNamara and George Bundy found US policies toward the Latin American military to have been 'effective in the goals set for them', namely 'improving internal security capabilities' and 'establishing predominant US military influence'. That was very important, because, in what they called the 'Latin American cultural environment' – remember, these are serious intellectuals, so they're interested in things like that – it is the task of the military 'to remove government leaders from office whenever, in the judgement of the military, the conduct of these leaders is injurious to the welfare of the nation'.

(Chomsky 1998: 7)

17.1 Introducing the idea of a geopolitical tradition

Whilst the term 'geopolitics' can refer to many things, it is important for students of geography (a subject with which geopolitics has been associated) to understand that geopolitics is commonly associated with a particular way of writing (and thinking) about space, states and the relations between them. Often this takes the form of mapping, emphasizing the strategic importance of particular places.

Geopolitics in this sense sees itself as a tradition. That is, something conscious of its unfolding historical development and with a sense of important founders (and certain key texts written by them). Those who write about this tradition or see themselves as working within or extending it usually trace its origins to the late nineteenth-century writings of a conservative Swedish politician Rudolf Kjellén. Kjellén is reputedly the first person to have used the term geopolitics in published writings (see Holdar 1992). But beyond this idea of a founding moment when the geopolitical 'tradition' begins with the first use of the term by Kjellén, things start to get complicated. The 'tradition' divides, fractures, multiplies and finds itself translated into many languages and cropping up in everything from the writings and speeches of American politicians to texts written by Brazilian generals and Russian journalists. All these reinvent and re-work the 'tradition' as they go along. As the introduction to a critical collection on 'rethinking geopolitics' explains:

> the word 'geopolitics' has had a long and varied history in the twentieth century, moving well beyond its original meaning in Kjellén's work. . . . Coming up with a specific definition of geopolitics is notoriously difficult, for the meaning of concepts like geopolitics tends to change as historical periods and structures of world order change.
>
> (Ó Tuathail 1998: 1)

In other words, exactly what is meant by the term 'geopolitics' has changed in different historical and geographical contexts. However, a good way to understand something about this idea of a tradition of geopolitics is to look at some of the people, places and ideas associated with one of its most significant appearances, in the ideology of the right-wing dictatorships that ruled many South American countries in the 1960s, 1970s and 1980s. Although today they are nearly all democracies, between the mid-1960s and the mid-1980s many South American countries were dominated by their armies. In most cases they had come to power through military coups and became more concerned with internal security (patrolling the towns and country as a kind of police force and running the wider government) than with fighting or preparing to fight wars with other countries. In virtually all cases the military regimes received a significant degree of support from the United States, which had often trained the military elites in counter-insurgency and provided weapons and logistical support. For in the US, Central and South America and the Caribbean were presented as its legitimate sphere of influence, America's 'backyard'. The US had first declared this to be the case back in 1823 – when it was termed the 'Monroe Doctrine' (so named after the fifth US President James Monroe, 1817–25) (Figure 17.1). However, the full significance of the Monroe Doctrine would only be fulfilled once the USA achieved superpower status after 1945. Alarmed by the presence of pro-Soviet 'communist' forces in Cuba following the successful revolution there in the late 1950s which overthrew a corrupt US-backed regime, the United States was determined to allow no more radical governments to emerge in the region. US political and economic interests would be protected – by whatever means necessary.

Figure 17.1 The Monroe doctrine.
Source: Skidmore and Smith (1988). Copyright 2004 Star Tribune

Chile is a good example, both of the US role and of the active presence of the geopolitical tradition. On 11 September 1973, the elected left-wing government of Chile (headed by the communist President Salvador Allende) was overthrown in a brutally violent coup led by the head of the armed forces General Augusto Pinochet (Plate 17.1). The left-wing regime that Pinochet overthrew had already faced American economic and political power and sanctions and the CIA backed the coup. Promising to impose order on a country whose politics had long been democratic but was wracked by left–right divisions, Pinochet and his generals ruled Chile with an iron fist. All opposition was crushed and thousands of people were rounded up by the armed forces. Many 'disappeared' into secret military jails. Hundreds of these prisoners were tortured and murdered. The use of torture became a routine instrument of state. For the next seventeen years, Chile was ruled by Pinochet and his cronies who imposed a new 'neoliberal' economic and political model (of the form that is described in Chapter 8) and murdered any significant opposition. Pinochet finally stepped down as head of state in 1990, when democracy was restored, but not before he had

entrenched himself as a senator (in the parliament) for life. He also guaranteed that the armed forces (which he continued to head) would continue to have a significant role in Chile's state structures and that they could not be prosecuted for any crimes committed during the coup and brutal years of military rule.

The ideology of the military regime in Chile (like those elsewhere in South America) was intensely nationalist. More importantly it was a particular conservative form of Chilean nationalism. The nation was held to be sacred and the military were rescuing this 'sacred body' from communists, subversives and so on (i.e. anyone who opposed the military vision of ultra-nationalism and order and the neoliberal economic and social model which was now imposed). It is in this nationalism and conception of the nation as a kind of sacred body that geopolitics enters the picture. A few years before the coup, Pinochet had published an army textbook on geopolitics. Intended mainly for use in Chilean military academies, the 1968 textbook indicated that Chile's future dictator took the subject deadly seriously. For Pinochet, geopolitics was a science of the state, a set of knowledges and programmes to perfect the art of statecraft, that is to strengthen the state in a continent and wider world in which it was held to be in competition with others. The starting point and the heart of Pinochet's textbook and of the wider bleak tradition of geopolitics that it presents is an organic theory of the state.

17.2 The organic theory of the state

This idea is at the core of most South American geopolitical writings, and of the wider geopolitical tradition. In summary, it holds that the state or country (and the sense of nation that goes with it), is best understood as being like a living being. Like any living organism, it therefore needs space to grow and it will be in competition with other living beings. An idea that the state needs 'living space' (a certain amount of resources and land) for the nation to thrive is particularly evident in South American geopolitics. A certain reading of Charles Darwin's idea of a struggle for the 'survival of the fittest' is therefore transferred to the realm of states. Drawing upon conservative German writings of the nineteenth and early twentieth centuries (which had been codified by the German academic Friedrich Ratzel) and with this organic notion of the state in mind, geopolitics claims to

Plate 17.1   General Pinochet: geopolitics in action. (Popperfoto/Reuters)

identify certain laws that govern state behaviour and that, once identified, can be a guide for those charged with furthering and protecting the 'national interest' (or at least a certain definition of the latter).

It is not difficult to see how this idea could appeal to South American military dictators, for it gives the military, equipped with the supposedly scientific study of geopolitics, a special mission. In other words it legitimizes their rule. Moreover, the organic idea was extended further to legitimize the extermination of those whom the military defined as enemies of the state. People labelled as communist or subversive, indeed all those who oppose the military dictatorship, can be compared with a disease or cancer which threatens the life-blood of the state-organism and is best 'cut out', that is eliminated. The otherwise unthinkable (the murder of thousands of people in the name of the wider welfare of the nation-state) is made to seem natural and a good thing, since it serves the longer-term interest of the health and power of the 'living being' that is the state (see Hepple 1992 for an exploration of this). With the 'cancer' of subversion and disloyalty eliminated the 'state-body' can go on to grow and thrive.

17.3 Brazilian national integration

Aside from Chile, one of the most important expressions of geopolitics in South America has been in Brazil. The military in Brazil overthrew the democratic government in 1964 and stayed in power for the next 20 years. Again the Brazilian military were supported by the USA, whose ambassador in Brazil termed the military coup 'the single most decisive victory of freedom in the mid-twentieth century' (cited in Chomsky 1998: 7). Although never as despotic as the regime led by Pinochet (or similar military governments in Argentina and Uruguay), the long years of military rule gave the Brazilian generals a chance to elaborate and impose a geopolitical vision on the country. One of the most evident aspects of Brazilian geopolitics is the idea that state security requires a measure of national integration. The vast scale of Brazil, the difficulty of travel across its Amazon 'heartland' (the world's largest tropical rain-forest) and the fact that it shares borders with every other South American country except Chile gave the geopolitics of the Brazilian generals a special sense of its national mission and an obsession with the potential for the country to be a great power (known in Portuguese as *Grandeza*).

The associated sense of the urgency of the integration of Brazil required the extension of a network of highways across Amazonia and the settlement of its lands by farmers and ranchers. It is this geopolitically inspired vision, combined with a highly corrupt system of patronage and favours to those close to the regime, that underlies the enclosure and division of Amazonia into private lands (some larger than European countries such as Belgium) and the accompanying transformation or destruction of the tropical rain-forest (see Hecht and Cockburn 1989) (Figure 17.2).

The consequences of this geopolitically motivated and economically profitable strategy have been disastrous for the indigenous peoples of Amazonia. They have found themselves forced off land that was traditionally theirs, sometimes murdered or attacked by ranchers, the state and settlers and disorientated by the arrival of a frontier culture of violence, destruction and consumption. Poor peasants, particularly from the impoverished north-east of Brazil, have also moved into the Amazon region in search of land and freedom. They are fleeing from the oppression and landlessness they themselves face in the north-east of Brazil, where the vast majority of the land is in the hands of a small elite. Of course, in the geopolitical visions of the Brazilian generals, this population movement is seen as positive, reinforcing the Brazilian population of the Amazon and redistributing what they regard as marginal 'surplus' people in the process. At the same time this would defuse political pressures for land reform in north-east Brazil. In all this we can see how Brazilian geopolitics is implicated in a complex of social and environmental transformations. The destruction of the Amazonian ecosystem is therefore not simply the result of 'population pressure' or abstract forces of development, rather it is to be understood as a social process and therefore as the consequence of particular political choices and (im)balances of power.

17.4 Antarctic obsessions

An interest, at times an obsession, with the last unexploited continent of Antarctica is evident in both Brazilian and Chilean geopolitics. Antarctica is also prominent in geopolitical writings from Argentina and Uruguay and also appears in Peruvian and Ecuadorian geopolitical texts (see Dodds 1997). To understand what this amounts to and what forms it takes necessitates some understanding of Antarctica's exceptional political geography. Antarctica has a unique

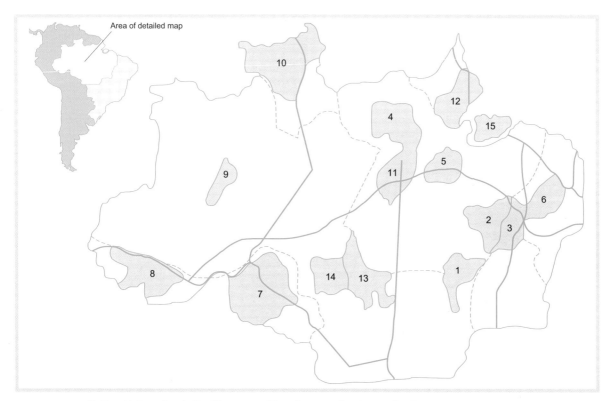

Figure 17.2   National integration in Brazilian geopolitics: Amazon development poles.
Source: Hecht and Cockburn (1989: 127)

territorial status, in so far as it has no recognized state on its territory. Everywhere else in the world forms part of the patchwork of states that we learn to be familiar with and take for granted. Antarctica is a stark reminder that there is nothing natural or inevitable about this. The absence of recognized states in Antarctica reflects the inhospitability and remoteness of the continent, the only one without an indigenous human population. Although the plankton-rich waters around Antarctica were exploited for seal and whale hunting in the nineteenth century, it was not until the twentieth century that exploration of the interior began. Even today, Antarctica has only a non-permanent population (at any time) of a few hundred scientists. This presence, plus the relative proximity of the continent to South America and the possibility of exploitable mineral resources, has led to a series of territorial claims. Although Argentinean writers had already represented Antarctica as an extension of the southern Argentine area of Patagonia, the British made the first formal claim in 1908. In turn, parts of this claim were 'granted' to Australia and New Zealand. France and Norway made claims in the 1930s and 1940s, followed shortly by Argentina and Chile. After 1945, the USSR and the USA also established a wide network of bases, although

without staking formal claims to territory. The claims made by Argentina, Chile and the UK overlapped – and foreshadowing the British–Argentine conflict over the Falklands/Malvinas British and Argentine forces exchanged fire in Antarctica in the late 1940s and early 1950s. In the context of this potential for conflict and growing possibility of Cold War confrontation (on the Cold War, see Section 17.7) a United Nations Treaty in 1959 agreed that all claims would be (forgive the pun) 'frozen' for at least 30 years, and the continent reserved for scientific (not commercial or military) use. The Treaty was extended in 1991. But prior to then, it was not clear what the future status of the continent would be. Even today, the Treaty merely defers the issue of claims.

Argentina is the South American country with the largest claim (Figure 17.3). The Argentine claim has also become central to geopolitical discourse there. As Child (1985: 140–1) explains in a study of geopolitics and conflict in South America:

For Argentine geopolitical writers, the subject of Antarctica is not only linked to tricontinental Argentina [a power in the South American continent and the South Atlantic], but also to . . .

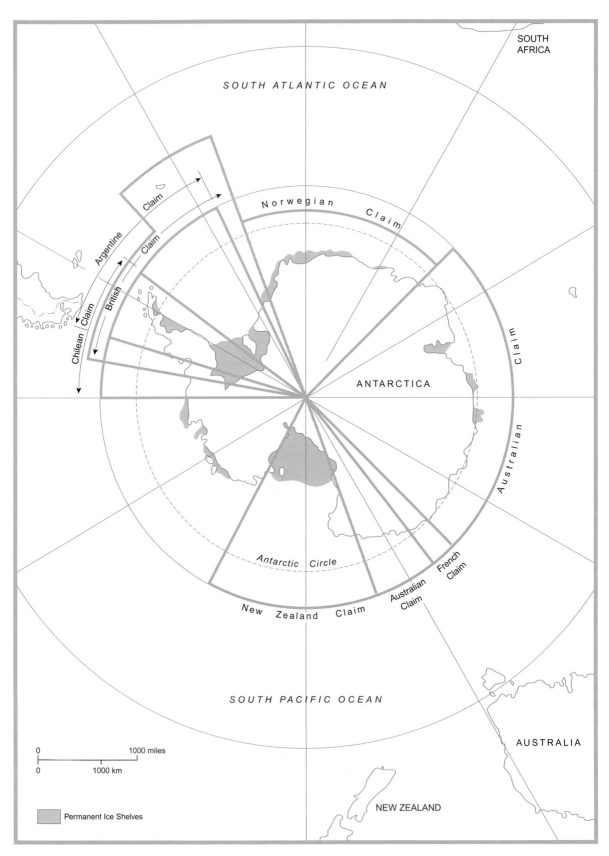

Figure 17.3 Claims on Antarctica.

Source: Adapted from M.I. Glassner, *Political Geography*, John Wiley & Sons Inc., 1993, p. 498. Copyright © 1993 John Wiley & Sons Inc. This material is used by permission of John Wiley & Sons Inc.

national sovereignty, patriotism, and pride. This is a particularly touchy combination after the humiliating defeat of the Malvinas/Falklands conflict. The Argentine National Antarctic Directorate has professors of Antarctic geopolitics on its staff. Through the media, maps, and postage stamps and the centralized educational system, Argentines are constantly taught and reminded that there is an Argentine Antarctic just as much are there are Argentine Malvinas. The need to assert Argentine rights in the Argentine Sea, islands, and Antarctica is linked to dreams and national projects of Argentine greatness.

17.5 Heartland

Although Antarctica and the South Atlantic are significant components of Argentine (along with Brazilian and Chilean) geopolitics, the former also looks beyond other Argentine frontiers. In particular, Argentine geopolitical writers have taken a particular interest in the security of Bolivia, together with some Brazilians and Chileans. They have scripted Bolivia (a relatively impoverished land-locked mountainous country, which has borders with Argentina, Brazil, Chile, Paraguay and Peru) and proximate areas of its neighbours as a key strategic continental 'heartland' (see Kelly 1997). Control of Bolivia, in this vision, would be a vital key to a relative dominance in the South American continent. That Bolivia has a strong revolutionary tradition and was for many years characterized by chronic political instability has reinforced the tendency of the other South American countries to meddle in Bolivian politics. Indeed, during the years (1976–82) when Argentina was last ruled by a geopolitically obsessed military junta, the Argentine armed forces were actively involved in supporting a Bolivian military government. This activity took the form of the kinds of brutal suppression and frequent murder of those (trade union leaders, dissidents, opposition members and leaders) who opposed the military government and its economic and social strategies. The idea of Bolivia as a 'heartland', control of which would be a kind of magic (geopolitical) key to domination of South America, links back to one of the best known genres in classical (European) geopolitical thought. For although the South American countries have seen some of the most significant expressions of geopolitical discourse of modern times, geopolitics originates in Europe and it

is to some examples of European geopolitics, including the idea of heartland, that we now turn.

The designation of a heartland was first made by the British geographer (and strongly pro-imperialist conservative politician) Halford Mackinder. In what has since become a widely cited (if less often read) article first published in 1904 following its presentation to the Royal Geographical Society (RGS), Mackinder argued that the age of (European) geographical exploration was drawing to a close. This meant that there were hardly any unknown 'blank' spaces left on European maps of the world. According to Mackinder, the consequence of this closing of the map, this end of the centuries-long task of exploration and discovery, was that political events in one part of the world would invariably affect all others, to a much greater extent than hitherto. There would be no more frontiers for Europeans to explore and conquer. Instead, the great powers would now invariably collide against each other. Mackinder called this end of European exploration 'the post-Columbian age' and the closing of frontiers, the emergence of a 'closed political system':

> From the present time forth, in the post-Columbian age, we shall again have to deal with a closed political system, and none the less that it will be one of world wide scope. Every explosion of social forces, instead of being dissipated in a surrounding circuit of unknown space and barbaric chaos, will be sharply re-echoed from the far side of the globe, and weak elements in the political and economic organism of the world will be shattered in consequence. There is a vast difference of effect in the fall of a shell into an earthwork and its fall amid the closed spaces and rigid structures of a great building or ship. Probably some half-consciousness of this fact is at last diverting much of the attention of statesmen in all parts of the world from territorial expansion to the struggle for relative efficiency.

(Mackinder 1904: 422)

Given that this was the case, Mackinder claimed to identify the places of greatest world-strategic significance, control of which would give any great power a key to world power. In his 1904 paper, he termed this the 'pivot area'. With Mackinder's address to the RGS and his subsequent article came a series of maps, the most frequently reprinted one of which claims to describe 'The Natural Seats of Power' (Figure 17.4). As Ó Tuathail's (1996: 25) critical account of Mackinder has recently argued:

Figure 17.4 'The Natural Seats of Power' according to Mackinder (1904).

Mackinder's January 25 04, address to the Royal Geographical Society, 'The Geographical Pivot of History', is generally considered to be a defining moment in the history of geopolitics, a text to which histories of geopolitics invariably point.

Mackinder himself did not use the term geopolitics. But his 1904 paper and its maps were later appropriated into other geopolitical traditions. Indeed, the map of 'The Natural Seats of Power':

> is perhaps the most famous map in the geopolitical tradition. . . . The map is meant to create the illusion of the unveiled geographical determinants of power. It is the global vision, the post-Columbian made possible, the pure panorama produced by the penetrative gaze of the global voyeur. Yet, interestingly, the map is not purely visual but is marked with text. Huge swathes of territory are stamped with a definitive positionality and function: 'Pivot Area', 'Inner or Marginal Crescent', and 'Lands of Outer or Insular Crescent'. These are macrogeographical identities around which Mackinder spatializes history and reduces it to formulaic equations and timeless east/west, land/sea conflicts. 'The social movements of all times', he concluded, 'have played around essentially the same physical features.'
>
> (Ó Tuathail 1996: 31–3)

A certain geographical determinism can be detected here, in which the physical environment ('essentially the same physical features' in Mackinder's words) provide the fixed, unchanging backdrop to and therefore something of the explanation of human history. This view, which ignores the social construction of resources (see Chapter 5) and other features of nature, is common to virtually all geopolitics (something which it shares with a great deal of twentieth-century geography). One of the key formulas in Mackinder's notion of geographical determinism is his identification of a 'pivot-area' or 'heartland zone' in east-central Europe, control of which would be a kind of magic key to world domination. Reworking the ideas of his 1904 article after the First World War, Mackinder refined this notion and penned a formula that encapsulated it:

> Who rules East Europe commands the Heartland.
> Who rules the Heartland commands the World-Island.
> Who rules the World-Island commands the world.
>
> (Mackinder 1919: 4)

According to this (simplistic) formula, 'rule' of the eastern portion of Europe offered the strategic path to that of the African-Asian-European continents (which together constitute what Mackinder terms the 'World-

Island') and hence a dominant position on the world scene.

In addition to the general backdrop of the emergence of the sense of closed political space, or in Mackinder's terms the completion of the age of exploration (which he called, after its best known figure, Christopher Columbus, the Columbian age), it is important to recognize that Mackinder, whilst claiming to offer a universal and objective model (with 'laws', for example the formula about 'Heartland') is in fact highly subjective. Heartland is another organic metaphor, invoking the idea that territory has a heart, like a living being. Moreover, Mackinder's writings reflect not only the moment at which they were penned, but also the place. Mackinder, writing from the vantage point of imperial Britain, is concerned to identify threats and dangers to British power. At the time when Mackinder presented his paper to the RGS, Britain was the pre-eminent world power. It still seemed that way to Mackinder in 1919 when he wrote about the 'Heartland'. But British imperial politicians, like Mackinder, were aware of the growing power of America, Germany and Russia. In fact, potential British imperial competition with the latter in Asia provided a key context to Mackinder's work. As Peter Taylor (1994: 404) explains:

> Behind every general model there is a specific case from which it is derived. For the heartland model this is particularly easy to identify. Throughout the second half of the nineteenth century Britain and Russia had been rivals in much of Asia. While Britain was consolidating its hold on India and the route to India, Russia had been expanding eastwards and southwards producing many zones of potential conflict from Turkey through Persia and Afghanistan to Tibet. But instead of war this became an arena of bluff and counter-bluff, known as the 'Great Game'. . . . Mackinder's presentation to an audience at the Royal Geographical Society would not have seemed so original as it appears to us reading this paper today. . . . Put simply, the heartland model is a codification and globalization of the Great Game: it brings a relatively obscure imperial contest on to centre stage.

Not only does this envisage the world in a particular way, as a 'stage', but it sees only select key actors as the significant figures at play. These are the European powers (plus Russia). Other peoples and places are merely the backdrop for action by White Men. The taken-for-granted racism of Mackinder's model, in

which only Europeans make history, is also that of European imperialism and that of the bulk of wider European geographical and historical writings of the time (see Chapter 3).

Yet, although Mackinder's 1904 paper is very much a product of its time and Mackinder's own conservative world view, it has proven durable and has been integrated into rather different contexts, which saved it from the relative obscurity that it deserves as a turn-of-the-century imperialist text. The transfer of the discourse of 'Heartland' to Bolivia by South American codifiers of geopolitics has already been noted. In addition, 'Heartland' was appropriated by German geopolitics in the 1930s and 1940s and formed part of the backdrop to Cold War American strategy from the late 1940s through to the last decade of the twentieth century. Taylor (1994: 405) notes:

> First in the inter-war years the heartland theory became an integral part of German geopolitics. It fitted the needs of those who advocated lebensraum [living-space], the policy of expanding into eastern Europe [and enslaving and murdering its indigenous populations], coupled with accommodating the U.S.S.R., as a grand continental policy for making Germany a great power again. Second with the onset of the Cold war after World War II, the heartland theory got another lease of life as the geostrategic basis of nuclear deterrence theory. The west's nuclear arsenal was originally justified in part as compensation for the U.S.S.R.'s 'natural' strategic advantage as the heartland power.

The next two sub-sections of this chapter will examine aspects of Nazi, Fascist and Cold War geopolitics in greater detail.

17.6 Nazi and Fascist geopolitics

The formal tradition of writing about space and power under the title of 'geopolitics' also found fertile contexts in Italy, Portugal, Spain and Japan. Influenced and supported by Nazi Germany, all these countries (together with Hungary and Romania) saw the rise and victory of ultra-nationalist or Fascist governments (often through violent struggle or full-scale civil war with democratic or communist forces) (see Chapter 3, pp. 68–71). In each case geopolitical debates were crucially negotiated through other cultural and political debates about race, nationalism, the colonial pasts and futures, supposed national 'missions' and destinies and the European and global political contexts. The German case has become the best known. In Germany, organic notions of the state had already been popularized by conservative nineteenth-century academics. Moreover, Germany was characterized by extreme political and economic turbulence in the decades following its defeat in the 1914–18 World War. This combination provided a fertile environment for the elaboration and circulation of a distinctive geopolitical tradition. In Ó Tuathail's (1996: 141) words:

> After the shock of military defeat and the humiliation of the dictated peace of Versailles, the Weimar Republic proved to be fertile ground for the growth of a distinct German geopolitics. Geopolitical writings, in the words of one critic, 'shot up like mushrooms after a summer rain'.

The main features of these writings (which they shared with a wider German nationalism, later codified in Nazism) were a critique of the established 'World Order', and of the injustices imposed on Germany by the victors. German claims were often presented graphically in maps that were widely circulated (see Herb 1989: 97).

Like the variants of the geopolitical tradition that were developed amongst right-wing and military circles in Italy, Portugal, Spain and Japan, German geopolitics also asserted an imperial destiny. Indeed, as Agnew and Corbridge (1995: 58–9) explain: 'The Nazi geopoliticians of the 1930s came up with formalized schemes for combining imperial and colonized peoples within what they called "Pan-Regions".'

In this vision, notions of racial hierarchy were blended with conceptions of state 'vitality' to justify territorial expansion of the Axis powers (see O'Loughlin and van der Wusten 1990) (Figure 17.5). In Europe, related conceptions of the need for an expanded German living-space were used as justification for the mass murder of occupied peoples and those who did not fit into the grotesque plans of 'racial/territorial' purity. The practical expression of these was the construction of a system of racial 'purification' and mass extermination. At least 6 million Jewish people were murdered in concentration camps together with millions of others: disabled people, gays and lesbians, gypsies and political opponents. Historical debates about the role and relative significance of German geopolitics within the Holocaust and within

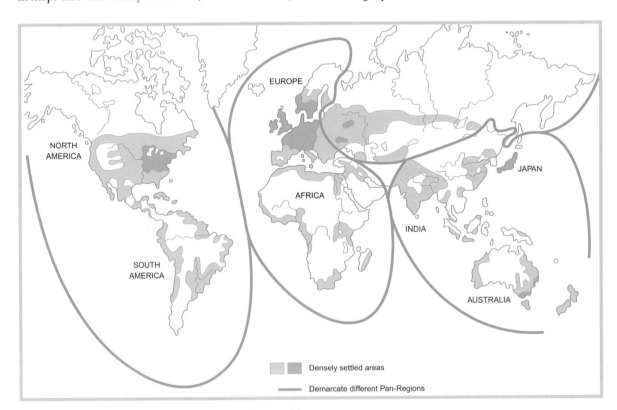

Figure 17.5 'Pan-Regions' as envisaged in Nazi geopolitics.

Source: O'Loughlin and Van der Wusten (1990) Political geography and the pan regions, *Geographical Review*, **80**, adapted with permission of the American Geographical Society. Copyright American Geographical Society, 1990

broader Nazi ideology and strategy continue (see Heske 1986: 87, Bassin 1987, Ó Tuathail 1996, Natter 2003). In the United States in particular, German geopolitics became the subject of lurid tales and depictions, cropping up in media, military and government 'explanations' of Nazi danger. Popular magazines, such as *Readers Digest*, would inform Americans of the 'scientists behind Hitler' describing German geopolitics as *the* key to Nazi strategy. And whilst the relative significance of the German geopolitical tradition in the wider genocidal ultra-nationalism of German Fascism was certainly overstated in such accounts, we should see geopolitics as a particular expression of *wider* academic and intellectual involvement and complicity in authoritarian state power, war-making and genocide. Much more widely, beyond the geopolitical tradition *per se*, academic geography was deeply implicated in these activities. Exploring this, Natter (2003: 188) notes how:

> The work of disciplinary historians of geography has demonstrated the extent to which the demarcation of geography seems inseparable from the history of war, imperialism and quests for

national identity. . . . Geopolitics, thus, would mark a particular, but in no way separable (and hence containable) geopolitical deployment of geo-power.

An example of this wider complicity was the way in which the models of 'Central Place Theory' (an abstract model of the ideal spatial distribution of towns), which were developed in Germany by Walter Christaller in the 1930s, were elaborated with the express purpose of providing planners with a model of German settlement to impose on conquered territories of Eastern Europe, once the indigenous Jewish and Gypsy people had been gassed to death and the majority Slavic population enslaved (see Rössler 1989).

The defeat of the Axis powers in 1945 (culminating in the use of American atomic weapons against the civilian populations of Hiroshima and Nagasaki) and the lurid wartime depictions of Nazi geopolitics in the United States dealt something of a blow to the formal tradition of geopolitics. However, it continued in subsequent decades, in both Spain and Portugal (see Sidaway 1999, 2000), where Fascist regimes remained in power to the mid-1970s and, as we have seen, in Latin America. These regimes (and plenty of others)

Plate 17.2 The gateway to the Auschwitz concentration camp. The Nazi regime used concentration camps to execute their policy of racial/territorial 'purification'. (Corbis/Ira Nowinski)

FOR
ROSE-MARIE

Dr. K. Haushofer

THE WORLD OF
GENERAL HAUSHOFER

Geopolitics in Action

BY ANDREAS DORPALEN

With an Introduction by
Colonel Herman Beukema, U.S.A.

FARRAR & RINEHART, INC.
New York • Toronto

Plate 17.3   German geopolitics: as represented in the USA.

were integrated into the US-led anti-communist network of allies. In a policy that became known as containment, the US aimed to encircle and block the potential expansion of Soviet power and influence beyond the immediate borders of the USSR and the pro-Soviet states installed by the USSR in the Eastern European territories that it had occupied during its Second World War battle for survival against Nazi Germany. In the allied countries (and as an expression of this strategy of containment), the functions of geopolitics, in particular its 'strategic vision' and claim of 'scientific' objectivity, continued to operate or were displaced into other disciplines and branches of knowledge, including geography and the expanding subject of International Relations. This displacement of strategic knowledge was particularly evident in the United States, which by 1945 was the greatest power that the world had ever seen, constituting over half of the world economy and (until the recovery of the Soviet Union and its own development of atomic weapons in 1949) possessing a virtually unrivalled military capacity.

17.7 Cold War geopolitics and the logics of containment

Like a fast-moving glacier, the Cold War cut across the global terrain in the 1940s, scraping and grinding solid bases, carving and levelling valleys, changing the topography. For nearly half a century the Cold War competition between the United States and the Soviet Union dominated the international system. . . . Like all retreating glaciers, the receding Cold War left deposits of debris and exposed an altered, scarred landscape that leaders in the 1990s struggled to measure and fathom.

(Paterson 1992: 2)

The Cold War is used as a shorthand description of the conflict (which to many seemed to be a conflict for world domination) between the 'communist' East, led by the USSR and the 'capitalist' West, led by the USA between around 1949 to around 1989. This was not simply a rift between two great powers, but a complex ideological conflict, which often appeared to be about different ways of life and contrasting social systems (the bureaucratic 'command economies' based on predominant state ownership of the means of production, with a ruling class drawn from the bureaucracy, and the capitalist economies based on predominantly private ownership with a ruling class drawn from the owners of capital). It was also a conflict in which *direct* military confrontation between the two great powers was avoided. Hence the metaphor of Cold, i.e. 'poor, frozen or frosty' relationships, but short of all-out 'hot' war, involving direct exchanges of missiles, bombs and so on. In part this avoidance of direct conflict arose because both powers were conscious of the enormous stakes, armed with nuclear, chemical and biological weapons as well as vast arsenals of increasingly high-tech 'conventional' armaments. By the 1960s, they were each capable of destroying virtually all human life several times over. Moreover, neither could attack the other without being sure that the enemy would still possess enough undamaged nuclear weapons to retaliate, a 'balance' of terror known, appropriately enough, as MAD (Mutually Assured Destruction). Instead, conflict took the form of the continual preparation for war, plus proxy wars in what became known as the (originally not pro-West, not pro-East, but contested) 'Third World' and many other political, economic and cultural forms of competition. That is, it was a conflict conducted by *every* means *except* direct military confrontation between the superpowers.

Of course there was no shortage of indirect confrontation and preparation for war. Europe was divided into two armed camps and split down the middle between Soviet and American zones of influence by a fortified military and ideological frontier that became known as the 'Iron Curtain'. Whilst Europe was characterized by an uneasy stability and balance of terror, by the 1970s the term geopolitics had been revived (or rediscovered) by US national security advisors Henry Kissinger and Zbigniew Brzezinski (see Hepple 1986, Sidaway 1998) to refer to the strategic vision deemed necessary to circumvent the sense of growing Soviet power, particularly in the Third World, where a wave of successful revolutions had brought left-wing, pro-Soviet governments to power. As has already been noted, one of these was in Cuba, but by the end of the 1970s, there were also pro-Soviet governments in the countries of Indochina (Vietnam, Laos and Kampuchea) where the US had been unable to defeat communist insurgencies despite a massive military effort, in the African countries of Angola and Mozambique, in the Arabian country of South Yemen and in Afghanistan as well as in the small Central American state of Nicaragua. Communist China and North Korea added to American concerns. In this

Plate 17.4 'Containment' explained: US Secretary of State for Defense briefing journalists during the Vietnam war. (Corbis UK/Bettmann)

world of superpower competition, geopolitics again found its moment and expression in the visions of the national security advisors and American generals.

But to focus on formal geopolitics would be too narrow. For the relative absence of explicit references to geopolitics in American strategic culture during the 1950s and 1960s did not mean that the kind of mapping of places as pacified/dangerous, ours/theirs, threatening/stable, strategic/secure, and the intersection of a geographical imagination with a military/strategic one that characterize the 'common-sense' understanding of the geopolitical tradition was absent. As has been noted, such understandings proliferated into other academic realms, practices, knowledges and disciplines. As well as cropping up in the subject of International Relations, this proliferation of quasi-geopolitics could be found in academic geography where, under the title of 'political geography', American geographers would proclaim their difference from the 'disreputable' tradition of geopolitics, and then proceed to engage in more or less the same kinds of strategic mapping and policy advocacy (see Smith 1984, 2003). More widely, American leaders and policy-makers declared that the

Soviet Union had to be *encircled* and *contained*. The metaphor of disease (containment) was mixed with that of 'dominoes' – if one country (say Vietnam or Cuba) had 'fallen' to Soviet control or influence, then it could (like a chain reaction) 'infect' other proximate ones. The irony of this (which is equally present in the formal geopolitical tradition) is that in the name of security and strategy, the real complexity of *human* geographies in the places (such as Vietnam or Chile, or Afghanistan, Nicaragua or the Persian Gulf) that are deemed strategic is sometimes obscured or erased. Forget about the complex details of the people, culture and society. What matters is the 'strategic value' of the place, or the political identity of its government as an ideological and strategic friend or foe. At times this could take quite extreme forms, as with US National Security Adviser Robert McNamara (previously head of Ford, and later head of the World Bank):

Robert McNamara was, of course, the leading specimen of *homo mathematicus* – i.e., men who behave and believe other men [and, we might suppose, women too] behave primarily in response to 'hard data', usually numbers (infiltration rates, 'kill ratios', bomb tonnage). Like the classic private eye on television, they are always looking for 'the facts', but usually the wrong ones. They miss reality, for they never get close enough or related enough to another society to do more than count things in it. If you relate to a country as a military target you do not need to know anything about it except details as are easily supplied by reconnaissance satellites, spy ships, secret agents, etcetera. You need never know who the victims of your attack were. Your task is merely to assess the results of what you have done and this is done by counting bodies, destroyed factories, enemy soldiers.

(Barnet 1973: 119)

Such logic is seemingly far removed from the prominence of maps in the formal geopolitical tradition. Yet we can see that, in *both* the kind of thought present in McNamara's mind and in classic geopolitics, are moments when the myriad complexity of the world is reduced to a simple black and white representation, be it numerical or cartographic:

By gathering, codifying, and disciplining the heterogeneity of the world's geography into the categories of Western thought, a decidable, measured and homogeneous world of geographical objects, attributes, and patterns is made visible, produced.

(Ó Tuathail 1996: 53)

These logics are both *colonial* and *colonizing*. They are colonial, in so far as they are rooted in the fact-gathering epoch of colonial exploration and share the arrogance of its classifications of peoples and places, which are then turned into a hierarchy of civilized to uncivilized, with the former (who are seen in racist terms) as the 'natural' leaders, or most developed, free or 'advanced'. They are also colonizing, in so far as they expand into and affect other seemingly distant fields of knowledge and social or cultural activities. Everything from scientific debates to art and cinema becomes caught up in geopolitics, and registers its spread, its dissemination.

17.8 Cold War geopolitics in art and culture

The art world and, in particular, the course of modern art in the second half of the twentieth century became intertwined with Cold War geopolitics. More generally, art and politics have frequently been intertwined. The Nazis (and their leader Adolf Hitler, who was a failed artist of minimal talents) detested modern art, condemning it as decadent. For them a gross parody of classical art, sculpture and architecture was all that was fit for the German Reich. Other dictators, such as Joseph Stalin, the leader of the Soviet Union between the mid-1920s and his death in 1953, were also wary of modern art and favoured a garish realist style, depicting heroic smiling workers and peasants. Meanwhile, and especially in the years between the two World Wars, Paris became the centre of production of modern art. But this was disrupted by the devastation of the Second World War and the rise of American economic, cultural and political powers. As Sylvester (1996: 405) explains:

In the wake of World War II and the catapulting of the United States into superpowerdom and the Cold War, there was parallel activity in US art circles. Media figures, such as Howard Devree and Alden Jewell, advocated a position for US art in the world that would mirror the international ascendancy of the US political economy. Devree maintained in the *New York Times* that the United States was the most powerful country and therefore needed to create art that was strong and virile to replace the art of Paris.

But this was not only words from influential figures. The American Central Intelligence Agency (CIA) stepped in to fund and promote American modern art, bankrolling major exhibitions and tours, to demonstrate the creative and in Devree's words, the 'virile' artistic culture of the United States. This assertion of 'virility' reveals too the masculinity of much American Cold War language and policy, differentiating the thrusting, powerful American society (later personified by people such as President John F. Kennedy) from a Europe portrayed as weak and effeminate and from the more brutal, less cultured men in charge of the Soviet Union. This strategy was more subtle than the vulgar anti-communism that pervaded Hollywood at the time. It reinforced more general arguments about the US as a bastion of 'free expression' and seemed to tally with the relative dynamism of the post-war American economy and the sense that the leadership of the Western World (or, as American commentators preferred to call it, the 'Free World') had decisively passed in military, political, economic and cultural terms to the USA. As Cockcroft (1974: 129) explains:

CIA and MOMA [the New York based Museum of Modern Art, which was funded by the multi-millionaire Rockefeller family] cultural projects could provide the well-funded and more persuasive arguments and exhibits needed to sell the rest of the world on the benefits of life and art under capitalism.

Of course, the Cold War was waged too in everything from science fiction comics and action movies and thrillers (in which good Americans, like Rambo, and the occasional heroic Brit, like James Bond, faced and defeated totalitarian, alien and frequently communist enemies) to Olympic sports (in which the Soviet, Chinese, East Germans and Bulgarians invested enormous resources in training and drug-enhanced muscles, to 'prove' the superiority of the communist system in the Olympic stadiums).

The important thing to understand here is that whilst the sense of a formal geopolitical tradition may be seen as a self-conscious and fairly direct expression of states, space, power and the relations between them, the latter also take a vast variety of other forms. At the same time, the backdrop of a geopolitical conflict (such as the Cold War) can provide the basis for the making of a coherent sense of a particular way of life, for bonding a country, emphasizing a shared conformist identity and a common enemy (see Tyler May 1989; Campbell 1992).

17.9 Conclusions: a new world order and postmodern geopolitics?

The notion of the Cold War faded with the collapse of the USSR and associated communist allies between 1989 and the early 1990s, leaving the United States as

Plate 17.5 Cold War in cinema: James Bond (007) saves the West from communists and criminals. (Topham Picturepoint)

Thematic Case Study 17.1

The continuing Cold War?

The Cold War in Asia was always a very complex mixture of ideological, territorial and national disputes. An extra dimension was added by fact that China became a nuclear power in the early 1960s at the same time that the Chinese communists split with the USSR on the grounds of ideological differences.

Moreover, whilst it is now more than a decade since proclamations of the 'end of the Cold War' began to multiply in Europe, in the 'Asia-Pacific' region remnants of the global Cold War confrontation remain in place. Despite substantial changes and accommodations, China, North Korea and Vietnam are still one-party 'communist' political systems. Moreover, a series of divisions and territorial disputes whose history is deeply rooted in the Cold War remain unresolved. Most notably, the Korean peninsula remains divided between mutually antagonistic regimes and both the People's Republic of China and the offshore island of Taiwan have both claimed to be the legitimate government of the whole of China and remain fundamentally opposed.

Arguably, therefore, there is a sense in which the narratives about 'the end of the Cold War' reflect the Euro-Atlantic experience rather than being globally valid. The 'end' may have happened in Europe and the Soviet Union may no longer exist, but Cold War legacies and logics are active in East Asia (see Hara 1999) and beyond.

Plate 17.6 North Korean soldiers patrol their side of the demilitarized zone between North and South Korea. (Corbis/Michel Setboun)

the sole effective global military superpower. Yet this Western and American 'victory' has also produced a certain sense of disorientation. This had been evident before, as the so-called 'bipolar' world of the early years of Cold War confrontation in which Washington and Moscow were the main political, economic and ideological points of orientation became progressively more 'multipolar' from the 1960s onwards. That is, as Western Europe and Japan recovered from wartime devastation and as Communist China split on ideological grounds with the USSR, the sense that the world was divided into just two superpower points of orientation lessened. But the sensation of geopolitical complexity was to grow precipitously once one of the established 'poles' (Soviet communism and pro-Soviet regimes in Central and Eastern Europe) collapsed at the start of the 1990s. Having a clearly demarcated and identifiable enemy – as both sides had in the Cold War – offered contrasting but apparently solid ideas of identity, purpose and common cause that has faded in a post-Cold War world. If the Cold War embodied a relatively coherent geopolitical map (East against West, with the allegiance of the Third World as one of the prizes), the post-Cold War world is characterized by a diversity of maps and scripts. A variety of interpretations and geopolitical 'models' now attempt to offer explanation and impose meaning on contemporary events.

Amongst the most influential of these has been the discourse of 'New World Order', articulated by US President George Bush in the early 1990s. In Bush's vision, what is called the 'international community' (the member states of the UN, led by the USA, but with financial and military support from key allies such as Germany, the UK and Japan) would act as a kind of global police force, intervening where and when they felt necessary to maintain or restore 'order'. Both the 1990–1 Gulf War and the 1999 conflict with Yugoslavia were justified and conducted (in part) in the name of building such a 'New World Order'. Critics have pointed out that, as Booth (1999: 49) puts it, the New World Order 'means the New World (gives out the) Orders'. That is, the United States seeks global military hegemony and with the Soviet resistance out of the way, there is little, save its own 'public opinion', to stop it. The 'Monroe Doctrine' mentioned in Section 17.1 is extended to virtually the entire world. Critics point out that this suits Western and especially American commercial interests. The United States' global military posture in the aftermath of the drama of 11 September 2001, in particular the extension of

US military power deeper into what were once the heartlands of the Soviet Union in Central Asia, is one key aspect of this. US strategy in the Middle East, Africa, Asia and the Americas points to a wider American imperial geopolitics, which extends the strategies and doctrines developed in the twentieth century and finessed in the Cold War (Smith 2003).

Others have stressed that such apparent American military superiority and rhetoric counts for less than might appear at first sight, and that what continues to be evident in the post-Cold War world is enhanced economic (and in some ways, cultural) competition between blocs or great powers. In this vision, American power is contested by an integrating European Union and by Pacific powers such as Japan and China. But the terms of the 'contest' are less in conventional military and strategic terms than in competition for markets, productivity and profit. A world of 'geopolitical' competition is, it is argued, being partially displaced by 'geoeconomic' competition (for an example, see Luttwak 1990). In the US, this has also been interpreted as an impending 'supposed clash of civilizations' (e.g. Huntington 1993), between that of the 'West' and, for example, societies orientated to alternative belief systems such as (Chinese) 'Confucianism' or 'Islam'. Although superficially quite different, notions of geoeconomic competition and a clash of civilizations both betray a sense that the relative military dominance of the USA at the end of the Cold War may not in itself be enough to preserve America's sense of leadership, power and world-historical destiny in the context of a complex 'multipolar' world.

Indeed despite early post-Cold War proclamations about the great prospects for a more peaceful and secure New World Order, the world now appears to American strategists as more 'disordered' and unpredictable. 'Enemy regimes' (at various times, Afghanistan, Iran, Iraq, Libya, North Korea, Sudan and Yugoslavia) have been targeted as the United States continues to declare that certain regimes must be contained, punished or changed. At the same time, high-tech weapons spread, disseminating greater destructive capacity into areas of local confrontation and conflict, such as South and East Asia and the Middle East. Manufacturers in the West, China and the former communist countries have found plenty of lucrative markets for weapons of all kinds, from fighter planes to machine guns. Such exports very often end up in relatively impoverished societies, including many in Africa, the Balkans and Asia that are awash with the destructive debris of the Cold War. And in East Asia a number of disputes and differences rooted in the Cold War endure (see Case study 17.1). With other significant geopolitical divides (for example, the rival claims made by India and Pakistan over the territory of Kashmir, and the Arab–Israeli conflict) that were sometimes overshadowed by the Cold War but have endured beyond it, the reality and future prospects for larger scale wars and conflicts persists, and with them the spectres of nuclear or other (chemical or biological) 'weapons of mass destruction'. Commenting on this, Barkawi and Laffey (2002: 125-6) note how:

> The end of the Cold War arguably made nuclear war *more* likely, especially given that Soviet weapons, nuclear materials and technical personnel are far from being concentrated under imperial control and indeed may even be available for purchase on the open market. The buyers may well be non-state actors such as al Qaeda who, on the evidence of 11 September, would be far more willing to use nuclear weapons of mass destruction than the leadership of a state with a vulnerable homeland. If India and Pakistan, among other possibilities, indicate that inter-state and even nuclear war cannot be assigned to the dustbin of history, al Qaeda and the 'War on Terror' are indicative of new forms of international and globalised war ...

Moreover, for the Western powers and for the USA in particular, no longer is the main 'enemy headquarters' so clearly demarcated and fixed behind a stable 'Iron Curtain' – as the USSR and its communist allies were. Following the hijacking of airliners and their use in attacks on highly symbolic economic and military targets in the United States on 11 September 2001 ('9/11'), this has been felt even more deeply. Although there is no absence of traditional forms of geopolitical discourse, which define a clear boundary of 'our' space against 'them', some have argued that today Western strategy operates in the frame of a new kind of 'postmodern geopolitics'. For example, Gearóid Ó Tuathail (1998: 31–2) examined recent US foreign policy debates and notes how they argue that:

> In contrast to the transcendent containment imperative and fixed posture of Cold War strategy, what is required [according to US strategists] in response to persistent territorial concerns and proliferating deterritorialized threats is a geo-

strategic doctrine premised on flexibility and speed. The 1995 National Military Strategy of the United States is subtitled 'a strategy of flexible and selective containment'. ... Threats are described as widespread and uncertain, the possibility of conflicts probable, but their geographic sites are too often unpredictable. This document describes the current strategic landscape as characterized by four principal dangers that the US military must address: 'regional instability; the proliferation of weapons of mass destruction; transnational dangers such as drug trafficking and terrorism; and the dangers to democracy and reform in the former Soviet Union, Eastern Europe and elsewhere'. ... What is remarkable about these threats is that none has a fixed spatial location; regional instability refers not only to the Middle East but also to Europe and Africa; proliferation and transnational dangers are global; even dangers to reform, which is the only danger explicitly linked to certain places, are potentially ubiquitous, as the 'elsewhere' indicates. Military strategy still has to negotiate territory and place but it has also, in interesting ways, become untethered from place and territory. Anywhere on the globe is now a potential battlefield. The document concedes as much, noting that 'global interdependence and transparency coupled with our worldwide security interests, make it difficult to ignore troubling developments almost anywhere on earth'.

Others have argued that the level of connections and reconfiguring of power amount to a new form of global empire (Hardt and Negri 2000), fundamentally different from the imperialisms of the past (such as those described in Chapter 2). Whilst the global maps and scripts of geopolitical alliances (friends and foes) and actions have changed quite dramatically since the Cold War declined, there are also some important continuities and many legacies. The shocking attacks of 11 September 2001 have multiple causes and consequences. However amongst the causes are links back to the superpower confrontation of the Cold War. Thus as Halliday (2002: 36) notes:

> ... the rise of [Islamic] fundamentalist groups [such as *al-Qaida* which claimed responsibility for the 9/11 attacks] is not subsequent to, but an integral result of the Cold War itself.
>
> The Cold War indeed contributed to this crisis and in particular to the destruction of Afghanistan from 1978 onwards, but in a way that should give

comfort for few. One can here suggest a 'two dustbins theory' of Cold War legacy: if the Soviet system has left a mass of uncontrolled nuclear, chemical and biological weapons and unresolved ethnic problems, the West bequeathed a bevy of murderous gangs, from National Union for the Total Independence of Angola (UNITA) [which waged a violent struggle against the pro-Moscow regime in that African country in the 1970s and 1980s and later continued to wage war in Angola to control resources like diamonds] and Cuban exiles in the Caribbean and Miami to the *mujahidin* of Afghanistan, who are now on the rampage [both of which were earlier funded and armed by the United States and its allies].

Not all such non-state and anti-statist movements are of the right, however, and Routledge (2003) is among those who celebrate the prospects of a radical 'geopolitics from below' and a counter- or anti-geopolitics from those movements (such as the Zapatistas of Mexico and the *Movimento Sem Terra* or Landless Peasants Movement of Brazil and other indigenous and social movements critical of 'globalization') that seek progressive alternatives. According to Routledge (2003: 236) therefore:

… myriad alternative stories can be recounted which frame history from the perspective of those who have engaged in resistance to the state and the practices of geopolitics.

Yet if it began with the modern age of geopolitics, announced by Kjellén and Mackinder, it might be argued that the twentieth century ended less in a proliferation of anti-geopolitics and more as an age of 'postmodern geopolitics'. Gone is the apparent simplicity of the great power and Cold War confrontations, let alone demarcated and fixed 'heartlands', replaced by the sense of a world of proliferating uncertainty and 'threat'. Scares, dangers and anxieties about terrorism since '9/11' have greatly reinforced this. And there is no easy or predictable end to this situation.

At least with all this complexity, and the preceding points in mind, we might well argue that a critical examination of 'geopolitics' should not confine itself to the tradition that goes under that name (also see Tesfahuney 1998, Sharp 1999 and Nagel 2002). To do so would be to miss much wider issues about how broader sets of ideas about space, power, order, society and the international scene are combined into common sense notions about 'us' and 'them' and the whole geographical order of what is presented as the good life.

Learning outcomes

Having read this chapter, you should understand:

- The term 'geopolitics' has been associated with a wide variety of texts and contexts.

- However, geopolitics is frequently associated with the sense of a self-conscious tradition of writing about and mapping the relations between states, geography and power.

- This is usually traced to the late nineteenth-century writings of Rudolf Kjellén.

- One of the key features of this tradition is the idea that the state resembles a living organism which requires living space and will compete with others (the organic theory of the state).

- The oppressive military regimes in South America (notably Argentina, Brazil and Chile) during the 1960s and 1970s elaborated and applied a variety of the geopolitical tradition.

- However, it also proliferated in a number of European countries and Japan, particularly between the two World Wars.

- Since this proliferation of geopolitics was most evident in Fascist states, the defeat of the Axis powers in 1945 signified its relative decline as a formal tradition, although it continued in Fascist Spain and Portugal and in South American military circles.

- Beyond the self-conscious tradition of geopolitics, similar forms of thinking about territory, states and power have proliferated in many disciplines.

- This has been particularly evident during the Cold War, when broader geopolitical discourses suffused many aspects of art, science, culture and daily life.

- Since the decline of the Cold War, there has been a proliferation of geopolitical visions of the 'New World Order', amongst them a more complex cartography of perceptions of threats and dangers.

Further reading

Agnew, J. (2003) *Geopolitics: Revisioning World Politics*, 2nd edition, Routledge, London and New York. A historical and contemporary guide which places geopolitics in comparative historical perspective.

Agnew, J. and **Corbridge, S.** (1995) *Mastering Space. Hegemony, Territory and International Political Economy*, Routledge, London and New York. Part II (Chapters 5, 6 and 7) is useful on aspects of the 'New World Order'.

Agnew, J., Mitchell, K. and **Toal, G.** (eds) (2003) *A Companion to Political Geography*, Blackwell, Oxford. Includes some very useful accounts of geopolitical traditions, amongst other aspects of past and present political geographies.

Campbell, D. (1992) *Writing Security: United States Foreign Policy and the Politics of Identity*, University of Minnesota Press, Minneapolis. A challenging but rewarding account of how foreign policy and the Cold War have been productive of America's sense of self-identity.

Dodds, K. (2000) *Geopolitics in a Changing World*, Prentice Hall, Harlow. An accessible introduction.

Halliday, F. (2002) *Two Hours that Shook the World. September 11, 2001: Causes and Consequences*, Saqi Books, London. Amongst the most nuanced of the dozens of books that have appeared on this subject.

Heffernan, M. (1998) *The Meaning of Europe: Geography and Geopolitics*, Arnold, London. A readable account of European geopolitics and much more.

Hyndman, J. (2003) Beyond either/or: a feminist analysis of September 11th, *ACME: An International E-Journal for Critical Geographies*, **2**(1), 1–13 (freely downloadable from www.acme-journal.org). Indicates how feminist approaches might problematize and counter dominant geopolitical interpretations.

O'Loughlin, J. (ed.) (1994) *Dictionary of Geopolitics*, Greenwood Press, Westport, CT. A useful source.

Ó Tuathail, G. (1996) An anti-geopolitical eye: Maggie O'Kane in Bosnia, 1992–3, *Gender, Place and Culture*, **3**(2), 177–85. Shows how commitment and orientation can produce alternative narratives amounting to an 'anti-geopolitics'. Suggestive and worth reflection.

Ó Tuathail, G. and **Dalby, S.** (eds) (1998) *Rethinking Geopolitics*, Routledge, London and New York. Useful examples of how geopolitics is being critically rethought at a more advanced level.

Ó Tuathail, G., Dalby, S. and **Routledge, P.** (eds) (1998) *The Geopolitics Reader*, Routledge, London and New York. A collection of short texts (readings) on different aspects of geopolitical discourse, including extracts from Mackinder, Haushofer and Kissinger and debates about the New World Order. Good clear editorial introductions.

For annotated, clickable weblinks and useful tutorials full of practical advice on how to improve your study skills, visit this book's website at www.pearsoned.co.uk/daniels

CHAPTER 18

Territory, space and society

DAVID STOREY

Topics covered

- Territorial strategies
- Social processes and spatial relations
- Contested spaces
- Geographies of resistance

Human spatial relations are the results of influence and power. Territoriality is the primary spatial form power takes.

(Sack 1986: 26)

18.1 Territoriality

In his important book *Human Territoriality. Its Theory and History*, published in 1986, Robert Sack rejects determinist views of human territoriality as a basic instinct and argues instead that it is a geographic and political strategy. He draws attention to the means through which territorial strategies may be used to achieve particular ends. In essence the control of geographic space can be used to assert or to maintain power, or, importantly, to resist the power of a dominant group. Territoriality, he suggests, is 'a device to create and maintain much of the geographic context through which we experience the world and give it meaning' (Sack 1986: 219). In this way territoriality, which is deeply embedded in social relations, can be viewed as a process linking space and society.

Sack (1986) argues that territoriality involves a classification by area whereby geographic space is apportioned. This spatial division needs to be communicated through various means, most notably through the creation and maintenance of boundaries. Through these mechanisms territories are created. Territoriality also tends to reify power so that it appears to reside in the territory itself rather than in those who control it. In this way attention is deflected away from the power relationships underpinning the maintenance of territories and their boundaries. Ultimately territoriality can be seen as 'a primary geographic expression of social power' (Sack 1986: 5).

It follows from this that territories themselves are social constructions rather than natural entities. They are created, contested, modified and destroyed (Paasi 2003). They are produced under particular conditions and serve specific ends. Once created, they can become the spatial containers in which people are socialized through various social practices and discourses. As Paasi (2003) suggests, a number of important dimensions of social life and social power are brought together in territory. There is a material component such as land, there is a functional element associated with control or attempts to control that space and there is also a symbolic component associated with people's social identity. People identify with territories in such a way that they can be seen 'to satisfy both the material requirements of life and the emotional requirements of belonging' (Penrose 2002: 282).

The creation of territories and the utilization of territorial strategies can be observed at a variety of spatial scales. While some of the most discussed expressions of territoriality are manifested at the level of the state (which is the focus of Chapter 19), many more micro-level examples of territorial control and territorial strategies may be observed. This chapter explores some of these examples, demonstrating how particular social practices are mapped onto space and therefore mark and configure territories.

There are five key foci here: class-based divisions, racialized space, gendered spaces, sexuality and space, and personal space. While the various topics discussed in this chapter are examined in relatively discrete sections, it should be abundantly clear that many of the issues raised are interrelated (and also overlap with earlier chapters in the book). People have more than one single identity: gender, sexuality and ethnicity cross-cut each other in a system of overlapping identities. Ethnic groups and classes are not immutable, they are social constructs. In other words, they vary in time and space and are not simply natural categories, but human (that is social) products. As such we need to be aware of the dangers of seeing them as rigidly defined.

18.2 Class-based divisions

As other chapters have pointed out, geographers and others involved in the study of urban areas have long drawn attention to spatial divisions in cities, linked to residential patterns, economic activities and land use (see Chapters 9 and 11). Most cities have distinct residential neighbourhoods, colloquially defined as 'rich' or 'poor', 'working class' or 'middle class'. Socio-economic differentiation may well be the most important cleavage within the urban landscape (Knox and Pinch 2000). Segregation along class lines is effected through various mechanisms such as the housing market, effectively determining who can afford to live where. This appears at its most formalized in US cities where the process of municipal incorporation means that better-off territorially defined urban areas can effectively secede from the larger city of which they are a part. In doing so they enjoy a degree of fiscal autonomy which means that residents do not have to support services, such as public transport, for poorer areas outside their own municipality (Johnston 1984, Knox and Pinch 2000). In this way, through a combination of land use zoning and territorially based funding of services, effective barriers prevent those deemed 'undesirable' from moving in. Economic apartheid results as an area opts out of the broader

Plate 18.1   Closed space: 'Gated' residential development, Worcester, UK.

urban environment and so residents shed themselves of any sense of collective responsibility for the poor. The incorporated municipality can decide on certain local regulations such as excluding industrial developments. It also has the power to enforce minimum lot sizes and prohibit such things as mobile homes, and so on. In this way it can effectively exclude poorer residents, thereby insulating the relatively affluent residents from potential declines in property prices. This is a clear example of the political manipulation of space with power mediated through a territorial process which may have serious racial as well as class connotations (see next section).

Other more subtle, but perhaps equally effective, processes contribute to territorially-based residential segregation (see Chapter 9, pp. 220–4, and Chapter 11, pp. 260–3). Gentrification is where parts of the urban area experience regeneration or renewal resulting in more affluent residents moving in and displacing the original predominantly working-class inhabitants. Driven in part by economic considerations and in part through consumer choice, it serves to reinforce economic divisions within society and thereby perpetuate the idea that some households do not belong in particular places (see N. Smith 1996). As Short (1989) suggests, the built environment reflects

the needs or perceived needs of different household types and social categories. It is also argued that the role of 'urban gatekeepers' (such as estate agents) may play a key role here in altering (or endeavouring to maintain) the social composition of particular areas (Knox and Pinch 2000). Gentrification reflects broader socio-economic processes and the resultant residential territorialization can be seen as an expression of the financial power of home-owners and the power of finance capital. In this way places like London's Docklands are transformed from manufacturing and working-class residential areas into service sector (particularly financial services) zones with a resident middle-class population. This, as Short (1989) points out, reflects more than a simple change in land use, it also reflects changes in the meaning of place. Docklands, like other 'regenerated' urban zones, has been transformed into a different place, with quite a different symbolic meaning.

The idea of the gated community is becoming increasingly evident, particularly in the United States. Davis (1990) drew attention to what he termed 'Fortress LA' where security guards patrol the perimeter of walled residential zones in an effort to exclude what are seen as 'undesirables'. This phenomenon is repeated in many north American cities (and increasingly so elsewhere) with private security firms patrolling more affluent urban areas in order to exclude those seen as undesirable, thereby maintaining the 'undefiled' nature of the neighbourhood. The increasing prevalence of apartment blocks and other new housing developments surrounded by security fences further illustrates this trend. More broadly M. Davis (1995) sees this as the destruction of democratic urban space and its replacement with privatized exclusionary space. Added to 'secure' shopping centres, office blocks and apartment buildings complete with gates and intercom systems, this exemplifies a trend towards socio-spatial design whereby territorial strategies associated with crime prevention effectively exclude those not wanted (Plate 18.1). City residents are effectively barred from parts of the city – a policy of territorial containment enforced through increased surveillance and architectural design features (Cozens *et al.* 1999).

These territorial strategies work in ways which ensure a particular residential mix and may well serve to link together both racial and class divisions. Conversely, many working-class housing estates, seen as being inhabited by an underclass, are often seen as unsafe 'no-go' areas. Such stigmatizing of place in itself becomes part of the problem, serving to reinforce class divisions and to reproduce various forms of social exclusion. Effectively this results in a form of ghettoization where the residents of particular areas become stigmatized with a whole series of negative consequences (such as employability) flowing directly from this.

The apparent consignment of poorer residents to certain areas can also of course provide the spatial framework for forms of resistance. It may facilitate the election (at local level at least) of political representatives for residents of those areas. Thus they are given a voice that might otherwise be denied them. It may facilitate the mobilization of people in support of, or in opposition to, issues of direct concern to them. In this way a territorial strategy can be utilized in order to defend the interests of those who, if more spatially scattered, would be unable to do so. The extent to which politicians elected in certain areas can be said to be representative of those who live there is another issue.

18.3 Racialized spaces

Just as class is mapped onto the urban fabric, so too is ethnicity (see Chapter 11, pp. 257–60). In this way we can speak of racialized spaces. Race and ethnic categories are, however, social constructions rather than simple biological realities. While race can be questioned as a problematic and often dubious form of social classification derived from past notions of hierarchy and domination (notably slavery and colonialism), there is no doubt that racism or 'race thinking' is a very real social phenomenon. Jackson (1989: 132–3) defines racism as

> the assumption, consciously or unconsciously held, that people can be divided into a distinct number of discrete 'races' according to physical, biological criteria and that systematic social differences automatically and inevitably follow the same lines of physical differentiation.

Although there may be little basis for a racial categorization of people, that does not prevent some people from behaving as though there was. While issues linked to 'race' are clearly social phenomena, they are often manifested spatially. The classic example of racialized space was that devised under the apartheid system in South Africa: a territorial system that enhanced and entrenched political, economic and social power of a minority white population over non-white populations. Both nationally and at the more localized level of individual urban areas, space was divided on racial lines (Figure 18.1). Although apartheid no longer officially exists, its legacy in South Africa means that a division of space into black and white 'territories' endures based on the racial and class lines reinforced during the apartheid era and which will take many years to dissolve.

Non-white people were 'placed' in locations not of their own choosing in order to enhance minority white power. In this way, there was a legal transposition of inequality onto geographical space. This spatial arrangement was designed to ensure greater degrees of control over the black population and is a classic example of the utilization of a territorial strategy to attain political objectives. At very localized levels, there was a racialization of space with buses, public toilets and other amenities reflecting this divide. A racial ideology was mapped onto the South African landscape. Such extreme racializations have by no means been confined to South Africa, however. More recently, controversy surrounded the construction of a wall (since demolished) around an area of housing occupied by gypsies in the town of Usti nad Lebem in the Czech Republic in the late 1990s. The generally negative stereotyping of gypsies in much of Europe, particularly in the east, has led to considerable discrimination with gypsies seen as an undesirable 'Other', as a consequence of which they are effectively de-territorialized; they are seen not to belong anywhere (Fonseca 1996).

Elsewhere, overt racial discrimination may be illegal but black people and other non-white ethnicities may well experience exclusion through the implementation of regulations excluding the poor (referred to in the previous section) who, in most US cities, for example, are disproportionately black or Hispanic. A variety of factors serve to support this ideology: fear of the unknown, a deeply felt hostility to the 'other', to those seen as different, economic fears associated with the belief that ethnic minorities are taking jobs, and so on. Either way, such patterns of exclusion and inclusion and attendant territorialities reflect the complex intersections of race, class and ideology.

In considering the evident spatial concentrations of ethnic groups in urban areas, it might be argued that individuals choose to locate in such areas for a variety of reasons. In brief there are a combination of 'positive' and 'negative' factors; for some there are

attractions such as 'being amongst one's own', while others may feel driven to seek sanctuary from a racist, hostile society. Following Knox and Pinch (2000), key reasons may be summarized as follows. Clustering affords *defence* against attack by the majority group. It also provides a degree of *mutual support* and bolsters a sense of belonging and community. This in turn may be a useful means through which group cultural norms and heritage may be preserved. Finally, clustering produces *spaces of resistance* whereby external threats, whether to cultural norms or of physical attack, may be reduced.

None of the above should detract from the fact that such residential clustering may be more a function of necessity rather than free and unconstrained choice. It should be borne in mind that the degree of choice available to many members of ethnic minority groups may be extremely limited. The idea that people may choose to cluster is to ignore the fact that quite often no viable alternatives are available. Discriminatory ideologies of race work to exclude people from particular milieus thereby

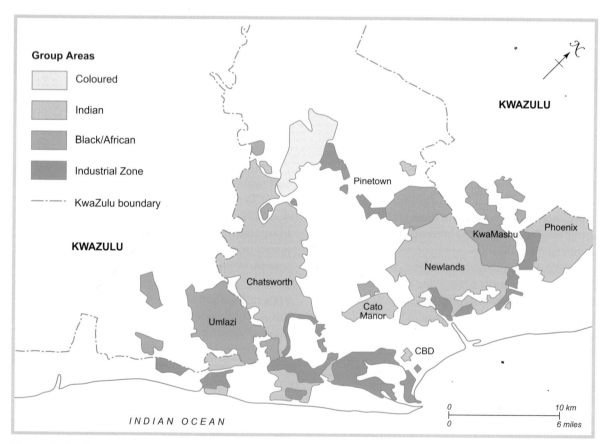

Figure 18.1 Racialised space: group areas, Durban, South Africa during apartheid.
Source: Smith (1990)

translating social exclusion into geographical exclusion. Thus as Wacquant (1995) argues, 'ghetto' areas are institutionally produced; they are not simply the product of individual choice.

It should also be borne in mind that such clustering itself further contributes to future rounds of marginalization and exclusion so that 'what was a spatial reflection of economic and social marginality becomes a spatial constraint on economic advance and social mobility' (S.J. Smith 1999: 18). This further demonstrates the idea that territorial practices serve to reproduce particular social outcomes in a complex relationship whereby society does not simply impact on space, but spatial arrangements in turn impact on society.

In considering so-called ghetto areas, there is a tendency to see them in quite negative terms. The ghetto is a territorial entity and it is one that evokes many negative connotations; the term is often seen as synonymous with 'slum', the juxtaposition leading to a stigmatizing of its residents. In such ways the hegemony of the dominant group is maintained and the 'other' (in this case, ethnic minorities) remains geographically, as well as socially and economically, marginalized. In Western cities, 'ghettos' and other 'inner city' areas are often seen as the home of an 'underclass' locked into a cycle of deprivation and disadvantage. Places such as south central Los Angles, for example are represented as dangerous areas to avoid (or as dominated by violent street gangs: see Case study 18.1). In this way the 'ghetto' is portrayed *as a problem* rather than a place whose residents *experience* problems. It is also depicted in a somewhat monolithic way that ignores its internal social, economic and cultural diversity.

Areas with high concentrations of ethnic minority groups can have much more positive connotations – areas like the so-called 'Banglatown', centred on Brick Lane in London, which is predominantly inhabited by Bengalis. As an area with a long history of Jewish and Irish immigration it can be seen as epitomizing a certain cosmopolitanism, which can act in a very positive way for local residents (Dwyer 1999). As a multi-cultural space, it offers the potential for people of diverse ethnic backgrounds to learn more about each other and to mix freely. (Whether this actually happens is another matter.) It also becomes commercialized with the area now being known for its Indian restaurants and featuring in tourist guide books of London. The area also hosts an annual festival celebrating cultural diversity (see

www.bricklanefestival.com/). However, this profile can also lead to a targeting of such areas by racist groups. In 1999 a nail-bomb attack was carried out in Brick Lane, an apparent assault on the area's non-white ethnic groups. Thus, the promoting of a multi-cultural or 'diasporic space' (Dwyer 1999) is met with resistance from those who wish to see such spaces remain ethnically 'pure'.

Such areas can be seen as territorial manifestations of social inequality reflecting the uneven distribution of power in society. Of course, like any other territorial entity, the ghetto can be used as a means of mobilizing residents, of providing the territorial frame within which people can operate with a view to improving their own conditions. Successful residents' groups often emerge and may become sufficiently well organized to be able to engage in lobbying, in forms of self-help, etc. Such forms of community action may well make a positive difference to the lives of ghetto residents. In this way, it can be argued that the ghetto, just as with any other territorial formation, can take on its own sense of identity and can become a mechanism for the expression of group interests. For example, Bozzoli (1999) shows how a South African township in the 1980s, a space designed and built by the dominant power to house subordinate groups, became transformed into a space of resistance. Spatial arrangements engender specific social relations as a consequence of human interaction. Spaces can take on different meanings from those originally intended: 'space develops over time a "hidden transcript" of its meaning to those who inhabit it, different from the "public transcript" of its meaning for those who rule' (Bozzoli 1999: 6). Residents of Alexandra township in Johannesburg worked to make the place ungovernable through a combination of tactics including an action committee and through symbolic territorial actions such as patrolling the streets and the actual renaming of streets to reflect their struggle (for example, one street was renamed ANC Street). The space that was Alexandra was being reshaped into something different from that originally intended. It became a space of resistance rather than merely a space of oppression.

Sometimes the places in which the poor and oppressed live start off as bounded and prison-like stalags. But there are times when rebellious inhabitants seek to transform the stalag into a space of their own upon which their meanings are imprinted and whose boundaries become the

Thematic Case Study 18.1

Territoriality on the streets

The activities of urban street gangs have received considerable media attention in recent years, particularly in the US. These gangs engage in occasionally quite violent conflict underpinned by a strong territorial component. Their *raison d'être* is the assertion of control over their 'turf'. Streets are contested spaces. Rival gangs and the police are not welcome on their patch. This territorial behaviour might be seen as reflecting some sort of innate territoriality, but it can also be interpreted as a consequence of the marginalization of many poorer young people in impoverished urban areas. In places such as south central Los Angeles, a combination of high unemployment, deprivation and political powerlessness creates a situation where territoriality may be a means of expressing power using the only resource available to them, the streets and neighbourhoods in which they live. These gangs may be linked to criminal behaviour and may be involved in controlling illegal activities in their patch, mirroring the behaviour of 'older' criminal gangs who also display a territorially based organizational structure. These phenomena contribute to a stigmatizing of place in which areas are seen as dangerous and to be avoided. Quite often the problems experienced by residents are seen to be ethnic or racial in origin rather than resulting from structural features. This further ghettoizes the area and its residents, cutting them off still further from the rest of society.

Street gangs tend to lay down territorial markers to indicate to others their 'ownership' of particular places. Graffiti on walls, bridges and buildings is one very visual method of claiming space. Markers are quite literally placed on the landscape to signal control of territory or 'turf ownership'. Work in Philadelphia indicated that graffiti became denser closer to the core of that gang's territory. Ley and Cybriwsky (1974) were able to demarcate reasonably accurately the spatial extent of gang control in the city. In this way, aspects of popular culture are translated into a territorial frame. The claiming of space is what is important. It may be a means by which marginalized youth make their claim to existence; through planting their mark on territory, that territory becomes theirs (see Scott and Brown 1993). Rather than pathologizing gang members, there is a need to look at the broader processes occurring in such areas in an attempt to understand these sorts of territorial conflicts.

---

defiant barricades which keep the authorities out, rather than the symbolic walls which keep the persecuted in.

(Bozzoli 1999: 40)

We should also be mindful of the fact that much discussion surrounding issues of 'race' and ethnicity tends to assume 'whiteness' as the norm. As McGuinness (2000) argues, much progressive research itself falls into this trap with a focus on non-white groups tending to deflect attention away from white ethnicity. One consequence of the pursuit of this 'new exoticism', as McGuinness terms it, is that relatively little attention is given to 'white spaces'. Ideas of white flight to the suburbs (in response to the evolution of 'black' ghettoes) and the creation of 'white' territories are themselves elements in the racialization of space. Similarly the construction of rural Britain as a 'white

space' reflects deeply embedded ideas associated with belonging and with national identity. Such constructions can have serious implications for those who are non-whites. These sorts of racialized spaces require further investigation in order to broaden and deepen our understanding of the connections between geography and ethnicity (Bonnett and Nayak 2003).

This section has demonstrated the ways in which racist and exclusionary ideologies are transposed onto space. It has also indicated ways in which those racist constructions are opposed. Just as particular power relations are refracted through a territorial frame, so those relations are contested through territorial strategies. The spaces to which people are consigned may provide the means through which they contest their marginalization.

A territorial base may serve as a means through which an ethnic identity or a class identity is reinforced and reshaped, in part at least in opposition to other identities. Other identities, such as religious affiliation may also provide the basis for parallel territorialities. Of course we need to be mindful that such identities may be deeply contested and are far from monolithic, though there may often be attempts through community representatives to portray them as such (Yuval-Davis 2003). Case study 18.2 provides an example of segregation based on religious affiliation.

18.4 Gendered spaces and the public–private divide

The growth of what became known as the 'feminist movement' from the 1960s onwards, building on earlier attempts to achieve equality for women, has been influential in gaining recognition in many societies for the unequal status of women and men in all dimensions of life; in the home, in (paid) workplaces, in the broader political and social arena. Feminist writers and activists have been instrumental in attempting to explain how patriarchal systems of power have tended to reinforce male dominance and how women have often been marginalized. Feminist geographers have highlighted the 'geography' of discrimination against women, particularly in drawing attention to the manner in which space and place are heavily gendered. Critical attention has been focused on divisions between the public and private domain and its spatial corollary of a division between what is seen as public space and private space. It is argued that patriarchal systems of power have led to a division between 'male' public and 'female' private space, resulting in social practices whereby certain activities and certain spaces are seen as male preserves. Issues of gender are mapped onto space in various ways. In its most simple form this is reflected in the idea that 'a woman's place is in the home'. The home has tended to be seen as a space of reproduction juxtaposed to the workplace as a space of production (Laurie *et al.* 1999). The implications of gender are seen to be as important as other political, social and economic factors in the structuring of spaces and places.

Underpinning this are ideas that distinguish between sex as a biological fact and sex as gender, which refers to the socially constructed roles of both male and female identities (Women and Geography Study Group 1984). In emphasizing the role of social conditioning, the argument is that as individuals we are not biologically predetermined to be more suited to some roles rather than to others.

One reason for the relative absence of women in particular places is overt discrimination or active discouragement in the sense of certain activities or pursuits not being deemed suitable for women (see Case study 18.3). Historically, women who transgressed these boundaries were often portrayed in a negative light, an idea reflective of notions of 'good' and 'bad' women. Women out alone at night might be seen as not conforming to what is expected of them. A crucial aspect of the relationship between women and place centres on the perception of some specific places as 'unsafe'. Many women do not feel safe in certain public places, most notably darkened streets. As Valentine (1989) investigates, women transfer a fear of male violence into a fear of certain spaces. This has profound implications for the ways in which men and women negotiate their way through urban spaces (Fell 1991). This is also contested. Clearly, the various strands of feminist thought and practice have resulted in significant advances with regard to equal rights for women. While this can be seen within the arena of legislation in many countries associated with equal pay and related issues, it is also reflected in terms of spaces. Thus, the heightened visibility of women in public space reflects the changing status of women. Phenomena such as 'reclaim the night' marches demonstrate the overt use of a spatial strategy to make a political and human point. While particular groups may find themselves excluded from certain spaces, those spaces can also be reclaimed (as an example see www.srcf.ucam.org/rtn/intro.html).

Where women enter the workforce, they may still encounter territorial divisions in the workplace. Thus, Spain (1992) alludes to the 'closed door' jobs of managers (mainly men) and the 'open floor' jobs of manual workers (who may be predominantly women in certain countries/regions/sectors). Many employers may even locate in particular localities (or countries) in order to take advantage of what they see as an appropriate workforce based on prevailing wage levels or skills and assumptions about gender roles. Hanson and Pratt (1995), for example, show how some firms located in particular parts of Worcester, Massachusetts in order to avail of what they saw as desirable characteristics based on stereotypical assumptions associated with gender and ethnicity. The latter related to jobs that the employer saw as suitable for women such as sewing, knitting etc. All of this implies that spatial or territorial strategies can be used in order to retain social control over women. Social processes reproduce attitudes that tend to 'naturalize' a gendered

Regional Case Study 18.2

Religious segregation in Northern Ireland

Linked to the idea of racialized spatial divisions, religious differences can also result in territorial separation within urban areas. One of the best known examples is that which occurs in Northern Ireland (see Case study 3.1). In urban areas in Northern Ireland, there are Catholic residential districts and Protestant districts, a reflection of the region's long-standing political problems and conflict between unionists (predominantly Protestant) who wish the region to remain part of the United Kingdom and Irish republicans (predominantly Catholic) whose aim is a united Ireland. This political divide is reflected in the social geography of Belfast with high levels of residential segregation along religious lines. However, it should be borne in mind that this reading of the conflict can be slightly misleading; not all Catholics are nationalist and similarly not all

Protestants are unionist. Nevertheless, it is apparent that there is considerable religious segregation even if this cannot, and should not, be seen to correspond completely to people's wider political beliefs. Once again, this spatial segregation highlights the ways in which political relationships are mapped onto space.

Religious segregation has been reinforced through population movements, both voluntary and forced. Catholic families have periodically been burned out of their homes in areas in which Protestants formed an overall majority. Equally, perceived influxes of the 'other' community have occasionally led to a drift away from certain areas by the 'other side'. This further reinforces pre-existing divisions, with housing estates becoming predominantly (and in some case exclusively) Catholic or Protestant. As with other forms of ghettoization, this can be seen to be partly through choice and partly through force of circumstance.

In Belfast, where the so-called Shankill–Falls divide separates the two groups in the west of the city,

Plate 18.2   Divided space: 'Peace' wall, Belfast.

Plate 18.3 Claiming space: Irish republican wall mural, Belfast.

a very high degree of segregation occurs, highlighted by the inappropriately named 'peace line' – a wall literally dividing roads in the area and designed to prevent confrontations (Plate 18.3). Parts of the city are seen as out-of-bounds to one side because it is the other's territory. Once again, this highlights the terri-

torial nature of political conflict and emphasizes the manner in which territory very easily comes to be seen in terms of 'ours' and 'theirs'. The territoriality of Belfast is such that a person's address is quite likely to reflect their place in a religious as well as a merely geographical sense.

Regional Case Study 18.2 continued

Plate 18.4 Claiming space: Ulster loyalist wall mural, Belfast.

Just as international borders have flags and other territorial markers to indicate their location, so also these segregated residential areas come complete with their own sets of boundary signifiers. Red, white and blue kerbstones indicate loyalist areas, while Irish flags and a variety of republican wall murals are highly visible symbols in nationalist areas (Plates 18.3 and 18.4). Similarly, graffiti orders 'Prods keep out', 'no pope here', 'Ulster is British' or 'tiocfaidh ar lá' ('our day will come', a common republican slogan). These markers serve both to reassure residents and to send out a clear message to the 'other side' that their presence is not welcome. For people who find themselves living in – or even passing through – the 'wrong' area, the consequences can be fatal. These areas are quite often patrolled, to some extent controlled, by members of paramilitary organizations. Certainly in the past the Provisional IRA has seen itself as a defender of Catholic areas of Belfast while organizations such as the UDA (Ulster Defence Association) and UVF (Ulster Volunteer Force) have portrayed themselves in a similar light in relation to Protestant areas. More recently in-fighting between different loyalist paramilitary groups has manifested itself as a 'turf war' within Protestant areas of Belfast. Once again territoriality allows a mobilizing process to take place. Support is garnered for both political parties and paramilitary organizations.

For more information on the Northern Ireland conflict see http://cain.ulst.ac.uk/. See also *Political Geography* **17**(2), 1998 (special issue: Space, place and politics in Northern Ireland)

division of labour in which women perform certain functions which are acted out in specific spaces, e.g. 'home-making' and child-rearing in domestic space. Behaviour is thus not just gendered but also spatialized.

Historically, this division of labour has tended to confine women to the private realm, leaving men to inhabit (much of) the public domain. This view of women as playing a subordinate role has in the past been reflected in discriminatory attitudes and practices, particularly in relation to women in the paid workforce with active discouragement through lower wages, if not actual exclusion, from many jobs. These views are predicated on the undesirability of women going out to work. Massey (1994) has argued that opposition to women going out to work was greater than opposition to women working *per se*. It can be argued that this ascribing of women's role, through delimiting the spaces in which women were encouraged to appear, is another spatial expression of power. In other words, the confining of women to domestic space, and their exclusion from male territories, was the key element in male control. McDowell (1983) has referred to urban areas as containing masculine centres of production and female centres of reproduction. With increasing female participation in the workforce and a raft of anti-discriminatory legislation in many countries, such a generalization may appear to have lost some of its validity. Nevertheless, the division between a (largely)

male public sphere and a (largely) female private sphere still has considerable resonance in many societies (although the extent of this is itself immensely geographically variable across the world).

If women have sometimes been seen as 'belonging' in or closely associated with the home, it is not the case that the space of the home is undifferentiated. Even within the home, territorial divisions take place, the most obvious being the notion of the kitchen as a female preserve. The relative neglect of such territorial behaviour in all its manifestations is commented on by Sibley:

> In geography, interest in residential patterns wanes at the garden gate, as if the private province of the home, as distinct from the larger public spaces constituting residential areas, were beyond the scope of a subject concerned with maps of places.
>
> (Sibley 1995: 92)

There has, however, been a growing interest in the domestic sphere. Daphne Spain (1992) provides many examples of sex segregation in the home from different cultural contexts and from different time periods. Groups such as the Berbers in Morocco or the Bedouin have often had a clear spatial division between men's and women's spaces (Figure 18.2). Similarly, the Jivaro indians in South America live in tents that have separate entrances for men and women. Shurmer-Smith and Hannam (1994) illustrate a similar point through reference to the Yoruba in Nigeria. Here they

Thematic Case Study 18.3

The gendered spaces of football

The geographer Doreen Massey (1994) recounts how, as a teenager, she used to get a bus into central Manchester on Saturdays and how she was struck by the fact that football pitches were spaces inhabited totally by males. In this way, she argues, there is a clear gendering of space whereby some spaces come to be seen as the preserve of one or other gender. To take Massey's example one step further, it can be argued that professional football stadia were, at least until quite recently, almost exclusively male preserves. With a resurgence of interest in football in the 1990s, this

phenomenon has altered slightly with an increase in the numbers of female supporters and clubs encouraging family attendance. The relative growth in popularity of women's football is another dimension to this. Nevertheless, this division reflects ideas about separate leisure activities and, hence, separate spaces for women.

Viewed from a feminist geographic perspective, such social practices are built upon ideas of what is or is not acceptable behaviour for women to engage in (built on socially or culturally constructed notions of masculinity and femininity); in other words, the everyday reproduction of patriarchy.

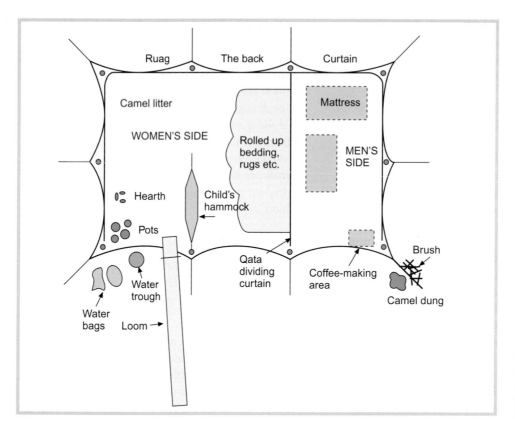

Figure 18.2
Gendered space:
Bedouin tent.
Source: Spain
(1992)

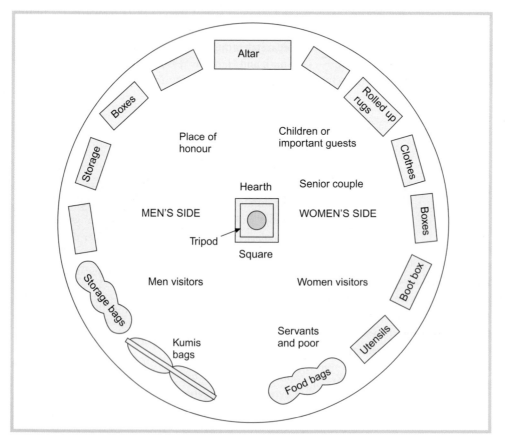

Figure 18.3
Gendered space:
Mongolian ger.
Source: Spain
(1992)

point to the non-existence of designated spaces for unmarried women or for children (Figure 18.3). Within Western societies too, deep-seated ideas about woman's role as home-maker, cook, cleaner, child-rearer, etc., meaning that women have often been presumed to 'belong' in some rooms more than others have often been challenged; yet they frequently endure.

All these reflect territorial expressions of power whereby the designation or apportionment of space within the domestic sphere reflects the relative status of the individuals concerned. Spain therefore reminds us that 'houses are shaped not just by materials and tools, but by ideas, values and norms' (1992: 111).

18.5 Sexuality and space

The idea of places territorialized by particular groups on the basis of sexual orientation has been widely discussed. At its most elementary level this has seen the mapping of gay and lesbian 'zones' in selected cities. It is fair to point out that such spaces are not as easy to identify as, say, areas inhabited predominantly by a particular ethnic group. It is equally obvious that, in the main, these are not strictly demarcated areas. Rather, they are zones where gay people may feel more at ease through being accepted rather than rejected, scorned or ignored (or worse) by their neighbours. There have been criticisms of such straightforward mapping of 'gay territories' in that it may be interpreted as a mapping of what some see as deviant behaviour, that it essentializes sexuality and that it reinforces a gay/straight dichotomy (Knopp 1995, T. Davis 1995). Nevertheless, the fact that gay and lesbian people do, in some instances, cluster in particular places suggests that territorial behaviour may be important. Within these areas, gay-owned businesses, bars, restaurants, and a general concentration of gay residents allows the creation of what Castells calls 'a space of freedom' (1997: 213). Areas such as the Castro district of San Francisco (Plate 18.5) and the more spatially confined 'gay village' in Manchester serve as important examples of this trend. The construction of such zones may arise for reasons similar to those associated with ghettos and other forms of segregated space. Castells (1997) has argued that there are two key factors: protection and visibility. The first of these is fairly obvious. Gays or lesbians will feel more comfortable and less vulnerable to attack when surrounded by other gays or lesbians. The idea of 'strength in numbers' may make people feel safer from

homophobic 'gaybashers'. The second reason, that of visibility, relates to the need for those who identify themselves as gay or lesbian to assert an identity within a culture that is predominantly straight (heterosexual) and in which a straight discourse dominates, a society in which other sexualities are often still seen by many as deviant or abnormal behaviours. Gay neighbourhoods are a means of asserting other sexual identities, a means by which those who identify as gays, lesbians or bisexual can proclaim 'we are here'. Harry Britt, a one-time key figure amongst San Francisco's gay community, once commented that 'when gays are spatially scattered, they are not gay, because they are invisible' (cited in Castells 1997: 213).

Rothenburg (1995) examined the development of lesbian spaces in cities where relatively high concentrations begin to drift into gentrified housing. In this way the place becomes associated with the creation of an identity. As a consequence of the clustering of people sharing that particular identity, the place becomes transformed. Hence a territorial identity emerges which acts as a powerful mechanism in transmitting a sense of the group and in making it easier for gay or lesbian people to feel comfortable with their sexuality.

The significance of San Francisco's gay area was reflected in that community's ability to gain political representation. In obtaining power over territory they also gained political representation and San Francisco has become recognized as something of a 'gay capital' of the United States with a somewhat more liberal attitude exemplified through the hosting of a public annual 'gay wedding day' involving hundreds of couples in a commitment ceremony. In this way, the designation of 'gay territories' plays a crucial role in raising awareness of gay people and issues and also provides a means by which some degree of power and self-confidence can be attained. Thus celebratory events such as 'gay pride' marches can appear as a staking of claim to territory.

The examples of gay and lesbian spaces suggest another important point, that of the temporality of territory. The longevity of these spaces may be quite brief as the 'scene' moves to somewhere else. Valentine (1995) has drawn attention to the fact that lesbian spaces may be very short-lived in time, whether due to the transience of lesbian bars/clubs or the even more short-term phenomena of lesbian or gay evenings.

Of course such places and such events may often be 'co-opted' to present an 'acceptable' image of the group concerned. The evolution of 'gay spaces' has often been associated with an economic imperative. The importance of the 'pink pound (or dollar)' in aiding urban regeneration has frequently added a strong commercial angle to these developments. Some activists have expressed disquiet over the appropriation of such events and their dislocation from their original social and cultural roots and from their original territorial base. The fact that Manchester's 'gay village' and San Francisco's Castro district are firmly on the tourist trails of their respective cities may be lauded as an acceptance of identities previously scorned but it can also be seen as a commercialization of that identity which may, to some extent, serve to further ghettoize it. It might be argued that it is acceptable to be openly gay in an area so designated but in other spaces and societies the pressure to keep this identity hidden may well persist (see www.sfguide.com/sights/neighborhoods/castro.htm). Here of course there are profound links between the small (often neighbourhood or urban) scale territorialities and the wider policies of the city or state concerned. Many states still criminalize gay sexuality and almost

everywhere sexual mores and regulations are highly territorialized.

18.6 Personal territories

At its most elementary level, the assertion of territoriality is reflected in claims to private property. Thus, people desire to own their own home, to adorn it in their chosen style and, in various ways, to mark it out as theirs. The geographer Jean Gottman suggests that 'civilised people . . . have always partitioned the space around them carefully to set themselves apart from their neighbours' (1973: 1). This manifestation is commonly interpreted as being symptomatic of our inherently territorial nature. Private property is regarded by many as a logical and necessary outcome of human territorial behaviour. It represents a claim to space that is reinforced by the legal system of many countries. However, as Alland points out, it might well be the case that 'private property is the child of culture and develops into a major preoccupation only with the evolution of complex society' (1972: 64). It follows that we need to be careful to avoid the trap of translating a need for personal space into an ideological claim for the sanctity of private property.

Even within buildings territorial behaviour can be recognized. As we have seen, the idea of the kitchen as a 'woman's place' is one example of this. Spain (1992) has collated details of how the domestic home in many different cultural contexts is often spatially divided, not just in terms of gender but also in terms of age, with certain spaces being designated for women or for children. The banning of children from some rooms and the proprietorial attitude towards one's own room in a house are other examples of this. In the home, space is even being claimed at the level of 'my chair', 'my place at the table' and so on.

Equally, in workplaces some areas and rooms can only be entered by staff of a certain level. Some zones are out of bounds to more junior staff. These can be interpreted as managerial strategies designed to ensure a particular outcome; staff know their 'place' and can be more effectively controlled. Hanson and Pratt (1995) reveal how companies reproduce social segregation through spatial practices within the workplace whereby different sets of workers inhabit different parts of the factory and rarely, if ever, meet. Thus, office staff may be located downstairs in 'cubicles' separated by room dividers, with sales staff and management upstairs in individual or shared

Plate 18.5   Gay space: Rainbow flag flies in the Castro District, San Francisco, USA.

offices while production staff are located in an entirely separate part of the building. Socializing between workers tends to reflect their spatial segregation even to the extent of each department having separate annual parties. Work hierarchies are reflected in the spatial arrangements of the workplace. These practices have clear outcomes. They may render it difficult for workers to organize through physically keeping them separate and through engendering a sense of difference between different sections of the workforce.

At a more elementary level, there are echoes of this in how people treat their workspace, whether it be an office, a workstation within an office, locker room, garden shed or whatever. In part of course this relates to notions of comfort and familiarity and the desire to protect our own personal space. The psychological need for space and privacy does not, however, take from the fact that what is happening is an attempt to wield power through a territorial mechanism.

Taking the idea of territory down to its most elementary level, the desire for personal space can be seen as a form of territorial behaviour. Humans like to have a pocket of space around them that is 'theirs' and they resent others 'invading' their space. This can be interpreted as a territorial claim to a portion of geographic space. While this might be taken as reflecting a natural tendency, it is worth noting that the amount of space needed appears to vary from one society to another, a fact noted long ago by Hall (1959). Nurture, culture, power and politics (and not just nature), all need to be considered where the complexities and diversities of human territorialities are concerned.

18.7 Conclusion

The key argument of this chapter is that social practices are reflected in space and that territorial strategies are utilized by both dominant and non-dominant groups in order to attain certain political ends. This may be to do with maintaining power or with resisting the imposition of power by a dominant group. The examples provided are evidence of the way in which social relations are expressed through spatial patterns and they highlight the ways in which this geography helps in turn to shape social relations.

Phenomena such as racism and sexism regularly embody a territorial component. Territorial strategies are often used to control the 'place' of women or of ethnic or racial groupings. In this way particular ideologies are transposed onto space. People are confronted with wider practices through their use of space or through the ways in which they are allowed to use space. Power relationships take on a spatial dimension, even at the most mundane and everyday level. The basis for territorial division will vary depending on the particular fault line within that society. The examples used here demonstrate the spatialization of wider ideas and they show how people are kept 'in their place' whether through legal means (apartheid), administrative practices (urban incorporation in the United States) or surveillance strategies (women's fear of crime). Social boundaries are being communicated through space. As Harvey has suggested, 'the assignment of place within a socio-spatial structure indicates distinctive roles, capacities for action, and access to power within the social order' (1990: 419).

However, just as dominant ideologies can be reinforced through territorial practices, they can also be resisted. Territorial strategies are useful mechanisms in the assertion of identity. Spatial concentrations within particular geographic areas make visible people and issues that might otherwise remain unseen. In doing so, this demonstrates the 'positive' and 'negative' dimensions to territoriality; it can be both a force for oppression and also one for liberation. Particular strategies can be used to assert an identity and territorially transgressive acts can be employed to reclaim space.

While many people do not necessarily freely choose their 'place', they may, nevertheless, identify with their immediate neighbourhood or locality. This sense of identity can in turn be converted into forms of action aimed at obtaining particular outcomes. The formation of community or residence groups reflects feelings of belonging or attachment to a particular place. It follows that notions of territory are connected with ideas of social power. The claiming of space is a political act whether it occurs in the public or private arena. It is important to be aware of 'how relations of power and discipline are inscribed into the apparently innocent spatiality of social life' (Soja 1989: 6).

Learning outcomes

Having read this chapter you should understand:

- How social issues linked to racism, ethnicity, gender and sexuality are mapped onto geographic space.

- The ways in which social inequalities are also associated with spaces and territoriality.

- How territorial strategies are used to exert or to resist control.

- The ways in which many aspects of everyday life are reflected in spatial arrangements and territoriality.

Further reading

Two social geography texts dealing with many of the concerns of this chapter:

Panelli, R. (2004) *Social Geographies. From Difference to Action*, Sage, London.

Valentine, G. (2001) *Social Geographies. Space and Society*, Prentice Hall, Harlow.

Texts dealing with territory and territoriality:

Sack, R. (1986) *Human Territoriality: Its Theory and History*, Cambridge University Press, Cambridge.

Storey, D. (2001) *Territory. The Claiming of Space*, Prentice Hall, Harlow.

A useful urban geography text touching on many of the issues dealt with here:

Knox, P. and **Pinch, S.** (2000) *Urban Social Geography. An Introduction*, 4th edition, Prentice Hall, Harlow.

On issues of race and ethnicity:

Jackson, P. (1989) *Maps of Meaning. An Introduction to Cultural Geography*, Routledge, London.

For discussions of geography and gender:

Massey, D. (1994) *Space, Place and Gender*, Polity Press, Cambridge.

Women and Geography Study Group (1997) *Feminist Geographies. Explorations in Diversity and Difference*, Longman, Harlow.

For a range of perspectives on gay and lesbian spaces see the various chapters in:

Bell, D. and **Valentine, G.** (eds) (1995) *Mapping Desire. Geographies of Sexualities*, Routledge, London.

For annotated, clickable weblinks and useful tutorials full of practical advice on how to improve your study skills, visit this book's website at www.pearsoned.co.uk/daniels

CHAPTER 19

The place of the nation-state

CARL GRUNDY-WARR AND JAMES SIDAWAY

> **Topics covered**
>
> - The ubiquity and geographical diversity of nations and states
> - Interpretations of nationalism as a territorial project
> - Relationships between nations, states and territory

The territory of a nation is not just a profane part of the earth's surface. It is a constitutive element of nationhood which generates plenty of other concepts and practices directly related to it: for example, the concept of integrity and sovereignty; border control, conflict, invasion and war. It defines and has some control over many other national affairs, such as the national economy, products, industry, trade, education, administration, culture and so on. Unarguably, the territory of a nation is the most concrete feature of a nation for the management of nationhood as a whole. For a theoretical geographer, it is the territoriality of a nation. . . . For people of a nation, it is a part of SELF, a collective self. It is a nation's *geo-body* . . . Geographically speaking, the geo-body of a nation occupies a certain portion of the earth's surface which can be easily identified. It seems to be concrete to the eyes and having a long history as if it were natural, and independent from technology or any cultural and social construction. Unfortunately, that is not the case . . . the geo-body of a nation is merely the effect of a modern geographical knowledge and it technology of representation, a map. The geo-body, the territoriality of a nation as well as its attributes such as sovereignty and boundary, are not only political but also cultural constructs.

(Winichakul 1996: 67)

19.1 Historical and geographical variability of states

Although it is also other things, notably a provider of services, a system of regulations, ideologies, legal regulations and police powers ('law and order'), flows of capital (budgets, taxation and government spending) backed up by the threat of discipline and violence (for example, armed forces), the state can be interpreted as a form of *community*. As earlier chapters (in particular Chapters 1, 2 and 3) have shown, forms of human community and their attendant territorialities (see Chapter 18) have been extremely historically and geographically variable. This variation is enormous, from the claims of spokespersons for a state (it could almost be any state) that it is a natural embodiment of a 'nation' with the right to a select segment of the world (a territory with boundaries) and an ancient history (so even archaeological remains of past societies somehow come to be classified as 'national monuments', like Stonehenge in England, the Acropolis in Greece, Angkor Wat in Cambodia, the Pyramids in Egypt and so on) to nomadic communities or complex tribal or dynastic ones with no or little sense of themselves as belonging to or requiring a national state, to those who would reject national and state identity in the name of what are seen as other or more worthy causes (and attendant territorial visions), for example a religious identification, local community or world revolution.

The contemporary system of states, in which all of the land surface of the earth (with the partial exception of Antarctica as detailed in Chapter 17) is divided into state units, whose outlines become familiar to us from maps and globes, is after all fairly new. Earlier in the twentieth century, the borders between many of today's states were only vaguely defined, and more recently large areas of the world were ruled by colonial empires or dynastic realms (the Austro-Hungarian in Central Europe, Ottoman in the Balkans and Eastern Mediterranean, and Ch'ing empire in China, for example). Alternatively, the concept of nation-statehood was simply not in the political vocabulary. In the latter cases, ethnically, linguistically or religious specific groups owed allegiance to the imperial order rather than to a defined nation. That is, loyalty would be *foremost* to the empire and any sense of national or ethnic identity would be a local or 'private' matter.

Such imperial visions are no longer dominant and today few formal colonies and no large-scale dynastic empires remain (see Chapter 3, pp. 75–9). Therefore the territories that were once ruled as part of, for example, a Japanese, British, French, Portuguese or Russian empire are today mostly divided into self-avowed and recognized *sovereign* states. They possess the same apparatus of statehood (leaders, flags, capital cities, administrations, postage stamps, seats at the United Nations and so on) that the former imperial powers have. Moreover, the global map of states continues to change. In recent decades, some have disappeared (like the former East Germany and South Yemen) as separate

Plate 19.1 The state as a system of organized violence: armed forces. (Trip/G Grieves)

423

states, whilst others have split into component parts (like the former USSR and Yugoslavia). All this reinforces the point that states, like other communities, particularly the 'nations' with which they are associated, are not to be taken at face value. The claims made for and on behalf of them deserve critical examination. Having said that, this is not an easy task. For, as Benjamin Akzin (1964) pointed out over 40 years ago, to discuss nations, states and nationalism is to enter a terminological maze in which one easily and soon becomes lost. This chapter will indicate some pathways into the maze. It begins with an account of 'nations' and 'nationalism' before returning to the relationship of these to states. This will also help to pave the way for the discussions of citizenship and political action in Chapter 20.

19.2 Nations as 'imagined' political communities

One of the most influential and suggestive critical studies of nations and nationalism is a book by Benedict Anderson (1983) entitled *Imagined Communities: Reflections on the Origin and Spread of Nationalism*. He begins with a reminder of the ubiquity of nations and nationalism:

> Almost every year the United Nations admits new members. And many 'old nations' once thought fully consolidated find themselves challenged by 'sub'-nationalisms within their borders – nationalisms which naturally dream of shedding this subness one happy day. The reality is quite plain: the 'end of the era of nationalism', so long prophesied, is not remotely in sight. Indeed, nationness is the most universally legitimate value in the political life of our time.
>
> (Anderson 1983: 3)

Although enormously variable between, say, Nepali, Israeli, Singaporean, Nicaraguan, Vietnamese, Eritrean, American, Greek, Turkish and Irish versions, the ideology of nationalism holds that everyone will have a primary identity with a particular 'nation'. Such communities should be able to express themselves in a state; that is, they should enjoy what is called 'sovereignty' within certain geographical boundaries. 'Sovereignty', which is a term of long vintage and was previously associated with royal dynasties (the sovereign monarch), shifts to the 'people' of a 'nation', and even if a royal figurehead is retained, she or he will have to become in some way a 'national' symbol. However, it is important to remember

that territorial sovereignty as depicted on the world political map of today is of relatively recent origin (see Section 19.3), that more differentiated forms of sovereignty and other territorialities have existed in the past, and that sovereignty is continually being challenged in various ways.

Examples of quite different conceptions of sovereignty in the past are abundant in many parts of the world, from medieval Europe to most parts of the pre-colonial world. The labyrinthine world of medieval times in Europe was at some levels intensely 'local', involving much smaller communities and political units than today, although these were usually a 'part of a complex hierarchy of political or cultural entities, such as the Church of Rome, the Hanseatic League, or the dynastic Habsburg Empire' (J. Anderson 1986: 115). Sovereignty was not rigidly territorial as it mostly is with modern nation-states (see Chapter 1, pp. 26, 30–1). As James Anderson (1995: 70) explains:

> Political *sovereignty* in medieval Europe was shared between a wide variety of secular and religious institutions and different levels of authority – feudal knights and barons, kings and princes, guilds and cities, bishops, abbots, the papacy – rather than being based on territory *per se* as in modern times. Indeed the territories of medieval European states were often discontinuous, with ill-defined and fluid frontier zones rather than precise or fixed borders. Then the term 'nation' meant something very different and *non*-political, generally referring simply to people born in the same locality. Furthermore, the different levels of overlapping sovereignty typically constituted *nested* hierarchies, for example parish, bishopric, archbishopric for spiritual matters; manor, lordship, barony, duchy, kingdom for secular matters. People were members of higher-level collectivities not directly but only by virtue of their membership of lower-level bodies.

In many parts of the pre-colonial world, sovereignty was not based upon fixed boundaries and territorial control *per se*. For instance, much of North Africa and the Middle East had very different forms of political sovereignty in pre-modern times. According to George Joffé (1987: 27):

> political authority was expressed through communal links and was of varying intensity, depending on a series of factors involving, *inter alia*, tradition, geographic location and political relationships. The

underlying consideration, however, was common throughout the region and involved a concept of political sovereignty that derived from Islamic practice. The essential condition was that ruler and ruled were bound together through a conditional social contract in which the ruler could expect loyalty in return for enduring the conditions in civil society for the correct practice of Islam.

Not surprisingly, many of the colonially inspired geometrical boundaries that define the modern states of the Middle East and North Africa have limited relation to pre-colonial political landscapes. This is also true of other parts of the world. In many pre-colonial Asian states, for example, the emphasis of sovereignty was not on the territorial limits of control 'but on pomp, ceremony and the sacred architecture of the symbolic centre' (Clarke 1996: 217). In the classical Indianized states of South and South-east Asia, sovereignty was often focused on rulers who claimed divinity, and further eastwards, emperors held the 'mandate of Heaven', and the mandarins' right to exercise their authority was derived from their being 'superior men' (Sino-Vietnamese, *quan-tu*; Chinese *chun-tzu*) 'who acted according to Confucian ideals' (Keyes 1995: 195). But it would be too simplistic to think of sovereignty purely in terms of emperors, kings, queens, chiefs and so on, as ruler–ruled/ state–society relations in pre-modern societies were often complex, hierarchical, shifting and not based on strict territoriality. O.W. Wolters (1982: 16–7) described the scheme of power relations in South-east Asia as *mandala* (a Sanskrit word that defies easy translation, but which refers to a political apparatus that was without fixed boundaries, but which rested on the authority of a central court):

> [The] *mandala* represented a particular and often unstable political situation in a vaguely definable geographical area without fixed boundaries and where smaller centers tended to look in all directions for security. Mandalas would expand and contract in concertina-like fashion. Each one contained several tributary rulers, some of whom would repudiate their vassal status when the opportunity arose and try to build up their own networks of vassals.

This system of tributary relationships carried its own forms of obligations, sanctions, and allegiance. Mandalas created complex geographies of power, 'a polycentric landscape-seascape' (Friend 2003: 18),

including smaller chiefdoms paying tribute to more than one 'overlord' at the same time. Initially, the multiple sovereignties of the region were very confusing to the European imperial powers in the region who were eagerly trying to carve out their own spheres of unambiguous control. As Theodore Friend (2003: 21) notes in relation to the making of Indonesia:

> The Netherlands required centuries to unify Indonesia in their own fashion: first for mercantilist advancement of trade and then for nineteenth-century motives of geographic empire ... How could so few succeed over so many? The answer: because only a handful of mandalas had to be overcome, each caring little about the others or knowing nothing of them. The Dutch brought a layer of assiduous modernity to political vacuums strung throughout a vast archipelago. Geographically disconnected and culturally discordant but now administratively centralized, the Netherlands East Indies was for the length of one human generation the first comprehensive empire that region had ever known.

Elsewhere in South-east Asia, there were frequent clashes in conceptions of space and power relations between indigenous polities and the colonial intruders. Thongchai Winichakul (1995: 70) recounts a typical conceptual clash between the British and the Siamese Court. The British in 1845 were seeking to delimit precise boundaries of 'uniform rule' from Chiangmai to the Kra isthmus, and so they sent an intimidating letter to the Court of Bangkok, advising them to 'issue strict orders along their frontiers so that all subordinate authorities may clearly understand the line of boundary'. In response, the Court's reply confounded the British, for instead of a boundary line the Siamese only recognized a zone of relatively fluid control incorporating 'teak forests, mountains upon mountains, muddy ponds where there were three pagodas ... and so forth'. As Thongchai explains, the Court saw the outer margins of sovereignty as more a matter for the local authorities to decide, and the Siamese Kingdom had discontinuous boundaries, including some forests and mountains where sovereignty was ambiguous, and some tributaries where sovereignty was shared. It was only by the late nineteenth century that the Siamese rulers began to adopt the concept of territorial sovereignty in the same sense as the British and French were doing in mainland South-east Asia.

Although significant vestiges of such territorial structures remain, today they have mostly been

Regional Case Study 19.1

Nations without states: the Kurdish case

Gerard Chailand (1993: 4) notes that: 'the Kurdish people have the unfortunate distinction of being probably the only community of over 15 million persons which has not achieved some form of national statehood, despite a struggle extending back several decades.'

The lands predominantly inhabited by Kurds are in fact divided between four states (see Figure 19.1): Iran, Iraq, Turkey and Syria (there are also Kurdish communities in the former Soviet Union and migrant populations in Western Europe, especially in Germany).

Despite early twentieth-century promises from the then imperial powers (including Britain and France), in the end it suited no great power to establish an

Figure 19.1 Regions historically inhabited by the Kurds.
Source: After Chailand (1993: ix)

Regional Case Study 19.1 continued

independent Kurdistan. As in much of Asia (and beyond), the borders were determined in large part by foreign imperialists, taking little or no account of the wishes and rights of local peoples. The rise of Turkish, Iranian and Arab nationalisms – themselves all partly responses to imperial domination – left little space for the Kurds.

Yet frequent attempts on the part of the states in the region to eradicate Kurdish identity and nationalism have not been able to quash either the widespread sense of Kurdishness or the often violent struggle in pursuit of a Kurdish nation-state.

displaced by nation-states with fixed (though sometimes disputed) boundaries. Through the twentieth century, the ideology of nationalism (and the associated idea of nation-states) became one of the dominant and most widespread influences on politics across the world. Occasionally nationalisms recognize that the state itself may be multi-national (as in British nationalism, which contains English, Scottish and other affiliations), but in so doing, the wish is usually expressed that somehow a more inclusive national identity will evolve or has evolved which coincides with the boundaries of the state. Britain, Switzerland, South Africa, Nigeria and the United States are all cases where different versions of the claim and goal of an inclusive 'national identity' that supposedly unites disparate 'sub-nations' or communities have been asserted. Often, the assertion of a particular dominant nationalism in a territory has required the suppression of or conflict with other national claims on the same territory (the emergence of the state of Israel is a clear example of this; and in turn the project of a national home for Jewish people in the land of Palestine is in part a reaction to the genocidal extremes associated with German and other European nationalisms earlier in the twentieth century). As a result many 'national' communities that assert a claim to statehood are denied this (see Case study 19.1, for example). Others remain contested. Burma, for example, has witnessed over five decades of protracted ethno-nationalist struggles, particularly between successive military regimes holding power in Rangoon and the predominantly ethnic-Burman heartlands, and various movements in the provinces that are seeking either greater political autonomy within a federal structure or complete independence from the fragile 'Union of Myanmar' (as Burma is named by the ruling military junta) (see Case study 19.2). Similarly a number of 'multi-national' communist countries, notably Yugoslavia and the Soviet

Union, sought to regulate nationalism by assigning citizens to one of a number of constituent national identities. This sometimes involved inventing nationalities to rationalize and simplify more complex tribal and religious identities, whilst asserting that these should be subservient to an overarching sense of Soviet or Yugoslav identity which coincided with the boundaries of the USSR or of Yugoslavia. So, for example, in the USSR, people could be declared on their identity documents as having one of a number of officially recognized nationalities (Uzbek, Latvian or Russian, for example), but they would also and supposedly above all be citizens of the USSR. In due course, it was in part because of local nationalist challenges to wider Soviet and Yugoslav affiliations that the USSR and Yugoslavia collapsed in the early 1990s.

Many of those self-identified nations without their own state (the Kurds are an example, see Case study 19.1; the Karens, Karenni, Shan and Kachin are other examples, see Case study 19.2) or, at the least, an allocated autonomous region within a broader (multi-national) state claim the right to one. And frequently, smaller nations within a multi-national state (such as Scotland in the UK or Quebec in Canada) become the basis for claims that they should enjoy full statehood. In many cases too, either the central state or some other community with another affiliation resists. There are numerous examples of this, including the complex case of Northern Ireland, where most Catholics (Republicans) would wish to see the province united with the rest of the Irish Republic (which itself successfully broke away from the British colonial empire earlier in the twentieth century). Most Protestants (Unionists), who claim descent from settler populations from England and Scotland, wish to remain part of a 'United Kingdom' (see Case study 3.1).

The Irish case is just one of dozens of situations where conflicting nationalist and confessional

Regional Case Study 19.2

Burma: a national geo-body or a mosaic of ethnic homelands?

According to Chao Tzang Yawnghwe, a son of the first President of the Union of Burma, who took an active part in the Shan resistance movement in the mid-twentieth century: 'I feel that the greatest flaw in current works dealing with post-1948 Burma is the confusion over the term "nation-building" in general, and more specifically, its connotations within the internationally political perimeter known as Burma,

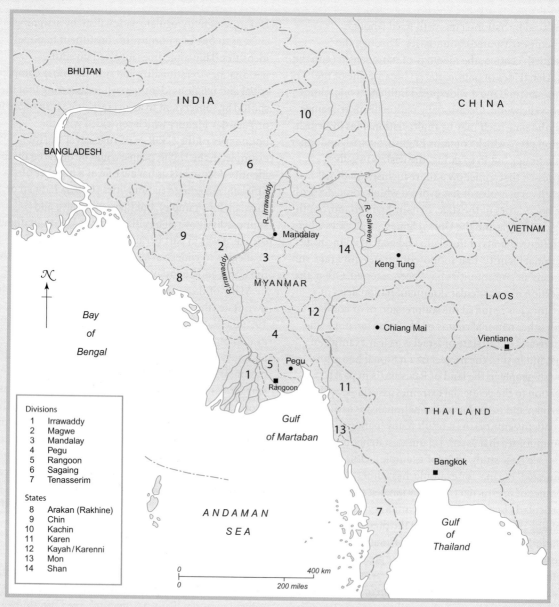

Figure 19.2 Burma: national geo-body or mosaic of ethnic homelands?

which, in reality is a composite of many homelands' (Yawnghwe 1987: ix).

From a geographical perspective we could say that Burma is composed of the broad plain lying on both sides of the Irrawaddy River that flows into the sea at the Gulf of Martaban and the Bay of Bengal; and surrounding this plain there are extensive border regions populated by different ethnic groups. In each of these 'ethnic minority' (in Burma as a whole) areas the 'majority' Burman people are hardly present. It is these border states that Yawnghwe (1987) refers to as the ancestral 'homelands' of different ethnic groups, such as the Karen, Karenni (Kayah), Kachin, Shan or Tai, Chin, Arakan, and others (see Figure 19.2).

During the nineteenth century, the British colonial rulers sought to try to manage the complex political and ethnic mosaic that came to be defined as Burma, by dividing the colonial administration into 'Burma Proper' (covering much of Lower Burma of the Irrawaddy) and the so-called 'Frontier Areas' (encompassing the Kachin, Chin, Karenni uplands, the Karen Salween district, and the Shan plateau). In fact, the 'Frontier Areas' included almost 50 per cent of the total land area and about 16 per cent of the total population of Burma. However, it was only 'Burma Proper' that was to be under a highly centralized and direct form of bureaucratic rule, whilst the 'Frontier Areas' were thinly administered with forms of indirect rule that left some of the tiny chiefdoms with more political autonomy. By imposing different degrees of administration on the lowland and hill people 'colonial modernity began the institutionalization of ethnic categories' (Lang 2002: 31), which was to become intensely politicized in the years following Burma's independence in 1948.

In the extensive border regions (Figure 19.2), there have been many battles for control over territory, resources and people between the Burma Army or *tatmadaw* and the 'sub-state' nationalist movements, such as the Karen National Union (KNU), the Karenni National Progressive Party (KNPP), various Shan factions, and the Kachin Independence Organization (KIO), to name a few. Each of the main ethno-national groups maintained armies and for much of the period since independence they controlled large parts of what they called their 'national homelands'. Ethnic groups with sizeable populations straddling international boundaries, such as the Kachin people across the Burma–China border, have also been able to develop strong social and cultural connections beyond these homelands. Recent years have witnessed ongoing conflict in many of these troubled border regions of Burma. Countermoves by the Burmese army have led to huge numbers being forcibly displaced and many have fled to Thailand or live in constant fear of the Burma Army (Lang 2002, Grundy-Warr and Wong 2002).

(religious–cultural) logics collide, often with violent consequences. There is clearly something very powerful going on, whereby nationalist visions are linked with particular territories and conceptions of state. Yet, as Anderson (1983: 3) recognizes:

But if the facts [of the existence of many and sometimes conflicting nationalisms] are clear, their explanation remains a matter of long-standing dispute. Nation, nationality, nationalism – all have proved notoriously difficult to define, let alone analyse. In contrast to the immense influence that nationalism has exerted on the modern world, plausible theory about it is conspicuously meagre. Hugh Seton-Watson, author of by far the best and most comprehensive English-language text on nationalism, and heir to a vast tradition of liberal historiography [theories of history writing] and social science, sadly observes: 'Thus I am *driven* to the conclusion that no "scientific definition has existed and exists." '

Readers may wish to 'prove' Seton-Watson's observation for themselves, by trying to come up with a universally valid definition of a nation. Faced with this task, students (and their teachers) will often work through a long list of characteristics ascribed to nationality. But it seems that exceptions can always be found. Language is a favoured criterion, but then many languages are spoken by more than one 'nation' (English is an example) and some 'nation-states' (Switzerland or Mozambique, for example) are divided into communities speaking different languages. Religion is sometimes chosen, but the same objections apply. Ethnicity and race turn out to be problematic criteria, favoured by racists of all stripes, and often part of the basis for national identities, but susceptible to the obvious and undeniable point that everywhere is much too mixed up historically and genetically for such categorizations to be watertight. Besides, some nationalisms have come to celebrate their multi-racial and multi-cultural composition, as in the 'melting pot' United States or the 'rainbow nation' of South Africa. 'Culture' usually crops up as a criterion. Yet as Chapter 12 has indicated, cultures are always (though of course to varying degrees) heterogeneous and contested. Think, for example, of age, gender, class and other

variations and the coexistence of multiple sub-cultures (which, as Chapter 18 has shown, are often related to distinctive 'local' territorialities) that characterize every supposedly 'national society'.

All this leads Anderson to declare that nations are in a sense 'imagined communities'. This imaginary status is not to deny that they are not in a sense real to those who feel they belong to them. Indeed, Anderson (1983: 6) feels that:

> In fact, all communities larger than primordial villages of face-to-face contact (and perhaps even those) are imagined. Communities are to be distinguished, not by their falsity/genuineness, but by the style in which they are imagined.

Plate 19.2 'Nation' imagined as race. (Corbis UK/Tim Page)

The style in which nations are imagined (and therefore the way that they become to seem real by those who belong to them) is, according to Anderson (1983: 7) as 'limited', 'sovereign' and as 'communities':

> The nation is *imagined* as limited because even the largest of them, encompassing perhaps a billion living human beings, has finite, if elastic boundaries, beyond which lie other nations.... The most messianic nationalists do not dream of a day when all the members of the human race will join their nation in the way that it was possible, in certain epochs for, say, Christians to dream of a wholly Christian planet....
>
> It is imagined as *sovereign* because the concept was born in an age in which Enlightenment and Revolution were destroying the legitimacy of the divinely-ordained, hierarchical dynastic realm ... nations dream of being free.... The gage and emblem of this freedom is the sovereign state.
>
> Finally, it is imagined as *community*, because regardless of the actual inequality and exploitation that may prevail in each, the nation is always conceived as a deep, horizontal comradeship. Ultimately it is this fraternity that makes it possible, over the past two centuries, for so many millions of people, not so much to kill, as willingly to die for such limited imaginings.

Within all this is a profound territorial link between the nation and the state. The state claims to be a sovereign expression of the nation – bound to it and to a particular place. That is, it has territorial limits. The nation-state has a *geography*, which is charted, demarcated, mapped and represented to the 'national population' in their school atlas and geography lessons. Such a system of 'national' geographical representation is always combined with an historical vision, a grand narrative of 'national history', often assuming that the nation is ancient, even primeval.

Yet Anderson and other critical accounts of nationalism stress that it is very much a modern ideology. For what makes mass nationalism possible are certain socio-economic and technical transformations, notably the arrival of media and national educational systems. Schooling, newspapers, and later radio and television all help to promote and popularize the idea that people belong to and share in the nation. Whilst others have emphasized the longer historical roots of many nationalisms in pre-modern ethnic affiliations (for example, Smith 1986), for Anderson and most other critical observers, what is striking is nationalism's

relative modernity. Even where nationalists imagine the nation as ancient, such an imagination is itself more often than not predominantly a nineteenth- and twentieth-century phenomenon. At a pinch, nationalism might be traced to the seventeenth century. But the ideology of the nation-state was not anywhere very evident before then. So whilst someone might talk about a thousand years (or more) of, for example, English or British or Indian national history, they conveniently forget that English or British and Indian nationalisms are relatively modern concepts, much less than a thousand years old.

The tendency for nationalists to 'reinvent the past', to pick out selective moments from the past, or to manipulate history, is one of the main arguments of historian Eric J. Hobsbawm (1996: 255), who argues:

> For historians are to nationalism what poppy-growers in Pakistan are to heroin addicts: we supply the essential raw material for the market. Nations without a past are contradictions in terms. What makes a nation *is* the past; what justifies one nation against others is the past and historians are the people who produce it.

Whilst history is undoubtedly significant, the nationalist brew would be incomplete without ancestral connections to 'homeland' and to particular places. In other words, *territory* is central to nationalism, and this is often why whenever there are divergent nationalist claims to one piece of land, extreme violence often follows. Indeed, the terrible forms of so-called 'ethnic cleansing' witnessed in the former Yugoslavia in the 1990s are illustrations of recent nationalist extremism and the significance of historical imaginations concerning territorial–political and cultural identity.

Nationalist narratives and claims are not always linked to such violent actions. At times they may be rather more rhetorical or theatrical. Graham E. Clarke (1996: 231–2) raises a particular Nepalese example:

> ... [I]n the 1960s in Kathmandu the sole national newspaper carried articles, no doubt read tongue in cheek by some educated Nepalese, arguing that since Lumbini, the birthplace of the historical Buddha (Gautama), was located some few five miles north of the current southern border, that Buddha was *therefore* Nepalese and not Indian.'

In a sense, the nationality claims to Buddha are no more or less absurd than the Macedonian and Greek counter-claims about the 'national' identity of a distinctly pre-nation-state historical figure – Alexander the Great. As Loring Danforth (1995: 171) writes:

> At the level of popular discourse both Greeks and Macedonians make extensive use of Alexander the Great as a powerful symbol of the historical and cultural continuity, which, from their perspectives, links them with their glorious ancestors, the ancient Macedonians. Demonstrations held throughout the world in the early 1990s by Greeks and Macedonians in opposition to or in support of international recognition of the Republic of Macedonia were led by men dressed as Alexander the Great, wearing a crested helmet, breast plate, and greaves and holding a shield and a spear. Each group was outraged that the other group had attempted to appropriate its own famous ancestor for such obviously political purposes.

All of this relates to the important issue of how we deal with things like space and identity in history. As Morris-Suzuki (1996: 42) writes:

> The nation ... casts a long shadow backwards on our vision of the past, and channels our perceptions into a particular spatial framework. In my bookcase, I have a volume on the history of Thailand since the tenth century, which, considering the repeated political and cultural realignments within the space we now label 'Thailand', seems only a little more bizarre than its neighbour on the shelf, a history of the Soviet Union from palaeolithic times to World War II.

Claims to nationhood frequently involve the blending of 'tangibles' and 'intangibles' in a unique brew that may contain such volatile ingredients as blood, language, race and religion. A fundamental goal of nationhood is to generate a strong *sense* of belonging associated with a particular territory. In practice it may prove difficult to distinguish between 'the nation' and other human collectivities, but as Walker Connor (1994: 93) puts it:

> ... what ultimately matters is not *what is* but *what people believe is.*

There has been a strong tendency to see nations and nationalisms as essentially being derived from American and European prototypes. It is important to challenge the idea that nationhood is a nationalist project originating in the West that was simply implanted onto the rest of the world without significant transformations in the various cultural, ethnic and religious contexts it touched. African, Middle Eastern or Asian variants of nationalism or constructions of statehood are far more than simply colourful (or failed) replicas of the European–American model(s). To be sure, imported idioms and motifs from Europe and America have been important; for example, the official state language in a fair number of former colonies is that of the old colonial power. And European (and less often United States) imperialism has provided an important backdrop to the trajectory of nationalisms in Africa and Asia, not least because the anti-colonial nationalists themselves were often educated in western institutions and adopted concepts such as self-determination, national liberation and territorial sovereignty in their political struggles against the colonial powers. Yet whilst 'globalizing' capitalism and especially imperialism are part of the picture, African, Middle Eastern and Asian nationalisms are also (like those of Europe and the Americas) rooted in *local* historical trajectories. In many cases these are of much longer vintage, often based on sophisticated cultural, religious and ethnic tapestries, some of the main patterns of which have subsequently been appropriated or (re)invented as pre-colonial 'national' histories.

Thus, we should perhaps think in terms of multiple histories and geographies of nations and states, and not just of particular European or American modular forms. The influential Bengali intellectual Partha Chatterjee (1993) has persuasively argued against the imagining of Indian national identity through the lens of the colonial power and stresses the 'essential' inner or spiritual domains of culture that were never colonized, never European. Furthermore, there are different coexisting national voices or 'fragments' – among women, peasants, elite, castes, outcasts, and so on – each with its separate discourse. As Stein Tønnesson and Hans Antlöv (1996a: 32) put it:

> When the national idea entered Asia it could not be implemented without mediation, hence transformation, by indigenous agencies in particular settings. There were existing and alternative ideas with which European-style nationalism interacted and intermixed ... (T)he nationalist ideas were invested with local qualities, meanings and nuances which could not be found in Europe. People had their own views on what constituted a legitimate social order, and such views could not be ignored by the modernizing leaders of the anti-colonial struggles.'

As well as different cultural 'forms of the nation', we may also consider different gendered perspectives of nationalism and nation-states. Although Chatterjee's (1993) arguments are quite specific to the South Asian case, they do reveal the coexistence of different voices, men and women, and this relates to the point about nationalism and constructions of nation-statehood invoking different conceptions of 'manhood' and 'womanhood' more widely. Although there are immense historical and geographical variations of this, there is a general tendency for women to be seen as particularly important transmitters of 'national culture' (for example, in the idea of a 'mother-tongue' language) (Yuval-Davis 1997) and as somehow the embodiment of the nation. McClintock (1994: 352) thus notes that:

> Nations are contested systems of cultural representation that limit and legitimize people's access to the resources of the nation-state, but despite many nationalists' ideological investment in the idea of popular *unity*, nations have historically amounted to the sanctioned institutionalization of gender *difference*. No nation in the world grants women and men the same access to the rights and resources of the nation-state'.

Moreover, particular ideas of the 'family' are central to most imagined communities:

> Nations are frequently figured through the iconography of familial and domestic space. The term nation derives from [the Latin] *natio*: to be born. We speak of nations as 'motherlands' and 'fatherlands'. Foreigners 'adopt' countries that are not their native homes and are naturalized into the national 'family'. We talk of the 'Family of Nations', of 'homelands' and 'native' lands. In Britain, immigration matters are dealt with at the Home Office; in the United States, the president and his wife are called the First Family. Winnie Mandela was, until her recent fall from grace, honoured as South Africa's 'Mother of the Nation'. In this way, despite their myriad differences, nations are symbolically figured as domestic genealogies.

In several European nationalisms (including 'British') one significance of this is that the sexist notion of women (usually grouped with children) supposedly as *naturally* inferior to men, or as requiring the protection of *naturally* more powerful men, frequently provided the prior backdrop to depicting hierarchies (between dominating and dominated classes, for example) *within* the nation as natural, like those of a family. A similar conception provided part of the racist justification for overseas colonies. Peoples subject to colonial domination were often presented as being equivalent to 'families' of black children ruled by a 'benevolent' white father.

19.3 Constructing boundaries

States come in diverse shapes and sizes. But they all have boundaries (which, if they have a coastline, also extend into the sea and are governed by international protocols: see Stienberg 2001). The extract from Winichakul (1996) quoted at the start of this chapter noted how the 'territoriality' of the modern nation-state is of a bounded space: 'a certain portion of the earth's surface which can be easily identified'. Premodern dynastic realms and empires could make do with loose boundaries. But modern nation-states have felt it necessary to *demarcate* their boundaries, to iron out perceived irrationalities and anomalies – sometimes by going to war. In short, they have sought to nationalize and unify 'their' space and 'their' (national) populations and render them into known, surveyed and defensible sovereign territory and communities. As Chapter 15 explores, the rise of geopolitics in the late nineteenth and early twentieth centuries represents a particular expression of just such a rationality – but it has wider forms and deeper roots. The nineteenth-century invention and refinement of statistics (*state-istics* = the science of the state) and demography (*demos* = people, *graphy* = writing) and the establishment of geography (*geo* = earth, *graphy* = writing) as a university discipline and field of research are all part of this regime of knowledge and power where people are surveyed and made into subjects of nation-states. For states with land boundaries the occupation and official demarcation of the frontiers is an important part of this process.

Consider the example of the border that is often recognized as the *oldest*, more or less stable, still-existing border in the world, that between Spain and Portugal. The mutual recognition of the border between Spain and Portugal is usually traced in modern Spanish and (particularly) Portuguese history to the Treaty of Alcañices signed on 12 September 1297 (in fact, the date on the Treaty is 12 September 1335, but amongst other things, our conventions for counting years have altered since). The Treaty has acquired something of the status of a foundational text. In other words, it is interpreted as a kind of proof of the ancient historical basis of the nations concerned.

Yet the treaty was *not* signed in the name of states or 'nations'. Named after the Templar castle in which it was sealed and witnessed, the text of the treaty of Alcañices begins with the words 'In the name of God. amen.' and is signed in the presence of the Templars and other holy orders by those who describe themselves as:

> by the grace of God the King of Portugal and the Algarve and by the grace of God the King of Castille and Leon,

and has as its subject:

> towns, castles and lands, town boundaries, divisions and orderings ... [disputes over which] have caused many wars, homicides and excesses, as a consequence of which the lands of both [kingdoms] have been looted, burnt and ruined, weighing heavily on God ... [and] because of our sins, risking the danger of losing them and them falling into the hands of our enemies in the faith and most gravely [causing] the violation of God's will and injury to the holy church of Rome and Christendom.
>
> (Cited in Martíñez 1997: 15; our translation)

Moreover, the actual demarcation of the Portuguese–Spanish borderline had to wait until the nineteenth and twentieth centuries. Two formal delimitation treaties set out an agreed frontier and established mechanisms for its physical demarcation on the ground. But gone are 'amens' and references to 'the will of God'. Instead, the treaties (the first of 1864 for the northern half of the border and the second of 1926 for the southern half of the border) make reference to the need to impose order and well-defined demarcations, to 'eliminate the anomalous situation in which, in the shadow of ancient feudal traditions' (Tratado de Limites 1866: 1, our translation), some frontier areas (including a number of villages) were recognized as *shared* or common lands with usage rights by communities of both states/'nationalities'. Following the treaties, such areas were divided and the border, where not demarcated by a river, was marked on the ground by rectangular boundary stones every few hundred metres (although it took a couple of decades to put in place all the boundary markers and resolve local conflicts and differences over the 'fine-scale' division of lands). In other words, the Portuguese–Spanish case indicates how the epoch of modern nationalism (i.e. the nineteenth and twentieth centuries) saw what James Anderson (1995: 71) calls:

> a territorialisation of politics, with a sharpening of differences at the borders of states and of nations between 'internal' and 'external', 'belonging' and 'not belonging', 'us' and 'them'.

By the same logic, or an extreme extension of it,

Plate 19.3
Representing the limits of the 'nation-state': boundary patrols. (© Michelle Gienow)

minorities who do not 'fit in' sometimes have to be moved, murdered or deported – or, in the euphemisms of our times, 'cleansed' (see Sibley 1995). This has been associated with, for example, Nazi racial territorial visions of a Europe cleansed of Jews and Gypsies and an expanded racially pure Germany. The Nazi Holocaust represents an extreme case. But 'cleansing' or 'purification' of national space, amounting to the expulsion or murder of people defined as 'foreign' or 'alien', has been widespread in the twentieth century. There are examples from all continents: the creation of modern Turkey whose Armenian population were slaughtered in the second decade of the twentieth century (and which still refuses to admit the rights of a Kurdish minority); the creation of Israel in the 1940s, which was accompanied by the flight of indigenous Palestinian Arabs from territories deemed part of the new Jewish

national state; the system of reservations for native Americans in the USA and Canada; the violent exchange of Muslim, Hindu and Sikh populations which accompanied the 1947 partition of India; the 1974 partition of Cyprus; the wars of the 1990s in the former Yugoslavia; and the partition of Ireland (see Spotlight box 19.1). These logics have also produced calls for the modification of boundaries, sometimes resulting in violence and war. No continent has been without examples of such conflict and violence and it is probably impossible to find a state which has not at some time expelled people or murdered them in the name of some nation or other. Yet each nationalism must also be unique, establishing itself as different from others (even if the idioms of flags and so on are similar), constructing a sense of self against others who are defined as outside the imagined community.

Spotlight Box 19.1

Irredentism and secessionism

Irredentism is defined by Mark Blacksell (1994: 299) as: 'The assertion by the government of a country that a minority living across the border in a neighbouring country belongs to it historically and culturally, and the mounting of a propaganda campaign, or even a declaration of war, to effect that claim.' Whilst Blacksell focuses on official irredentism, sometimes irredentist parties and organizations promote the claim without it necessarily becoming their government policy. Often the irredentist claim includes areas where populations are mixed, making for potentially violent conflict. The collapse of Yugoslavia and the Soviet Union have been accompanied by irredentist claims. But there are dozens of other examples around the world.

Often closely related to irredentism is the phenomenon of secessionism, whereby a minority in one state seeks the transfer of a section of the state's territory to another new or existing state. Although it has also had its own local cultural and political dynamics, the conflict in and over Northern Ireland exemplifies irredentism and secessionism. The Irish Republic was

born of a secession from the British Empire (whilst at the same time being seen as the first of many national liberations for colonized peoples) in the 1920s. The Irish constitution long claimed sovereignty over the whole of Ireland, thus making an irredentist claim on that part which remained under British jurisdiction, whilst nationalists in what they often called the north of Ireland (refusing the term Northern Ireland which signified a separate jurisdiction) sought secession and a unified Ireland.

Current schemes for the peaceful resolution of the conflict include visions for a measure of joint sovereignty over what (to be as neutral as possible) is called 'the province' (which could be a province of the United Kingdom and/or of the Irish Republic and/or Ulster; a place apart).

Meanwhile, James Anderson (1997: 216) suggests that: 'Northern Ireland is the residue of failures in nation- and state-building, whether viewed from either a British or an Irish perspective. The disputed labels for territory – repeated at a local level in, for example, Derry/Londonderry (or, ironically, 'Stroke City') – reflect the rival national identities and suggest that the ideal of the nation-state is unachievable.'

19.4 Nation-states as symbolic systems

As we have seen, the state is the bureaucratic expression of nationalism. This is not to say that a widespread sense of nationalism inevitably precedes the state. Indeed, what is often termed the state apparatus (everything from regulations governing schools, to tax inspectors and politicians) seeks to foster national subjectivities out of the frequently ambivalent and disparate array of identities contained by its boundaries. If, as has been argued by many, nationalism is akin to religion, then the state becomes its symbolic structure. When examining the superficially quite different cases of Australia and Sri Lanka, Bruce Kapferer (1988) claims that nationalism is itself a religion, owing to the fact that, as with most religious-like beliefs, nationalism demands the recognition (his word is 'reification', which means something more than this) of an immanent and all-encompassing entity (the nation). The nation fulfils the role of a sacred cause, something greater than any individual and something which may be worth dying for. The nation is the God or deity of the religion of nationalism and the state is its theology or temple. Moreover, as Herzfeld (1992: 37) notes:

> Every bureaucratic action affirms the theology of the state. Just as nationalism can be viewed as religion, bureaucratic actions are its most commonplace rituals. There are other such everyday

rituals: Hegel saw the reading of the morning newspaper as the secular replacement of prayer.

There are indeed many public rituals of nationality and statehood: coronations and remembrance days, military parades, national holidays, national prowess (or the lack of it) at football or the Olympics, swearing in of governments, state funerals. (Football in the UK is also revealing of the complexity of British nationalisms and the survival of separate Scottish, English, Welsh and Northern Irish squads is a testimony to the limits to 'Britishness' and the survival or 'revival' of other national affiliations.) In all these rituals, the nation is reaffirmed and the state performed and made to seem omnipresent, historical and real. Dramatic cultural or political events covered by the media greatly reinforce senses of national community. The events of 11 September 2001 were thus narrated in the United States as an 'attack on America'. It is also said, for example, that virtually every adult *American* alive at the time can recall where they were on hearing the news of the assassination of US President John F. Kennedy in 1963. Whilst his funeral, like that of Diana Spencer, became a global media spectacle, it was represented and felt most acutely as a kind of national loss. 'Goodbye *England's* Rose . . .' as Elton John put it at Diana's funeral. (It should be noted that there was some controversy over the lyrics of Elton John's tribute song, in so far as Diana was not only the Princess of *Wales*, but a former member of the avowedly *British* monarchy. Rather like

Plate 19.4 Declaring the nation, performing the state: arrivals at Changi Airport, Singapore. (P. Vanriel/Robert Harding)

the case of football, this bears out the complexities and unresolved tensions of British nationality.)

But the nation-state also demands (like all effective religions) personal commitment and more minor ritualistic acts, many of which rest on a geographical imagination of inside and outside, belonging and otherness (see Taussig 1997). The anthropologist Michael Herzfeld (1992: 109) examines this theme and is worth quoting at some length:

> Nationalist ideologies are systems of classification. Most of them are very clear about what it takes to be an insider. That, at least, is the theory. In practice, however, divergent interpretations give the lie to such essentialist claims, as to take one prominent and current example, in the debate currently waging in Israel about the definition of a Jew. Such taxonomic [naming] exercises ... are central to the very existence of the nation-state. All other bureaucratic classifications are ultimately calibrated to the state's ability to distinguish between insiders and outsiders. Thus ... one can see in bureaucratic encounters a ritualistic enactment of the fundamental principles upon which the very apparatus of state rests. Seen in these terms, arguments about the number on a lost driver's license or an applicant's entitlement to social security do not simply challenge or reinforce the power of particular functionaries of state. They rehearse the logic of the state itself.

Not only that, but the whole exercise of state power gets taken for granted as the natural order of things. Only when many of those activities which are ascribed

Plate 19.5 War-making and with it a sense of shared struggle can be productive of national identity. (Corbis UK/Bill Ross)

to the state are no longer carried out (in situations of war, for example) or when a person finds themselves on the wrong side of a state-sanctioned category (the wrong side of the boundary, the wrong side of the law) is the power (a power over life and death and thousands of lesser things) and the universality of the symbolic order of 'the state' revealed. The state claims the monopoly over these things. And for others to exercise judgment and to punish or to kill is 'to take the law into their own hands'. These things are reserved for the territorial state.

19.5 Conclusions: the place of the nation-state?

There have never been so many states as there are today. In the nineteenth century a wave of new states (countries such as Argentina, Mexico, Bolivia) emerged in the Americas in revolt against Spanish and Portuguese empire. These American prototypes and the USA itself (which dates from 1776) provided an example to nationalists elsewhere in the colonial world. But it was not until the twentieth century that most of Africa and Asia could escape direct colonial domination. This often required violent 'national-liberation' struggles against entrenched resistance from the colonial powers and white settler populations. In the decades between about 1945 and 1975, dozens of new 'sovereign' states emerged in place of the old colonial map. And if, for example, the Tamil nationalists of Sri Lanka, the Karen, Karenni and Shan of Burma, the Palestinian and Kurdish nationalists of the 'Middle East' or the Basque, Scottish and Corsican nationalists in Western Europe get their way, there will be even more.

Yet, for many years the *demise* or decline of the state has been discerned or predicted – and such claims have of late become even more common in discourses about 'globalization'. These usually argue that the growing scale and power of transnational flows, particularly of capital (but also of people, ideas and religious affiliations, technologies and so on) is *subverting* the capacity of the state and *weakening* national identities. The nation-state is often described as being 'hollowed-out' or 'eroded'. In this view, the state no longer has the power to command, for example, the society and economy inside its boundaries that was once attributed to it. And such a 'hollowing out' is sometimes seen to point the way to a post-national world (or some kind of *shared* global culture in which national cultures are

replaced by a more hybrid, but common global mixture).

Some writers such as Arjun Appadurai have argued that there are new 'postnational cartographies' emerging. An illustration of this is given by Appadurai (2003: 342–3):

> The case of Khalistan is particularly interesting. Khalistan is the name of the imagined home that some Sikhs in India (and throughout the world) have given to the place that they would like to think of as their own national space, outside of the territorial control of the Indian state. Khalistan is not simply a separatist, diasporic nationalism.... Rather, Sikhs who imagine Khalistan are using spatial discourses and practices to construct a new, postnational cartography in which ethos and demos are unevenly spread across the world and the map of nationalities cross-cuts existing national boundaries and intersects with other translocal formations. This topos of Sikh 'national' identity is in fact a topos of 'community' (*qom*), which contests many national maps (including those of India, Pakistan, England, and Canada) ...

Appadurai (2003: 243) also mentions 'translocal affiliations' that are 'global or globalizing', such as Islamic, Christian, and Hindu fundamentalisms; as well as 'racial and diasporic' ones that 'cross-cut existing national boundaries' and intersect 'with other translocal formations.' The hyper-mobility of people, refugees, exiles, migrants and others, plus old and new identities, associations, networks and affiliations that are other than 'national', mean that the idea of the 'nation-state' is being challenged from all directions. Yuval-Davis (1997, reprinted in Brenner *et al.* 2002: 322) has pointed to the lack of congruence between nations and states, arguing that there is often a lack of 'overlap between the boundaries of state citizens and "the nation"', which requires us to have a much more 'multi-layered' notion of people's citizenship needs (see Chapter 20) 'because people's membership in communities and polities is dynamic and multiple.'

However, others object that transnational forces of 'globalization' are really nothing new and that capitalism in particular has shown itself not only able to coexist with states and nations but in symbiotic relationship with them. In other words, they reinforce each other. The same goes for many organized religious movements that coexist or support nation-states. (This is not to say that many movements with religious inspirations do not see themselves as anti-statist, even radically and violently so, as in the case of some fundamentalist Christian militias in the USA and Islamicist movements in Asia. But in so doing of course they ascribe to the state a certain power and centrality – thereby making it more real. A parallel to help readers understand is the atheist statement: 'There is no God', which cites something called God in order to deny its existence!) Moreover, 'globalization' is not only massively uneven (as we have seen in earlier chapters) but as likely to produce local backlashes as a universal culture of, say, the *same* fast food, drinks and soap operas and political orientations everywhere in the world. That is, the technologies of capitalism (particularly media) provide the preconditions for strengthening, rather than undermining, imagined communities of nationalism, while states still act to fine tune the regulatory frameworks for continued capital accumulation (resorting where needed to force to suppress opposition). States still enact laws about business, trade unions, property rights and so on. Everywhere buying a property or land or setting up a legal business require some kind of registration with the state. No amount of globalization has ended this. However, instead of further entering the large and complex debates about the impacts of 'globalization' on nations and states, for which Anderson (1995) is a good introductory guide, let us return to the conception of the state as a symbolic system, and as a complex of *representations*. The historian-philosopher and activist Michael Foucault (1979: 29) argued in one of his most famous essays that:

> We all know the fascination which the love, or horror, of the state exercises today; we know how much attention is paid to the genesis of the state, its history, its advance, its power and abuses, etc. . . . But the state, no more probably today than at any other time in its history, does not have this unity, this individuality, this rigorous functionality, nor, to speak frankly, this importance; maybe after all the state is no more than a composite reality and a mythicised abstraction, whose importance is a lot more limited than many of us think.

Hence, to recognize the state as, in part at least, a symbolic system is also to recognize that, as Rose and Miller (1992, 172) argue: ' "the state" itself emerges as an historically variable linguistic device for conceptualising and articulating ways of ruling'. In other words, things like the 'nation' and the 'state' are *made* real mainly in certain words, texts (including maps) and deeds, that is, in language and action. Think again of the staging of those *national* sporting, political and cultural occasions, which bring the nation home, and the more mundane or bureaucratic acts of state, such as the display of maps in schools and public buildings or the action of showing a passport or filling in a form with your *national* (insurance, registration or identity) number. So perhaps nation-states exist above all as systems of actions and beliefs – an 'imagi*nation*' if you like – which must be continually re-enacted, re-narrated and re-imagined as territorial sovereign spaces in order to seem important and real to us. Just as they have been here.

Plate 19.6   Identifying citizenship. (Alan Thomas)

Learning outcomes

Having read this chapter you should understand:

- Nations, nationalism and states are complex historically and geographically variable phenomena.

- Nationalism is an ideology (a system of beliefs) which holds that people have a primary identity to a particular nation and that such communities should be able to express themselves in a *geographically* defined state.

- Nations can be understood as a kind of imagined community. They are imagined because not all members of a national community can know each other.

- Such national imaginaries contain a geography, a mental map of national space and its boundaries.

- National imaginaries are also gendered, for example in the idea of mothers as key transmitters of 'national culture' to the next generation.

- States can be understood as complex symbolic systems.

Further reading

Agnew, J., Mitchell, K. and **Toal, G.** (eds) (2003) *A Companion to Political Geography*, Blackwell, Oxford. Contains several chapters relevant to the material covered here.

Anderson, B. (1983) *Imagined Communities. Reflections on the Origin and Spread of Nationalism*, Verso, London. An expanded second edition from 1991 is available. Likely to be challenging, but worthwhile.

Anderson, J., Brook, C. and **Cochrane, A.** (eds) (1995) *A Global World? Re-ordering Political Space*, Oxford University Press and the Open University, Oxford. Chapter 2 on 'The exaggerated death of the nation-state' is a good place to follow up debates about the impact of globalization mentioned above in Section 19.5.

McDowell, L. and **Sharp, J.P.** (eds) (1997) *Space, Gender and Knowledge: Feminist Readings*, Arnold, London. Section 7 on 'Gender, Nation and International Relations' contains extracts of some challenging articles and lots of suggestions for further reading.

McNeill, D. (2004) *New Europe: Imagined Spaces*, Arnold, London. Contains material on European nation-states and their relationships to alternative scales of politics (such as the European Union and the municipalities).

Seton-Watson, H. (1977) *Nations and States: An Enquiry into the Origins of Nations and the Politics of Nationalism*, Westview Press, Boulder, Colorado. There are hundreds of case studies of particular nationalisms. Select a nation and you will usually find many to choose from. Although now out of print, this book can often be found in libraries and Seton-Watson remains a good starting point.

Taylor, P.J. and **Flint, C.** (2000) *Political Geography: World-Economy, Nation State and Locality*. Prentice Hall, Harlow. Now in its fourth edition, this text may be consulted for further ideas on most political geography topics.

Yuval-Davis, N. (1997) *Gender and Nations*, Sage, London. An accessible and rewarding survey of its theme.

Finally, two very different studies of European state boundaries and national identities around them:

Berdahl, D. (1998) *Where the World Ended: Re-unification and Identity in the German Borderland*, University of California Press, Berkeley.

Paasi, A. (1996) *Territories, Boundaries and Consciousness: The Changing Geographies of the Finnish–Russian Frontier*, Wiley, Chichester.

For annotated, clickable weblinks and useful tutorials full of practical advice on how to improve your study skills, visit this book's website at www.pearsoned.co.uk/daniels

CHAPTER 20

States, citizenship and collective action

MURRAY LOW

Topics covered

- The history and geography of citizenship
- The relationships between political parties, social movements, states and territory

20.1 Inside the border

From birth, we find ourselves thrown into a world already divided into mutually exclusive territorial compartments each with their own state organizations. A huge number of consequences flow from the apparently arbitrary fact of our being born inside one of these rather than another. Inside each compartment, people find themselves confronted with states that can differ markedly from others in their capacity to intervene in and shape society, their openness to popular needs and demands, and the degree to which they rely on violence to control their populations. They may or may not have access to state services, be able to exercise political rights such as voting or running for office, or be liable to conscription by the government in periods of warfare. These inherited state structures are extremely difficult to replace, or even to dramatically reshape. Revolutionary episodes, in which mass pressure can lead to overall changes in the way states are organized, and where new combinations of social interests can come to control state organizations, are rare in the contemporary world. Furthermore, because of immigration laws, it is usually not easy for people to move to a different territorial state compartment in search of political institutions that are less oppressive or more capable of responding to their needs or interests.

Yet much of the time, we take our membership of particular political communities for granted. We also, within limits, usually accept, or take for granted the right of the political institutions responsible for our own territorial state to make laws, raise taxes, regulate economic and social institutions such as firms and families, and deal with other states on our behalf. This is usually taken to mean that most people accord 'their' states a certain amount of *legitimacy*, meaning that they accept that the political organizations which govern them have the right to do so, and that, on the whole, their laws and policies should be accepted as valid by the populations under their jurisdiction.

In reality, the legitimacy of states and their activities is ambiguous and contested. Many people in the world would rather live under alternative arrangements because established borders and state organizations, while perhaps legitimate in the eyes of some, do not allow for the expression of their national identities. States in some regions of the world, such as central and western Africa (e.g. Sierra Leone, Liberia,

the Congo, Rwanda), the Middle East and Central Asia (e.g. Iraq, Israel/Palestine, Afghanistan) or the Balkans (Bosnia and Serbia) have in recent years shown that violent forms of contestation of states' authority, legitimacy and overall character is an on-going process affecting the lives of millions of people. Moreover, virtually everywhere, particular fields of government activity are subject to fierce contestation by groups calling the legitimacy of policies into question, from contemporary 'pro-democracy' movements in, for example, Burma, China or Cuba to movements in the United States that dispute the right of the Federal government to levy taxes, interfere in local matters, and to strive for a monopoly on the use of effective organized force (e.g. via control of gun ownership). Furthermore, the absence of noisy or visible opposition to states and their activities does not mean the latter have firm foundations in popular consent. Many people are of necessity too concerned with economic survival or everyday life to register much interest, positive or negative, in what 'their' states are doing. Most people have limited knowledge of state activities and the taken for granted status of existing arrangements, often central to an appearance of legitimacy, can be quickly undermined with the public revelation of hitherto unknown information about the workings of government or the effects of particular policies.

There are, of course, institutional and ideological mechanisms by means of which modern states can derive some popular legitimacy, however unstable this may prove to be in particular circumstances. Nationalism is an ideology which, amongst other things, tells us which political communities we ought to belong to. It has played a primary part in making states and their borders seem justifiable to most of their inhabitants (as well as being a means for contesting these) and was considered in Chapter 19. The state organizations dividing up the modern political map have also come to be accepted because of the rights and material benefits they claim to guarantee to their inhabitants including, in many cases, the right to influence state policy through democratic processes. This chapter begins by considering *citizenship*, a status as a member of a particular, usually geographically demarcated, community that allows claims to be made on its state organizations. Citizenship has developed in socially and geographically uneven ways. Like nationalism, it seems necessarily to have exclusionary implications for those who do not fit dominant ideas about who belongs inside a particular state. Nevertheless, citizenship has helped force states to have a wide range of internal and external policies and has facilitated the political mobilization of their populations in pursuit of policy change.

Geographers – and other social scientists – have often characterized these domestic processes as involving relations between states and various types of social actors. The second section of this chapter discusses two important types of collective actor shaping conflicts over domestic policies, political parties and social movements, stressing the interdependence between these two channels of citizen influence on states. In conclusion, it considers some of the limits to these established political geographies of sovereignty and citizenship, raising the possibility that the institutions and compromises underpinning our taken-for-granted view of politics inside states may not survive indefinitely.

The contested division of the world into territorial compartments neatly segregating people into communities defined by common citizenship is possibly the most basic fact of the political geography of the past century. So is the division between 'domestic' institutions and policies from 'international' ones that has largely structured thinking about politics in universities and public life. It is so important that, paradoxically, it has often been taken for granted by geographers and other social scientists. Sometimes, this has led to the assumption that 'society' is a phenomenon defined by national state boundaries, or that states as we know them are somehow the 'natural' basic units of politics (see Agnew and Corbridge 1995). It has also, at times, led geographers to focus on urban or local states and international or transnational political processes and organizations in the sense that these are somehow 'more geographical' than the politics going on at the national scale. The 'domestic' political geographies of citizenship and collective action discussed in this chapter are of course geographical in the sense that they vary across the globe and within countries. Organizations like political parties and social movements, moreover, operate simultaneously at local, national and broader scales. But because it is possible to see 'national' level institutions such as citizenship and political parties as interesting political, but not very geographical, topics the basic geographical structure of a world divided into states which makes them possible should be kept in mind throughout this chapter.

20.2 Citizenship: an unevenly developed institution

Citizenship is a central institution underpinning the legitimacy claimed by modern states. It has two dimensions that make this possible:

- Citizenship makes the inhabitants of territorially defined compartments 'members' of political communities. It gives them certain rights of influence over the personnel and policies of state organizations. In return, citizens have certain responsibilities to obey laws, to pay taxes to fund state activity and to mobilize in episodes of warfare.

- Citizenship, partly because of the political pressure it allows citizens to exert, helps generate and guarantee the provision by states of benefits and services that can give men and women a material stake in established political arrangements.

The first of these dimensions shows that the appeal of citizenship as an institution has a lot to do with its connections with democracy, and relates to a kind of bargain where, in return for political rights, citizens are obliged to give their states material and moral support. In this sense, citizenship has a long history. It has, nonetheless, undergone significant transformations. In the city-states of classical Athens or renaissance Florence, membership of the city could mean an obligation to participate directly in political decision-making. Many people today find this participatory model of citizenship appealing. Historically, however, such participation was restricted to minorities of usually relatively well-to-do males: being a citizen was a privilege marking out some individuals as politically more equal than others.

The crucial innovation in modern models of citizenship was to make it possible to imagine citizenship as an *equalizing* status, creating political equality even if people were unequal in other respects. In the late eighteenth century, the writings of 'Enlightenment' philosophers such as Rousseau and Kant, and the experiences of political and social revolution in America and France, connected this sense of citizens as political equals with the idea that citizens are ideally self-governing, or *autonomous*. This meant that they should be able to imagine themselves as the authors of what states do. So, in these models, membership of a modern state should permit everyone, as equal citizens, to mobilize politically and influence state policy. However, the greater scale and complexity of modern states has usually been said to make the direct participation of citizens in decision-making awkward or impossible, and to have led to the development and diffusion of elections and representative democracy.

This vision of citizens as equals in the political sphere, governing themselves through representative political institutions, is at the heart of the ideals of liberal democracy, and political *liberalism* more generally. Liberal political theory places a great deal of emphasis on the *rights* regulating the relationship between individual citizens and their states. Some of these citizenship rights, such as the right to vote, or to free speech and assembly, connect states and their citizens in ways designed to make politics legitimate by freeing channels of communication and influence between states and societies. Others are designed to create barriers between citizens and states by specifying what states cannot do to their citizens, such as seize their property, invade their homes or arrest them without proper legal justification. Modern citizenship rights thus form part of a liberal model of politics that depends on or assumes a division between the state and the private sphere of the economy and households. In the modern world, this model suggests, the ongoing economic inequality of individuals and families in this private sphere is compatible with an historically

unprecedented emphasis on the equality of individuals in politics.

One of the key dimensions of contemporary globalization is the supposed generalization of this political model around the world. When everyone lives under it, we have famously been told by Fukuyama (1992), we will have arrived at 'the end of history'. Later in this section, a series of questions will be raised about how far the model, and the citizenship rights it rests on, can really can be seen as 'universal', applicable in *principle* to everyone, everywhere. Here, we should simply note that it has always been criticized for being in many respects out of step with the *facts* of political power. Although the idea that everyone, however unequal in other respects, has the same political weight has been crucial in campaigns to eliminate particular political inequalities, in practice women and the poor have been excluded from democratic arrangements in most countries until well into the twentieth century (see Case study 20.1). The underlying reason is obviously that other forms of social and economic inequality affect the degree to which political equality can be institutionalized. The innovations necessary to create liberal citizenship rights are often only partially implemented, can be reversed, and in many political–geographical contexts do not exist at all. Most notably perhaps, the proportion of states in the world

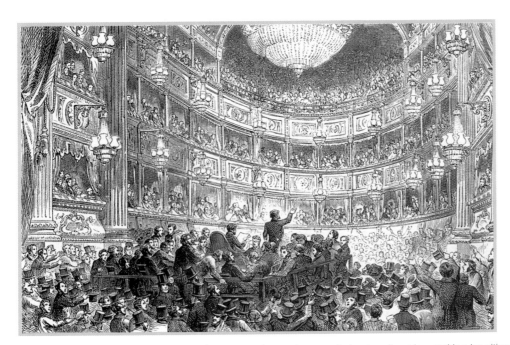

Figure 20.1 Modern representative democracy does not generally involve direct law-making by citizens, but it does generate political participation of other kinds. This free-trade meeting in mid-nineteenth-century London illustrates this kind of direct involvement of citizens in meetings, states and other political institutions. (Mary Evans Picture Library)

Thematic Case Study 20.1

The uneven introduction of women's suffrage

Political citizenship rights have tended to develop in ways that reflect and help reinforce existing gender inequalities. Table 20.1 shows how, in many 'first world' contexts, the right of women to vote was only achieved a long time after that for men. In other parts of the globe, some countries, such as those in South America that gained independence in the nineteenth century, show a similar pattern. In many African and Asian countries which were decolonized during the twentieth century, universal suffrage for men and women was introduced simultaneously.

In some contexts, the issue of women's voting rights has been complicated by the existence of other social divisions. In Peru, for example, although women were granted the right to vote in 1955, literacy tests disenfranchised most of the country's peasant population, men and women (about a third of the total) until 1979. It has proved possible for the right to vote to be granted in a regionally unequal manner within a state. In Nigeria, women in the south of the country were eligible to vote from 1957, while women in the predominantly Islamic north were enfranchised only in 1978.

Table 20.1 shows only the first dates from which all men and all women were entitled to vote for selected countries. In some countries, subsequent political history has meant that these rights have been suspended, or rendered meaningless by the absence of other prerequisites for democracy, such as freedom of speech and association, or by the suppression of political opposition.

Table 20.1 Dates of introduction of universal voting rights for men and women

Country	Votes for all men	Votes for all women
Belgium	1893	1948
Finland	1906	1906
France	1848	1944
Ireland	1918	1918
Spain	1869	1976
Switzerland	1919	1971[1]
UK	1918	1928
USA	1870[2]	1920[2]
New Zealand	1879	1893
Mexico	1917	1953
Uruguay	1916	1932
India	1950	1950
Japan	1925[3]	1945
South Korea	1948	1948
Morocco	1959	1959
Gambia	1961	1961
Kenya	1963	1963

Notes:

1 Switzerland has a Federal system, in which the individual Cantons play an important role in people's lives. It was only in 1990 that women acquired the right to vote on local or cantonal issues in the Canton of Appenzell.

2 These are the dates at which formally universal suffrage was complete. But through the use of poll taxes, literacy tests and intimidation, most African-Americans of both sexes were unable to vote in the southern states until after the passage of the Voting Rights Act of 1965.

3 Men of 25 years and over only.

Sources: Derbyshire and Derbyshire (1989), Lane *et al.* (1991), Krieger (1993), Nelson and Chowdhury (1994)

which can in some minimal, liberal sense be called 'democratic' has remained, despite some ups and downs, fairly consistently low through most of the last century. Despite globalization, then, the global geography of democracy is an uneven one (see Figure 20.1). In many states, public criticism of, and protest against, government actions are criminalized (and in relatively long-established democratic contexts such as the United Kingdom, legislation has recently imposed new and significant restrictions on public demonstrations). Court systems with powers to block government activity that contravenes citizen rights, and that are deliberately insulated from overt forms of governmental pressure, have emerged to try and regulate state actions and to preserve the abilities of citizens to act politically on their own behalf. Yet such systems seem to merely introduce new forms of power in which judges and processes of litigation which are not resolved by democratic rules increasingly decide important issues. To the extent that modern citizens govern themselves at all, they only do so 'at a distance', through elected representatives. Liberal democracy on this model is therefore often thought to be at best an approximation to popular rule. There have always been

Figure 20.2
This British poster from 1910, arguing for the right of women to vote, emphasizes how the granting of formal political citizenship to excluded groups can lead to pressure for changes in state policies. Not surprisingly, established political forces often resist the inclusion of such groups because they anticipate such changes. (Mary Evans Picture Library)

critics who have argued for more direct or participatory decision-making in politics to reduce the influence of necessarily unreliable representatives and to bring the scale of politics closer to the scale of people's lives. Frequently such critics draw on the examples of the lost (but reimagined) ancient and medieval city-states or cite more recent tribal and nomadic communities as 'models' of participation.

The second dimension of modern citizenship, the provision of benefits and services by states, reminds us that states have to *do* things to become important to their inhabitants, and for their powers and activities to enjoy some level of popular legitimacy. Historically, states have been organizations that have specialized in violence, and even now at a very basic level being a citizen of a modern state means sharing with other citizens the apparatus it provides (or sometimes fails to provide) for protection from supposed internal and external threats. Tilly (1985) has described the violent processes through which the modern map of European states took shape, suggesting that states are like rival criminal 'protection rackets', exacting payments in the form of taxes or military service from those who find

themselves on their territory. In return, they are able to provide a degree of security to the latter because of their specialization in violence. He underlines the fact that more 'developed' or diverse forms of citizenship have been fairly late in arriving. European states as recently as the 1780s have plausibly been described as 'little more than elaborated networks of drill sergeants, recruiting officers, impressment gangs and attendant tax officials' (Mann 1993: 371). Put this way, it is easy to see why the legitimacy enjoyed by most states in history has been rather grudgingly given and fragile.

The story of the development of modern citizenship, as it is usually told, involves the expansion of political membership rights which have drawn many states into new fields of activity inside their borders. This story, whatever its flaws (see Case study 20.2), draws attention to the fact that citizenship has never entirely been about formal rights to political influence, protection from government, and *political* equality. Citizenship has unleashed a dynamic through which formal political equality is used to make a political issue of inequalities in the private sphere and creates pressures to change these. As a result of this dynamic,

Degree of
political freedom
(1=high, 7=low)

◼ 1

◼ 2 & 3

☐ 4 & 5

☐ 6 & 7

Figure 20.3 The uneven development of political citizenship, 2002. Political rights remain highly variable across the globe. This map is based on 'political freedom'. Scores produced by Freedom House, an American organization that monitors political and civil rights (for details visit http://freedomhouse.org/). Measuring political rights is difficult and controversial (since the criteria of what is right is itself contested), you should interpret the map with caution.

being a citizen has become much more central to people's life-chances than in the past, and in the twentieth century there was an uneven but nonetheless massive expansion of state activities in the fields of economic and social policy in most parts of the globe. Large-scale efforts at poverty relief, the provision of income for unemployed people, minimum wages for workers, and health services and education at a range of levels are largely twentieth-century phenomena.

As with the first dimension, links with democracy are central to citizenship's capacity to legitimate modern states. Ideas about popular rule are insufficient

to create pressures for democratization: people and movements around the world have struggled for democratic institutions because of the improvements in life-chances more inclusive political regimes seem to promise. In turn, democratic regimes acquire popular support because of their results in terms of socially and economically interventionist policy initiatives. It is not a coincidence that the minimal forms of democracy that have been established in some places in the last hundred years have coincided with the development of welfare states and 'mixed' economies, with states intervening to shape social and economic

---

Thematic Case Study 20.2

A simple model of the development of modern citizenship

It has been tempting to characterize the expansion of citizenship through simple evolutionary models. The best known of these is that of T.H. Marshall (1973) in which the development of citizenship can be broken down into several stages, each of which can be thought of as building on rights already established:

- *Civil citizenship*: legal and civil rights, such as rights of access to the courts, security of property and freedom from arbitrary arrest. These emerged in the seventeenth and eighteenth centuries for selected segments of the population in a few countries.
- *Political citizenship*: rights to free speech (including the public dissemination of views on political questions in various media), of assembly and protest, and the right to vote. The expansion of political citizenship is largely a nineteenth-century phenomenon.
- *Social citizenship*: rights to state guarantees of socio-economic well-being, such as rights to employment, healthcare, pensions or family benefits. These have been characteristic of the uneven expansions of citizenship in the twentieth century.

Marshall mapped the stages of his model onto specific centuries because it was primarily intended as background to a discussion of the role and future of the British welfare state. Nevertheless, critics have been quick to point out how hard it is to generalize his optimistic scheme to the rest of Europe, let alone the rest of the world (e.g. Mann 1988). In some contexts, such as Germany, social rights have appeared well ahead of meaningful political rights. In others certain 'stages' (notably that of political citizenship) have failed to materialize.

The expansion of citizenship has, moreover, rarely had the linear and progressive character which Marshall envisaged. Even at the level of political citizenship, democratization processes have led to fragile results and have commonly been reversed, giving rise to patterns of oscillation over time between more and less empowering bundles of citizenship rights for much of the world's population.

Marshall's model is also rather descriptive: it is not very useful in *explaining* how citizenship rights emerge and become institutionalized, that is stable over time as a result of their acceptance and enforcement by political and legal institutions. There is much disagreement on this last point among researchers. Some view citizenship rights as the product of struggle by subordinate groups such as the working class and their political organizations (e.g. Korpi 1983). Other, more sceptical scholars see them as innovations sponsored by the economically or politically powerful which are really designed to make the task of governing easier by 'buying off' popular opposition to normal politics (e.g. Piven and Cloward 1971).

developments in the interests of their populations. One does not really seem possible in the long run without the other.

Like those characterizing the first dimension of modern citizenship, the policy innovations here have been uneven. The establishment of certain social and economic rights, especially those entailing increased government expenditure, is dependent on a series of economic and political conditions which vary in space and time. In many Third World countries, for example, attempts to create indigenous versions of European welfare states or 'socialism' have been economically difficult and have also challenged the economic and social status of established elites. Both economic development and the changing balance of power and numbers between dominant and subordinate socio-economic groups therefore remain central to explaining the uneven development of citizenship benefits (see Rueschemeyer *et al.* 1992).

Like other citizenship rights, those associated with 'social' citizenship have developed in highly *gendered* ways. Access to many benefits, such as unemployment benefits and pensions, has often been conditional on participation in the formal labour force. Many women have therefore been excluded from important programmes altogether or have only been able to access them through their husbands. The kinds of benefits available have often reflected a model of society in which male 'breadwinners' are imagined as deserving of state aid, as the chronic inadequacy of childcare programmes in most countries highlights, and women have often been central in images of welfare 'dependency' used to attack the principle of social citizenship, as in the attempted demonization of 'single mothers' in the US or UK in recent years. In discussing this gendering of citizenship, Walby (1994) concluded that although social citizenship rights have been important as a means through which many women have achieved some level of freedom from private domination in the household by men, welfare states have generally been anything but gender-neutral.

Most models of modern citizenship imply that, however spatially and historically uneven its development may be, the rights to which the modern citizen is entitled are in principle applicable to everyone, wherever they happen to be, and whatever cultures or social categories they belong to. Much discussion and criticism of political regimes which have not introduced, or have eliminated, many of the rights discussed above, have drawn on a *universal* set of standards based on the experiences of liberal

democracies in Western countries. In recent years, the very idea of 'universal' citizenship, in which sets of rights can be imagined as in principle applicable to all the individuals inhabiting states across the globe, has come under increasing criticism. These criticisms centre on the abstraction from important differences between social groups and societies in such universal models:

1. Contemporary *communitarians* view the emphasis in these models on citizen rights, as opposed to responsibilities, as undermining the sense of obligation citizens should have to the communities and societies in which they live. They suggest that the expansion of state activities in the past century, fuelled by abstract ideas about individual rights, has undermined the role of moral and political civic action rooted in geographically and historically specific communities (e.g. Etzioni 1996). The widespread attempt to reduce state intervention in societies characteristic of many countries in recent decades has been accompanied by escalating criticism of the language of individual citizenship rights and of the resulting high expectations people have of what states can do for them. This, in turn, has led to communitarian ideas becoming more important in policy discussions. So, increasingly emphasis has been placed on the role of voluntary sector organizations at the local level in coping with a range of problems from poverty to crime prevention. Likewise, community-based solutions to 'development' issues have become increasingly important in recent years, in part because they better reflect local goals and cultures than policies developed by national governments and international organizations, but also because of the recent emphasis by international institutions on restructuring the role of the state. In post-colonial contexts, alternative concepts of citizenship and democracy based on more 'communitarian' models deriving from tribal and other indigenous political forms are widely debated, particularly where, as in many African states (such as Uganda, South Africa or Rwanda), 'Western' forms of individual citizenship and electoral democracy have coexisted with such 'indigenous' forms of organization under colonialism and more recently (Mamdami 1996). Controversies abound here because, while the more consensual, community orientated forms of politics associated with non-Western political practices in Africa and elsewhere are appealing in many ways,

and not least for their home-grown character, non-electoral forms of democracy and the maintenance of 'traditional' communal hierarchies and inequalities can also serve to cement the political dominance of elites.

2. From early-modern contention over the regulation of food prices to twentieth-century struggles over social policy, women have reshaped citizenship in important ways (Skocpol 1992, Thompson 1994). Yet in practice citizenship rights such as the rights over property and person, the right to vote, or access to welfare benefits, have been granted in highly unequal ways to men and women. This inequality of access, which has already been alluded to, is no accident, feminist scholars have argued. 'Universal' citizens have, in the Western liberal tradition, been imagined on the model of *male* individuals leaving the clearly-defined 'private' spaces of the home to vote, work and fight (see Prokhovnik 1998). The liberal citizenship model can only pretend to be a universal one, therefore, by systematically concealing relevant gender differences from view, gender differences which might call into question the idea that one model can fit everybody. In this sense, universal citizenship may not only be weak as a solution to gender inequalities, but may also be a source of gender oppression by blocking forms of political representation better able to advocate and decide issues concerning women. Although the issues involved are often very different, this sort of criticism is also useful in considering the relevance of ideas of universal citizenship to political conflicts involving class, ethnicity, 'race' and sexuality (see Young 1989 for a general discussion).

3. The emphasis on universal ideas about citizenship can also be criticized on the grounds that it is 'eurocentric', meaning that differences between political and cultural traditions in the developed countries of the West and those elsewhere are ignored in the assertion by Western governments and international institutions of catalogues of 'human' rights deriving from European Enlightenment models. This has already been touched on in discussing community-based forms of citizenship. One key example here, however, is of models of political membership and of state–society relationships that stress religious criteria, such as those in some predominantly Islamic countries (for example, the ruling family of Saudi Arabia claim

divine sanction and the 'constitution' of the country comprises the sacred texts of Islam). These seem incompatible with universal lists of rights which have a predominantly secular bias and which, for example, may depend on the separation of political and religious institutions. Elsewhere, in one-party political systems (the People's Republic of China is a prominent example), the ruling party rejects western liberal democracy in the name of stability, order and the claim that the party represents popular interests.

4. The internal territorial structure of states, particularly the division of powers between central and local governments, poses some deep problems for universal models of the citizen. The varying activities of 'local' states are rooted in the regionally differentiated problems experienced by their local citizens (see, for example, Duncan and Goodwin 1988). By the same token, local constraints and challenges and the extent of local state financial resources spell uneven access to the benefits of supposedly universal citizenship rights. In this century, many central state organizations have achieved greater importance than local governments in domestic policy through their role in overcoming such inequalities through inter-regional resource redistribution and the standardization of services. Yet disparities remain, as do tensions between national and local priorities, particularly in contexts such as the United States (and other federations) where decentralized patterns of government and decision-making are formalized within federal arrangements. This appears to be another more or less empirical question about the uneven development of citizenship, or the resources necessary to sustain it, but it also relates to important questions of principle about local versus larger-scale democracy. It is currently fashionable to argue for greater power and responsibilities for local governments, and the enhancement of local and regional forms of citizenship, on the grounds that decisions taken closer to the people have greater democratic legitimacy than those taken by central governments (see Borja and Castells 1997). Yet, greater decentralization of policy-making to regional and local governments is likely to make these problems of geographical inequality more acute, and place our inherited concepts of universal citizenship under even more strain (Staehli *et al.* 1997, Rodriguez-Pose and Gill 2003).

5. Most of the above issues are about challenges to dominant Western ideas about universal citizenship from groups and regions who do not necessarily fit the universal pattern. Yet, even inside the dominant institutions of 'the West', the criteria regulating political membership are not really universal at all. Immigration policy reveals very quickly the limits to abstract, universal ideals about what citizenship should be. As modern citizenship has deepened, guaranteeing citizens more and more enforceable rights and often costly services, there has been in many countries a 'narrowing' of access to its benefits. Since the nineteenth century, the problem for states of deciding who is or who is not one of their citizens has become more acute, and they have engaged far more actively in regulating admission to the communities within their borders. In many countries there are sizeable populations of persons, resident for work purposes, who do not enjoy many basic citizenship rights. The formal and informal criteria states use to exclude migrants from citizenship rights (birth, nationality, kinship relations with existing citizens and so on) show how citizenship is in reality entwined with questions of nationality and 'race' (Brubaker 1990). In Germany, for example, opening citizenship to those who are

Plate 20.1 Jean-Marie Le Pen of the French Front National, one of the most consistently successful of many European political parties seeking to gain electoral support by defining citizenship in ways which exclude immigrants and resident populations viewed as culturally or 'racially' different. (Associated Press/Michael Euler)

not 'blood relations' of Germans has been a hotly contested issue, and in most countries a casual nationalism whereby people 'like us' can be citizens and others generally cannot is very much the order of the day. Thus it is difficult to view citizenship as being fundamentally different from, or at least untainted by, other kinds of belonging. It has in practice proved just as exclusionary as national or 'racial' ideas of community for those people who are seen as not fitting into a given territorial space (see Kofman 2003).

This section has reviewed the relationship between citizenship and questions about the legitimacy of states, emphasizing the difficulties involved in assuming that standard modern models of citizenship rights can provide solutions to problems of legitimacy across different historical or geographical contexts or across social divisions such as gender or 'race'. Questions about citizenship are central to political geography because of regional differences within states or between states in what citizenship means, and the differences it makes to people's lives. Citizenship raises important questions about the difficulties, or even the desirability, of coming up with global prescriptions for how people and states ought to relate to each other.

20.3 Collective action and policy: citizens and states

States have become subject to enormous pressures from the societies within their borders to transform their actions in different ways. States become sites of contention as soon as their activities impinge on the everyday lives of their subjects. The central interventions characteristic of states over much of their history – particularly their taxing and conscripting activities – have been struggled over intensively and often violently. Many of the citizenship rights discussed in the previous section owe their origins and shape to protracted conflicts between different social groups over state policy and the expanded social, economic, and environmental activities of twentieth-century states have in turn generally promoted high levels of contention over a much wider range of policies than in the past.

Some of these struggles are largely *distributional*, concerned with state provision of resources and services or with the distribution of burdens such as taxes or pollution. Other conflicts have more to do with *regulatory* matters such as labour relations, workplace health and safety, environmental protection and the possible risks associated with technology. Others primarily have to do with the recognition and protection of identity, from that of the national or ethnic minority groups already considered to the gay and lesbian sexual identities (see Brown 1997, who links the politics of sexuality and identity to ideas about citizenship). Geographers have long been interested in the relationships between states at various scales and social groups or social divisions, so as to be better able to explain state activities. To this end, they have often drawn on one of a number of theories. For example, pluralist theories of the state explain policies as the outcomes of conflict between relevant interest groups. Marxist theories of the state, by contrast, view the relationships and conflicts between different social classes as central to understanding why states act the way they do. The use of such theories to understand geographical and historical variations in, for example, the economic policies of states or the power of organized business or landowning elites, is beyond the scope of this chapter (see Brenner *et al.* 2003, Judge *et al.* 1995). But all of these approaches, although they do not always emphasize it, depend on understanding how citizens (or firms or other relevant participants in policy conflicts) are able to act collectively in pursuit of their political goals. The second part of this chapter is about such *collective action*, and in keeping with the general theme of citizenship, deals with popular, rather than elite, mobilization.

Such organized mobilization can take many forms. *Voting* is a form of collective action organized in such a way, and on a large enough scale, that participants cannot really coordinate their action effectively. In most contemporary societies, political parties are the organizational forms by means of which this important but diffuse form of collective action is coordinated around sets of contrasting policy aims. *Political parties* are usually general in their appeals. They mobilize coalitions which are usually socially heterogeneous during limited and infrequent electoral episodes. Many *social movements*, on the other hand, involve the direct participation of highly motivated groups in political action aimed at achieving specific results. All these have complex geographies.

20.3.1 Political parties and the geography of electoral coalitions

In those states with political systems open to popular influence, a set of institutions (e.g. parties, elections, representative governments) and practices (e.g. periodic voting, ongoing public political debate) channel political conflict through regulated organizations and routines. Marshall (Case study 20.2) related these institutions and practices to the idea of political citizenship rights. Whether the existence of these institutions and practices is enough to make a state democratic is a crucial question. States are subject to all kinds of pressures and influences from economically powerful individuals and business organizations, as well as from movements and pressure groups of a more popular character, which operate through channels other than elections. In many countries, elections seem as much influenced by the money channelled to candidates and parties as they are by popular preferences. Moreover, even in liberal democracies with functioning multiparty systems and

active political contests, the connections between the detail of what governments do and the preferences of electorates are often obscure, leading to the suspicion that electoral democracy is merely a device used by the powerful to help make often unpopular government activities seem legitimate and to channel popular discontent into easily managed and infrequent episodes. All of these worries are, to varying degrees, on target. They also help explain the apparent unease with established political mechanisms felt by many citizens in countries with declining voter turnout in elections. Such problems and qualifications are part of what makes electoral democracy a matter of great concern. It is, on the other hand, hard to deny that the mass participation in elections has been crucial in making states pay much more attention to popular life-chances than they had in the past.

In contemporary liberal democracies, political parties are the main organizations responsible for persuading people to vote, for structuring the choices available to them when they do, and for developing sets of appeals and policy positions which, if they gain

Plate 20.2 The powerful symbolic as well as practical significance of the right to vote is brought home forcefully by this image of voters queuing in the South African elections in 1994, when the majority black population were first able to participate. (Associated Press/Denis Farrell)

sufficient support, form the basis of governmental programmes. They are thus central to explaining the ways in which people vote, and crucial actors shaping the relationships between states and their citizens. The practices of representation in which parties engage form the central political mechanism in electoral democracies. As mentioned earlier, it is usually viewed as the crucial political innovation permitting some measure of popular influence over the fairly large states characteristic of the modern world (Dahl 1989, Manin 1997). What do the parties represent? Different parties do represent different social interests, but we should not take this to mean that voters come neatly pre-divided into clearly identifiable 'voting blocks' based, for example, on class, ethnicity, occupation or gender. If this were the case, we would know what the outcomes of elections would be in advance, simply by looking up census tables! Citizens often have positions on political issues which make sense in terms of their social background, and some parties do become strongly identified with particular social groups. But most citizens have multiple social positions and identities, which can push their political identities and preferences in different directions, and make electoral politics less predictable both for citizens and for politicians aspiring to governmental power.

Moreover, political parties, even those such as labour parties or nationalist parties which seem to appeal to a particular group, can rarely afford to reflect one particular interest. In most contemporary societies there is not an identifiable social group of sufficient numerical strength and internal coherence on which parties can rely to win elections and legitimate their actions in government. A good example of this problem is the situation faced by many socialist parties in European countries. The core interest these parties were founded to represent, the industrial working class, has been steadily declining as a proportion of most electorates, and has rarely formed an electoral majority in any case (Przeworski 1986). Even allowing for definitions of the working class which would include workers in the expanding tertiary and quaternary sectors of the economy, it is clear that building winning electoral coalitions has necessarily involved these parties attempting to attract a variety of social groups simultaneously. These are likely to see themselves as having diverse interests and policy preferences. Likewise, those parties that are strongly identified with business interests, such as the conservative or Christian Democratic parties of Europe or Latin America, cannot aspire to governmental

power without finding ways of appealing through their programmes, policies and rhetoric to large non-business constituencies, including parts of the working class.

In consequence, much of the most interesting work of political parties goes into actively shaping and reshaping partisan political identities in attempts to weld together heterogeneous electoral coalitions. This is achieved, however precariously, through face-to-face contact, mobilization at rallies and other public events, and the extensive use of print and electronic media. The creation of new broad electoral coalitions offering otherwise different groups a shared political identity, and suppressing their differences, has often been the crucial factor in mobilizing sufficient support for reorganizations and expansions of state activity. In the twentieth century, the development of some of the most radical democratic sets of policies and institutions designed to protect ordinary people's interests and redistribute social wealth, in Scandinavia, the United States, and in many Asian and Latin American contexts, have been backed up by electoral coalitions uniting rural small farmers and urban workers who hitherto had not seen themselves as falling into the same political category (see Esping-Anderson 1985). To compensate for declining working-class populations and rural support, the parties concerned have, in recent decades, become more and more reliant on adding voters who define themselves as 'middle class' to their electoral coalitions. In contemporary Europe, the competition for middle-class support has therefore become in many ways the key factor sustaining or undermining established patterns of state intervention. The dilemma for left-of-centre parties is, of course, that increasing reliance on this support often involves diluting their commitments to the very policies they needed middle-class votes to sustain. We can see here how the ability of electoral democracy to deliver material benefits to modern citizens depends not only on the presence of elections and political citizenship rights, but also on the skill and will of parties in creating supportive shared political identities in sometimes rapidly changing circumstances.

This alliance-building activity is, in turn, likely to involve holding together different *regional* bases of support. In the case of the 'farmer–labour coalitions' just mentioned, this is easy to see. They involve constructing electoral bridges between urban and rural, or central and peripheral, regions of the space-economy. The United States example is especially clear.

Presidential Election of 1896

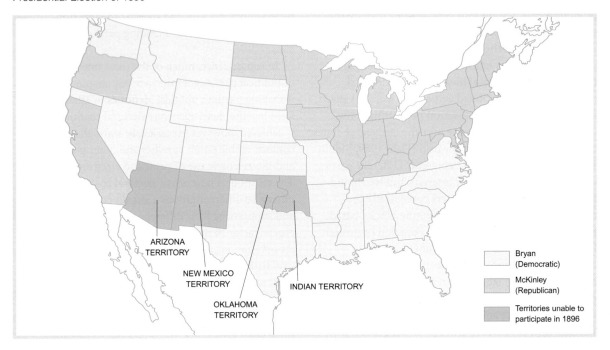

Presidential Election of 1932

Figure 20.4   The formation of a cross-regional coalition: the United States in the 1930s. Bryan, the Democratic Party presidential candidate, lost the important 1896 election despite winning the support of voters in the southern and western agricultural periphery. Franklin Roosevelt's 1932 victory, and subsequent 'New Deal' reforms, were made possible by linking the Democrat's rural base with the votes of northeastern industrial workers.

Source: Based on Nash *et al.* (1990: 670, 623)

The creation of the modern Federal (central) government, with its array of institutions and policies in the fields of industrial regulation, social policy and agricultural support, was propelled by a striking change in the regional electoral appeal of the Democratic party during the Great Depression of the 1930s (Figure 20.4). The 'New Deal coalition' that resulted embraced both the traditional southern rural base of the party and working- and middle-class support in the core industrial areas of the north-east and mid-west, a formula that provided unprecedented legitimacy for popular reforms (Bensel 1986). In many Latin American and Asian contexts, where industrial and service workers form a small and spatially concentrated proportion of the electorate, the building of farmer–labour coalitions has been central to efforts to create legitimacy for inclusive, non-military, governments (Collier and Collier 1991). This sort of coalition is extremely difficult to construct, and has usually been held together by promises of redistributive social policies and land reform. It has, therefore, often been viewed as a threat by urban and land-holding elites. In several Latin American cases, such as the regimes of Juan Perón in 1940s and 1950s Argentina, of Salvador Allende in early 1970s Chile, or of Getúlio Vargas and João Goulart in 1950s–1960s Brazil, attempts to govern on the basis of such cross-regional 'populist' alliances have been an important factor provoking elite-supported military coups.

It is possible for parties to govern on the basis of coalitions which marginalize particular regions on a long-term basis, but this is likely to provoke controversy and, in countries with strong regionalist or nationalist traditions, may even threaten national integrity. Particularly (but not only) where boundaries have been created originally by foreign imperial administrations, for example, in much of Africa, the Balkans, the Middle East, or South-east Asia, the presence of national (or other distinctive ethnic, confessional or linguistic) minorities may foster the formation of party systems organized at the regional or sub-state scale rather than the level of the territorial state as a whole. Concerns about 'national integration' have been important in sustaining one-party (or military) regimes in many such countries, and efforts at sustaining democracy usually have to cope with the threat of one-region rule. In Nigeria, for example, attempts have been made to legally bar the development of parties with concentrated regional bases of support, and one of the thorniest issues for post-apartheid South Africa was the degree to which

politics should be decentralized, allowing regionally based organizations like the Zulu Inkatha Freedom Party to potentially 'rule their own roosts' (see Tordoff 1997: 120). In the United Kingdom in the 1980s and 1990s, there was much commentary about the Conservative government's apparent reliance on a 'two-nation' political strategy involving the use of relatively affluent and populous southern England as an electoral base (Jessop 1989, Gamble 1994). With three parties (the Conservatives, Labour and the Liberal Social Democrat Alliance) dividing the vote, it was possible for the Conservatives to retain national power with less than an overall majority, and crucially with low levels of electoral support elsewhere in the country (see Johnston et al. 1988 for details). The possibility of such regionally polarized rule and the resentments it provokes has been an important reason for the recent devolution of power to Scotland and Wales and is one factor behind pressures for the UK to adopt a different, less geographically based, electoral system (see Case study 20.3).

High levels of regional differentiation and inequality can make the construction and maintenance of nationwide coalitions within parties a difficult task. In Italy, a country with deep interregional economic and social inequalities, the Christian Democrats became the dominant party over much of the post-World War Two period by forging links and compromises between northern business and southern agrarian interests, a strategy which was successful in keeping the Communist party out of government, and which involved the support of both the Vatican and the United States government (see Ginsborg 1990). But during a wave of political scandals in the 1990s it became only too clear that a range of dubious and illegal techniques had become routine parts of holding this alliance together, from the corrupt funnelling of government contracts and finance to linkages with southern Italian organized crime (Stille 1997). As these practices were progressively uncovered, the Christian Democrat-based interregional alliance was 'de-legitimated'. The party collapsed and the entire party system became transformed, along with the 'rules of the game' of the Italian political system. One symptom of the problem faced by the Christian Democrats was the rise of regionalist political forces such as the Northern League seeking to undermine the established geographical organization of the system and sever some of the links built up between North and South. This may seem an extreme case, but the coalition-building in which parties have to engage creates, in

Thematic Case Study 20.3

The spatial organization of electoral democracy: the power of boundaries

Most democracies around the world have adopted electoral systems that make it difficult for a political party to govern alone on the basis of less than a majority share of the overall vote. In these, the numbers of seats allotted to parties in legislatures and other elected bodies are to varying degrees *proportional* to the share of the votes they win nationally (or regionally). The United States and the United Kingdom stand out because in these countries key legislative elections are won by parties accumulating victories for single candidates in a large number of geographically delimited constituencies or districts. This system can be defended in that it maintains a strong representative link between local constituencies and government that is weaker or lacking in other contexts. It can also be criticized, not only for its unfair consequences in terms of translating vote shares into law-making power (particularly through its bias against smaller parties), but also because the construction of the boundaries of electoral constituencies or districts creates the potential for important additional forms of bias in electoral arrangements (see Taylor and Johnston 1979, Morrill 2004).

A key dimension of liberal political citizenship is an *equal* capacity of citizens to influence political outcomes, often expressed using the phrase 'one person, one vote.' The experiences of the US and UK (as well as of other countries, such as Japan, France and Australia, that have used or do use geographically defined constituencies as a component of their systems) demonstrate that those controlling the processes by which electoral boundaries are drawn, redrawn or left as they are, are able to ensure that the votes of some groups in society are more important than those cast by others. One mechanism through which this can be achieved, *malapportionment*, is easy to spot, but for political reasons has proved hard to reform. Where electoral boundaries create constituencies containing unequal numbers of voters that

nonetheless each elect one representative, it is clear that the votes of the residents of the constituencies with the smaller electorates have more impact than those of their fellow citizens. Malapportionment has typically arisen from a reluctance to change constituency or district boundaries in response to population changes, particularly those associated with urbanization and industrialization. In nineteenth century Britain, and until well into the second half of the twentieth century in the US, this resulted in a form of bias where rural areas were hugely over-represented in legislatures, given the size of their electorates, and urban areas correspondingly under-represented. This situation was not unwelcome from the point of view of established elites, rural and urban. As a result, the successive nineteenth century Reform Acts in the British Parliament that were, among other things, designed to address urban under-representation, were fiercely contested, and it took the intervention of the Federal Court system in the US to finally force Congress and the State legislatures to redraw their district boundaries to give rural and urban votes more equal political weight.

Periodic reapportionment in the US, where electoral boundaries now have to be systematically redrawn to reflect population changes, has become a protracted, complex affair. *Gerrymandering*, drawing electoral boundaries so as to deliberately engineer particular political outcomes, typically works by concentrating likely opposition voters into as small a number of districts as possible, where their votes are essentially 'wasted' by forming very large majorities in these few places, or dispersing them as harmlessly as possible into a large number of districts that can be won by the party the boundary drawers favour. It is harder to detect than malapportionment and, because drawing the boundaries *always* involves creating socially different constituencies, may be an inherent liability of US-type electoral practices.

In the past, gerrymandering was particularly prevalent in southern States attempting to minimize the political impact of African-American voters. In the contemporary US, debates have raged around the deliberate creation of Congressional districts, often

Thematic Case Study 20.3 continued

with (in conventional terms) very peculiar geograph-ical forms, designed to concentrate African-American and other minority voters and guarantee the election of minority representatives (see Forest 2001). Because there are no really politically 'neutral' boundaries to be found, elections in systems like this are always likely to be not only about how voters vote, but about how they are divided into electorates in the first place.

general, all sorts of interesting tensions, regional as well as social, within party organizations.

The heterogeneity of partisan coalitions, and the policy fudges and vacillations this helps produce, is one reason why parties in electoral democracies are rarely viewed admiringly by the public or democratic theorists. Another reason is that parties are not simply sustained by popular votes, but also by financial support from business and other elite sources. Inevitably, then, parties have to be seen as organizations effecting compromises of several kinds, both between different social and geographical interests and between elites and citizen majorities. However, it may be that the stitching together of unstable combinations of social forces in which parties specialize has been a key precondition for some degree of democratic government in the past hundred years or so, and for the resulting reorientation of state domestic activities around some popular concerns. The formation of competing partisan identities may even be an important, and underemphasized, antidote to more monolithic forms of political identity, such as nationalism. In any case, parties are never the only vehicles of political collective action in modern societies. When we view them as part of a range of forms of citizen mobilization which also includes social movements, potential correctives to their compromising character are not hard to find.

20.3.2 Social movements: linking formal and informal politics

Democracy is usually thought of as a form of politics involving the institutionalization of parties and voting, but it has never taken a purely electoral form. Social movements routinely involve people more directly in collective action to influence state policy or to exert pressure on other organizations, such as firms or parties. The 'formal' political sector laid out in constitutional arrangements and laws regulating

elections and political parties has always been accompanied by a lively informal sector, in which citizens march, demonstrate, protest, occupy public and private spaces or more quietly fund lobbying and direct action of various sorts by specialized social movement organizations. Social movements are often written about as if they are a more 'progressive', or at least modern, form of political mobilization than parties. One of the themes of this section will be that, exciting as they are, we can only begin to understand movements if we let go of the temptation to make them seem necessarily more innovative or radical than other forms of political action. Social movements are interesting to geographers because of their rootedness in particular places and contexts (although they may sometimes have greater geographical flexibility than political parties).

A social movement is a specialized group formed outside the state and party system in an attempt to promote or prevent social change. Social movements may act to influence government policy, as in the case of groups campaigning against fox-hunting in the United Kingdom, or those struggling for greater press freedom in Iran. They may also act without going through political channels to achieve their aims or make symbolic points. This sort of strategy has been made highly visible in recent times by Greenpeace interposing its ship, the *Rainbow Warrior*, between whaling vessels and their prey, by Operation Rescue blocking the entrances to abortion clinics in the United States, and earlier by 1960s African-American civil rights activists occupying 'whites-only' spaces on buses, at lunch counters and in department stores. The blockading of ports by animal rights activists seeking to prevent the shipment of livestock, the dumping of imported meat into the sea by farmers, and the uprooting of genetically modified crops are other recent examples from the British context.

These practices are not institutionalized and often unruly, but it is hard to see how contemporary politics could work without them. By themselves, votes are a

Plate 20.3 Members of the Falun Gong sect in China, in April 1999, occupying public space to put pressure on the government to recognize and decriminalize their organization. The image is a reminder that common images of social movements as 'modern', 'progressive' or even 'left-wing' may be misleading. This movement is difficult to understand using such conventional Western political categories. (Associated Press/Greg Baker)

notoriously blunt way of communicating opinion to those with influence or potential influence over what states do. Participating in social movements is one way in which citizens can more directly 'voice' their desires and beliefs, and express their political and social identities. Moreover, because social movements are by nature more specialized in their concerns and constituencies than political parties, they are a means by which pressure can be exerted on the powerful by groups concerned with specific policy issues. In this sense, social movements are related to 'interest groups' such as business organizations, which seek to influence the details of policy-making outside electoral channels. In electoral-democratic systems characterized by majority decision-making, the interests of non-elite minorities are often badly served. Some of the most important social movements of the past 50 years, such as the American civil rights movement, gay and lesbian organizations in many countries, or the movements for autonomy or secession, have formed around the protection of, and expansion of opportunities for, minority groups. For all these reasons, and despite the contentiousness and disorder that are often intrinsic to their activities, social movements can be useful to established parties and government policy-makers.

They furnish potentially valuable channels of communication between those in authority and specific groups in civil society as well as indicate the intensity of popular concern about specific policy issues.

Social movement activity usually requires much greater sacrifices of time, energy and resources than voting. Looked at from an economic point of view, collective action of this sort can seem irrational: an individual will almost always see his or her contribution to such activity as so marginal that it will hardly seem worth the effort (Olson 1965). Even if individuals are very concerned about a particular social or political issue they are, the story goes, usually happy to rely on others to do all the work to bring about the changes they want. This is one reason why political parties, which represent rather than directly involve the vast majority of the people in their coalitions, might be thought to be predictably dominant when it comes to linking citizens to policy. Yet levels of popular involvement in voluntary group activities, including politically oriented social movements, can be higher than turnout in elections: a comprehensive study (Verba *et al.* 1995) has suggested, for example, that this is indeed the case in the United States, a country with often remarkably low levels of voter turnout in recent years.

How do people and groups get involved in social movements? It is true that feeling strongly about something does not necessarily lead to doing anything about it. Unlike the situations imagined in most economic models, however, movement mobilization does not take place in a vacuum where independent individuals cautiously weigh up the costs and benefits of their potential contributions. Most social movements grow out of established networks of social relationships where cultures of involvement in political action have taken shape, and where people feel a strong sense of obligation to cooperate in activities organized by those they either know or respect. The American civil rights movement, which in some ways became the model for many other social movements in the US and elsewhere in the last quarter of the twentieth century, grew out of pre-established social networks among urban middle-class African-Americans, churchgoers and students in segregated institutions of higher education (McAdam 1983). Successive waves of activism around educational, nuclear power and environmental issues in Germany have similarly developed out of established sub-cultural networks involving young people and especially students (Markovits and Gorski 1993).

This 'rootedness' of social movements means that in many cases there are strong links between movement development and particular places or localities, and attendant territorialities (of the forms discussed in Chapter 18). The pre-existing cultures and social networks from which movements draw strength are likely, in many cases, to be highly localized. But it is important to set against this the fact that social movements mobilize at a variety of geographical scales. National mobilizations involving demonstrations, often in capital cities and in the full glare of the media, are perhaps the most obvious, although such movements typically develop from smaller-scale mobilizations around more local issues. As Tilly (1984, 1986) shows in the case of France, social movements have become 'nationalized' in step with the development of centralized state power and policy-making. Much attention in recent years has been given to 'transnational' movement mobilization, around issues concerning the environment in particular (Keck and Sikkink 1998, Miller 2004). The effects of such movement activity can be exaggerated, but clearly some organizations such as Greenpeace and many non-governmental organizations in the

development field operate across borders and in recent years there has been much coordinated pressure on governments and international organizations by activists and organizations: the mobilizations around French nuclear weapons testing in the South Pacific in the mid-1990s, the campaign to cancel Third World debt, and especially the diverse activities and strategies of the 'anti-globalization' movement (Plate 20.4) may well be harbingers of greater international networking and pressures by the social movements of the future.

Are we living in a period in which social movements are replacing parties as the main form of popular political mobilization? In the context of a declining working-class base for popular challenges to the state, with the accompanying dilution of socialist party politics and increasing popular antipathy to party politics as such, so-called *new social movements* have often been viewed as an emerging alternative form of radical politics (Touraine 1981, Dalton *et al*. 1990). The new social movements that have sprung up since the late 1960s, such as the women's, environmental, anti-nuclear, gay and lesbian and student movements, can be seen as representing a 'new politics' in which direct action and specialized struggles form an alternative to established democratic mechanisms.

New movements are said to differ from previous social movements, particularly the labour movement, in several ways. 'Old' movements were concerned with the advancement of the material interests of their members, as shown by the achievements in economic policy and in the expansion of social citizenship rights of nineteenth- and early twentieth-century worker mobilization. New movements are said to be more concerned with identity and life-style issues. The post-1968 feminist movement, for example, has often emphasized consciousness-raising among women and the celebration of gender difference as much as or more than established concerns over equality of opportunity in politics and labour markets. Environmental activism is difficult to relate to material concerns: so much so that the label 'post-material' politics has been coined to capture the sense that increasing affluence has enabled publics in developed countries to focus more on quality of life issues than they have in the past (Inglehart 1990). New movements are also said to operate more independently of established political channels than old movements and, as a result, to be far more 'radical' in their tactics, employing a variety of direct-action techniques to achieve their goals.

Plate 20.4   Protestors against the World Trade Organization (WTO) in Geneva in November 1999. The movement highlighting the role, power, and lack of democratic accountability of organizations such as the WTO and the International Monetary Fund has a definite trans-national character. (Popperfoto/Reuters)

There has been an important 'wave' of social movement activity worldwide dating roughly from the mid-1960s. Recent movements have staked out new areas for contesting policy and have often shown a remarkable capacity for innovation in the tactics they have employed. Moreover, as Dalton and Cotten (1996) show in the case of South Korea, where labour militancy has been supplemented by more diverse mobilizations around issues of citizenship, women's rights, environmentalism and poverty, such movements have not been confined only to Western contexts (see also Hipscher 1998). Indeed, labour, environmental and religious-based mobilizations have occurred on all continents. Nevertheless, two important critical points need to be made, which suggest that there is a difference between noting the emergence and proliferation of collective action around 'new' issues, and arguing that this marks a radical break from previous forms of movement and from established politics in general.

Firstly, some of the interest in new movements has rested on a series of flawed comparisons with labour movements. Nineteenth-century labour movements were as concerned with the establishment and assertion of working-class identities as they were with 'bread and butter' material issues. Similarly, non-economic lifestyle- and identity-asserting movements are not recent phenomena. Calhoun (1995) discusses a series of 'counter-cultural' religious and utopian community movements in pre-twentieth-century America that fit the new social movement pattern rather well. But the key counter-example here is probably nationalism, which is, among other things, a far from new social movement that is very much concerned with 'non-material', or at any rate non-economic, issues about language, identity, life-style and collective expression.

Secondly, the 'formal' political sector of elections and parties and the 'informal' sector of movement activism are related in quite fundamental ways. Part of the interest in new social movements was stimulated by the thought that political parties were perhaps things of the past. Most commentators would now suggest that parties have been transforming rather than declining (see Maisel 1994, Mair 1997). More attention is correspondingly being paid to the connections and interactions between movements, parties and national 'systems' of parties (Kitschelt 1990, Müller-Rommel 1990). Some parties grow out of movements. Green parties in Germany, Belgium and elsewhere have emerged out of pre-existing networks connecting environmental and anti-nuclear activism and 'new left' politics (Kitschelt 1989, Markovits and Gorski 1993). The right-wing anti-immigration parties dotting the present-day European political landscape have mostly developed out of background webs of racist organizations, which in some cases are the remnants of earlier attempts at something like party organization. Such party–movement linkages are not new. Earlier waves of party formation in developed countries had been fuelled by workers' and farmers' movements. Many important political parties in African and Asian societies, such as the Indian Congress party, derive from anti-colonial liberation movements.

Even if they originated in other circumstances, established parties often draw on movement support. This is especially obvious in cases where parties are loose, decentralized, and comparatively 'weak' in organizational terms. The United States is a good example of this sort of situation. There, for example, the Democratic party has established often fruitful relationships with many social movements, including the women's movement and civil rights groups in more recent decades. This sort of relationship can be a dangerous game for established party structures and elites, as there is sometimes a fine line between party–movement interactions and the invasion and 'capture' of party organizations by social movement organizations. This can lead to parties being transformed in new directions, and is often a key factor in creating new political possibilities from old political moulds. The periodic capture of some levels of the Democratic party by farmer and labour organizations was a major force for political change in America in the first part of the twentieth century, just as the influence of Christian organizations on the 'religious right' within the Republican party at local and national levels has had important repercussions in its closing decades.

Movements, more specialized vehicles for collective action than parties, are among the most exciting 'domestic' political phenomena. As some of the debates about new social movements suggest, the contrast with parties should not be taken too far. The common romantic image of movements as popular champions and vehicles of direct action is in many cases well deserved but they, like parties, can become as bureaucratic and reliant on financial donations to carry out their activities as many other political organizations. Major environmentalist organizations are often large and professionally staffed, like the business organizations they oppose (McCann 1986). Others, such as Greenpeace, fund direct action by essentially full-time employees by accumulating donations from their mostly passive memberships.

Indeed, like parties, social movements often claim to be acting on behalf of constituencies who are not directly taking part in their activities. None of these tendencies are inevitable: one of the most successful mass movements, the African National Congress in South Africa, has become a ruling party, but there is as yet little clear evidence that this has substantially lessened movement mobilization in that country (Klandermans *et al.* 1998). Nevertheless, social movements, like parties and other political collective actors, need to be critically evaluated from the point of view of both how they really operate and how representative they are of the wider groups their activities affect.

This section has discussed what social movements are and the kinds of activities they engage in: this has involved exploring their geographical rootedness in place and their capacity to organize across geographical scales that makes them so interesting as organizations that potentially operate across the boundaries of domestic and international politics. Their relationship to other forms of political collective action has been examined, and the importance of looking at their interactions with other organizations such as parties has been emphasized, as has the difficulty of clearly distinguishing between old and new social movements on the basis of interest-based versus identity politics. None of these debates about social movements make them any less important in shaping contemporary political geographies. On the contrary, by stressing the long-term nature of many recent movement goals and characteristics and showing how pervasive they are throughout the political spectrum, we can see just how important they are in making and contesting our political worlds.

20.4 Conclusions: limits of territorial politics

This chapter has discussed two dimensions of the relationship between modern states and the people inside their territories, a relationship structured by the basic spatial division of the contemporary world into pockets of 'domestic' political space. Modern citizenship, in conjunction with nationalism, has fostered the development of a certain degree of consent by domestic populations to be ruled by particular state organizations in particular territorial areas. Political parties and social movements have formed links between popular demands and state activity which have enhanced states' legitimacy by subjecting their policies to varying degrees of citizen control. They also create political linkages and alliances across geographical space, inconclusively binding together political constituencies at different scales and acting, particularly in the case of social movements, as mediators between domestic and global politics. It would be wrong to suggest that any of these mechanisms driving modern domestic politics are the best that could be imagined or achieved, that they exhaust our aspirations for what 'democracy' could be, or that they are operative in all parts of the globe. But it would be equally wrong to ignore the improvement of life circumstances that this recent, albeit geographically uneven, 'domestication' of hitherto generally violent and predatory state organizations has brought about for many people.

Given its recent emergence, it seems appropriate in closing to ask how stable this political geography is likely to be. One theme in the discussion of political parties was the importance of broad coalitions, such as those between industrial workers and other relatively underprivileged constituencies such as small farmers, in creating pressures for states to shift their activities in directions favourable to popular interests. Yet, as we have seen, this sort of reformist alliance has become harder and harder to assemble in industrialized countries in recent decades, as economic changes have eroded its social basis. De-industrialization, social fragmentation and middle-class concern about taxation have put pressure on established patterns of social and economic intervention and the idea of ever-expanding citizenship rights has come under fire, not only from conservatives, but also from the inheritors of socialist traditions. Inclusive democracy as it has been institutionalized in developed countries may depend on the delivery of real improvements to ordinary people's circumstances. It is an open question how far governments can go in withdrawing or modifying various citizenship benefits before serious legitimacy problems for states re-emerge.

Furthermore, the picture of citizenship and citizen mobilization presented here assumes a fairly clear distinction between states and their societies. Another reason why citizenship and established patterns of citizen mobilization may experience dramatic transformations lies in the possibility that traditional govern*ment*, theoretically responsive to popular influence through parties and movements, is being replaced in many contexts by new forms of govern*ance*. In these, states devolve power to non-governmental organizations, or to 'partnerships' between parts of the state and private actors such as business organizations or voluntary organizations. While these new arrangements, which are often temporary in character and highly focused in their aims, may be effective in tackling particular problems, they are rarely subject to direct democratic control. Furthermore, the restructuring and 'downsizing' of states that results may, in the long term, make it harder and harder for them to rely on the kinds of popular legitimacy which clearly identifiable *state* benefits and policies make possible. One key aspect of this problem has to do with geographical scale. As the previous chapter noted, the established 'national' states which have formed the main subject of this chapter may cease to be the key organizers of political life, as a result of a general redistribution of power resulting from globalization. If this is true, important consequences for citizenship, and other 'national' vehicles of popular mobilization, would follow, as well as for the imagined communities to which citizens see themselves belonging.

There are a whole series of important theoretical and practical challenges ahead in these shifting geopolitical contexts. What might 'post-national' citizenship look like? Could it be more effective than its forerunners in bridging important social divisions and cultures, or is it necessary to imagine multiple forms of citizenship, each with a certain validity in relation to particular regions or societies, or offering different sets of rights and/or benefits to social groups experiencing different forms of oppression or having different needs? How can governance, as opposed to government, be influenced democratically? How can international organizations be democratized? Should there be international elections, for example, or are conventional diplomatic channels or pressures from social movements or non-governmental organizations enough to give the activities of the United Nations, the World Bank or the World Trade Organization some minimal level of legitimacy? It has been argued (e.g. Linklater 1998) that, after the Second World War and with the development of the United Nations Charter of Human Rights and the concept of 'crimes against humanity', a form of global civil citizenship has been 'under construction'. Is it possible to construct some form of *political* citizenship across borders, languages and cultures? How meaningful could these new citizenships be without forms of non-national governmental activity providing more than formal rights and protections and making some real positive difference to everyday lives?

Modern states are supremely ambiguous social institutions. They are capable of large-scale action to shape societies and the life-chances of their inhabitants through the provision of health care and education, through economic management and the regulation of risks and external threats. They are also, because of their capacity for organized violence, the most dangerous organizations in human history. For a small part of the last millennium, states in some parts of the world have reached accommodations with their populations which have assured their legitimacy in return for higher levels of responsiveness to popular needs. It may be that these accommodations have depended on certain social and geographical conditions which will not prevail far into the new millennium. Whether this is a pessimistic conclusion or not will depend on the political imaginations and struggles of those shaping the political geographies that might displace them.

Learning outcomes

After reading this chapter, you should know that:

- To be a citizen of a state is to be a 'member' of a particular spatially bounded political community.

- The membership benefits associated with modern citizenship are usually said to comprise certain civil and political rights, as well as state benefits and services. The development of access to these rights and benefits has varied greatly across social divisions and geographical space.

- The idea that we can come up with a list of universal citizenship rights, which we think should be available to everyone and used to evaluate the actions of governments across the world, has been increasingly criticized. The ideal of universal citizenship, it is argued, cannot deal with problems of cultural differences between places and communities, or with social divisions such as gender, ethnicity or 'race'.

- Political parties are important organizations through which citizens are mobilized to influence state activities. They are capable of constructing very large coalitions of people for electoral purposes, but these coalitions are heterogeneous and often unstable, because of the social and geographical tensions they contain.

- Social movements are specialized forms of popular collective action, generally aiming to promote change in a particular policy area. Social movements and political parties are interrelated in important ways, and it is unlikely that social movements can replace parties as the main form of citizen mobilization.

- The institutions creating and organizing domestic political geographies in the last century or so may be under threat from government restructuring and a general spatial reorganization of world politics ('globalization'). This may lead to the development of post-national citizenships.

Further reading

Agnew, J. (2002) *Making Political Geography*, Arnold, London. In part an account of how the field of political geography has evolved with the twentieth century, this short text contains material relevant to this chapter and others in this section.

Barnett, C. and **Low, M.** (eds) (2004) *Spaces of Democracy*, Sage, London. The writers in this collection explore a variety of themes raised in this chapter, including electoral geography, citizenship, social movements and democratization.

Kofman, E. (2003) Rights and citizenship, in Agnew, J., Mitchell, K. and Toal, G. (eds) *A Companion to Political Geography*, Blackwell, Oxford. A useful geographic discussion of citizenship, emphasizing several of the themes raised here including the relationships between citizenship and migration.

Johnston, R. and **Pattie, C.** (2003) Representative democracy and electoral geography, in Agnew, J., Mitchell, K. and Toal, G. (eds) *A Companion to Political Geography*, Blackwell, Oxford. A good introduction to the geography of political parties and elections.

Jones, M., Jones, R. and **Woods, M.** (2004) *An Introduction to Political Geography: Space, Place and Politics*, Routledge, London. Useful on geographics of citizenship, nations and states.

Mouffe, C. (1993) Citizenship, in Krieger, J. (ed.), *The Oxford Companion to Politics of the World*, Oxford University Press, Oxford. A clear and concise discussion of the concept of citizenship.

Painter, J. (1995) *Politics, Geography and 'Political Geography': A Critical Perspective*, Arnold, London. A very good introduction to many of the topics considered in this chapter. Chapter 3 is on liberal democracy and the welfare state; Chapter 6 is on social movements.

Taylor, P. and **Flint, C.** (2000) *Political Geography: World-Economy, Nation State and Locality*, 5th edition, Prentice Hall, Harlow. This book has always been wonderful at connecting local and national 'domestic' political topics to a global framework. See Chapter 6 for an interesting discussion of the geographies of elections and political parties.

Walby, S. (1994) Is citizenship gendered? *Sociology*, 28(2), 379–95. A clear feminist analysis of citizenship, emphasizing inequalities between men and women in terms of access to modern citizenship rights.

For annotated, clickable weblinks and useful tutorials full of practical advice on how to improve your study skills, visit this book's website at www.pearsoned.co.uk/daniels

Challenges and promises

PETER DANIELS

MICHAEL BRADSHAW

DENIS SHAW

JAMES SIDAWAY

We routinely make sense of places, spaces, and landscapes in our everyday lives – in different ways and for different purposes – and these 'popular geographies' are as important to the conduct of social life as are our understandings of (say) biography and history.

(Gregory 1994: 11)

After all, things do not happen outside of space and time, and always take *place*. There will always be a place for 'thinking geographically'.

(Hubbard *et al.* 2002: 239)

The root of the word geography combines *geo* (earth) and *graphy* (writing). To engage in geography is to write about the earth (which includes its lands and seas, resources, places and peoples) or, more widely perhaps, to represent the earth in text (which includes maps: some of the most complexly crafted of all texts). Of course, many other branches of knowledge such as history, anthropology, sociology, oceanography, politics, and geology (from which geography draws) are also in some sense or other about the world. Traditionally, what has been distinctive about human geography is that it puts an emphasis on people in places and spaces, on landscapes modified by human interventions and on complex spatial connections: relations, discourses and imaginations of proximity, distance and territory. Today, human geography embraces issues such as spaces of identity, consumption and sexuality which are fundamental parts of the human geographies of our interconnected, contradictory, complex, conflict-ridden and fantastically diverse planet. We would argue that they are no less 'real world', nor necessarily more or less 'abstract' topics, than some of those longer established geographical issues of, for example, resources, population, cities or the changing distribution of economic activity. Indeed, we would argue that some of the most interesting, but tricky, challenges for human geography arise from its ambition to take account of human experiences, achievements and transformations. In other words, human geography is not just about 'out there', it is also about 'in here'. Thus, you all create your own personal geographies derived, for example, from where you live, where you work or when you travel. Moreover, through such myriad social-cultural, political and economic geographies, we inhabit a world, as John Pickles (2004: 5) points out:

> ... that has, in large part, been made as a geo-coded world; a world where boundary objects have been inscribed, literally written on the surface of the earth and coded by layer upon layer of lines drawn on paper.

Such complexity and challenges (as well as diverse 'geo-codes') are evident when we consider 'globalization'. The latter is something that is often considered a defining characteristic of the times in which we live. As we edited the first edition, we were conscious of, and regularly argued amongst ourselves about, what the term 'globalization' takes for granted, what kind of stances we should encourage the contributors to take towards it, and how far we could

agree once we saw the arguments of the chapters and sub-sections emerging. Such discussions were repeated in preparing this new edition. Indeed, as many of the chapters have spelled out, there is certainly a significant sense in which a combination of technical, political, cultural and economic transformations throughout the twentieth century enhanced the sense of global interconnection. Take the case of this book. The example that you are reading may well have been printed far from where you picked it up by a multinational publishing company, whose ownership and 'home' location are difficult to discern. Its production involved dozens of floppy disks, CDs, desktop and laptop computers (themselves usually manufactured in Asia), and software written and distributed by the one of the world's largest (by value) transnational companies (Microsoft). In the case of the first edition, numerous drafts of the chapters and sections were e-mailed between authors and editors, sometimes between offices next door to each other, or sometimes across Europe. In turn, the second edition has been prepared with one of the editors based in Singapore and the three others in the English Midlands, and contributors in a range of locations. And the most convenient place for the editors to meet to finalize the contents of these conclusions and other outstanding matters turned out to be in Philadelphia, whilst we were all attending the centenary convention of the Association of American Geographers in March 2004. Just two decades ago the writing and production process would have been very different. There would have been much less dependence on electronic media, more extensive use of surface or airmail to exchange pieces of the manuscript, and the production of the book, copy-editing, proofreading and printing largely completed locally. To describe the contemporary technological, economic, cultural or political tendencies as 'globalization' is, we agreed, to invoke a certain *geographical imagination*, a vision of the growing significance of a global scale of action, of the world as a single place. Yet, we know that much of the world still remains relatively marginalized from these supposedly hypermobile ways of living and working.

These experiences are made real and meaningful in the language that geographers (and others) adopt. Without the term 'globalization', we might use other terms: 'imperialism', or 'capitalism', for example. Each carries particular connotations. In this way, 'globalization' serves as a particular concept that is used to make sense of the world and one on which a certain geographical imagination (of an increasingly connected and 'shrinking' world, for example), is emphasized. Yet just because something is imagined and interpreted in particular ways and through references to particular geographies does not make it any less real to those caught up in it. Globalization is seen by some as a positive liberalizing force, breaking down barriers, making capitalism more efficient, and spreading its benefits throughout the world. For others it is a negative force enabling another round of exploitation, often with the further destruction of local cultures and identities. Obviously, globalization is contested, in terms of both the meanings attributed to it and the evaluation of its consequences. Judge for yourself, but do not make the mistake of assuming that everyone everywhere shares your experiences of the world.

Again, consider where you were at the moment of the new millennium or when the events that have become known as '9/11' took place in September 2001. The particular construction of space and time that saw the calendar New Year 2000 sweep across the time zones of the earth was also based on a specific geographical and historical imagination. Indeed, global time zones are demarcated with reference to a Royal Observatory at Greenwich in the capital (London) of what was once a great colonial power. However, the coming of the millennium was represented to us (and lived and made by us all) as an enormous global collective experience, likened by some to the moment when the Apollo 8 spacecraft in 1968 captured that first image of the earth as a whole from space.

Moreover, the idea of a twenty-first century, which frames the title of this text, has (as was noted in our Introduction) its roots in a Western (Christian) timetable. According to the calendars of China, Islam and Judaism this is not the twenty-first century, and Islamic, Hindu, Jewish and Chinese 'New Years' do not coincide with the 1st January. Furthermore, the calendar that produced 2000 as a millennium year was established first in the regions that we now refer to as Europe and the Middle East and then refined and disseminated over what we now call 'centuries'. In England the calendar now used was not established as official until 1751. And those seemingly arcane controversies over whether 2000 or 2001 marked a 'new century' are in part rooted in the fact that Western mathematics had no concept of zero, until it was imported, via the Arab–Islamic civilization, from India to Europe. So whilst the year 2000 is a Western construct, it turns out to rely on the idea of zero whose origins are elsewhere. One of the lessons here is that

the West owes much of its identity and knowledge to ideas, people and artefacts originating outside what have been considered its geographical limits. Many observers subsequently claimed that a new age (and perhaps the geopolitics of a new century) began on a sunny New York morning in September 2001, when the drama of hijacked civilian aircraft being used as weapons rapidly became a global media event. Yet the many causes and consequences of '9/11' are far from new. As geographers you will be able to explore their links to flows of resources, power and geopolitics and to the technologies, cultures and economies of mobility that enabled the use of airliners as weapons to strike symbolic targets. Moreover, how the events and meanings of '9/11' are represented, experienced and understood in different places varies enormously.

Yet as was also noted in the Introduction, our academic geographies (particularly as researched and taught in Western Europe and North America) frequently still tend to be rooted in a Western vision, often taking this for granted and assuming that what is the norm is the Western 'developed country' experience. Geography is only at the beginning of what promises to be a difficult but promising process of moving beyond such assumptions and outgrowing its modern roots as a Western science tied up with colonial knowledges of the 'rest' of the world. As it does so, assumptions, categories and taken-for-granted maps will come under scrutiny and challenge. This book has set out some of the ways in which this is taking place as geographers rethink what it means to represent the complexity and diversity and (uneven) interconnectedness of spaces, places and landscapes. As readers think more about this, it may be instructive to compare this book – whose contributors are mostly based in British universities – with three other introductory human geography texts (Le Heron *et al.* 1999, Wait *et al.* 2000, Knox and Marston 2001) written (and, like *An Introduction to Human Geography*, published by multinational information, media and publishing companies) from Australia, New Zealand and the United States respectively. Consider how the contents and style of this text, like those others, is marked by where it was written.

So, where else do you go from here? As a geography student (or any other kind of student for that matter), you will be faced with choices, for example between human and physical geography or between more specialized subjects within one of those traditions. It is in the nature of academic disciplines to sub-divide and to compartmentalize knowledge. Geography is unusual in that it sits at the intersections between the humanities, social sciences and natural sciences. This means that it internalizes both the disadvantages of disciplinarity and the potential advantages of multidisciplinarity. Much of the promise of geography lies on the margins between the sub-disciplines, between human and physical geography or, for example, between economic and cultural geography. In your studies you should make the most of this promise, seek out the space on the margins and look for the connections between the human and natural environments, the economic, the social and the cultural. Many of the challenges that face humankind in the new century also occupy these margins. The emphasis on places and spaces and how they are interconnected, imagined and represented, the hallmark of human geography, is itself promising. We hope that the chapters and arguments brought together in this book will (when supplemented by further reading, critical thought, writing and conversation) encourage our readers to continue to engage with this promise.

A

Abjection the process by which we seek to repress that which we regard as unclean, improper, impure or dangerous. In geographic terms, abject material tends to be located in marginal spaces which become repositories for those things which are regarded as threatening the social body.

Actor-network human actors in interaction through non-human intermediaries and power relations.

Agglomeration the concentration of productive activities in a particular region.

Agribusiness large-scale, capital-intensive, agricultural businesses incorporating supply, production and processing capacities.

Agro-food system the highly integrated system of agricultural production in the developed world which involves both upstream (e.g. suppliers) and downstream (e.g. processing) industries.

Apartheid the policy of spatial separation on racial grounds employed in South Africa under National Party rule between the late 1940s and early 1990s.

Appropriationism the replacement of agricultural inputs with industrial alternatives. It forms a key process in the industrialization of the agro-food system.

B

Balance of payments (BOP) the difference in value between a country's inward and outward payments for goods, services and other transactions.

Balkanization progressive subdivision of a region into small political units.

Bands societies of hunter gatherers typically numbering up to 500 individuals. See Spotlight box 1.1.

Billion thousand million.

Biodiversity the variability among living organisms from all sources including terrestrial, marine and other aquatic ecosystems and the ecological complexes of which they are part; this includes diversity within species, between species and of ecosystems.

Biomass plant and animal residue burnt to produce heat.

Bottom-up development economic and social changes brought about by activities of individuals and social groups in society rather than by the state and its agents.

Bounded applied to the behaviour of decision-makers whose access to information, for example, is constrained by financial resources or time. They also have a limited capacity to process the information that they are able to obtain and they will also be constrained (bounded) by the environment within which their behaviour is taking place. **Bounded** cultures may be regarded as reactions to perceived threats to local cultures, entailing a strong assertion of the latter. These can create a level of local fragmentation, with a parochial, nostalgic, inward-looking sense of local attachment and cultural identity. Bounded cultures generally involve very clear definitions of 'insiders' and 'outsiders' in the creation of a sense of belonging.

Brundtland Report published in 1987, it adopted the position that it was possible to pursue economic growth without compromising the environment and introduced the first widely used definition of sustainable development: 'development which meets the needs of the present without compromising the ability of future generations to meet their own needs'.

C

CAP Common Agricultural Policy.

Capital money put into circulation or invested with the intent of generating more money. See Spotlight box 15.2.

Capitalism an historically specific economic system in which production and distribution are designed to accumulate capital and create profit. The system is characterized by the separation of those who own the means of production and those who work for them.

Chiefdom social formation based on societies characterized by the internal and unequal differentiation of both power and wealth and organized around the principle of kinship. See Spotlight box 1.3.

City-states urban regions having political jurisdiction and control over a specific territory. A form of social organization associated with some of the earliest states.

Civilization refers to an advanced form of social development characterized by such things as urban life, commercial activity, writing systems and philosophical thought.

Class social distinctions between groups of people linked to their material conditions and social status.

Climate change a change in climate that is attributed directly or indirectly to human activity that alters the composition of the global atmosphere and which is in addition to natural climate variability observed over comparable time periods. Includes temperature rise, sea-level rise, precipitation changes, droughts and floods.

Cold War a period extending from the end of the 1950s to the late 1980s during which two ideologically opposed blocs emerged within the world system, both with nuclear capabilities. The first was headed principally by the USA while the other was dominated by the Soviet Union.

Command economy an economic system characterized by the state-led central planning of economic activity combined with the simultaneous suppression of market-type relations.

Commodity a good or service produced through the use of waged labour and sold in exchange for money.

Commodity fetishism the process whereby the material and social origins of commodities are obscured from consumers.

Communism a political theory attributed to the works of Karl Marx and Friedrich Engels. Communism is characterized by the common ownership of the means of production. The foundations for full communism were supposed to be laid during a transitional period, known as socialism. The Soviet Union was the first state to be ruled by a communist government, from 1917 to 1991.

Communitarianism a moral and political philosophy based on the extension and preservation of community. It is the opposite of liberalism which is often regarded as insufficiently sensitive to the social / community sources of self-worth.

Comparative advantage in economic geography, refers to an advantage held by a nation or region in the production of particular set of goods or services. See Spotlight box 13.3.

Competitive advantage exists when a firm can deliver the same benefits as competitors but at lower cost (cost advantage). A firm may also deliver benefits that are greater than those of competing products (differentiation advantage)

Conditional reserve those (mineral or ore) deposits that have already been discovered but that are not economic to work at present-day price levels with currently available extraction and production technologies.

Contact zone the space in which transculturation takes place – where two different cultures meet and inform each other, often in highly asymmetrical ways.

Containment the Western strategy of encircling the Soviet Union and its allies during the Cold War.

Core according to Wallerstein, the core refers to those regions of the capitalist world economy characterized by the predominance of core processes associated with relatively high wages, advanced technology and diversified production. See Spotlight box 2.1.

Counterurbanization population increases in rural areas beyond the commuting range of major urban areas.

Crude birth rate number of births in one year per 1,000 population.

Crude death rate number of deaths in one year per 1,000 population.

Culture a system of shared meanings often based around such things as religion, language and ethnicity that can exist on a number of different spatial scales (local, regional, national, global, among communities, groups, or nations). Cultures are embodied in the material and social world, and are dynamic rather than static, transforming through processes of cultural mixing or **transculturation**. See Spotlight box 12.1.

D

DDT a chemical pesticide banned by many countries in the 1970s as a pollutant which is a probable human carcinogen and causes damage to internal organs.

Deforestation   the removal of forests from an area.

Deindustrialization   refers to a relative decline in industrial employment. It may also refer to an absolute decline in industrial output as well as employment.

Demarginalization   the process whereby a marginal or stigmatized space becomes 'normalized', and its population incorporated into the mainstream.

Demographic transition model   traces the shift from high birth and death rates to low ones. See Case study 4.3.

Depopulation   the reduction of population in an area through out-migration or a reduction in the birth rate below the death rate.

Diaspora   literally, the scattering of a population; originally used to refer to the dispersal of Jews in AD 70, now used to refer to other population dispersals, voluntary and non-voluntary. Evokes a sense of exile and homelessness.

Diaspora space   the spaces inhabited not only by those who have migrated and their descendants, but also by those who are conceptualized as indigenous or 'native'. Similar to contact zone – a meeting point of different cultures.

Disorganized capitalism   a form of capitalism identifiable since the mid-1970s. It is characterized by a relative decline in the importance of extractive and manufacturing industries, together with a relative increase in the importance of services. There is also an increasing tendency for the production process to be dominated by small-scale and flexible forms of organization. See Case study 3.3.

Displaced persons   refugees who have no obvious homeland.

Division of labour   the separation of tasks in the production process and their allocation to different groups of workers.

Doubling time   time taken for a population to double in size. See Case study 4.1.

E

Earth Summit   the UN Conference on Environment and Development held in Rio de Janeiro in 1992. 178 countries negotiated a global strategy centrally concerned with sustainable development.

Ecocatastrophism   the prediction of impending environmental disasters, often as a result of human actions.

Economies of scale   occur when mass production of a good (product) results in a lower average cost for each item. Economies of scale occur within a firm, such as using expensive equipment more intensively, or outside the firm as a result of its location, such as the availability of a local pool of skilled labour or good transport links.

Edge city (exopolis)   a term referring to an area with city-like functions usually arising on the edge of an already urbanized area or conurbation and heavily dependent on fast communications systems. Edge cities are regarded as symptomatic of the most recent phase of urbanization.

Emerging markets   refers to those markets that are perceived to have a substantial growth potential. The term is often used in relation to countries of the former Soviet Union, Asia, Latin America and Africa.

Empire   an extended territorial political unit or state built up, often by force, under a supreme authority. Empires usually involve rule over alien or subject peoples.

Energy mix   the balance between various sources of energy in primary energy consumption.

Energy ratio   the relationship between energy consumption and economic growth in an economy:

$$\text{Energy ratio} = \frac{\text{Rate of change in energy consumption}}{\text{Rate of change in economic growth (GDP)}}$$

A value greater than 1 indicates that the amount of energy required to create an additional unit of GDP is increasing; a value less than 1 suggests the reverse.

Enlightenment   a philosophy or movement that emerged during the eighteenth century and was based on the idea of social progress through the application of reason.

Environmental Non-Governmental Organization (ENGO)   an organization formed by members of the public which has no government connections and is specifically concerned with environmental issues.

Environmentalism   a broad term incorporating the concerns and actions in aid of the protection and preservation of the environment.

Epidemiological transition   switch in the predominant causes of death noted in the movement from a pre-transition to a post-transition society, primarily from infections to degenerative diseases.

Ethnicity   refers to the process through which groups are recognized as possessing a distinct collective cultural identity.

EU European Union.

Exchange the process of interchange of goods and services between individuals, groups and/or organizations, whether involving money or not. Can also refer more broadly to social interactions.

Export processing zone a small closely defined area which possesses favourable trading and investment conditions created by a government to attract export-orientated industries.

F

Factors of production refers to those elements necessary for the effective functioning of the production process and typically includes land, labour and capital.

FAO Food and Agriculture Organization.

Fascism a term used particularly to describe the nationalistic and totalitarian regimes of Benito Mussolini (Italy, 1922–45), Adolf Hitler (Germany, 1933–45) and Francisco Franco (Spain, 1939–75).

Feudalism a hierarchical social and political system common in Europe during the medieval period. The majority of the population were engaged in subsistence agriculture while simultaneously having an obligation to fulfil certain duties for the landholder. At the same time the landholder owed various obligations (fealty) to his overlord. See Case study 1.2.

Financial exclusion the processes that prevent disadvantaged social groups from gaining access to the financial system.

Flexible specialization new methods of production such as multi-purpose machines and small-batch production. The term is frequently associated with agglomerations of firms, or industrial districts.

Food regimes distinct relationships discerned between patterns of international food production and consumption and the developing capitalist system. See Case study 7.1.

Fordism a regime of accumulation involving mass production and consumption. Named after Henry Ford (1863–1947), Fordism is known for a differentiated division of labour, assembly-line production and affordable mass-produced consumer goods.

Fourth World according to Castells (1998) a term referring to those regions effectively excluded from participation in the 'new' world economy. Importantly, it is of relevance not only to regions and countries of the recognized Third World but also to regions within the developed world. Also sometimes used to refer to indigenous or native peoples.

G

Gated communities residential developments protected by a range of mechanisms such as security gates, walls, private security guards and intercom systems.

GATT General Agreement on Tariffs and Trade. A United Nations agency founded in 1948 with one of its main purposes being to facilitate free trade through the gradual reduction of national tariffs on goods and services.

Gemeinschaft a form of community said to be common in traditional societies (as distinct from industrial societies) and associated with notions of stability and informal personal contact.

Gender refers to socially constructed ideas of difference between men and women.

Gender division of labour a division of labour constructed around gender in which particular tasks and occupations are deemed to be male or female activities.

Gendered space the ways in which certain spaces are seen to be occupied exclusively or predominantly by either males or females

Genetic Modification (GM) human manipulation of genetic material (plant, animal and human) to create altered organisms.

Gentrification the process by which middle- and upper-class incomers displace established working-class communities. Often associated with new investment in the built environment, gentrification may be small-scale and incremental (i.e. instigated by individual incomers), or be associated with major redevelopment and regeneration schemes.

Geographies of exclusion the spatial processes by which a powerful grouping consciously seeks to distance itself from other less powerful groupings.

Geopolitics a term that has been used to refer to many things, including a tradition of representing space, states and the relations between them; also emphasizing the strategic importance of particular places.

Gesellschaft   a form of association common in urban-based industrial societies (as distinct from traditional societies) and associated with non-permanent and utilitarian social relationships.

Ghetto   refers to very high concentrations of people drawn from a particular ethnic or cultural background living in specific parts of an urban area. The term is now commonly associated with notions of deprivation, unemployment and social exclusion.

Global cities   those cities performing a dominant role in the world economy and characterized by specialized service-type functions such as financial markets.

Global Environmental Facility (GEF)   an independent clearing house funding organization that disburses funds to developing countries for environmental and sustainable development projects. Managed by the World Bank, UNDO and UNEP.

Global production system   refers to the dominant form of production identifiable at a global scale. It has evolved over many centuries into its current form, referred to as post-Fordism.

Global triad   this term describes how the world economy is essentially organized around a tripolar, macroregional structure whose three pillars are North America, Europe, and East and South-east Asia.

Global warming   an increase in the temperature of the Earth's surface caused by trapping infrared radiation in carbon dioxide, increased amounts of which are produced by burning fossil fuels.

Globalization   a contested term relating to transformation of spatial relations that involves a change in the relationship between space, economy and society.

Glocalization   the interaction between the particular character of places or regions and the more general processes of change represented by globalization.

Green revolution   most commonly refers to the period during the late 1950s and 1960s when new high-yielding forms of wheat and rice were introduced to significant parts of the developing world in an attempt to boost output levels.

Gross domestic product (GDP)   the value left after removing the profits from overseas investments and those profits from the economy that go to foreign investors.

Gross national product (GNP)   a broad measure of an economy's performance; it is the value of the final output of goods and services produced by the residents of an economy plus primary income from non-residential sources.

H

Heartland   identified by the British geographer Halford Mackinder (1904) as the zone in East-Central Europe, control of which would be a key to world domination. The term has since been appropriated by Latin American geopolitics.

Hegemony   term derived from the work of Antonio Gramsci which refers to the ability of a dominant group to exert or maintain control through a combination of overt and subtle mechanisms.

Horizontal integration   occurs when two companies, within the same industry and at the same stage of production, merge.

High-value foods   includes such foods as fruit, vegetables, poultry and shellfish. World trade in high-value foods has increased markedly during the last two decades.

Hybridity   refers to groups as a mixture of local and non-local influences; their character and cultural attributes are a product of contact with the world beyond a local place. See Case study 12.5.

Hypothetical resources   those resources that might be expected to be found in the future in areas that have only been partially surveyed and developed.

I

IMF (International Monetary Fund)   international financial institution that originated from the 1944 Bretton Woods Conference. Its main roles include regulating international monetary exchange and controlling fluctuations in exchange rates in order to alleviate balance of payments problems.

Imaginary geographies   the ideas and representations that divide the world into spaces with particular meanings and associations. These exist on different scales (e.g. the imaginaries that divide the world into a developed core and less developed peripheries, or the imagined divide between the deprived inner city and the affluent suburbs).

Imperialism   a relationship of political, and/or economic, and/or cultural domination and subordination between geographical areas.

Indigenous peoples   those peoples native to a particular territory that was later colonized, particularly by Europeans.

Industrial revolution   a term that is often taken to refer to the marked transformation of productive forces, initially within the British economic system, between the mid-eighteenth and mid-nineteenth centuries. It resulted in the movement of Britain from a largely rural-based economy to one that was dominated by manufacturing and industrial production. Such a transformation resulted in substantial social, political as well as economic changes.

Infant mortality rate (IMR)   number of deaths of persons aged under 1 per 1,000 live births.

International division of labour   a term referring to the tendency for particular countries and regions of the globe to specialize in particular types of economic activity.

Islam   a monotheistic religion founded by the prophet Muhammad in the seventh century AD. Today there are two predominant divisions within Islam, the Sunni and Shi'i.

K

Kondratieff cycle   a term used to describe the cycles of boom and bust evident within the capitalist system since the mid-eighteenth century. Named after the Russian scholar N.D. Kondratieff. See Table 2.1.

Kyoto Protocol   the agreed outcome of a meeting of 160 nations in Kyoto, Japan, in 1997 whereby developed nations agreed to limit their greenhouse gas emissions, relative to the levels emitted in 1990.

L

Legitimacy   with regard to nation-states, a term meaning that the majority of people accept the rule of law of the governing political organizations.

Less developed countries (LDCs)   countries at a disadvantage in today's global competitive environment because their comparative advantage in cheap labour or natural resource endowments has become subordinated to knowledge-based factors. For example, knowledge is the key to innovations in production that ultimately make products more competitive. LDCs suffer from poor productive capacities and competitiveness

related, for example, to weak linkages between manufacturing, services and infrastructural sectors such as telecommunications; insufficiently developed human resources; deficiencies in physical infrastructure; and an inability to generate adequate resources to invest in reducing economic and social problems in ways that would help to improve productive capacity. Examples are Indonesia, Nigeria and Ethiopia.

Liberalism   a political philosophy distinguished by the importance it attaches to the civil and political rights of individuals. May also refer to belief in the importance of the free market.

Life expectancy (at birth)   average number of years of life expected on the basis of age-specific mortality schedules for the specified year.

Limits to Growth   the belief that there are natural limits to possible growth which if exceeded will lead to environmental catastrophe.

Local Agenda 21 (LA21)   the implementation by local administrations of sustainable development practices as defined in Agenda 21, an outcome of the Earth Summit, Rio de Janeiro, 1992.

Locality   a place or region of sub-national spatial scale.

M

Market-based states   modern states where the market is the dominant means by which land, labour, capital and goods are exchanged and has a major influence over social and political organization.

Marxism   a form of socialism and mode of analysis derived from the teachings of Karl Marx (1818–83). Marxism regards capitalism as an inherently unjust system with the capitalists (those who own the means of production) exploiting the proletariat (those who must sell their labour in order to live). It aims to replace capitalism with a fairer system, socialism maturing into communism. See Spotlight box 3.2.

Merchant capitalism   refers to an early phase of capitalist industrial development dominant in the larger urban regions of Europe from the late fifteenth century. Merchants were the principal actors engaged in both the provision of capital and the movement and trade of goods (predominantly bulky staples such as grain and manufactured goods). See Spotlight box 2.2.

Mergers   occur when two firms agree to form a new company.

Modernism   a term typically associated with the twentieth-century reaction against realism and romanticism within the arts. More generally, it is often used to refer to a twentieth-century belief in the virtues of science, technology and the planned management of social change.

Modernity   refers to a period extending from the late sixteenth and early seventeenth centuries (in the case of Europe) to the mid to late twentieth century characterized by the growth and strengthening of a specific set of social practices and ways of doing things. It is often associated with capitalism and notions such as progress.

Monopoly   in theory, exists in an industry when one firm produces all the output of a maket; in practice varies between countries. In the UK, for example, any one firm that has 25 per cent of the market is considered to hold a monopoly.

Moral panic   a term describing periodic episodes of concern about the threat of a particular group to the nation-state. Moral panics are normally fuelled by sensationalist media reporting, and are generally diffused by the state through policies which aim to counteract this imagined threat.

More developed countries (MDCs)   countries with significant competitive advantages in today's globalizing economy. They have well-developed, increasingly knowledge-based and strongly interconnected manufacturing and service sectors that provide a significant proportion of employment and contribute to significant national and individual wealth. Indices such as literacy levels, incomes and quality of life are high and these countries exercise considerable political influence at the global scale. Examples are the UK, the US, Germany and France.

Multiculturalism   refers to a belief or policy that endorses the principle of cultural diversity and supports the right of different cultural and ethnic groups to retain distinctive cultural identities. It has often been criticized for being too symbolic and not politically radical enough in challenging racism.

Multiple deprivation   usually refers to a situation in which an individual or group suffers a series of disadvantages, including poverty, poor health, exposure to criminality, inadequate housing and so on. Geographically often used to denote those denied easy spatial access to a range of services, such as housing, health care, education and also transport.

N

Nationalism   the ideology and sentiment of belonging to a 'nation' and the claim that the 'nation' should be expressed in a 'state'.

Nation-state and **State**   a symbolic system of institutions claiming sovereignty over a bounded territory.

Natural increase   surplus of births over deaths in a population.

Neoliberalism   an economic doctrine promoting market-led growth, deregulation and the privatization of state-owned enterprises.

Neo-Malthusian   the belief that environmental problems are a consequence of population growth, following the arguments set out by Thomas Malthus in the late eighteenth century.

New economic geography   an economic geography that recognizes the importance of culture as an influence on economic processes and outcomes. In this way it draws attention to the culturalization of the economy in contrast to the economization of culture.

New industrial districts (NIDs)   areas that specialize in a particular industry because of external economies which result from specialization.

New international division of labour (NIDL)   the global shift of economic activity that occurs when the process of production is no longer constructed primarily around national economies.

New World Order   the notion of a Western-led post-Cold War structure of global power, which sanctions intervention.

Newly agriculturizing countries (NACs)   refers to the handful of low–middle income countries which have come to specialize in the export of high-value foods. These include Brazil, Mexico, China, Argentina and Kenya.

Newly industrialized countries (NICs)   countries where there has been a relatively recent and significant shift away from primary activities towards manufacturing production. In some cases the proportion of manufacturing production is similar to that of the UK or the US. Examples are South Korea and Mexico.

Niche markets   refers to the existence of consumer groups with identifiable tastes and life-styles.

Non-governmental organization (NGO)   an organization formed by members of the public and one that has no government connections.

Non-renewable (stock) resources   those resources, mainly mineral, that have taken millions of years to form. Their availability is therefore finite as there is no possibility of their stock being replenished on a time scale of relevance to human society.

O

Organized capitalism   a form of capitalism that reached its zenith during the period 1945 to 1973. It was characterized by the dominance of extractive and manufacturing industries, large-scale manufacture, mass-production and a significant level of state involvement. See Spotlight box 3.1.

Overpopulation   used to suggest that the finite resources of a particular area will run out if the population expands beyond a given point. Similar to the idea of a carrying capacity, that there is a limit beyond which environmental degradation occurs.

P

Palaeolithic   the stage in the development of human society when people obtained their food by hunting, fishing and gathering wild plants, as opposed to engaging in settled agriculture. Also referred to as the Early Stone Age.

Pastoral nomadism   a form of social organization that is based on livestock husbandry for largely subsistence purposes. Pastoral nomads are characterized by a high level of mobility which allows them continually to search for new pastures in order to maintain their herds of animals.

Pathways of farm business development   refers to the different types of agricultural land uses to be initiated by the post-productivist transition. See Case study 7.2.

Patriarchy   system of gendered power relations through which men exercise power over women.

Peasant   the term has had a number of different interpretations over the years and in different parts of the world. In general it usually refers to those individuals whose livelihood is largely dependent on the land and on rural subsistence-type activities. The term peasant is usually reserved for those living in organized states, thus distinguishing them from band and tribal members, etc.

Periphery   according to Wallerstein, refers to those regions of the world capitalist economy characterized by low wages, simple technology and limited production.

Personal space   the apparent desire by humans to have a pocket of space around them and into which they tend to resent others intruding.

Political economy   an approach to social study that emphasizes the political/social construction and consequences of economic activity.

Post-Fordism production system   comprises a mix of different ways of organizing production at a number of spatial scales. Common to all these types of production is some form of flexible production.

Postmodernism   a philosophy that holds that the traits associated with twentieth-century modernism, such as belief in the possibility of managing social change according to sets of agreed principles, are now in retreat in the face of increasing individualism, pluralism and eclecticism.

Post-productivist transition   a term used to describe the movement away from productivist agricultural systems. This new phase of agricultural production is characterized by extensive and diversified patterns of farming and the growing importance of non-agricultural activities in the countryside, for example recreation. See Case study 7.2.

Poverty   the condition of possessing an income insufficient to maintain a minimal standard of living. Definitions of poverty are culturally specific, and thus relative to the social norms and expectations endemic to a given nation-state. However, the condition of absolute poverty (i.e. lacking the income to maintain a minimum diet) is acknowledged worldwide.

Power geometry   the ways in which different social groups/individuals are placed in relation to the forces of globalization, enabling some to benefit and others to be disadvantaged.

Prehistoric societies   societies that have left no written records.

Primary energy   the energy in the basic fuels or energy sources used, e.g. the energy in the fuel fed into conventional power stations.

Productivist agriculture   strongly associated with North America. It is characterized by the development of agro-industrial complexes based on grain-fed livestock. It was at its most influential during the period extending from the 1920s to the 1970s. See Case study 7.1.

Proto-industrialization   refers to the early phase of capitalist industrial development in Europe.

Characteristics of this period include a significant level of rural-based industrial activity and a low level of technological application in the production process. See Spotlight box 2.3.

Proven (proved) reserves   those deposits of a resource that have already been discovered and are known to be economically extractable under current demand, price and technological conditions.

Q

Quality assurance schemes   a series of checks for ensuring the origin and quality of (usually) agricultural foodstuffs typically associated with retailers and supermarkets.

R

Race   refers to the division of human beings into supposedly recognizable groups based predominantly on physical characteristics.

Racialized space   the ways in which certain spaces are seen to be occupied exclusively or predominantly by a particular 'racial' or ethnic group.

Racism   practices and attitudes that display dislike or antagonism towards people seen as belonging to particular ethnic groups. Social significance is attached to culturally constructed ideas of difference.

Ranked society   society in which there is an unequal division of status and power between its members, where such divisions are based primarily on such factors as family and inherited social position.

Reciprocity   a system of mutual exchange common in primitive societies. Often involved the mutual exchange of gifts and tokens in order to maintain good external relations or internal social relations.

Redistribution   in pre-capitalist societies a system whereby tribute and taxes were taken from subject peoples by a central ruler or other organization and used to pay officials, soldiers, for the ruler's own consumption, or for other purposes.

Regulated states   social formations organized on the basis of territory. The regulated, or pre-modern, state was not dominated by the market system with political, social and religious considerations usually being of greater importance. See Spotlight box 1.4.

Renewable energy   energy sources such as winds, waves and tides that are naturally replenished and cannot be used up.

Renewable (flow) resources   those resources that are naturally renewed within a sufficiently short time span to be of use to human society. The continued availability of such resources is increasingly dependent upon effective management.

Rent gap   a term describing the difference between the current ground rental for a building and its potential ground rent. As such, **gentrifiers** seek to exploit the rent gap by investing in property and real estate in areas where they perceive rentals to be below their potential market value.

Replacement rate fertility   number of babies that an average woman should have to replace her generation, given prevailing levels of mortality (2.1 per woman in post-transition societies).

Reserves/production ratio (R/P)   if the reserves remaining at the end of any year are divided by the production in that year, the result is the length of time those remaining reserves would last if production were to continue at that level.

Residential segregation   the ways in which, most obviously in urban areas, housing patterns can be observed where people live in areas divided along class or ethnic lines. In some instances this is conscious policy, in others it results from the interaction of social and economic processes.

Resource   a substance in the physical environment that has value or usefulness to human beings and is economically feasible and socially acceptable to use.

Rural services   social goods that comprise the fabric of non-productive life in rural areas (e.g. education, health, transport, shops).

Rurality   functionalist, critical political economy and social representation approach for distinguishing between rural and urban economy and society.

Rust Belt   a region of the north-eastern USA that suffered substantial industrial decline, especially after the Second World War.

S

Second industrial revolution   a term sometimes used to account for the profound technological and accompanying social changes that affected industrial capitalism from the late nineteenth century. During this period Britain's industrial might was challenged by Germany and the USA.

Semi-periphery   according to Wallerstein, refers to those regions of the world capitalist economy which, while exploiting the periphery, are themselves exploited by the core countries. Furthermore, they are characterized by the importance of both core and peripheral processes.

Sense of place   the feelings, emotions and attachments to a locality by residents (past or present), which may be articulated in art, literature, music or histories, or may become part of individual or group memory.

Services   work done as an occupation or business for other individuals or businesses that brings about a change in the condition of a person or of a good belonging to some economic unit which does not produce or modify physical goods.

Sexuality   refers to social differences linked to sexual identity and behaviour.

Simulacrum   a copy without an original. An example might be Disneyland's 'Main Street', which represents an ideal American high street, but is not modelled on an original.

Social exclusion   the various ways in which people are excluded (economically, politically, socially, culturally) from the accepted norms within a society.

Social movement   comprised of individuals, groups and organizations united by a common purpose or goal.

Social representation   lay and academic discourses that describe an event or context.

Spatial interaction   a term used to indicate interdependence between geographical areas. Covers the movement of people, goods, information and money between places.

Spatial relations   the ways in which people are connected across geographic space through economic, social, cultural or political processes.

Speculative resources   those resources that might be found in unexplored areas which are thought to have favourable geological conditions.

State   a political unit having recognizable control (claiming supreme power) over a given territory. Unlike earlier social formations like bands and tribes, states have always based their power on their ability to control a specified territory and its inhabitants.

Stigma   a term describing the condition of possessing a 'spoiled' or discredited identity.

Stratified society   society within which there is an unequal division of material wealth between its members.

Structural adjustment   loans designed to foster structural adjustment in LDCs by promoting market-led growth and a reduction in government intervention in the economy.

Sub-culture   a subdivision of a dominant culture or an enclave within it with a distinct integrated network of behaviour, beliefs and attitudes.

Substitutionism   refers to the increased use of non-agricultural raw materials characterized by the replacement of food and fibre with industrial alternatives. It forms a key process in the industrialization of the agro-food system.

Suburbanization   refers to the movement of middle- and skilled working-class people into residential areas located some distance away from their paid employment.

Sun Belt   refers to the major growth areas of the southern and western parts of the USA during recent years in contrast to the contracting and declining industrial base of the north-east. Can also be used in other parts of the developed world to describe dynamic regions, e.g. the M4 corridor in England.

Sustainable development   a vague yet highly influential term popularized by the publication of the Brundtland Report in 1987. At its simplest it requires contemporary societal development to be undertaken bearing in mind the needs and aspirations of future generations. It has been criticized for failing to represent any real challenge to the prevailing Western development model.

T

Takeover   occurs when one company purchases a controlling interest in a second company against the wishes of the latter's directors.

Territoriality   a term used to describe an expression of ownership and control by an individual, group or state over a particular area of land in order to achieve particular ends.

Territory   a recognizable region (area of land or sea) occupied and controlled by an individual, group or state.

Third World   a rather vague term used to describe those regions of the world in which levels of development, as understood by such measures as GDP, are significantly below those of the economically more advanced regions. The term is increasingly seen as an inadequate description of the prevailing world situation since it disguises a significant amount of internal differentiation.

TNCs   transnational corporations. See Spotlight box 13.2.

Top-down development   economic and social changes brought about by activities of the state and its agencies.

Total fertility rate (TFR)   in general terms equal to the average number of live births per woman within a population.

Traceability   in economic geography refers to the ability to determine the origin of agricultural foodstuffs.

Trade bloc   a group of nation-states that act together in order to further their influence over world trade, e.g. the European Union (EU) or the North American Free Trade Area (NAFTA).

Transculturation   the ways in which subordinated or marginal groups select and invent from dominant cultures; although such groups cannot control what emanates from the dominant culture, they do determine to varying extents what they absorb into their own and what they use it for.

Transgression   a term describing actions that breach social expectations of what is appropriate in a particular place. Deliberate transgression may thus constitute an act of resistance.

Transnationalism   multiple ties and interactions linking people or institutions across the borders of nation-states and measured, for example, as flows of capital, people, information, and images.

Tribe   a type of social formation usually said to be stimulated by the development of agriculture. Tribes tend to have a higher population density than bands and are also characterized by an ideology of common descent or ancestry. See Spotlight box 1.2.

U

Underclass   a term referring to poorer, more marginalized groups in society who are seen to experience multiple deprivation.

United Nations (UN)   organization officially formed in 1945. A collaboration of over 50 countries. Its aim is to maintain international peace and security, tackle economic, social, cultural and humanitarian problems, and promote human rights. This is coordinated through a number of councils and through a wider family of specialized agencies and funds.

Urban village   a residential sector of a city (often inner-city) believed to possess social characteristics typical of a village-type community.

Urbanization   refers to the increasing importance of the urban relative to the total population. It is initially stimulated by the movement of people from rural to urban areas.

V

Vertical integration   firms at different stages of the production chain merge together.

Vital trends   changes in fertility and mortality.

W

World Bank   international financial institution established in 1944. Its main role is to provide development funds to LDCs in the form of loans and technical assistance.

World Summit on Sustainable Development (WSSD) held in Johannesburg in 2002. A mega-conference similar to that held in Rio a decade earlier. Global leaders negotiated 30 global environmental protection targets in the Plan of Implementation.

World systems theory   Immanuel Wallerstein's conceptualization of the changing nature of the world socio-economic system into three distinct historical categories.

BIBLIOGRAPHY

A

Abrahamsen, R. (2001) *Disciplining Democracy: Development Discourse and Good Governance in Africa*, Zed Books, London.

ActionAid (1995) *Listening to Smaller Voices: Children in an Environment of Change*, ActionAid, Chard.

Adams, W.M. (1990) *Green Development: Environment and Sustainability in the Third World*, Routledge, London.

Adams, W.M. (1993) Sustainable development and the greening of development theory, in Schurmann, F. (ed.) *Beyond the Impasse: New Directions in Development Theory*, Zed Books, London.

Aghajanian, A. (1991) Population change in Iran, 1966–86: a stalled demographic transition? *Population and Development Review*, **17**, 703–15.

Aglietta, M. and Breton, R. (2001) Financial systems, corporate control and capital accumulation, *Economy and Society*, **30**: 433–66.

Agnew, J. (1988) *Geopolitics: Revisioning World Politics*, Routledge, London and New York.

Agnew, J. (2002) *Making Political Geography*, Arnold, London.

Agnew, J. (2003) *Geopolitics: Revisioning World Politics*, 2nd edition, Routledge, London and New York.

Agnew, J. and Corbridge, S. (1995) *Mastering Space: Hegemony, Territory and International Political Economy*, Routledge, London.

Agnew, J., Livingstone, D. and Rogers, A. (eds) (1996) *Human Geography: An Essential Anthology*, Blackwell, Oxford.

Agnew, J. Mitchell, K and Toal, G (eds) (2003) *A Companion to Political Geography*, Blackwell, Oxford.

Ahluwalia, P. (2001) *Politics and Post-colonial Theory: African Inflections*, Routledge, London.

Ake, C. (1996) *Democracy and Development in Africa*, Brooks, Washington.

Akzin, B. (1964) *State and Nation*, Hutchinson, London.

Alland, A. (Jr.) (1972) *The Human Imperative*, Columbia University Press, New York.

Allen, J. (1995a) Global worlds, in Allen, J. and Massey, D. (eds) *Geographical Worlds*, Oxford University Press and Open University, Oxford, 105–44.

Allen, J. (1995b) Crossing borders: footloose multinationals?, in Allen, J. and Hamnett, C. (eds) *A Shrinking World? Global Unevenness and Inequality*, Oxford University Press, Oxford, 55–102.

Allen, J. (1999) Cities of power and influence: settled formations, in Allen, J., Massey, D. and Pryke, M. (eds) *Unsettling Cities*, Routledge, London, 182–227.

Allen, J. and Hamnett, C. (1995) *A Shrinking World? Global Unevenness and Inequality*, Oxford University Press, Oxford.

Allen, J. and Massey, D. (1995) *Geographical Worlds*, Oxford University Press, Oxford.

Allen, J., Massey, D. and Pryke, M. (eds) (1999) *Unsettling Cities*, Routledge, London.

Allen, T. and Thomas, A. (2001) *Poverty and Development into the Twenty-first Century*, Oxford University Press and Open University, Oxford.

Almas, R and Lawrence, G. (eds) (2003) *Globalization, Localization and Sustainable Livelihoods*, Ashgate, Aldershot.

Amin, A. (ed.) (1994) *Post-Fordism: A Reader*, Blackwell, Oxford.

Amin, A. and Graham, S. (1999) Cities of connection and disconnection, in Allen, J., Massey, D. and Pryke, M. (eds) *Unsettling Cities*, Routledge, London, 7–47.

Amin, A. and Thrift, N. (1994) Living in the global, in Amin, A. and Thrift, N. (eds) *Globalization, Institutions and Regional Development in Europe*, Oxford University Press, Oxford, 1–22.

Amin, A. and Thrift, N. (2002) *Cities: Reimagining the Urban*, Polity Press, Cambridge.

Anderson, B. (1983) *Imagined Communities: Reflections on the Origin and Spread of Nationalism*, Verso, London.

Anderson, J. (1986) Nationalism and geography, in Anderson, J. (ed) *The Rise of the Modern State*, Harvester, Brighton, 113–26.

Anderson, J. (1995) The exaggerated death of the nation-state, in Anderson, J., Brook, C. and Cochrane, A. (eds) (1995) *A Global World? Re-ordering Political Space*, Oxford University Press and the Open University, Oxford, 65–112.

Anderson, J. (1997) Territorial sovereignty and political identity: national problems, transnational solutions?, in Graham, C. (ed.) *In Search of Ireland: A Cultural Geography*, Routledge, London, 215–36.

Anderson, K. (1991) *Vancouver's Chinatown: Racial Discourse in Canada 1875–1980*, McGill-Queens University Press, Montreal.

Anderson K. (2000) 'The beast within': race, humanity and animality, *Environment and Planning D – Society and Space*, **18**, 301–20.

Anderson, J., Brook, C. and Cochrane, A. (eds) (1995) *A Global World*, Oxford University Press, Oxford.

Anderson, K., Domosh, M., Pile, S. and Thrift, N. (eds) (2003) *Handbook of Cultural Geography*, Sage, London.

Agnew, J., Mitchell, K. and Toal, G. (eds) (2003) *A Companion to Political Geography*, Blackwell, Oxford.

An Atlas of Cyberspace. Online, available at: www.cybergeography.org/atlas/census.html

Anderson (1983) *Imagined Communities: Reflections on the Origins and Spread of Nationalism*, Verso, London.

Anonymous (1996) Ubuntu and democracy in South Africa, *Government Gazette*, Pretoria, No. 16943, 18.

Anonymous (1999) South Africa's middle-class pink currency, *Johannesburg Mail and Guardian*, 18 June, available at: www.globalarchive.ft.com.

Appadurai, A. (1990) Disjuncture and difference in the global cultural economy, *Theory, Culture and Society*, **7**, 295–310.

Appadurai, A. (2003) Sovereignty without territoriality: notes for a postnational geography, in Low, S.M. and Lawrence-Zúñiga, D. (eds) *The Anthropology of Space and Place*, Blackwell, Oxford, 337–49.

Ardener, S. (1995) Women making money go round: ROSCAs revisited, in Ardener, S. and Burman, S. (eds) *Money Go-Rounds: The Importance of Rotating Savings and Credit Associations for Women*, Berg, Oxford.

Ardener, S. and Burman, S. (eds) (1995) *Money Go-Rounds: The Importance of Rotating Savings and Credit Associations for Women*, Berg, Oxford.

Asian Age (1999) We will drive Monsanto out of the country, 6 March, 9.

Askew, I. (1983) The location of service facilities in rural areas, *Regional Studies*, **17**, 305–14.

Atkins, P. and Bowler, I. (2001) *Food in Society: Economy, Culture, Geography*. Arnold, London.

Atkinson, D. and Cosgrove, D. (1998) Urban rhetoric and embodied identities: city, nation and empire at the Vittorio Emanuele II monument in Rome, 1870–1945, *Annals of the Association of American Geographers*, **88**, 28–49.

Auty, R. (1979) World within worlds, *Area*, **11**, 232–5.

Auty, R. (1993) *Sustaining Development in Mineral Economies: The Resource Case Thesis*, Routledge, London.

Auty, R. (2001) *Resource Abundance and Economic Development*, Oxford University Press, Oxford.

B

Baker, B. (1998) The class of 1990: how have the autocratic leaders of sub-Saharan Africa fared under democratisation, *Third World Quarterly*, **19**, 115–27.

Bakhtin, M. (1984) *Rabelais and his World* (trans. Helen Iswolsky), Indiana University Press, Bloomington, IN.

Banister, D. and Norton, F. (1988) The role of the voluntary sector in the provision of rural services: the case of transport, *Journal of Rural Studies*, **4**, 57–71.

Baran, P.A. and Sweezy, P.M. (1966) *Monopoly Capitalism*, Pelican, Harmondsworth.

Barkawi, T. and Laffey, M. (2002) Retrieving the imperial: empire and international relations, *Millennium*, **31**, 109–27.

Barke, M. and Newton, M. (1997) The EU LEADER initiative and endogenous rural development, *Journal of Rural Studies*, **13**, 319–41.

Barnet, R.J. (1973) *Roots of War: The Men and Institutions behind US Foreign Policy*, Penguin, Harmondsworth.

Barnett, C. and Low, M. (eds.) (2004) *Spaces of Democracy*, Sage, London.

Barrett, H. and Browne, A. (1996) Export horticultural production in sub-Saharan Africa: the incorporation of The Gambia, *Geography*, **81**, 47–56.

Barrett, H., Ilbery, B., Browne, A. and Binns, T. (1999) Globalisation and the changing networks of food supply: the importation of fresh horticultural produce from Kenya into the UK, *Transactions of the Institute of British Geographers*, **24**, 159–74.

Barrow, C.J. (1995) *Developing the Environment: Problems and Management*. Longman, Harlow.

Barthes, R. (1972) *Mythologies*, Paladin, London.

Barton, J.R. (1997) *A Political Geography of Latin America*, Routledge, London and New York.

Bassin, M. (1987) Race contra space: the conflict between German Geopolitik and National Socialism, *Political Geography Quarterly*, **6**, 115–34.

Bater, J.H. (1986) Some recent perspectives on the Soviet city, *Urban Geography*, **7**, 93–102.

Bayley, C. (1989) *Atlas of the British Empire*, Hamlyn, London.

BBC (1999) *GM Foods: An Evaluation*. Online, available at: www.bbc.co.uk/science [accessed 12 December 2002].

Beardsworth, A. and Keil, T. (1992) Putting the menu on the agenda, *Sociological Review*, **40**, 253–93.

Bedford, T. (1999) Ethical Consumerism, Consumption, Identity and Ethics, unpublished PhD thesis, University of London, London.

Bell, D. and Valentine, G. (eds) (1995) *Mapping Desire: Geographies of Sexualities*, Routledge, London.

Bell, D. and Valentine, G. (1997) *Consuming Geographies: We are Where we Eat*, Routledge, London.

Bell, M. (1994) Images, myths and alternative geographies of the Third World, in Gregory, D., Martin, R. and Smith, G. (eds) *Human Geography: Society, Space and Social Science*, Macmillan, London, 174–99.

Bennett, J. (1998) Internal displacement: protecting the dispossessed, in *World Aid '96*, The Winchester Group, London.

Bensel, R.F. (1986) *Sectionalism and American Political Development 1880–1980*, University of Wisconsin Press, Madison, WI.

Benton, L. (1995) Will the real/reel Los Angeles please stand up? *Urban Geography*, **16**, 144–64.

Berdahl, D. (1998) *Where the World Ended: Re-unification and Identity in the German Borderland*, University of California Press, Berkeley.

Beresford, M.W. (1988) *East End, West End: The Face of Leeds during Urbanisation 1684–1842*, Thoresby Society, Leeds.

Berger, M.T. (1994) The end of the 'Third World'?, *Third World Quarterly*, **15**, 257–75.

Berger, M.T. (2001) The post-cold war predicament: a conclusion, *Third World Quarterly*, **22**, 1079–85.

Bergeson, A. and Schoenberg, R. (1980) Long waves of colonial expansion and contraction, 1415–1969, in Bergeson, A. (ed.) *Studies of the Modern World System*, Academic Press, New York.

Berman, M. (1983) *All That is Solid Melts into Air: The Experience of Modernity*, Verso, London.

Bernstein, H., Crow, B. and Johnson, H. (1992) *Rural Livelihoods: Crises and Responses*, Oxford University Press, Oxford.

Berry, B.J. (1976) *Urbanisation and Counterurbanisation*, Sage, Beverly Hills, CA.

Betjeman, J. (1937) *Continual Dew*, John Murray, London.

Betts, V. (2002) Geographies of youth, religion and identity in (post-)socialist Cuba, Unpublished PhD thesis, University of Birmingham.

Bhabha, H.K. (1994) *The Location of Culture*, Routledge, London.

Bianchini, F. and Parkinson, M. (eds) (1993) *Cultural Policy and Urban Regeneration. The West European Experience*, Manchester University Press, Manchester.

Bigg, T. (2003) *The World Summit on Sustainable Development: Was it Worthwhile?* IIED, London, available at: www.iied.org/docs/wssd/wssdreview.pdf

Birch, D.L. (1987) *Job Creation in America,* Free Press, New York.

Black, R. (1998) *Refugees, Environment and Development*, Longman, London.

Black, R. and Robinson, V. (1993) *Geography and Refugees*, Belhaven, London.

Blacksell, M. (1994) Irredentism, in Johnson, R.J., Gregory, D. and Smith, D.M. (eds) *The Dictionary of Human Geography,* 3rd edition, Blackwell, Oxford, 299.

Blaut, J.M. (1993) *The Colonizer's Model of the World: Geographical Diffusionism and Eurocentric History*, Guilford Press, New York and London.

Blench, R. (1998) Biodiversity Conservation and its Opponents. *ODI Natural Resource Perspectives* 32, available at: www.odi.org.uk/nrp/32.html

Block, F. (1987) *Revising State Theory*, Temple University Press, Philadelphia.

Blunden, J. (1995) Sustainable resources? in Sarre, P. and Blunden, J. (eds*) An Overcrowded World: Population, Resources and Environment*, Oxford University Press and Open University, Oxford, 161–213.

Bobek, H. (1962) The main stages in socio-economic evolution from a geographical point of view, in Wagner, P.L. and Mikesell, M.W. (eds) *Readings in Cultural Geography*, University of Chicago Press, Chicago, 218–47.

Bonfante, J. (1998) A German requiem, *The Economist*, 6 July, 25.

Bongaarts, J. (1996) Global trends in AIDS mortality, *Population and Development Review*, **22**, 21–46.

Bonnett, A. and Nayak, A. (2003) Cultural geographies of racialization – the territory of race, in Anderson, K., Domosh, M., Pile, S. and Thrift, N. (eds) *Handbook of Cultural Geography*, Sage, London, 300–12.

Booth, K. (1999) Cold wars of the mind, in Booth, K. (ed.) *Statecraft and Security: The Cold War and Beyond*, Cambridge University Press, Cambridge, 29–55.

Borden, I. (2001) *Skateboarding, Space and the City: Architecture and the Body,* Berg: Oxford.

Borgelt, C., Ganssauge, K. and Keckstein, V. (1987*) Mietshaus im Wandel, Wohnungen der behutsamen Stadterneuerung*, S.T.E.R.N., Berlin.

Borja, J. and Castells, M. (1997) *Local and Global: Managing Cities in the Information Age*, Earthscan, London.

Bornstein, D. (1996) *The Price of a Dream: The Story of the Grameen Bank and the Idea that is Helping the Poor to Change their Lives*, University of Chicago Press, Chicago.

Boserüp, E. (1965) *The Conditions of Agricultural*

Growth: The Economics of Agrarian Change under Population Pressure, Aldine Press, Chicago.

Boserüp, E. (1990) *Economic and Demographic Relationships in Development*, Johns Hopkins University Press, Baltimore

Bourdieu, P. (1984) *Distinction*, Routledge & Kegan Paul, London.

Bowler, I. (1992) Sustainable agriculture as an alternative path of farm business development, in Bowler, I., Bryant, C. and Nellis, D. (eds) *Rural Systems in Transition: Agriculture and Environment*, CAB International, Wallingford, 237–53.

Bowler, I. and Ilbery, B. (1997) The regional consequences for agriculture of changes to the Common Agricultural Policy, in Laurent, C. and Bowler, I. (eds) *CAP and the Regions*, INRA, Versailles, 105–15.

Bowler, I., Clark, G., Crockett, A., Ilbery, B. and Shaw, A. (1996) The development of alternative farm enterprises: a study of family labour farms in the north Pennines of England, *Journal of Rural Studies*, **12**, 285–95.

Bozzoli, B. (1999) Space and identity in rebellion. Power, target, resource. Paper presented at Regional Conference on Social Movements and Change, Cork.

BP (various years) *BP Statistical Review of World Energy*, BP, London (available at www.bp.com)

Brah, A. (1996) *Cartographies of Diaspora: Contesting Identities*, Routledge, London.

Brandt, W. (1980) *North-South: A Programme for Survival*, Pan, London.

Braudel, F. (1984) *Civilization and Capitalism 15th–18th Century V III: The Perspective of the World*, English translation, Collins/Fontana Press, London.

Bray, K. and Kilian, T. (1999) God does not pay for it: the political economy of cyberspace, paper presented at the 95th Annual Meeting of the Association of American Geographers, Honolulu.

Breckenridge, C.A. (ed.) (1995) *Consuming Modernity: Public Culture in a South Asian World*, University of Minnesota Press, Minneapolis.

Brenner, N., Jessop, B., Jones, M. and MacLeod, G. (eds) (2003) *State/Space: A Reader*, Blackwell, Oxford.

Bridge, G. and Watson, S. (eds) (2000) *A Companion to the City*, Blackwell, Oxford.

Brindle, D. (1999) Fathers under pressure to become superdads, *Guardian*, 16 June, 9.

Broome, R. (1994) *Aboriginal Australians: Black Responses to White Dominance, 1788–1994*, 2nd edition, Allen & Unwin, St Leonards, NSW.

Brown, D.J., Kaldenberg, D.O. and Browne, B.A. (1992) Socio-economic status and playing the lotteries, *Sociology and Social Research*, **76**, 161–7.

Brown, K. (2002) Innovations in conservation and development, *The Geographical Journal*, **168**, 6–17.

Brown, M. (1997) *Replacing Citizenship: AIDS Activism and Radical Democracy*, Guilford Press, New York.

Brubaker, R. (1990) *Citizenship and Nationhood in France and Germany*, Harvard University Press, Cambridge, MA.

Bruinsma, J. (2003) *World Agriculture: Towards 2015/2030*. FAO, Rome.

Bryant, C. and Johnston, T. (1992) *Agriculture in the City's Countryside*, University of Toronto Press, Toronto.

Bryden, J.M. (ed.) (1994) *Towards Sustainable Rural Communities*, School of Rural Planning and Development, University of Guelph, Ontario.

Bryman, A. (1995) *Disney and his Worlds,* Routledge, London.

Bryson, J.R. (2000) Spreading the message: management consultants and the shaping of economic geographies in time and space, in Bryson, J.R., Daniels, P.W., Henry, N.D. and Pollard, J.S,, *Knowledge Space, Economy*, Routledge, London, 157–75.

Bryson, J., Daniels, P.W. and Henry, N. (1996) From widgets to where? A region in economic transition, in Gerrard, A.J. and Slater, T.R. (eds) *Managing a Conurbation: Birmingham and its Region,* Brewin Books, Birmingham, 56–168.

Bryson, J.R., Daniels, P.W. and Warf, B. (2004) *Service Worlds: People, Organizations, Technologies*, Routledge: London

Bryson, J.R., Henry, N., Keeble, D. and Martin, R. (eds) (1999) *The Economic Geography Reader: Producing and Consuming Global Capitalism,* John Wiley, Chichester.

Buck, D., Getz, C. and Guthman, J. (1997) From farm to table: the organic vegetable commodity chain of northern California, *Sociologia Ruralis*, **37**, 3–19.

Buckley, R. (1995) *The United Nations: Overseeing the New World Order*, Understanding Global Issues 96/6, Understanding Global Issues Ltd., Cheltenham.

Budd, L. and Whimster, S. (eds) (1992) *Global Finance and Urban Living*, London, Routledge.

Bunce, M. (1998) Thirty years of farmland preservation in North America: discourses and ideologies of a movement, *Journal of Rural Studies*, **14**, 233–47.

Burnham, W.D. (1981) The system of 1896: an analysis, in Kleppner, P. *et al.* (eds) *The Evolution of American Electoral Systems*, Greenwood Press, Westport, CT.

Butlin, R.A. (1993) *Historical Geography, Through the Gates of Time and Space*, Edward Arnold, London.

C

Caldeira, T. (1996) Building up walls: the new pattern of social segregation in São Paulo, *International Social Science Journal*, **147**, 55–65.

Caldeira, T. (2001) *City of Walls: Crime, Segregation and Citizenship in Sao Paulo*, University of California Press, Berkeley.

Caldwell, M.L. (2004) Domesticating the French Fry: McDonald's and consumerism in Moscow, *Journal of Consumer Culture*, **4**, 5–26

Calhoun, C. (1995) New social movements of the early nineteenth century, in Traugott, M. (ed.) *Repertoires and Cycles of Collective Action*, Duke University Press, Durham, NC.

Calvocoressi, P. (1991) *World Politics since 1945*, Longman, London.

The Cambridge History of Islam (1970), Cambridge University Press, London.

Campbell, B. (1993) *Goliath: Britain's Dangerous Places*, London, Methuen.

Campbell, C. (1987) *The Romantic Ethic and the Spirit of Modern Consumerism*, Blackwell, Oxford.

Campbell, D. (1992) *Writing Security: United States Foreign Policy and the Politics of Identity*, University of Minnesota Press, Minneapolis, MN.

Carapico, S. (1985) Yemeni agriculture in transition, in Beaumont, P. and McLachlan, K. (eds) *Agricultural Development in the Middle East*, John Wiley & Sons, Chichester, 241–54.

Carneiro, A. de M. (1990) *Descrição da Fortaleza de Sofala e das mais da Índia*, Fundação Oriente, Lisbon.

Carroll, R. (1998) The chill east wind at your doorstep, *Guardian*, 28 October, 2–3.

Carter, N (2001) *The Politics of the Environment: Ideas, Activism, Policy*, Cambridge University Press, Cambridge.

Caspian Revenue Watch (2003) *Caspian Oil Windfalls: Who Will Benefit?*, The Open Society Institute, New York. Available at: www.revenuewatch.org/reports

Castells, M. (1989) *The Informational City: Information Technology, Economic Restructuring and the Urban and Regional Process*, Blackwell, Oxford.

Castells, M. (1996) *The Information Age: Economy, Society and Culture*, vol. 1, *The Rise of the Network Society*, Blackwell, Oxford.

Castells, M. (1997) *The Power of Identity*, Blackwell, Malden, MA.

Castells, M. (1998) *The Information Age: Economy, Society and Culture*, vol. 3, *End of Millennium*, Blackwell, Oxford.

Castells, M. and Hall, P. (1994) *Technopoles of the World*, Routledge, London.

Castles, S. and Miller, M.J. (2003) *The Age of Migration: International Population Movements in the Modern World*, Palgrave Macmillan, Basingstoke.

Caulkin, S. (1998) Credit where credit's overdue, *Observer*, 15 November, 2.

Cavazzani, A. and Fuller, A. (1982) International perspectives on part-time farming, *GeoJournal*, **6**, 383–90.

Central Intelligence Agency (2003) *CIA World Factbook*, CIA, Washington, DC.

Chaliand, G. (1993) Introduction, in Chaliand, G. (ed.) *A People without a Country: The Kurds and Kurdistan*, 2nd edition, Zed, London, 1–10.

Chaliand, G. and Rageau, P. (1985) *Strategic Atlas – World Geopolitics*, Penguin, London.

Champion, A.G. (ed.) (1988) *Counterurbanisation: The Changing Pace and Nature of Population Deconcentration*, Edward Arnold, Sevenoaks.

Champion, A.G. (1994) International migration and demographic change in the Developed World, *Urban Studies*, **31**, 653–77.

Chandler, J. (ed.) (1993) *John Leland's Itinerary*, Sutton, Stroud.

Chaney, D. (1990) Subtopia in Gateshead, the MetroCentre as a cultural form, *Theory, Culture and Society*, **7**, 49–68.

Chaney, D. (1996) *Lifestyles*, Routledge, London.

Chapman, K. and Walker, D. (1987) *Industrial Location: Principles and Policies*, Blackwell, Oxford.

Chatterjee, P. (1993) *The Nation and its Fragments: Colonial and Postcolonial Histories*, Princeton University Press, Princeton, NJ.

Child, J.C. (1985) *Geopolitics and Conflict in South America: Quarrels among Neighbours*, Praeger, New York.

Chomsky, N. (1998) Power in the global arena, *New Left Review*, **230**, 3–27.

Chossudovsky, M. (1997) *The Globalisation of Poverty*, Zed Books, London.

Christaller, W. (1966) *Central Places in Southern Germany* (trans. by C.W. Baskin), Prentice Hall, Englewood Cliffs, NJ.

Chung, R. (1970) Space-time diffusion of the demographic transition model: the twentieth century patterns, in Demko, G.J., Rose, H.M. and Schnell, G.A. (eds) *Population Geography: A Reader*, McGraw-Hill, New York.

Clapham, C. (1985) *Third World Politics*, Croom Helm, London.

Clark, W. (2000) Monocentric to polycentric: new urban forms and old paradigms, in Gridge, G. and Watson, S. (eds) *A Companion to the City*, Blackwell, Oxford, 141–54.

Clarke, G.E. (1996) Blood, territory and national identity in Himalayan states, in Tønnesson, S. and Antlöv, H. (eds) *Asian Forms of the Nation*, Curzon, Richmond, 205–36.

Clark, W.C. (1989) Managing planet Earth. *Scientific American*, **261**, 19–26.

Clifford, J. (1992) Travelling cultures, in Grossberg, L., Nelson, C. and Treichler, P. (eds) *Cultural Studies*, Routledge, London.

Cliquet, R.L. (1991) *The Second Demographic Transition: Fact or Fiction?*, Council of Europe, Strasbourg.

Cloke, P. (1985) Counterurbanisation: a rural perspective, *Geography*, **70**, 13–23.

Cloke, P. (1996) Rural life-styles: material opportunity, cultural experience, and how theory can undermine policy, *Economic Geography*, **72**, 433–49.

Cloke, P., Crang, P. and Goodwin, M. (1999) *Introducing Human Geographies*, Arnold, London.

Cloke, P. and Edwards, G. (1986) Rurality in England and Wales 1981: a replication of the 1971 index, *Regional Studies*, **20**, 289–306.

Cloke, P. and Little, J. (eds) (1997) *Contested Countryside Cultures: Otherness, Marginalisation and Rurality*, Routledge, London.

Coakley, J. (1998) *Sport in Society: Issues and Controversies*, McGraw-Hill, New York.

Cochrane, A. (1995) Global worlds and worlds of difference, in Anderson, J., Brook, C. and Cochrane, A. (eds) *A Global World*, Oxford University Press, Oxford, 249–80.

Cochrane, A. (1999a) Administered cities, in Pile, S., Brook, C. and Mooney, G. (eds) *Unruly Cities? Order/Disorder*, Routledge, London, 299–344.

Cochrane, A. (1999b) Redefining urban politics for the twenty first century, in Jonas, A. and Wilson, D. (eds) *The Urban Growth Machine: Critical Perspectives Two Decades Later*, State University of New York Press, Albany, NY.

Cochrane, A. and Jonas, A. (1999) Reimagining Berlin: world city, national capital or ordinary place?, *European Urban and Regional Studies*, **6**, 145–65.

Cochrane, A., Peck, J. and Tickell, A. (1996) Manchester plays games: the local politics of globalization, *Urban Studies*, **33**, 1317–34.

Cockcroft, E. (1974) Abstract expressionism, weapon of the Cold War, *Artform*, **12**, 39–41.

Coffey, W.J. (1996) The 'newer' international division of labour, in Daniels, P.W. and Lever, W.F. (eds) *The Global Economy in Transition*, Longman, Harlow, 40–61.

Cohen, I (1997) *Green Fire*, HarperCollins, Australia.

Cohen, R. (1998) Recent trends in protection and assistance for IDPs, in Earthscan (1998) *Internally Displaced People: A Global Survey*, Earthscan, Guildford, 3–9.

Coleman, D. (1996) New patterns and trends in European fertility: international and sub-national comparisons, in Coleman, D. (ed.) *Europe's Population in the 1990s*, Oxford University Press, Oxford.

Collier, R.B. and Collier, D. (1991) *Shaping the Political Arena: Critical Junctures, the Labour Movement, and Regime Dynamics in Latin America*, Princeton University Press, Princeton, NJ.

Connor, W. (1994) *Ethnonationalism: The Quest for Understanding*, Princeton University Press, Princeton, New Jersey.

Conzen, M.P. (1990) *The Making of the American Landscape*, Unwin Hyman, Boston.

Cook, I. and Crang, P. (1996) The world on a plate: culinary culture, displacement and geographical knowledges, *Journal of Material Culture*, **1**, 131–53.

Cook, I., Crang, P. and Thorpe, M. (1999) Eating into Britishness: multicultural imaginaries and the identity politics of food, in Roseneil, S. and Seymour, J. (eds) *Practising Identities*, Macmillan, London, 223–48.

Cook, I., Crang, P. and Thorpe, M. (2000a) Regions to be cheerful, culinary authenticity and its geographies, in Cook, I., Crouch, D., Naylor, S. and Ryan, J. (eds) *Cultural Turns/Geographical Turns*, Longman, London, 109–39.

Cook, I., Crouch, D., Naylor, S. and Ryan, J. (eds) (2000b) *Cultural Turns/Geographical Turns*, Prentice Hall, London.

Cook, I. *et al.* (2000c) Social sculpture and connective aesthetics: Shelley Sacks' 'Exchange Values', *Ecumene: A Journal of Cultural Geographies*, **7**, 337–44.

Cooke, P. (ed.) (1995) *The Rise of the Rust Belt*, UCL Press, London.

Coquery-Vidrovitch, C. (1996) *African Women: A Modern History*, Westview Press, Oxford.

Corbridge, S. (ed.) (1995) *Development Studies: A Reader*, Edward Arnold, London.

Corbridge, S., Martin, R. and Thrift, N. (eds) (1994) *Money, Power and Space*, Blackwell, Oxford.

Corner, J. and Harvey, S. (eds) (1991) *Enterprise and Heritage: Crosscurrents of National Culture*, Routledge, London.

Cosgrove, D. (1990) Environmental thought and action: pre-modern and post-modern. *Transactions of the Institute of British Geographers (NS)*, **15**, 344–58.

Council for the Protection of Rural England (1993) *The Regional Lost Land*, CPRE, London.

Council for the Protection of Rural England (1995)

Measuring the Unmeasurable: Twenty Indicators for the Countryside, CPRE, London.

Cowen, M.P. and Shenton, R.W. (1996) *Doctrines of Development*, Routledge, London.

Cox, K (2002) *Political Geography: Territory, State and Society*, Blackwell, Oxford.

Cox, K. and Mair, A. (1991) From localised social structures to localities as agents, *Environment and Planning A*, **23**, 155–308.

Cozens, P.M., Hillier, D. and Prescott, G. (1999) Crime and the design of new-build housing, *Town and Country Planning*, **68**, 231–3.

Crang, M. (1998) *Cultural Geography*, Routledge, London.

Crang, P. (1997) Cultural turns and the (re)constitution of economic geography, in Lee, R. and Wills, J. (eds) *Geographies of Economies*, Arnold, London, 3–15.

Crang, P. and Jackson, P. (2000) Consuming geographies, in Morley, D. and Robins, K. (eds) *British Cultural Studies,* Oxford University Press, Oxford.

Crawford, M. (1992) The world in a shopping mall, in Sorkin, M. (ed.) *Variations on a Theme Park: The New American City and the End of Public Space*, Noonday, New York.

Cresswell, T. (1996) *In Place/Out of Place: Geography, Ideology and Transgression*, Minneapolis, University of Minnesota Press.

Crewe, L. and Gregson, N. (1998) Tales of the unexpected, exploring car boot sales as marginal spaces of consumption, *Transactions of the Institute of British Geographers,* **23**, 39–53.

Crone, P. (1986) The tribe and the state, in Hall, J.A. (ed.) *States in History*, Basil Blackwell, Oxford, 48–77.

Crone, P. (1989) *Pre-Industrial Societies*, Blackwell, Oxford.

Cronon, W. (1991) *Nature's Metropolis: Chicago and the Great West*, W.W. Norton, New York.

Cross, M. and Waldinger, R. (1992) Migrants, minorities and the ethnic division of labour, in Fainstein, S., Gordon, I. and Harloe, M. (eds) *Divided Cities: New York and London in the Contemporary World*, Blackwell, Oxford.

Crouch, D.P., Garr, D.J. and Mundigo, A.I. (1982) *Spanish City Planning in North America*, MIT Press, Cambridge, MA.

Crush, J. (1995) Imagining development, in Crush, J. (ed.) *Power of Development*, Routledge, London, 1–26.

Cummins, S. and MacIntyre, S. (1999) Food deserts, evidence and assumptions in health policy-making, *British Medical Journal*, **325**, 436–8.

Cybriswksy, R. and Ford, L. (2001) Jakarta, *Cities*, **18**, 199–210.

D

Dahl, R. (1989) *Democracy and its Critics*, Yale University Press, New Haven, CT.

Daily Star (1995) Korma out for a meal, *Daily Star*, 5 December.

Dalton, B. and Cotton, J. (1996) New social movements and the changing nature of political opposition in South Korea, in Rodan, G. (ed.) *Political Oppositions in Industrialising Asia*, Routledge, London.

Dalton, R.J., Kuechler, M. and Bürklin, W. (1990) The challenge of new movements, in Dalton, R.J. and Kuechler, M. (eds) *Challenging the Political Order: New Social and Political Movements in Western Democracies*, Oxford University Press, Oxford.

Danforth, L.M. (1995) *The Macedonia Conflict: Ethnic Nationalism in a Transnational World*, Princeton University Press, Princeton, New Jersey.

Daniels, P.W. (1975) *Office Location: An Urban and Regional Study*, G. Bell, London.

Daniels, P.W. (1999) Overseas investment by US service enterprises, in Taylor, P. and Slater, D. (eds) *The American Century: Consensus and Coercion in the Projection of American Power,* Blackwell, Oxford, 67–83.

Daniels, P.W. and Lever, W.F. (eds) (1996) *The Global Economy in Transition*, Longman, Harlow.

Daniels, S. and Lee, R. (eds) (1996) *Exploring Human Geography: A Reader*, Arnold, London.

Dankelman, I. and Davidson, J. (1988) *Women and Environment in the Third World*, Earthscan, London.

Daugherty, H.G. and Kammeyer, K.C.W. (1995) *An Introduction to Population,* 2nd edition, Guilford Press, New York.

Davies, G. (1994) *A History of Money: From Ancient Times to the Present Day*, University of Wales Press, Cardiff.

Davis, J. and Goldberg, R. (1957) *A Concept of Agribusiness*, Harvard Business School, Boston, MA.

Davis, M. (1990) *City of Quartz: Excavating the Future in Los Angeles*, Verso, London.

Davis, M. (1992) *Beyond Blade Runner. Urban Control: The Ecology of Fear*, Magazine Pamphlet Series, Pamphlet No. 23, Open Media, Westfield, NJ.

Davis, M. (1995) Fortress Los Angeles: The militarization of urban space, in Kasinitz, P. (ed.) *Metropolis: Centre and Symbol of our Times*, Macmillan, Basingstoke, 355–68.

Davis, M. (1998) *Ecology of Fear. Los Angeles and the Imagination of Disaster*, Metropolitan Books, New York.

Davis, T. (1995) The diversity of queer politics and the redefinition of sexual identity and community in urban spaces, in Bell, D. and Valentine, G. (eds) *Mapping*

Desire: Geographies of Sexualities, Routledge, London, 284–303.

Dean, K. (2003) The Kachin spatiality in the modern state system, unpublished paper for the SSRC Workshop, *Borders and Regional Markets, Economies and Cultures*, Chiang Mai, Thailand, July 4–6, 2003.

Dennis, R. (1984) *English Industrial Cities of the Nineteenth Century*, Cambridge University Press, Cambridge.

Department for International Development (DFID) (1997) *White Paper on Eliminating World Poverty: A Challenge for the Twenty-first Century*, HMSO, London.

Derbyshire, J.D. and Derbyshire, I. (1989) *Political Systems of the World*, Chambers, London.

Desai, V and Potter, R (eds) (2001) *The Arnold Companion to Development Studies,* Arnold, London.

Desforges, L. (1998) 'Checking out the planet': global representations/local identities and youth travel, in Skelton, T. and Valentine, G. (eds) *Cool Places: Geographies of Youth Cultures*, Routledge, London, 174–91.

Desforges, L. (1999) Travel and tourism, in Cloke, P., Crang, P. and Goodwin, M. (eds) *Introducing Human Geographies,* Arnold, London, 296–304.

Destexhe, A. (1995) *Rwanda and Genocide in the Twentieth Century*, Pluto, London.

Dicken, P. (1992) *Global Shift: The Internationalization of Economic Activity*, Paul Chapman, London.

Dicken, P. (1998) *Global Shift: Transforming the World Economy*, 3rd edition, Paul Chapman, London.

Dicken, P. (2003) *Global Shift: Reshaping the Global Economic Map in the 21st Century*, 4th edition, Sage, London.

Dicken, P. and Lloyd, P.E. (1990) *Location in Space: Theoretical Perspectives in Economic Geography*, Harper & Row, New York.

Dickenson, J., Gould, B., Clarke, C., Mather, S., Prothero, M., Siddle, D., Smith, C. and Thomas-Hope, E. (1996) *A Geography of the Third World*, Routledge, London.

Dobb, M. (1946; 1963) *Studies in the Development of Capitalism*, Routledge & Kegan Paul, London.

Dodd, N. (1994) *The Sociology of Money: Economics, Reason and Contemporary Society*, Continuum Publishing Company, New York.

Dodds, F. (ed.) (1996) *The Way Forward: Beyond Agenda 21*, Earthscan, London.

Dodds, K. (1997) *Geopolitics in Antarctica: Views from the Southern Ocean Rim*, Wiley, Chichester.

Dodds, K. (2000) *Geopolitics in a Changing World*, Prentice Hall, Harlow.

Dodgshon, R.A. (1987) *The European Past: Social Evolution and Spatial Order*, Macmillan, Basingstoke.

Dodgshon, R.A. (1998) *Society in Time and Space. A Geographical Perspective on Change*, Cambridge University Press, Cambridge.

Dodgshon, R.A. and Butlin, R.A. (eds) (1990) *An Historical Geography of England and Wales*, 2nd edition, Academic Press, London.

Doherty, B (2002) *Ideas and Actions in the Green Movement*, Routledge, London.

Dolukhanov, P.M. (1996) *The Early Slavs*, Longman, London.

Donald, J. (1997) This, here, now: imagining the modern city, in Westwood, S. and Williams, J. (eds) *Imagining Cities: Scripts, Signs, Memory*, British Sociological Society and Routledge, London.

Döös, B.R. (1994) Why is global environmental protection so slow?, *Global Environmental Change*, **4**, 179–84.

Dorfman, A. and Mattelart, A. (1975) *How to Read Donald Duck: Imperialist Ideology in the Disney Comic*, International General Editions, New York.

Dorling, D. and Rees, P. (2003) A nation still dividing: the British census and social polarisation 1971–2001, *Environment and Planning A*, **35**, 1287–313.

Dorling D. and Shaw, M. (2000) Life chances and lifestyles, in V. Gardiner and H. Matthews (eds) *The Changing Geography of the UK*, London, Routledge, 230–60.

Dorpalen, A. (1942) *The World of General Haushofer: Geopolitics in Action*, Farrar & Rinehart, New York.

Douglas, M. (1966) *Purity and Danger*, Routledge and Kegan Paul, London.

Doyle, T (2000) *Green Power: The Environment Movement in Australia*. University of New South Wales Press, Sydney.

Driver, F. (1999) Imaginative geographies, in Cloke, P., Crang, P. and Goodwin, M. (eds) *Introducing Human Geographies*, Arnold, London, 209–16.

du Gay, P. (1996) *Consumption and Identity at Work*, Sage, London.

Dudrah, R.K. (2001) Drum N Dhol: British Bhangra, *European Journal of Cultural Studies* (forthcoming).

Duijzings, G. (2000) *Religion and the Politics of Identity in Kosovo*, Columbia University Press, New York.

Duncan, S. and Goodwin, M. (1988) *The Local State and Uneven Development*, Polity Press, Cambridge.

Dunn, K.M. (2001) Representations of Islam in the politics of mosque development in Sydney, *Tijdschrift voor Economische en Sociale Geografi*, **92**, 291–308.

Dunning, J.H. (1981) *International Production and the Multinational Enterprise*, Allen & Unwin, London.

Dwyer, C. (1999) Migrations and diasporas, in Cloke, P., Crang, P. and Goodwin, M. (eds) *Introducing Human Geographies*, Arnold, London, 287–95.

Dwyer, C. and Crang, P. (2002) Fashioning ethnicities: the commercial spaces of multiculture, *Ethnicities*, **2**, 410–30.

Dwyer, C. and Jackson, P. (2003) Commodifying difference: selling EASTern fashion, *Environment and Planning D: Society and Space*, **21**, 269–91.

Dyson, T. (1996) *Population and Food: Global Trends and Future Prospects*, Routledge, London.

E

Earthscan (1992) *Facing the Future: Beyond the Earth Summit*, Earthscan, Guildford.

Earthscan (1997) *The Reality of Aid 1997–1998: An Independent Review of Development Cooperation*, Earthscan, Guildford.

Earthscan (1998) *Internally Displaced People: A Global Survey*, Earthscan, Guildford.

Edwards, L.E. (1983) Towards a process model of office-location decision making, *Environment and Planning A*, **15**, 1327–42.

Edwards, M. and Hulme, D. (1992) *Making a Difference: NGOs and Development in a Changing World*, Earthscan, Guildford.

Ehrlich, P., Ehrlich, A. and Daily, G. (1993) Food security, population and the environment, *Population and Development Review*, **19**, 1–32.

Elliott, D. (2003) *Energy, Environment and Society*, 2nd edition, Routledge, London.

Elson, D. and Pearson, R. (1981) Paying for nimble fingers: the exploitation of women in the third world, *Feminist Review*, **7**, 87–107.

Emerson, T. (1995) Fool's gold, *Newsweek*, October, 32–7.

Engardio, P. (2001) Smart bombs, so why no smart aid?, *BusinessWeek*, 24 December, 58.

Engels, F. (1845) *The Condition of the Working Class in England*, 1987 edition, Penguin, London and New York.

Escobar, A. (1995) *Encountering Development*, Princeton University Press, Princeton, NJ.

Esping-Anderson, G. (1985) *Politics and Markets*, Princeton University Press, Princeton, NJ.

Espiritu, C. (1986) Transnational agribusinesses in the Third World, in Dembo, D., Dias, C., Morehouse, W. and Paul, J. (eds) *The International Context of Rural Poverty in the Third World*, Council of International and Public Affairs, New York.

Esposito, J.L. (ed.) (1995) *The Oxford Encyclopedia of Islam in the Modern World*, Oxford University Press, New York.

Esteva, G. (1992) Development, in Sachs, W. (ed.) *The Development Dictionary*, Zed, London.

Etzioni, A. (1996) *The New Golden Rule*, Touchstone Books, New York.

European Commission (1999) *Sustainable Urban Development in the European Union: A Framework for Action*, Office for Official Publications of the European Communities, Luxembourg.

Evans, D. (1992) *A History of Nature Conservation in Great Britain*, Routledge, London.

Evans, N., Morris, C. and Winter, M. (2002) Conceptualising agriculture: a critique of post-productivism as the new orthodoxy. *Progress in Human Geography*, **26**, 313–32.

F

Fairbairn, D. and Darling, D. (1997) Mapping and map-making: new approaches to the teaching of cartography, Proceedings of the 18th International Cartography Conference, Stockholm, 23–27 June, 4, 1955–1962.

Featherstone, M. (1995) *Undoing Culture: Globalization, Postmodernism and Identity*, Sage, London.

Featherstone, M., Lash, S. and Robertson, R. (eds) (1995) *Global Modernities*, Sage, London.

Fejes, F. (1981) Media imperialism, an assessment, *Media, Culture and Society*, **3**, 281–9.

Fell, A. (1991) Penthesilea, perhaps, in Fisher, M. and Owen, U. (eds) *Whose Cities?*, Penguin, London, 73–84.

Ferguson, J. (1990) *The Anti-Politics Machine: 'Development', Depoliticization and Bureaucratic Power in Lesotho*, Cambridge University Press, Cambridge.

Fieldhouse, D.K. (1999) *The West and the Third World*, Blackwell, Oxford.

Fielding, A.J. (1989) Migration and urbanisation in Western Europe since 1950, *Geographical Journal*, **155**, 60–9.

Finer, A. and Savage, G. (eds) (1965) *The Selected Letters of Josiah Wedgwood*, Cory, Adams & MacKay, London.

Fisas, V. (1995) *Blue Geopolitics: The UN Reform and the Future of the Blue Helmets*, Pluto Press, London.

Fisher, T. (2003) Differentiation of growth processes in the peri-urban region: an Australian case study, *Urban Studies*, **40**, 551–65.

FitzSimmons, M. (1986) The new industrial agriculture: the regional differentiation of speciality crop production, *Economic Geography*, **62**, 334–53.

FitzSimmons, M. (1997) Restructuring, industry and regional dynamics, in Goodman, D. and Watts, M. (eds) *Globalising Food*, Routledge, London, 158–68.

Florida, R. (1996) Regional creative destruction: production organisation, globalization, and the economic transformation of the Midwest, *Economic Geography*, **72**, 314–34.

Flowerdew, R. and Martin, D. (1997) *Methods in Human Geography: A Guide for Students Doing a Research Project*, Longman, Harlow.

Foley, R. (1995) *Humans Before Humanity: An Evolutionary Perspective*, Blackwell, Oxford.

Fonseca, I. (1996) *Bury me Standing*, Vintage, London.

Forbes, D.K. (1984) *The Geography of Underdevelopment*, Johns Hopkins University Press, Baltimore, MD.

Forbes, D. and Thrift, N. (1987) *The Socialist Third World: Urban Development and Territorial Planning*, Blackwell, Oxford.

Forest, B. (1995) West Hollywood as symbol, the significance of place in the construction of a gay identity *Environment and Planning D – Society and Space*, **13**, 133–57.

Forest, B. (2001) Mapping democracy: racial identity and the quandary of political representation, *Annals of the Association of American Geographers*, **91**, 143–66.

Forty, A. (1995) *Objects of Desire: Design and Society Since 1750*, Thames & Hudson, London.

Foster, R.F. (ed.) (1989) *The Oxford History of Ireland*, Oxford University Press, Oxford.

Fothergill, S. and Gudgin, G. (1982) *Unequal Growth: Urban and Regional Employment Change in the UK*, Heinemann, London.

Foucault, M. (1979) On governmentality, *Ideology and Consciousness*, **6**, 5–29.

Frank, A.G. (1969) *Capitalism and Underdevelopment in Latin America*, Monthly Review Press, New York.

Frank, A.G. (1997) The Cold War and me, *Bulletin of Concerned Asian Scholars*, **29**, available at http://csf.colorado.edu/bcas/symmpos/syfrank.htm [accessed 8 January 2003].

Freeman, C. (2000) *High Tech and High Heels in the Global Economy*, Duke, Durham.

Freeman, M. (1986) Transport, in Langton J. and Morris, R.J. (eds) *Atlas of Industrializing Britain, 1780–1914*, Methuen, London and New York.

Freidberg, S. (2003) Cleaning up down South: supermarkets, ethical trade and African horticulture, *Social and Cultural Geography*, **4**, 27–44.

Friedland, W. (1984) The new globalisation: the case of fresh produce, in Bonnano, A., Busch, L., Friedland, B., Gouveia, L. and Mingione, E. (eds) *From Columbus to Conagra: The Globalisation of Agriculture and Food*, University of Kansas Press, Lawrence, 210–32.

Friedman, J. (1994) Globalization and localization, in Friedman, J. (ed.) *Cultural Identity and Global Process*, Sage, London, 102–16.

Friedman, J. and Rowlands, M.J. (1977) Notes towards an epigenetic model of the evolution of civilization, in Friedman, J. and Rowlands, M.J. (eds) *The Evolution of Social Systems*, Duckworth, London, 201–76.

Friedman, T. (1999) It's a small world, *Sunday Times*, 28 March, Section 5.

Friedmann, H. (1982) The political economy of food: rise and fall of the postwar international food order, *American Journal of Sociology*, **88**, Supplement, 246–86.

Friedmann, H. (1993) The political economy of food, *New Left Review*, **197**, 29–57.

Friedmann, H. and McMichael, P. (1989) Agriculture and the state system: the rise and decline of national agricultures, 1870 to the present, *Sociologia Ruralis*, **29**, 93–117.

Friedmann, J. (1966) *Regional Development Policy: A Case Study of Venezuela*, MIT Press, Cambridge, MA.

Friedmann, J. (1992) *Empowerment: The Politics of Alternative Development*, Blackwell, Oxford.

Friedmann, J. (1997) Global crises, the struggle for cultural identity and intellectual porkbarrelling: cosmopolitans versus locals, ethnics and nationals in an era of de-hegemonisation, in Werbner, P. and Modood, T. (eds) *Debating Cultural Hybridity. Multi-Cultural Identities and the Politics of Anti-Racism*, Zed, London, 70–89.

Friend, T. (2003) *Indonesian Destinies*, The Belknap Press of Harvard University Press, Cambridge, MA.

Frouws, J. (1998) The contested redefinition of the countryside. An analysis of rural discourses in the Netherlands, *Sociologia Ruralis*, **38**, 54–68.

Fujita, K. and Child Hill, R. (1995) Global Toyotaism and local development, *International Journal of Urban and Regional Research*, **19**, 7–22.

Fukuyama, F. (1992) *The End of History and the Last Man*, Penguin, London.

Furnham, A. and Argyle, M. (1998) *The Psychology of Money*, Routledge, London.

Furtado, C. (1964) *Development and Underdevelopment*, University of California Press, Berkeley.

Furuseth, O. (1992) Countryside in revolt: rural response to a proposed hazardous waste facility, in Gilg, A.W. (ed.) *Restructuring the Countryside*, Avebury, Aldershot, 159–77.

Fyfe, N. (2004) Zero Tolerance, maximum surveillance? Deviance, difference and crime control in the late modern city, in Lees, L. (ed.) *The Emancipatory City: Paradoxes and Possibilities*, Sage, London.

G

Gad, G. and Holdsworth, D.W. (1987) Corporate capitalism and the emergence of the high-rise office building, *Urban Geography*, **8**, 212–31.

Gamble, A. (1994) *A Free Economy and a Strong State: The Politics of Thatcherism*, 2nd edition, Macmillan, Basingstoke.

Gandy Jr, O.H. (1995) It's discrimination, stupid!, in Brook, J. and Boal, I.A. (eds) *Resisting the Virtual Life*, City Lights Books, San Francisco.

Gans, H. (1962) *The Urban Villagers*, Free Press, New York.

Gans, H. (1968) *People and Problems: Essays on Urban Problems and Solutions*, Basic Books, New York.

Garmise, S., Morgan, K. and Rees, G. (1996) *Networks and Local Economic Development*, Papers in Planning Research 160, Department of City and Regional Planning, University of Wales, Cardiff.

Garreau, J. (1991) *Edge City: Life on the New Frontier*, Doubleday, New York.

Gary, I. and Karl, T.L. (2003) *Bottom of the Barrel: Africa's Oil Boom and the Poor*, Catholic Relief Services. Available at: www.catholicrelief.org/africanoil.cfm

Gbadegesin, A. (1991) Farming in the urban environment of a developing nation: a case study from Ibadan metropolis in Nigeria, *Environmentalist*, **11**, 105–11.

Gelbard, A., Haub, C. and Kent, M.M. (1999) World population beyond six billion, *Population Bulletin*, **54**, 1–44.

Geobusiness Magazine (2001) *The World Bank*, Geobusiness, Geneva.

German, T. and Randall, J. (eds) (1993) *The Reality of Aid*, Action Aid, London.

Geyer, H. (1996) Expanding the theoretical foundation of differential urbanization, *Tidjschrift voor Economische en Sociale Geografie*, **87**, 44–59.

Giddens, A. (1990) *The Consequences of Modernity*, Polity Press, Cambridge.

Giddens, A. (1991) *The Constitution of Society*, Polity Press, Cambridge.

Giles, C. and Goodall, I.H. (1992) *Yorkshire Textile Mills, 1770–1930*, HMSO, London.

Gilg, A. and Battershill, M. (1998) Quality farm food in Europe: a possible alternative to the industrialised food market and to current agri-environmental policies: lessons from France, *Food Policy*, **23**, 25–40.

Gilroy, P. (1993) *The Black Atlantic: Modernity and Double Consciousness*, Verso, London.

Ginsborg, P. (1990) *A History of Contemporary Italy*, Penguin, Harmondsworth.

Glassner, M.I. (1993) *Political Geography*, John Wiley, New York.

Gleeson, B. (1998) The social space of disability in colonial Melbourne, in Fyfe, N. (ed) *Images of the Street*, London, Routledge, 93–110.

Glennie, P. and Thrift, N. (1992) Modernity, urbanism and modern consumption, *Environment and Planning D, Society and Space*, **10**, 423–43.

Godlewska, A. and Smith, N. (eds) (1994) *Geography and Empire*, Blackwell, Oxford.

Goetz, A.M. and Gupta, R.S. (1996) Who takes the credit? Gender, power, and control over loan use in rural credit programs in Bangladesh, *World Development*, **24**, 45–63.

Goldman, R. and Papson, S. (1998) *Nike Culture: The Sign of the Swoosh*, Sage, London.

Gomez Pompa, A. And Kaus, A. (1992) Taming the wilderness myth, *BioScience*, **42**, 271–9.

Goodland, R.J.A., Daly, H.E., El Serafy, S. (1993) The urgent need for rapid transformation to global environmental sustainability, *Environmental Conservation*, **20**, 122–30.

Goodman, D. and Watts, M. (eds) (1997) *Globalising Food*, Routledge, London.

Goodman, D., Sorj, B. and Wilkinson, J. (1987) *From Farming to Biotechnology: A Theory of Agro-Industrial Development*, Blackwell, Oxford.

Goodwin, M. (1995) Poverty in the city, you can raise your voice, but who is listening?, in Philo, C. (ed.) *Off the Map: The Social Geography of Poverty in the UK*, Child Poverty Action Group, London, 134–54.

Goss, J. (1993a) Placing the market and marketing place, tourist advertising of the Hawaiian Islands,1972–92, *Environment and Planning D, Society and Space*, **11**, 663–88.

Goss, J. (1993b) The magic of the mall, form and function in the retail built environment, *Annals of the Association of American Geographers*, **83**, 18–47.

Goss, J. (1999a) Consumption, in Cloke, P., Crang, P. and Goodwin, M. (eds) *Introducing Human Geographies*, Arnold, London, 114–21.

Goss, J. (1999b) Once upon a time in the commodity world, an unofficial guide to the Mall of America, *Annals of the Association of American Geographers*, **89**, 45–75.

Gottdiener, M. (1982) Disneyland, a Utopian urban space, *Urban Life*, **11**, 139–62.

Gottdiener, M. (1997) *The Theming of America: Dreams, Visions and Commercial Spaces*, Westview Press, Boulder, CO.

Gottman, J. (1973) *The Significance of Territory*, University Press of Virginia, Charlottesville.

Grabher, G. (1993) *The Embedded Firm: On the Socioeconomics of Industrial Networks*, Routledge, London.

Graham, S. (2002) Bridging urban digital divides? Urban polarisation and information and communications technologies (ICT), *Urban Studies*, **39**, 33–56.

Green, A. (1997) Income and wealth, in Pacione, M.

(ed.) *Britain's Cities: Geographies of Division in Urban Britain*, London, Routledge, 78–102.

Green, F. and Sutcliffe, B. (1987) *The Profit System: The Economics of Capitalism,* Penguin, London.

Greenwald, J. (1998) Sticky currency, *Time*, 28 September, 82–3.

Gregory, D. (1982) *Regional Transformation and Industrial Revolution: A Geography of the West Yorkshire Woollen Industry*, Macmillan, London.

Gregory, D. (1994) *Geographical Imaginations*, Blackwell, Oxford.

Gregory, D., Martin, R. and Smith, G. (eds) (1994) *Human Geography: Society, Space and Social Science*, Macmillan, London, 146–73.

Griffiths, I.L. (1994) *The Atlas of African Affairs*, 2nd edition, Routledge, London.

Grigg, D. (1985) *The World Food Problem 1950–1980*, Blackwell, Oxford.

Groves, P.A. (1987) The northeast and regional integration, 1800–1860, in Mitchell, R.D. and Groves, P.A. (eds) *North America: The Historical Geography of a Changing Continent*, Hutchinson, London.

Grundy-Warr, C. (2002) Forced migration and contested sovereignty along the Thai-Burma/Myanmar border, in Schofield, S., Newman, D., Drysdale, A. and Brown, J.A. (eds) *The Razor's Edge: International Boundaries and Political Geography*, Kluwer Law International, London, 337–83.

Grundy-Warr, C. and Wong, S.Y. (2002) Geographies of displacement: the Karenni and the Shan across the Myanmar-Thailand border, *Singapore Journal of Tropical Geography*, **23**, 93–122.

Gutkind, E.A. (1964) *Urban Development in Central Europe*, Macmillan, London and New York.

Gwynne, Robert N, Klak, Thomas and Shaw, Denis J B (2003) *Alternative Capitalisms: Geographies of Emerging Regions*, Arnold, London.

H

Halfacre, K. (1993) Locality and social representation: space, discourse and alternative definitions of the rural, *Journal of Rural Studies*, **9**, 23–37.

Halfacre, K. (1996) Out of place in the country: travellers and the 'rural idyll', *Antipode*, **18**, 42–72.

Hall, C., Tharakhan, P., Hallock, J., Cleveland, C. and Jefferson, M. (2003) Hydrocarbons and the evolution of human culture, *Nature*, **426**, 318–22.

Hall. E.T. (1959) *The Silent Language*, Doubleday, Garden City.

Hall, P. (ed.) (1966) *Von Thünen's Isolated State*, Pergamon, Oxford.

Hall, P. (1996) *Cities of Tomorrow*, updated edition, Blackwell, Oxford.

Hall, P. (1998) *Cities in Civilization: Culture, Innovation and Urban Order*, Weidenfeld & Nicolson, London.

Hall, R. (1995) Households, families and fertility, in Hall, R. and White, P. (eds) *Europe's Population: Towards the Next Century*, UCL Press, London.

Hall, S. (1991) Brave new world, *Socialist Review*, **91**, 57–8.

Hall, S. (1992) The West and the rest: discourse and power, in Hall, S. and Gieben, B. (eds) *Formations of Modernity*, Open University Press, Milton Keynes, 275–331.

Hall, S. (1995) New cultures for old, in Massey, D. and Jess, P. (eds) *A Place in the World? Places, Cultures and Globalization*, Oxford University Press, Oxford, 175–213.

Hall, S. (1996) The meaning of new times, in Morley, D. and Chen, K.-H. (eds) *Stuart Hall: Critical Dialogues in Cultural Studies*, Routledge, London, 223–37.

Halliday, F. (2002) *Two Hours that Shook the World. September 11, 2001: Causes and Consequences*, Saqi Books, London.

Hamelink, C. (1983) *Cultural Autonomy in Global Communications*, Longman, New York.

Hamnett, C. (1995) Controlling space: global cities, in Allen, J. and Hamnett, C. (eds) *A Shrinking World?*, Oxford University Press, Oxford, 103–42.

Hamnett, C. (2003) *Unequal City: London in the Global Arena*, Routledge, London.

Hannah, L., Lohse, D., Hutchinson, C., Carr, J.L. and Lankerani, A. (1994) A global assessment of global impact, *Ambio,* 23, 246–50.

Hannerz, U. (1992) The global ecumene, in Hannerz, U. (ed.) *Cultural Complexity: Studies in the Social Organization of Meaning*, Columbia University Press, New York, 217–67.

Hanrou, H. and Obrist, H.-U. (1999) Cities on the move, in Bradley, F. (ed.) *Cities on the Move: Urban Chaos and Global Change. East Asian Art, Architecture and Film Now*, Hayward Gallery, London, 10–15.

Hanson, S. and Pratt, G. (1995) *Gender, Work and Space*, Routledge, London.

Hara, K. (1999) Rethinking the 'Cold War' in the Asia-Pacific, *Pacific Review*, **12**, 515–36.

Harcourt, W. (ed.) (1999) *Women@Internet: Creating New Cultures in Cyberspace*, Zed Books, London.

Hardin, G. (1968) The tragedy of the commons, *Science*, **162**, 1243–8.

Hardin, G. (1972) The survival of nations and civilisations, *Science*, **172**, 129.

Hardt, M. and Negri, A. (2000) *Empire*, Harvard University Press, Cambridge, MA.

Harris, D.R. (1989) An evolutionary continuum of plant-animal interaction, in Harris, D.R. and Hillman, G.C. (eds) *Foraging and Farming: The Evolution of Plant Exploitation*, Unwin Hyman, London, 11–26.

Harris, D.R. (1996a) Introduction: themes and concepts in the study of early agriculture, in Harris, D.R. (ed.) *The Origins and Spread of Agriculture and Pastoralism in Eurasia*, UCL Press, London, 1–9.

Harris, D.R. (1996b) The origins and spread of agriculture and pastoralism in Eurasia: an overview, in Harris, D.R. (ed.) *The Origins and Spread of Agriculture and Pastoralism in Eurasia*, UCL Press, London, 552–70.

Harrison, P. (1973) *The Third Revolution: Environment, Population and a Sustainable World*, Penguin Books, Harmondsworth.

Harrison, S. (1998) *Japanese Technology and Innovation Management*, Edward Elgar, Cheltenham.

Hart, G. (2001) Development critiques in the 1990s: culs-de-sac and promising paths, *Progress in Human Geography*, **25**, 649–58.

Hart, K. and Wilson, G. (1998) UK implementation of agri-environmental Regulation 2078/92/EEC: enthusiastic supporter or reluctant participant?, *Landscape Research*, **23**, 255–72.

Hartwick, E. (1998) Geographies of consumption, a commodity-chain approach, *Environment and Planning D, Society and Space,* **16**, 423–37.

Harvey, B. and Hallett, J.D. (1977) *Environment and Society: an Introductory Analysis*, Macmillan, London.

Harvey, D (1974) Population, resources and the ideology of science, *Economic Geography*, **50**, 256–77.

Harvey, D. (1982) *The Limits to Capital*, Blackwell, Oxford.

Harvey, D. (1985a) *The Urbanization of Capital: Studies in the History and Theory of Capitalist Urbanization*, Blackwell, Oxford.

Harvey, D. (1985b) Money, time, space and the city, in Harvey, D. (ed.) *The Urban Experience*, Johns Hopkins University Press, Baltimore, MD.

Harvey, D. (1989) *The Condition of Postmodernity: An Enquiry into the Origins of Cultural Change*, Blackwell, Oxford.

Harvey, D. (1990) Between space and time: reflections on the geographical imagination, *Annals of the Association of American Geographers*, **80**, 418–34.

Harvey, D. (1997) Contested cities: social process and spatial form, in Jewson, N. and MacGregor, S. (eds) *Transforming Cities: Contested Governance and New Spatial Divisions*, Routledge, London, 19–27.

Haskell, T.L. (1985a) Capitalism and the origins of the humanitarian sensibility, part 1, *American Historical Review*, **90**, 339–61.

Haskell, T.L. (1985b) Capitalism and the origins of the humanitarian sensibility, part 2, *American Historical Review*, **90**, 547–66.

Hastings, A. and Dean, J. (2003) Challenging images: tackling stigma through estate regeneration, *Policy and Politics*, **31**, 171–84.

Hayter, R. and Watts, H.D. (1983) The geography of enterprise: a re-appraisal, *Progress in Human Geography*, **7**, 157–81.

Healey, M.J. and Ilbery, B.W. (1990) *Location and Change: Perspectives on Economic Geography*, Oxford University Press, Oxford.

Healey, P., Davoudi, S., Graham, S. and Madani-Pour, A. (eds) (1995) *Managing Cities: The New Urban Context*, John Wiley, London.

Healey, P., McNamara, P., Elson, M. and Doak, A. (1988) *Land Use Planning and the Mediation of Urban Change*, Cambridge University Press, Cambridge.

Hebdige, D. (1979) *Subculture: The Meaning of Style*, Methuen, London.

Hebdige, D. (1987) Object as image: the Italian scooter cycle, in *Hiding in the Light: On Images and Things,* Comedia/Routledge, London, 77–115.

Hecht, S. and Cockburn, A. (1989) *The Fate of the Forest: Developers, Destroyers and Defenders of the Amazon*, Verso, London and New York.

Heffernan, M. (1998) *The Meaning of Europe: Geography and Geopolitics*, Arnold, London.

Heilbronner, R.L. (1972) *The Economic Problem*, 3rd edition, Prentice Hall, Englewood Cliffs, NJ.

Held, D., McGrew, D., Goldblatt, D. and Perraton, J. (1999) *Global Transformations: Politics, Economics and Culture*, Polity, Cambridge.

Henry, N.D. and Passmore, A. (1999) Rethinking 'global' city centres: the example of Birmingham, *Soundings: A Journal of Politics and Culture*, **13**, 60–6.

Henry, N., McEwan C. and Pollard J.S. (2002) Globalization from below: Birmingham – postcolonial workshop of the world?, *Area*, **34**(2), 118–27.

Hepple, L. (1986) The revival of geopolitics, *Political Geography Quarterly*, **5**, 521–36.

Hepple, L. (1992) Metaphor, geopolitical discourse and the military in South America, in Barnes, T.J. and Duncan, J.S. (eds) *Writing Worlds*, Routledge, London, 136–54.

Herb, G.H. (1989) Persuasive cartography in Geopolitik and National Socialism, *Political Geography Quarterly*, **8**, 289–303.

Herb, G.H. (1997) *Under the Map of Germany: Nationalism and Propaganda 1918–1945*, Routledge, London.

Herf, J. (1984) *Reactionary Modernism: Technology, Culture and Politics in Weimar and the Third Reich*, Cambridge University Press, Cambridge.

Herzfeld, M. (1992) *The Social Production of Indifference: Exploring the Symbolic Roots of Western Bureaucracy*, University of Chicago Press, Chicago.

Heske, H. (1986) German geographical research in the Nazi period, *Political Geography Quarterly*, **5**, 267–81.

Heske, H. (1987) Karl Haushofer: his role in German geopolitics and Nazi politics, *Political Geography Quarterly*, **6**, 135–44.

Hettne, B. (1995) *Development Theory and the Three Worlds,* 2nd edition, Longman, London.

Hildyard, N. (1994) The big brother bank, *Geographical Journal*, June, 26–8.

Hildyard, N. (1997) *The World Bank and the State: A Recipe for Change?*, N. Hildyard, Bretton Woods Project, London.

Hindess, B. and Hirst, P. (1975) *Pre-Capitalist Modes of Production*, Routledge & Kegan Paul, London.

Hinrichs, C. (2003) The practice and politics of food system localization, *Journal of Rural Studies,* **19**, 33–45.

Hipscher, P. (1998) Democratic transitions as protest cycles: social movement dynamics in democratising Latin America, in Meyer, D. and Tarrow, S. (eds) *The Social Movement Society*, Rowman & Littlefield, Lanham, MD.

Hirschmann, A.O. (1958) *The Strategy of Economic Development*, Yale University Press, New Haven, CT.

Hirst, P. and Thompson, G. (1996) *Globalisation in Question*, Polity Press, Cambridge.

Hobsbawm, E.J. (1991) *Nations and Nationalism Since 1780*, Cambridge University Press, Cambridge.

Hobsbawm, E. (1995), *The Age of Extremes: The Short Twentieth Century, 1914–1991*, Abacus, London.

Hobsbawm, E.J. (1996) Ethnicity and nationalism in Europe today, in Balakrishnan, G. (ed.) *Mapping the Nation*, Verso, London, 255–66.

Hodges, R. (1988) *Primitive and Peasant Markets*, Blackwell, Oxford.

Hoggart, K. (1990) Let's do away with rural, *Journal of Rural Studies*, **6**, 245–57.

Holbrook, B. and Jackson, P. (1996) The social milieux of two North London shopping centres, *Geoforum,* **27**, 193–204.

Holdar, S. (1992) The ideal state and the power of geography: the life-work of Rudolf Kjellen, *Political Geography*, **11**, 307–23.

Holloway, L. and Hubbard, P. (2001*) People and Place: The Extraordinary Geographies of Everyday Life,* Prentice Hall, London.

Homburger, E. (1994) *The Historical Atlas of New York City*, Henry Holt, New York.

Honey, R. (1983) The social costs of space: providing services in rural areas, *Geographical Perspectives*, **51**, 24–37.

Hoogevelt, A. (1997) *Globalisation and Postcolonialism*, Macmillan, London.

hooks, b. (1984) *Feminist Theory: From Margins to Centre*, South End Press, Boston.

hooks, b. (1992) Eating the other, desire and resistance, in hooks, b. (ed.) *Black Looks: Race and Representation,* South End Press, Boston, MA, 21–39.

Hooper, J. and Connolly, K. (2001) Berlusconi breaks ranks over Islam, *Guardian*, 27 September.

Horticultural Crops Development Authority (HCDA) (1997) *Performance of the Horticultural Industry*, HCDA, Nairobi.

Hosking, G. (1992) *A History of the Soviet Union, 1917–1991,* final edition, Fontana Press, London.

Howard, E. (1902/1965) *Garden Cities of Tomorrow*, edited by Osborn, F.J., with a preface by Mumford, L., Faber and Faber, London.

Hubbard, P. (1999*) Sex and the City: Geographies of Prostitution in the Urban West*, London, Ashgate.

Hubbard, P. Kitchin, R, Bartley, B. and Fuller, D. (2002) *Thinking Geographically: Space,Theory and Contemporary Human Geography*, Continuum, London.

Hughes, A. (2000) Retailers, knowledges and changing commodity networks: the case of the cut flower trade, *Geoforum*, **31**, 175–90.

Huillet, C. (1991) New management for rural services, *OECD Observer*, **172**, 17–19.

Hulme, D. and Murphree, M. (1999) Communities, wildlife and the 'new conservation' in Africa, *Journal of International Development*, **11**, 227–85

Humphreys, J. (1988) Social provision and service delivery: problems of equity, health and health care in rural Australia, *Geoforum*, **19**, 323–38.

Huntington, S.P. (1993) The clash of civilizations?, *Foreign Affairs*, **72**, 22–49.

Hutnyk, J. (1997) Adorno at Womad: South Asian crossovers and the limits of hybridity-talk, in Werbner, P. and Modood, T. (eds) *Debating Cultural Hybridity: Multi-Cultural Identities and the Politics of Anti-Racism*, Zed, London, 106–36.

Hyndman, J. (2003) Beyond either/or: a feminist analysis of September 11th, *ACME: An International E-Journal for Critical Geographies*, **2**, 1–13 (freely downloadable from www.acme-journal.org).

I

Ilbery, B. (ed.) (1998) *The Geography of Rural Change*, Longman, Harlow.

Ilbery, B. and Bowler, I. (1996) Industrialisation and

world agriculture, in Douglas, I., Huggett, R. and Robinson, M. (eds) *Companion Encyclopaedia of Geography*, Routledge, London, 228–48.

Ilbery, B. and Bowler, I. (1998) From agricultural productivism to post-productivism, in Ilbery, B. (ed.) *The Geography of Rural Change*, Longman, London, 57–84.

Ilbery, B. and Holloway, L. (1997) Responses to the challenge of productivist agriculture, *Built Environment*, **23**, 184–91.

Ilbery, B. and Kneafsey, M. (1998) Product and place: promoting quality products and services in the lagging rural regions of the European Union, *European Urban and Regional Studies*, **5**, 329–41.

Ilbery, B. and Kneafsey, M. (2000a) Producer constructions of quality in regional speciality food production: a case study from south west England, *Journal of Rural Studies*, **16**, 217–30.

Ilbery, B. and Kneafsey, M. (2000b) Registering regional speciality food and drink products in the United Kingdom: the case of PDOs and PGIs, *Area*, **32**.

Ilbery, B., Clark, D., Berkeley, N. and Goldman, I. (1995) Telematics and rural development, *European Urban and Regional Studies*, **2**, 55–68.

Ilbery, B., Healey, M. and Higginbottom, J. (1997) On and off-farm business diversification by farm households in England, in Ilbery, B., Chiotti, Q. and Rickard, T. (eds) *Agricultural Restructuring and Sustainability: A Geographical Perspective*, CAB International, Wallingford, 135–51.

Independent on Sunday (1998) March of the child soldiers, 6 December, 15.

Info Box (1996) *Info Box. The Catalogue*, Dirk Nishen, Berlin.

Inglehart, R. (1990) *Culture Shift in Advanced Industrial Societies*, Princeton University Press, Princeton, NJ.

International Broadcasting Trust (IBT) (1998) *The Bank, the President and the Pearl of Africa*, IBT/Oxfam, Oxford.

International Council for the Exploration of the Sea (ICES) (2003) ICES advises zero catches of cod and other fish stocks. ICES Press Release 20 October 2003. Available at www.ices.dk/aboutus/pressrelease/ACF Mautumn2003.pdf.

International Council for Local Environmental Initiatives (2002) *Second Local Agenda 21 Survey*. Published as UN Dept of Economic and Social Affairs, Background Paper 15 for the 2nd Preparatory Conference for the World Summit on Sustainable Development, DESA/DSD/PC2/BP15, 29pp. Available at www.iclei.org/rioplusten/final_document.pdf.

International Data Corporation (1999) Internet Commerce Market Model, March.

International Labour Office (1999) *Key Indicators of the Labour Market 1999 (KILM)*, ILO, Geneva.

International Monetary Fund (1996) *International Capital Markets: Developments, Prospects and Key Policy Issues*, IMF, Washington, DC.

Iyer, P. (1989) *Video Nights in Kathmandu: Reports from the Not-so-far East*, Black Swan, London.

J

Jackson, P. (ed.) (1987) *Race and Racism: Essays in Social Geography*, Allen and Unwin, London.

Jackson, P. (1989) *Maps of Meaning: An Introduction to Cultural Geography*, Routledge, London.

Jackson, P. (1993) Towards a cultural politics of consumption, in Bird, J., Curtis, B., Putnam, T., Robertson, G. and Tickner, L. (eds) *Mapping the Future: Local Cultures, Global Change*, Routledge, London.

Jackson, P. (1999) Commodity culture, the traffic in things, *Transactions of the Institute of British Geographers*, NS, **24**, 95–108.

Jackson, P. and Taylor, J. (1996) Geography and the cultural politics of advertising, *Progress in Human Geography*, **19**, 356–71.

Jackson, P. and Thrift, N. (1995) Geographies of consumption, in Miller, D. (ed.) *Acknowledging Consumption*, Routledge, London, 204–37.

Jackson, R. (1994) Demand for lottery products, *Journal of Consumer Affairs*, **28**, 313–25.

Jacobs, J. (1961) *The Death and Life of Great American Cities*, Penguin, Harmondsworth.

Japan Institute for Social and Economic Affairs (1996) *Japan 1997: An International Comparison*, Keizai Koho Center, Tokyo.

Jarosz, L. (1992) Constructing the dark continent: metaphor of geographic representation of Africa, *Geografiska Annaler*, **74**(2), 105–15.

Jessop, B. (1989) *Thatcherism: A Tale of Two Nations*, Blackwell, Oxford.

Jezkova, P. (1995) Changes in textiles: implications for women, in Mitter, S. and Rowbotham, S. (eds) *Women Encounter Technology: Changing Patterns of Employment in the Third World*, Routledge, London, 93–110.

Joffé, G. (1987) Frontiers in North Africa, in Blake, G.H. and Schofield, R.N. (eds) *Boundaries and State Territory in the Middle East and North Africa*, MENAS Press, Berkhamsted.

Johnston, D. (1998) Why territorial development matters, *OECD Observer*, **210**, February–March.

Johnston, R.J. (1984) *Residential Segregation, the State and Constitutional Conflict in American Urban Areas*, Academic Press, London.

Johnston, R.J. (1997) *Geography and Geographers: Anglo–American Human Geography since 1945*, 5th edition, Arnold, London.

Johnston, R. and Pattie, C. (2003) Representative democracy and electoral geography, in J. Agnew, K. Mitchell and G. Toal (eds) *A Companion to Political Geography*, Blackwell, Oxford.

Johnston, R.J., Gregory, D., Pratt, G. and Watts, M. (eds) (2000) *The Dictionary of Human Geography*, 4th edition, Blackwell, Oxford.

Johnston, R.J., Pattie, C.J. and Allsopp, J.G. (eds) (1988) *A Nation Dividing? The Electoral Map of Great Britain 1979–1987*, Longman, London.

Jones, G. and Hollier, G. (1997) *Resources, Society and Environmental Management*, Paul Chapman, London.

Jones, H. (1990) *Population Geography*, 2nd edition, Paul Chapman, London.

Jones, H., Caird, W., Berry, W. and Dewhirst, J. (1986) Peripheral counterurbanisation: findings from an integration of census and survey data in northern Scotland, *Regional Studies*, **20**, 15–26.

Jones, M., Jones, R. and Woods, M. (2004) *An Introduction to Political Geography: Space, Place and Politics*, Routledge, London.

Jones, P.S. (2000) Why is it alright to do development 'over there' but not 'here'? Changing vocabularies and common strategies across the 'First' and 'Third Worlds', *Area*, **32**, 237–41.

Joseph, A. and Phillips, D. (1984) *Accessibility and Utilization: Geographical Perspectives on Health Care Delivery*, Harper & Row, New York.

Judah, T. (1997) *The Serbs: History, Myth and the Destruction of Yugoslavia*, Yale University Press, New Haven.

Judge, D., Stoker, G. and Wolman, H. (eds) (1995) *Theories of Urban Politics*, Sage, London.

Jurgens, U. and Gnad, M. (2002) Gated communities in South Africa – experiences from Johannesburg, *Environment and Planning B: Planning and Design*, **29**, 337–53.

K

Kabeer, N. (1994) *Reversed Realities: Gender Hierarchies in Development Thought*, Verso, London.

Kahn, J.S. (1995) *Culture, Multiculture, Postculture*, Sage, London.

Kaminsky, G. and S. Schmukler (2002) Emerging markets instability: do sovereign ratings affect country risk and stock returns?, *World Bank Economic Review*, **16**, 171–95.

Kapferer, B. (1988) *Legends of People, Myths of State: Violence, Intolerance and Political Culture in Sri Lanka and Australia*, Smithsonian Institution Press, Washington, DC.

Kasson, J.F. (1978) *Amusing the Million: Coney Island at the Turn of the Century*, Hill & Wang, New York.

Katz, R.S. and Mair, P. (eds) (1994) *How Parties Organize: Change and Adaptation in Party Organizations in Western Democracies*, Sage, London.

Kaye, H.J. (1984) *The British Marxist Historians*, Polity Press, Cambridge.

Kearns, G. and Philo, C. (eds) (1993) *Selling Places: The City as Cultural Capital, Past and Present*, Pergamon, Oxford.

Keck, M.E. and Sikkink, K. (1998) *Activists Beyond Borders: Advocacy Networks in International Politics*, Cornell University Press, Ithaca, NY.

Keeble, D. (1993) Small firm creation, innovation and growth and the urban-rural shift, in Curran, J. and Storey, D. (eds) *Small Firms in Urban and Rural Locations*, Routledge, London, 54–78.

Keith, M. (2000) Walter Benjamin, urban studies, and the narratives of city life, in Bridge, G. and Watson, S. (eds) *A Companion to the City*, Blackwell, Oxford, 410–29.

Kennedy, P. (1988) *The Rise and Fall of the Great Powers*, Random House, New York.

Kershaw, I. (1993) *The Nazi Dictatorship: Problems and Perspectives of Interpretation*, 3rd edition, Arnold, London.

Keyes, C.F. (1995) *The Golden Peninsula: Culture and Adaptation in Mainland Southeast Asia*, University of Hawaii Press, Honolulu.

Khan, M. (1997) Aligarh city: urban expansion and encroachment, *Asian Profile*, **25**, 47–51.

Kipling, R. (1912) *Rudyard Kipling's Verse: Definitive Edition*, Hodder & Stoughton, London.

Kitschelt, H. (1989) *The Logic of Party Formation: Structure and Strategy of Belgian and West German Ecology Parties*, Cornell University Press, Ithaca, NY.

Kitschelt, H. (1990) New social movements and the decline of party organization, in Dalton, R.J. and Kuechler, M. (eds) *Challenging the Political Order: New Social and Political Movements in Western Democracies*, Oxford University Press, Oxford.

Klandermans, B., Roefs, M. and Olivier, J. (1998) A movement takes office, in Meyer, D. and Tarrow, S. (eds) *The Social Movement Society*, Rowman & Littlefield, Lanham, MD.

Kneafsey, M. and Ilbery, B. (2001) Regional images and the promotion of speciality food and drink products: initial explorations from the 'West Country'. *Geography*, **86**, 131–40.

Knights, D. and Odih, P. (1995) It's about time! The

significance of gendered time for financial services consumption, *Time and Society*, **4**, 205–31.

Knopp, L. (1995) Sexuality and urban space: a framework for analysis, in Bell, D. and Knox, P. and Pinch, S. (2000) *Urban Social Geography: An Introduction*, 4th edition, Prentice Hall, Harlow.

Knox, P. and Agnew, J. (1994) *The Geography of the World Economy*, 2nd edition, Edward Arnold, London.

Knox, P. and Agnew, J. (1998) *The Geography of the World Economy*, 3rd edition, Edward Arnold, London.

Knox, P., Agnew, J. and McCarthy, L. (2003) *The Geography of the World Economy: An Introduction to Economic Geography*, 4th edition, Hodder Arnold, London.

Knox, P. and Marston, S.A. (1988) *Human Geography: Places and Regions in Global Context*, Prentice Hall, Upper Saddle River, NJ.

Knox, P.L. and Marston, S.A. (2001) *Human Geography: Places and Regions in Global Context*, 2nd edition, Prentice Hall, Upper Saddle River, NJ.

Knox, P.L. and Marston, S.A. (2004) *Human Geography: Places and Regions in Global Context*, 3rd edition, Prentice Hall, Upper Saddle River, NJ.

Knox, P. and Pinch, S. (2000) *Urban Social Geography: An Introduction*, 4th edition, Prentice Hall, Harlow.

Kobayashi, A. and Peake, L. (1994) Unnatural discourse: 'race' and gender in geography, *Gender, Place and Culture*, **1**, 225–44.

Koczberski, L. (1998) Women in development: a critical analysis, *Third World Quarterly*, **19**, 395–409.

Kofman, E. (2003) Rights and citizenship in J. Agnew, K. Mitchell and G. Toal (eds) *A Companion to Political Geography*, Blackwell, Oxford.

Kondratieff, N.D. (1925) The major economic cycles, *Voprosy Konjunktury*, **6**, 28–79; an English translation is to be found in *Lloyds Bank Review* (1978), **129**.

Kondratieff, N.D. (1935) The long waves in economic life, *Review of Economic Statistics*, **17**, 105–15.

Korpi, W. (1983) *The Democratic Class Struggle*, Routledge & Kegan Paul, London.

Korten, D.C. (1998) Your mortal enemy, *Guardian*, 21 October, 4–5.

Koter, M. (1990) The morphological evolution of a nineteenth-century city centre: Łódź, Poland, 1825–1973, in Slater, T.R. (ed.) *The Built Form of Western Cities*, Leicester University Press, Leicester and London.

Kothari, U. and Minogue, M. (eds) (2002) *Development Theory and Practice: Critical Perspectives*, Macmillan, London.

Krieger, J. (ed.) (1993) *The Oxford Companion to Politics of the World*, Oxford University Press, Oxford.

L

Lacey, R. (1986) *Ford: The Men and the Machine*, Heinemann, London.

Laïdi, Z. (1988) Introduction: What use is the Soviet Union?, in Laïdi, Z. (ed.), *The Third World and the Soviet Union*, Zed Books, London, 1–23.

Lakha, S. (1994) The New International Division of Labour and the Indian computer software industry, *Modern Asian Studies*, **28**, 381–408.

Lall, S. (1983) The rise of multinationals from the third world, *Third World Quarterly*, **5**, 618–26.

Landes, D.S. (1969) *The Unbound Prometheus*, Cambridge University Press, Cambridge.

Lane, J.-E., McKay, D. and Newton, K. (1991) *Political Data Handbook: OECD Countries*, Oxford University Press, Oxford.

Lang, H.J. (2002) *Fear and Sanctuary: Burmese Refugees in Thailand*, Southeast Asia Program Publications, Cornell University, New York.

Langley, P. (2003) *The Everyday Life of Global Finance*, International Political Economy Group Working Paper No. 5, University of Manchester, Manchester.

Langton, J. (1984) The industrial revolution and the regional geography of England, *Transactions of the Institute of British Geographers*, NS, **9**, 145–67.

Langton, J. and Morris, R.J. (eds) (1986) *Atlas of Industrializing Britain, 1780–1914*, Methuen, London and New York.

Lash, S. and Urry, J. (1987) *The End of Organized Capitalism*, Polity Press, Cambridge.

Lash, S. and Urry, J. (1994) *Economies of Signs and Space*, Sage, London.

Latour, B. (1993) *We Have Never Been Modern*, Harvester Wheatsheaf, Brighton.

Laulajainen, R. (1998) *Financial Geography*, School of Economics and Commercial Law, Göteborg University.

Laurie, N., Dwyer, C., Holloway, S. and Smith, F. (1999) *Geographies of New Femininities*, Longman, Harlow.

Lawton, R. (1989) Introduction: aspects of the development and role of great cities in the Western world in the nineteenth century, in Lawton, R. (ed.) *The Rise and Fall of Great Cities*, Belhaven, London.

Laxton, P. (1986) Textiles, in Langton, J. and Morris, R.J. (eds) *Atlas of Industrializing Britain, 1780–1914*, Methuen, London and New York, 106–13.

Le Corbusier (1929/1987) *The City of Tomorrow*, Architectural Press, London.

Lee, R.B. (1968) What hunters do for a living, or how to make out on scarce resources, in Lee, R.B. and DeVore, I. (eds) *Man the Hunter*, Aldine Press, Chicago, 30–48.

Lee, R. (1999) Local money: geographies of autonomy and resistance?, in Martin, R.L. (ed.) *Money and the Space Economy*, Wiley, Chichester.

Lee, R. (2000) Radical and postmodern? Power, social relations and regimes of truth in the social construction of alternative economic geographies, *Environment and Planning A*, **32**, 991–1009.

Lees, L. (2000) A re-appraisal of gentrification: towards a geography of gentrification, *Progress in Human Geography*, **24**, 389–408.

Lefebvre, H. (2003) *The Urban Revolution* (first published 1970), University of Minnesota Press, Minneapolis.

LeGates, R.T. and Stout, F. (eds) (1996) *The City Reader*, Routledge, London.

Le Heron, R. (1993) *Globalised Agriculture: Political Choice*, Pergamon, Oxford.

Le Heron, R., Murphy, L., Forer, P. and Goldstone, M. (eds) (1999) *Explorations in Human Geography: Encountering Place*, Oxford University Press, Auckland.

Lemon, J.T. (1987) Colonial America in the eighteenth century, in Mitchell, R.D. and Groves, P.A. (eds) *North America: The Historical Geography of a Changing Continent*, Hutchinson, London.

Lester, A. (1999) in Graham, B. and Nash, C. (eds) *Modern Historical Geographies*, Prentice Hall, Harlow.

Lester, A. (2002) Obtaining the 'due observance of justice': the geographies of colonial humanitarianism, *Environment and Planning D: Society and Space*, **20**, 277–93.

Lewis, G. (1990) Community through exclusion and illusion: the creating of social worlds in an American shopping mall, *Journal of Popular Culture*, **24**, 121–36.

Lewis, G. (2000) Changing places in a rural world: the population turnaround in perspective, *Geography*, **85**, 157–65.

Lewis, M. (1990) *Liar's Poker: Rising Through the Wreckage on Wall Street*, London, Penguin Books.

Ley, D. and Cybriwsky, R. (1974) Urban graffiti as territorial markers, *Annals of the Association of American Geographers*, **64**, 491–505.

Leyshon, A. (1995a) Geographies of money and finance, *Progress in Human Geography*, **19**, 531–43.

Leyshon, A. (1995b) Annihilating space? The speed up of communications, in Allen, J. and Hamnett, C. (eds) *A Shrinking World?*, Oxford University Press, Oxford.

Leyshon, A. (1996) Dissolving difference? Money, disembedding and the creation of 'global financial space', in Daniels, P.W. and Lever, W.F. (eds) *The Global Economy in Transition*, Longman, London.

Leyshon, A. and Thrift, N. (1994) Access to financial services and financial infrastructural withdrawal: problems and policies, *Area*, **26**, 268–75.

Leyshon, A. and Thrift, N. (1997) *Money Space: Geographies of Monetary Transformation*, Routledge, London.

Liebes, T. and Katz, E. (1993) *The Export of Meaning: Cross-Cultural Readings of Dallas*, 2nd edition, Polity Press, Cambridge.

Ling, P.J. (1990) *America and the Automobile: Technology, Reform and Social Change*, Manchester University Press, Manchester.

Linklater, A. (1998) Citizenship and sovereignty in the post-Westphalian European state, in Daniele, A., Held, D. and Köhler, M. (eds) *Re-imagining Political Community: Studies in Cosmopolitan Democracy*, Polity Press, Cambridge.

Livingstone, D. (1992) *The Geographical Tradition*, Blackwell, Oxford.

Loewenson, R. (1992) *Modern Plantation Agriculture*, Zed Books, London.

Logan, J. and Molotch, H. (1987) *Urban Fortunes: The Political Economy of Place*, University of California Press, Berkeley.

Lowe, P., Buller, H. and Ward, N. (2002) Setting the next agenda? British and French approaches to the second pillar of the CAP, *Journal of Rural Studies*, **18**, 1–17.

Lowe, P., Murdoch, J. and Cox, G. (1995) A civilised retreat? Anti-urbanism, rurality and the making of an Anglo-centric culture, in Healey, P., Cameron, S., Davoudi, S., Graham, S. and Madani-Pour, A. (eds) *Managing Cities: The New Urban Context*, John Wiley, London.

Lowe, P., Murdoch, J., Marsden, T., Munton, R. and Flynn, A. (1993) Regulating the new rural spaces: the uneven development of land, *Journal of Rural Studies*, **9**, 205–22.

Lowe, P., Ray, C., Ward, N., Wood, D. and Woodward, R. (1998) *Participation in Rural Development: A Review of European Experience*, Centre for Rural Economy, University of Newcastle upon Tyne, Newcastle upon Tyne.

Löytönen, M. (1995) The effects of the HIV epidemic on the population of Europe, in Hall, R. and White, P. (eds) *Europe's Population: Towards the Next Century*, UCL Press, London.

Luke, T. (1991) The discourse of development: a genealogy of developing nations and modernity, *Current Perspectives in Social Theory*, **11**, 271–93.

Lury, C. (2000) The united colors of diversity: essential and inessential culture, in Franklin, S., Lury, C. and Stacey, J. (eds) *Global Nature, Global Culture*, Sage, London, 146–87.

Luttwak, E. (1990) From geopolitics to geoeconomics, *National Interest*, **20**, 17–24.

M

McAdam, D. (1983) *Political Process and the Development of Black Insurgency*, University of Chicago Press, Chicago.

McCann, M. (1986) *Taking Reform Seriously: Perspectives on Public Interest Liberalism*, Cornell University Press, Ithaca, NY.

McCarthy, C., Rodriguez, A., Buendia, E., Meacham, S., David, S., Godina, H., Supriya, K. and Wilson-Brown, C. (1997) Danger in the safety zone: notes on race, resentment and the discourse of crime violence and suburban security, *Cultural Studies*, **11**, 274–95.

McClintock, A. (1994) *Imperial Leather: Race, Gender and Sexuality in the Colonial Context*, Routledge, London.

McCormick, J (1989) *The Global Environmental Movement*, Belhaven Press, London.

McDowell, L. (1983) Towards an understanding of the gender division of urban space, *Environment and Planning D: Society and Space*, **1**, 59–72.

McDowell, L. (1994) The transformation of cultural geography, in Gregory, D., Martin, R. and Smith, G. (eds) *Human Geography. Society, Space and Social Science*, Macmillan, London, 146–73.

McDowell, L. (1997) Capital *Culture: Gender at Work in the City*, Blackwell, Oxford.

McDowell, L. and Sharp, J.P. (eds) (1997) *Space, Gender and Knowledge: Feminist Readings*, Arnold, London.

McEwan, C. (forthcoming) 'Walking the walk'? The state and gender transformations in the 'new' South Africa, *South African Geographical Journal*.

McGuinness, M. (2000) Geography matters? Whiteness and contemporary geography, *Area*, **32**, 225–30.

McNeill, D. (2004) *New Europe: Imagined Spaces*, Arnold, London.

Machel, G. (1996) *UN Study on the Impact of Armed Conflict on Children*, UNICEF, online, available at: www.unicef.org/graca/women/html.

Mackay, H. (ed.) (1997) *Consumption and Everyday Life*, Sage, London.

Mackinder, H. (1904) The geographical pivot of history, *Geographical Journal*, **23**, 421–44.

Mackinder, H. (1919) *Democratic Ideals and Reality: A Study in the Politics of Reconstruction*, Constable, London.

McLaughlin, B. (1986) The rural deprivation debate: retrospect and prospect, in Lowe, P., Bradley, T. and Wright, S. (eds) *Deprivation and Welfare in Rural Areas*, Geo Books, Norwich, 43–54.

McLaughlin, E. and Muncie, J. (eds) (1999) Walled cities: surveillance, regulation and segregation, in Pile, S., Brook, C. and Mooney, G. (eds) *Unruly Cities? Order/Disorder*, Routledge, London.

McLoughlin, J. (1991) *The Demographic Revolution*, Faber and Faber, London.

McLuhan, M. (1964) *Understanding Media: Extensions of Man*, Routledge & Kegan Paul, London.

McLuhan, M. (1975) The implications of cultural uniformity, in Bigsby, C.W.E. (ed.) *Superculture, American Popular Culture and Europe*, Bowling Green University Popular Press, Bowling Green, OH, 43–56.

McMichael, P. (1996) Globalisation: myths and realities, *Rural Sociology*, **61**, 25–55.

Madon, S. (1997) Information-based global economy and socio-economic development: the case of Bangalore, *Information Society*, **13**, 227–43.

Maffesoli, M. (1996) *The Time of the Tribes: The Decline of Individualism in Mass Society*, translated by Don Smith, Sage, London.

Mahon, R. (1987) From Fordism to ?: new technology, labour markets and unions, *Economic and Industrial Democracy*, **8**, 5–60.

Mair, P (1997) *Party System Change: Approaches and Interpretations*, Oxford University Press, Oxford.

Maisel, S.L. (ed.) (1994) *The Parties Respond: Changes in American Parties and Campaigns*, Westview, Boulder, CO.

Maisels, C.K. (1993) *The Emergence of Civilization*, Routledge, London.

Maitland, B. (1985) *Shopping Malls, Planning and Design*, Construction Press, London.

Malbon, B. (1997) Clubbing, consumption, identity and the spatial practices of every-night life, in Skelton, T. and Valentine, G. (eds) *Cool Places*, Routledge, London, 266–86.

Malbon, B. (1999) *Clubbing*, Routledge, London.

Mamdani, M (1996) *Citizen and Subject: Contemporary Africa and the Legacy of Late Colonialism*, Princeton University Press, Princeton.

Manin, R. (1997) *The Principles of Representative Government*, Cambridge University Press, Cambridge.

Mann, M. (1988) *States, War and Capitalism: Studies in Political Sociology*, Blackwell, Oxford.

Mann, M. (1993) *The Sources of Social Power, Volume II: The Rise of Classes and Nation-States, 1760–1914*, Cambridge University Press, Cambridge.

Marchand, M.H. and Parpart, J.L. (1995) *Feminism, Postmodernism, Development*, Routledge, London.

Marcuse, H. (1964) *One-Dimensional Man*, Abacus, London.

Marcuse, P. (2000) The language of globalization,

Monthly Review, **52**, 1–4, available at: www.monthlyreview.org/700marc.html

Markovits, A.S. and Gorski, P.S. (1993) *The German Left: Red, Green and Beyond*, Polity, Cambridge.

Marsden, T. (1997) Creating space for food: the distinctiveness of recent agrarian development, in Goodman, D. and Watts, M. (eds) *Globalising Food*, Routledge, London, 169–91.

Marsden, T. (1999) Rural futures: the consumption countryside and its regulation, *Sociologia Ruralis*, **39**, 501–20.

Marsden, T. and Arce, A. (1995) Constructing quality: emerging food networks in the rural transition, *Environment and Planning A*, **27**, 1261–79.

Marsden, T., Banks, J. and Bristow, G. (2000) Food supply chain approaches: exploring their role in rural development, *Sociologia Ruralis*, **40**, 424–38.

Marsden, T., Lowe, P. and Whatmore, S. (eds) (1990) *Rural Restructuring: Global Processes and Local Responses*, Wiley, London.

Marsden, T., Murdoch, J., Lowe, P., Munton, R. and Flynn, A. (1993) *Constructing the Countryside*, UCL Press, London.

Marshall, T.H. (1963) Citizenship and social class, in Marshall, T.H., *Sociology at the Crossroads*, Heinemann, London.

Martin, R.L. (1994) Stateless monies, global financial integration and national economic autonomy: the end of Geography?, in Corbridge, S., Martin, R. and Thrift, N. (eds) *Money, Power and Space*, Blackwell, Oxford, 253–78.

Martin, R.L. (1999) The new economic geography of money, in Martin, R.L. (ed.) *Money and the Space Economy*, Wiley, Chichester, 3–27.

Martin, R.L. and Minns, R. (1995) Undermining the financial basis of regions: the spatial structure and implications of the UK pension fund system, *Regional Studies*, **29**, 125–44.

Martínez, M. (1997) *Olivenza y el tratado de Alcanices*, Ayuntamiento de Olivenza.

Marx, K. (1867) *Das Kapital*; translated into English as *Capital*, 3 vols, International Publishers, New York (1967) and Penguin, London (1976).

Marx, K. (1973) *Grundrisse*, Penguin, London.

Mason, C. and R. T. Harrison (2002) The geography of venture capital investments in the UK, *Transactions of the Institute of British Geographers*, **27**, 427–51.

Massey, D. (1981) The geography of industrial change, in Potter, D., Anderson, J., Clarke, J., Coombes, P., Hall, S., Harris, L., Holloway, C. and Walton, T. (eds) *Society and the Social Sciences*, Open University Press, Milton Keynes, 302–13.

Massey, D. (1988) What's happening to UK manufacturing?, in Allen, J. and Massey, D. (eds) *The Economy in Question*, Sage, London, 45–90.

Massey, D. (1993) Power-geometry and a progressive sense of place, in Bird, J., Curtis, B., Putnam, T., Robertson, G. and Tickner, L. (eds) *Mapping the Futures: Local Cultures, Global Change*, Routledge, London, 59–69.

Massey, D. (1994) *Space, Place and Gender*, Polity, Cambridge.

Massey, D. (1995a) Imagining the world, in Allen, J. and Massey, D. (eds) *Geographical Worlds,* Oxford University Press, Oxford, 5–52.

Massey, D. (1995b) *Spatial Divisions of Labour: Social Structures and the Geography of Production*, Macmillan, London.

Massey, D. (1996) Space/power, identity/difference, tensions in the city, in Merrifield, A. and Swyngedouw, E. (eds) *The Urbanisation of Injustice*, Blackwell, London, 190–205.

Massey, D. (1999) Cities in the world, in Massey, D., Allen, J. and Pile, S. (eds) *City Worlds*, Routledge in association with the Open University, London, 99–175.

Massey, D. and Jess, P. (eds) (1995a), *A Place in the World? Places, Cultures and Globalization*, Oxford University Press, Oxford.

Massey, D. and Jess, P. (1995b) The contestation of place, in Massey, D. and Jess, P. (eds) *A Place in the World? Places, Cultures and Globalization*, Oxford University Press, Oxford, 133–74.

Massey, D. and Jess, P. (1995c) Places and cultures in an uneven world, in Massey, D. and Jess, P. (eds) *A Place in the World? Places, Cultures and Globalization*, Oxford University Press, Oxford, 215–39.

Massey, D., Allen, J. and Pile, S. (eds) (1999) *City Worlds*, Routledge, London.

Mather, A.S. and Chapman, K. (1995) *Environmental Resources*, Longman, London.

Mattelart, A. (1979) *Multinational Corporations and the Control of Culture*, Harvester Press, Brighton.

Maxwell, S. (2002) More aid? – Yes – and use it to reshape aid architecture, ODI opinions (3), available at: www.odi.org.uk/opinions/

Maxwell, S. and Singer, H. (1979) Food aid to developing countries: a survey, *World Development*, **7**, 225–47.

May, J. (1996) A little taste of something more exotic, the imaginative geographies of everyday life, *Geography*, **81**, 57–64.

Meadows, D.H., Meadows, D.I., Randers, J. and Behrens III, W.W. (1972) *The Limits to Growth: A Report to The Club of Rome*, Universe Books, New York.

Mehrotra, R. (1977) *Harvard Design Magazine*, Winter/Spring, 25–7.

Meinig, D.W. (1986) *The Shaping of America. A Geographical Perspective on 500 Years of History: Atlantic America, 1492–1800*, Yale University Press, New Haven and London.

Meyer, D.R. (1990) The new industrial order, in Conzen, M.P. (ed.) *The Making of the American Landscape*, Unwin Hyman, London and Winchester, MA.

Meyer, W.B. (1996) *Human Impact on the Earth*, Cambridge University Press, Cambridge.

Middleton, N. (2003) *The Global Casino*, 3rd edition, Arnold, London.

Mill, J. (1821) *Elements of Political Economy*, Baldwin, Craddock and Joy, London.

Miller, B. (2004) Spaces of mobilization: transnational social movements, in Barnett, C. and Low, M. (eds) *Spaces of Democracy: Geographical Perspectives on Citizenship, Participation and Representation*, Sage, London, New Delhi and Thousand Oaks, CA.

Miller, D. (1992) The young and the restless in Trinidad: a case study of the local and the global in mass consumption, in Silverstone, R. and Hirsch, E. (eds) *Consuming Technologies: Media and Information in Domestic Spaces*, Routledge, London, 163–82.

Miller, D. (1994) *Modernity, an Ethnographic Approach: Dualism and Mass Consumption in Trinidad*, Berg, Oxford.

Miller, D. (1995) Consumption as the vanguard of history, a polemic by way of an introduction, in Miller, D. (ed.) *Acknowledging Consumption: A Review of New Studies*, Routledge, London, 1–57.

Miller, D. (1997a) Coca-Cola, a black sweet drink from Trinidad, in Miller, D. (ed.) *Material Cultures*, UCL Press, London.

Miller, D.L. (1997b) *City of the Century: The Epic of Chicago and the Making of America*, Simon & Schuster, New York.

Miller, D. (1998) *A Theory of Shopping*, Polity, Cambridge.

Millington, A.C., Mutiso, S.K., Kirkby, J. and O'Keefe, P. (1989) African soil erosion: nature undone and the limitations of technology, *Land Degradation and Rehabilitation*, **1**, 279–90

Milner, M. (1998) IMF puts UK jobs at risk, *Guardian*, 17 July, 23.

Mintel (1996) *Nightclubs and Discotheques*, Mintel, London.

Mitchell, D (1996) Public space and the city, *Urban Geography*, **17**, 127–31.

Mitchell, J.B. (ed.) (1962) *Great Britain: Geographical Essays*, Cambridge University Press, Cambridge.

Mitchell, R.D. and Groves, P.A. (eds) (1987) *North America: The Historical Geography of a Changing Continent*, Hutchinson, London.

Mitchell, W.C. (1937) *The Backward Art of Spending Money and Other Essays*, McGraw-Hill, New York.

Mohan, J. (2000) Geographies of welfare and social exclusion, *Progress in Human Geography*, **24**, 291–300.

Mohanty, C.T. (1991) Under Western eyes: feminist scholarship and colonial discourses, in Mohanty, C.T., Russo, A. and Torres, L. (eds) *Third World Women and the Politics of Feminism*, Indiana University Press, Bloomington, 51–80.

Molina, M.J. and Rowland, F.S. (1974) Stratospheric sink chloroflouromethanes: chlorine atom catalysed destruction of ozone. *Nature*, **249**, 810–14.

Momsen, J. (1991) *Women and Development in the Third World*, Routledge, London.

Moore, R.I. (ed.) (1981) *The Hamlyn Historical Atlas*, Hamlyn, London and New York.

Mormont, M. (1990) Who is rural? Or, how to be rural: towards a sociology of the rural, in Marsden, T., Lowe, P. and Whatmore, S. (eds) *Rural Restructuring: Global Processes and their Responses*, David Fulton, London, 21–45.

Morrill, R. (2004) Representation, law and redistricting: the case of the United States, in Barnett, C. and Low, M. (eds) *Spaces of Democracy*, Sage, London.

Morris, C. (2000) Quality assurance schemes: a new way of delivering environmental benefits in food production?, *Journal of Environmental Planning and Management*, **43**, 433–48.

Morris, C. and Potter, C. (1995) Recruiting the new conservationists, *Journal of Rural Studies*, **11**, 51–63.

Morris, C. and Young, C. (1997) *A Preliminary Analysis of UK Farm Quality Assurance Schemes: 'New Deals' for the Rural Environment*, Anglo-French Conference of Rural Geographers, Nantes.

Morris, C. and Young, C. (2000) 'Seed to shelf', 'teat to table', 'barley to beer' and 'womb to tomb': discourses of food quality and quality assurance schemes in the UK, *Journal of Rural Studies*, **16**, 193–205.

Morris-Suzuki, T. (1996) The frontiers of Japanese identity, in Tønnesson, S. and Antlöv, H. (eds) *Asian Forms of the Nation*, Curzon, Richmond, 41–66.

Mortishead, C. (1999) Bangalore's program for success, *The Times*, 24 March, 29.

Moser, C. (1993) *Gender, Planning and Development: Theory, Practice and Training*, Routledge, London.

Mosse, G.L. (1975) *The Nationalization of the Masses*, H. Fertig, New York.

Mouffe, C. (1993) Citizenship, in Krieger, J. (ed.) *The Oxford Companion to Politics of the World*, Oxford University Press, Oxford, 138–9.

Mountjoy, A.B. (1976) Worlds without end, *Third World Quarterly*, **2**, 753–7.

Müller-Rommel, F. (1990) New politics parties and

new social movements in Western Europe, in Dalton, R.J. and Kuechler, M. (eds) *Challenging the Political Order: New Social and Political Movements in Western Democracies*, Oxford University Press, Oxford.

Munslow, B. and Ekoko, F. (1995) Is democracy necessary for sustainable development?, *Democratisation*, **2**, 158–78.

Murdoch, J. (1995) Actor-networks and the evolution of economic forms, *Environment and Planning A*, **27**, 731–57.

Murdoch, J. (1998) *Counterurbanisation and the Countryside: Some Causes and Consequences of Urban to Rural Migration*, Papers in Environmental Planning Research 15, Cardiff University, Cardiff.

Murdoch, J. (2000) Networks: a new paradigm of rural development, *Journal of Rural Studies,* **16**, 407–19.

Myers, N. (1991) *Population, Resources and the Environment: The Critical Challenges*, United Nations Population Fund, New York.

Myrdal, G. (1957) *Economic Theory and Underdeveloped Areas*, Duckworth, London.

N

Nagel, C. (2002) Geopolitics by another name: immigration and the politics of assimilation, *Political Geography*, **21**, 971–87.

Nash, G.B. *et al.* (1990) *The American People: Creating a Nation and a Society*, 2nd edition, Harper & Row, New York and London.

National Agricultural Lands Study (1980) *Where Have All the Farmlands Gone?*, NALS, Washington, DC.

National Research Council (2000) *Beyond Six Billion: Forecasting the World's Population*, National Academy Press, Washington, D.C.

National Statistics Online: www.statistics.gov.uk.

Natter, W. (2003) Geopolitics in Germany 1919–1945, in Agnew, J. Mitchell, K. and Toal, G. (eds) *A Companion to Political Geography*, Blackwell, Oxford, 197–203.

Nayek, A. (2003) Last of the Real Geordies? White masculinities and the subcultural response to deindustrialization, *Environment and Planning D: Society and Space*, **21**, 7–25.

Nelson, B. and Chowdhury, N. (eds) (1994) *Women and Politics Worldwide*, Yale University Press, New Haven, CT.

Netherlands Scientific Council for Government Policy (1992) *Ground for Choices: Four Perspectives for the Rural Areas in the European Community*, Report 42, NSCGP, The Hague.

New Internationalist (1999) *The A–Z of World Development*, New Internationalist, London.

Newman, O. (1972) *Defensible Space: People and Design in the Violent City*, Architectural Press, London.

Nixon, S. (1997) Circulating culture, in du Gay P (ed.) *Production of Culture/Cultures of Production,* Sage, London, 177–234.

Nostrand, R.L. (1987) The Spanish borderlands, in Mitchell, R.D. and Groves, P.A. (eds) *North America: The Historical Geography of a Changing Continent*, Hutchinson, London.

Nove, A. (1987) *The Soviet Economic System*, 3rd edition, Allen & Unwin, London.

Nye, D.E. (1992) *Electrifying America: Social Meanings of a New Technology*, MIT Press, Cambridge, MA.

O

O'Brien, R. (1992) *Global Financial Integration: The End of Geography*, Royal Institute of International Affairs, Pinter Publishers, London.

O'Connor, A. (1976) Third World or one world, *Area*, **8**, 269–71.

Odell, P.R. (1989) Draining the world of energy, in Johnston, R.J and Taylor, P.J. (eds) *A World in Crisis?*, 2nd edition, Blackwell, Oxford.

Ogborn, M. (1999) Historical geographies of globalisation, *c.*1500–1800, in Graham, B. and Nash, C. (eds) *Modern Historical Geographies*, Prentice Hall, Harlow.

Ogden, P.E. (1985) Counterurbanisation in France: the results of the 1982 population census, *Geography*, **70**, 24–35.

Ogilvy, J. and Schwartz, P. (2000) *China's Futures*, Wiley, London.

Ohmae, K. (1990) *The Borderless World: Power and Strategy in the Interlinked Economy*, Collins, London.

Ohnuki-Tierney, E. (1997) McDonalds in Japan: changing manners and etiquette, in Watson, J.L. (ed.) *Golden Arches East: McDonalds in East Asia*, Stanford University Press, Stanford, CA, 161–82.

O'Loughlin, J. (ed.) (1994) *Dictionary of Geopolitics*, Greenwood Press, Westport, CT.

O'Loughlin, J. and van der Wusten, H. (1990) Political geography and panregions, *Geographical Review*, **80**, 1–20.

Oldfield, J., Pollard, J.S. Randalls, S. and Thornes, J.E. (2004) *Weather Derivatives: Crossing the Borders of Human and Physical Geography*. Paper presented at the Meeting of the International Geographical Congress, Glasgow, 20 August.

Olson, M. (1965) *The Logic of Collective Action: Public Goods and the Theory of Groups*, Harvard University Press, Cambridge, MA.

Ong, A. (1999) *Flexible Citizenship: The Cultural Logics of Transnationality*, Duke University Press, Durham, USA.

O'Neill, M. and Black, M. (1996) Current quality issues in the Northern Ireland tourism sector, *Total Quality Magazine*, **8**, 15–19.

Oppong, J. and Hodgson, M. (1994) Spatial accessibility to health care facilities in Suhum District, Ghana, *Professional Geographer*, **46**, 199–209.

Organization for Economic Co-operation and Development (1991) *New Ways of Managing Services in Rural Areas*, OECD, Paris.

Organization for Economic Co-operation and Development (1993) *What Future for our Countryside?*, OECD, Paris.

Organization for Economic Co-operation and Development (1998) *OECD in Figures 1998,* OECD, Paris.

Organization for Economic Co-operation and Development (1999) *Environment in the Transition to a Market Economy*, OECD, Paris.

Origo, I. (1988) *The Merchant of Prato*, Penguin Books, Harmondsworth.

O'Riordan, T. (1976) *Environmentalism*, Pion, London.

O'Riordan, T. (ed.) (2000) *Environmental Science for Environmental Management*, 3rd edition, Prentice Hall, Harlow.

Østergaard, L. (1992) *Gender and Development: A Practical Guide*, Routledge, London.

Ó Tuathail, G. (1993) The effacement of place: US foreign policy and the spatiality of the Gulf crisis, *Antipode*, **25**, 4–31.

Ó Tuathail, G. (1994) Critical geopolitics and development theory: intensifying the dialogue, *Transactions of the Institute of British Geographers*, NS, **19**, 228–38.

Ó Tuathail, G. (1996) *Critical Geopolitics: The Politics of Writing Global Space*, Routledge, London.

Ó Tuathail, G. (1998) Postmodern geopolitics? The modern geopolitical imagination and beyond, in Ó Tuathail, G. and Dalby, S. (eds) *Rethinking Geopolitics*, Routledge, London, 16–38.

Ó Tuathail, G. and Dalby, S. (eds) (1998) *Rethinking Geopolitics*, Routledge, London.

Ó Tuathail, G., Dalby, S. and Routledge, P. (eds) (1998) *The Geopolitics Reader*, Routledge, London.

Owens, B. (1996) *The Case for Aid: A Manifesto*, Confederation of Aid Organisations, London.

Oxfam (1993) *Africa: Make or Break: Action for Recovery*, Oxfam, Oxford.

Oxfam (1995) *A Case for Reform: Fifty Years of the IMF and World Bank*, Oxfam, London.

Oxford Dictionary of Current English (1985) Penguin, Harmondsworth.

P

Paasi, A. (1996) *Territories, Boundaries and Consciousness: The Changing Geographies of the Finnish-Russian Frontier*, Wiley, Chichester.

Paasi, A. (2003) Territory, in Agnew, J., Mitchell, K. and Toal, G. (eds) *A Companion to Political Geography*, Blackwell, Malden, 109–22.

Packard, V. (1977) *The Hidden Persuaders*, Penguin, Harmondsworth.

Pahl, J. (1989) *Money and Marriage*, Macmillan, London.

Painter, J. (1995) *Politics, Geography, and 'Political Geography': A Critical Perspective*, Arnold, London.

Panelli, R. (2004) *Social Geographies: From Difference to Action*, Sage, London.

Pantazis, C., Gordon, D and Townsend, P. (1999) *The Necessities of Life in Britain*, Working Paper, Poverty and Social Exclusion Survey, University of Bristol.

Parasuraman, S. (1995) Economic marginalisation of peasants and fishermen due to urban expansion, *Pakistan Development Review*, **34**, 121–38.

Park, C. (1994) *Sacred Worlds: An Introduction to Geography and Religion*, Routledge, London.

Parker, G. (1987) French geopolitical thought in the interwar years and the emergence of the European idea, *Political Geography Quarterly*, **6**, 145–50.

Parrott, N., Wilson, N. and Murdoch, J. (2002) Spatializing quality: regional protection and the alternative geography of food, *European Urban and Regional Studies*, **9**, 241–61.

Past Worlds: The Times Atlas of Archaeology (1988) Times Books, London.

Paterson, T.G. (1992) *On Every Front. The Making and Unmaking of the Cold War*, revised edition, W.W. Norton, New York and London.

Peach, W.N. and Constantin, J.A. (1972) *Zimmerman's World Resources and Industries*, 3rd edition, Harper & Row, London.

Pearce, F. (2003) Saving the world, plan B, *The New Scientist*, 13 December, 6–7.

Pearson, R. (1995) Gender perspectives on health and safety in information processing: learning from international experience, in Mitter, S. and Rowbotham, S. (eds) *Women Encounter Technology: Changing Patterns of Employment in the Third World*, Routledge, London, 278–302.

Peck, J.A. and Yeung, H.W. (ed) (2003) *Remaking the Global Economy*, Sage, London.

Peet, R. (1989) The destruction of regional cultures, in Johnston, R.J. and Taylor, P.J. (eds) *A World in Crisis?*, 2nd edition, Blackwell, Oxford, 150–72.

Peet, R. with Hartwick, E. (1999) *Theories of Development*, Guilford Press, London.

Penrose, J. (2002) Nations, states and homelands: territory and territoriality in nationalist thought, *Nations and Nationalism* **8**, 277–97.

Pepper, D. (1996) *Modern Environmentalism: An Introduction*, Routledge, London.

Perry, R., Dean, K. and Brown, B. (eds) (1986) *Counterurbanisation: International Case Studies of Socio-Economic Change in Rural Areas*, Geo Books, Norwich.

Perry, S. and Schneck, C. (eds) (2001) *Eye-to-eye: Women Practising Development Across Cultures*, Zed Books, London.

Philip L. and Shucksmith M. (2003) Conceptualizing social exclusion in rural Britain, *European Planning Studies*, **11**, 461–80.

Philo, C. (1992) Neglected rural geographies: a review, *Journal of Rural Studies*, **8**, 193–207.

Philo, C. (1995) Where is poverty? The hidden geography of poverty in the United Kingdom, in Philo, C. (ed.) *Off the Map: The Social Geography of Poverty in the UK*, Child Poverty Action Group, London, 1–22.

Philo, C. (2000) Social exclusion, in Johnston, R., Gregory, D., Pratt, G. and Watts, M. (eds) *Dictionary of Human Geography*, Oxford, Blackwells, 751–2.

Pickles, J. (2004) A *History of Spaces: Cartographic Reason, Mapping and the Geo-coded World*, Routledge, London and New York.

Pieterse, J.N. and Parekh, B. (eds) (1995) *Decolonization of the Imagination: Culture, Knowledge and Power*, St Martins Press, London.

Pile, S. (1999) The heterogeneity of cities, in Pile, S., Brook, C. and Mooney, G. (eds) *Unruly Cities? Order/Disorder*, Routledge, London, 7–52.

Pile, S., Brook, C. and Mooney, G. (eds) (1999) *Unruly Cities? Order/Disorder*, Routledge, London.

Pinch, S. (1985) *Cities and Services: The Geography of Collective Consumption*, Routledge & Kegan Paul, London.

Pinch, S. (1994) Social polarization: a comparison of evidence from Britain and the United States, *Environment and Planning A*, **25**, 779–95.

Pinch, S. and Henry, N. (1999) Paul Krugman's geographical economics, industrial clustering and the British motor sport industry, *Regional Studies*, **33**, 815–27.

Piven, F.F. and Cloward, R. (1971) *Regulating the Poor: The Functions of Public Welfare*, Vintage, New York.

Pletsch, C. (1981) The three worlds or the division of social scientific labour 1950–1975, *Comparative Studies in Society and History*, **23**, 565–90.

Polanyi, K. (1968) *Primitive, Archaic and Modern Economies*, Beacon Press, Boston.

Policy Commission on the Future of Farming and Food (2002) *Farming and Food: A Sustainable Future*, HMSO, London.

Political Geography (1998) Space, *Place and Politics in Northern Ireland*, special issue of *Political Geography*, **17**(2).

Pollard, J.S. (1996) Banking at the margins: a geography of financial exclusion in Los Angeles, *Environment and Planning A*, **28**, 1209–32.

Pollard, J.S. (1999) Globalisation, regulation and the changing organisation of retail banking in the United States and Britain, in Martin, R.L. (ed.) *Money and the Space Economy*, Wiley, Chichester, 49–70.

Pollard, J.S. and Sidaway J.D. (2002) Euroland: economic, political and cultural geographies, *Transactions of the Institute of British Geographers*, **27**, 7–10.

Poole, R., Clarke, G. and Clarke, D. (2002) Growth, concentration and regulation in European food retailing. *European Urban and Regional Studies*, **9**, 167–86.

Popke, E.J. (2001) Modernity's abject space: the rise and fall of Durban's Cato Manor, *Environment and Planning A*, **33**, 737–52.

Poplawska, I. and Muthesius, S. (1986) Poland's Manchester: 19th-century industrial and domestic architecture in Łódź, *Journal of Social and Architectural History*, 148–60.

Porter, G. (1995) Scenes from childhood, in Crush, J. (ed.) *Power of Development*, Routledge, London, 63–86.

Porter, G. (1997) Mobility and inequality in rural Nigeria: the case of off-road communities, *Tijdschrift voor Economische en Sociale Geografie*, **88**, 65–76.

Porter, M.E. (1990) *The Competitive Advantage of Nations*, Free Press, New York.

Porter, P.W. and Sheppard, E.S. (1998) *A World of Difference: Society, Nature, Development*, Guilford Press, New York.

Potter, C. (1998) Conserving nature: agri-environmental policy development and change, in Ilbery, B. (ed.) *The Geography of Rural Change*, Longman, London, 85–105.

Potter, R., Binns, T., Elliott, J.A. and Smith, D. (1999) *Geographies of Development*, Longman, London.

Potter, R.B. (1979) Perception of urban retailing facilities: an analysis of consumer information fields, *Geografiska Annaler*, **61B**, 19–27.

Pounds, N.J.F. (1947) *An Historical and Political Geography of Europe*, Harrap, London.

Pounds, N.J.G. (1990) *An Historical Geography of Europe*, Cambridge University Press, Cambridge.

Powe, N. and Whitby, M. (1994) Economies of settlement size in rural settlement planning, *Town Planning Review*, **65**, 415–34.

Power, A. and Tunstall, R. (1997) *Dangerous Disorder:*

Riots and Violent Disturbances in Thirteen Areas of Britain, 1991–1992, Joseph Rowntree Foundation, York.

Power, M. (2003) *Rethinking Development Geographies*, Routledge, London.

PRB (2003) *World Population Data Sheet 2003*, Population Reference Bureau, Washington DC.

Preston, P.W. (1996) *Development Theory: An Introduction*, Blackwell, Oxford.

PriceWaterhouseCoopers (1999) *Technology Forecast 1999*, 10th Anniversary Edition, PriceWaterhouseCoopers, London.

Priestley, J.B. (1937) *English Journey,* cheap edition, William Heinemann, London.

Pritchard, W. (1998) The emerging contours of the third food regime: evidence from Australian dairy and wheat sectors, *Economic Geography*, **74**, 64–74.

Prokhovnik, R. (1998) Public and private citizenship: from gender invisibility to feminist inclusiveness, *Feminist Review*, **60**, 84–104.

Prunier, G. (1995) *The Rwanda Crisis 1959–1994: History of a Genocide*, Hurst & Co., London.

Przeworski, A. (1986) *Capitalism and Social Democracy*, Cambridge University Press, Cambridge.

Public Broadcasting Service (USA) (2004) Web-site on the Hoover Dam, includes films, maps, and accounts, available at: www.pbs.org/wgbh/amex/hoover/.

R

Raban, J. (1990) *Hunting Mr Heartache*, Collins Harvill, London.

Radcliffe, S.A. (1999) Re-thinking development, in Cloke, P., Crang, P. and Goodwin, M. (eds) *Introducing Human Geographies*, Arnold, London, 84–92.

Radcliffe, S. and Westwood, S. (1996) *Remaking the Nation: Place, Identity and Politics in Latin America*, Routledge, London.

Rahnema, R. (1997) Introduction, in Rahnema, R. and Bawtree, V. (eds) *The Post-Development Reader*, Zed Books, London.

Rahnema, R. and Bawtree, V. (1997) (eds) *The Post-Development Reader*, Zed Books, London.

Rathgeber, E. (1990) WID, WAD, GAD: trends in research and practice, *Journal of Developing Areas*, **24**, 489–502.

Rau, B. (1991) *From Feast to Famine: Official Cures and Grassroots Remedies in Africa's Food Crisis*, Zed, London.

Ravenhall, Y. (1999) *Women's Investing*, Part 1, online, available at: www.fool.co.uk/personalfinance.

Rawcliffe, P (1998) *Environmental Pressure Groups in Transition*, Manchester University Press, Manchester.

Ray, C. (1998) Culture, intellectual property and territorial rural development, *Sociologia Ruralis*, **38**, 1–19.

Redclift, M. (1984) *Sustainable Development: Exploring the Contradictions*, Methuen, London.

Reddish, A. and Rand, M. (1996) The environmental effects of present energy policies, in Blunden, J. and Reddish, A. (eds) *Energy, Resources and Environment*, Open University and Hodder & Stoughton, London, 43–91.

Rees, J. (1985) *Natural Resources: Allocation, Economics and Policy*, 2nd edition, Routledge, London.

Rees, J. (1991) Resources and environment: scarcity and sustainability, in Bennett, R. and Estall, R. (eds) *Global Change and Challenge: Geography in the 1990s*, Routledge, London, 5–26.

Reeves, R. (1999) Do we really need gay florists?, *Observer*, 8 August.

Renfrew, C. (1996) Language families and the spread of farming, in Harris, D.R. (ed.) *The Origins and Spread of Agriculture and Pastoralism in Eurasia*, UCL Press, London, 70–92.

Renner, M. (2001) Breaking the link between resources and repression, Worldwatch Institute, *State of the World 2002*, Earthscan, London, 149–73.

Ribeiro, A. (1935) Últimas conquistas e definação territorial, in Peres, D. (ed.) *História de Portugal*, vol. 2, Portucalense Editoria, Porto, 251–73.

Ricardo, D. (1817) *On the Principles of Political Economy and Taxation*, John Murray, London.

Richards, P.W. (1995) *Indigenous Agricultural Revolution,* Hutchinson, London.

Righter, R. (1991) *Utopia Lost: The United Nations and the World Order,* Twentieth Century Fund Press, London.

Rist, G. (1997) *The History of Development*, Zed, London.

Ritzer, G. (2000) *The McDonaldization of Society: An Investigation into the Changing Character of Contemporary Social Life*, Pine Forge Press, California.

Ritzer, G. (2004) *The Globalization of Nothing*, Pine Forge/Sage, London.

Robert, S. and Randolph, W. (1983) Beyond decentralization: the evolution of population distribution in England and Wales, 1961–81, *Geoforum*, **14**, 75–102.

Robins, K. (1991) Tradition and translation: national culture in its global context, in Corner, J. and Harvey, S. (eds) *Enterprise and Heritage: Crosscurrents of National Culture*, Routledge, London, 21–44.

Robins, K. (1995) The new spaces of global media, in Johnston, R.J., Taylor, P. and Watts, M. (eds) *Geographies of Global Change*, Blackwell, Oxford, 248–62.

Robinson, G. (1997) Greening and globalising: agriculture in 'the new times', in Ilbery, B., Chioti, Q. and

Rickard, T. (eds) *Agricultural Restructuring and Sustainability: A Geographical Perspective*, CAB International, Wallingford.

Robinson, G. (2003) *Geographies of Agriculture*, Pearson, London.

Robinson, G. and Ilbery, B. (1993) Reforming the CAP: beyond McSharry, in Gilg, A. (ed.) *Progress in Rural Policy and Planning*, vol. 3, Belhaven Press, London, 197–207.

Robinson, J. (1997) The geopolitics of South African cities: states, citizens, territory, *Political Geography*, **16**, 365–86.

Robinson, J. (1999) Divisive cities, in Pile, S., Brook, C. and Mooney, G. (eds) *Unruly Cities? Order/Disorder*, Routledge, London, 149–200.

Rodriguez-Pose, A. and Gill, N. (2003) The global trend towards devolution and its implications, *Environment and Planning C: Government and Policy*, **21**, 333–51.

Rogers, A., Viles, H. and Goudie, A. (eds) (1992) *The Student's Companion to Geography*, Blackwell, Oxford.

Rogers, R. and Gumuchdjian, P. (1997) *Cities for a Small Planet*, Faber and Faber, London.

Rollins, W.H. (1995) Whose landscape? Technology, Fascism and environmentalism on the National Socialist Autobahn, *Annals of the Association of American Geographers*, **85**, 494–520.

Rooum, D. (1995) *What is Anarchism?: An Introduction,* Freedom Press, London.

Rose, N. and Miller, P. (1992) Political power beyond the state: problematics of government, *British Journal of Sociology*, **43**, 173–205.

Ross, M. (2001) *Extractive Sectors and the Poor: An Oxfam America Report*, Oxfam. Available at: www.oxfamamerica.org/cirexport/index.html.

Rössler, M. (1989) Applied geography and area research in Nazi society: central place theory and planning, *Environment and Planning D: Society and Space*, **7**, 419–31.

Rostow, W.W. (1960) *The Stages of Economic Growth: A Non-Communist Manifesto*, Cambridge University Press, Cambridge.

Rostow, W.W. (1971) *The Stages of Economic Growth*, 2nd edition, Cambridge University Press, Cambridge.

Rostow, W.W. (1975) *How it All Began: The Origins of the Modern Economy*, Methuen, London.

Rothenberg, T. (1995) 'And she told two friends': lesbians creating urban social space, in Bell, D. and Valentine, G. (eds) *Mapping Desire: Geographies of Sexualities*, Routledge, London, 165–81.

Routledge, P. (2003) Anti-geopolitics, in Agnew, J., Mitchell, K. and Toal, G. (eds) *A Companion to Political Geography*, Blackwell, Oxford, 236–48.

de Ru, N. (2004) Hollocore, in Koolhas, R., Brown, S. and Link, J. (eds) *Content*, Taschen, Cologne, 336–49.

Rubenstein, J.M. (1992) *The Changing US Auto Industry: A Geographical Analysis*, Routledge, London.

Rubin Hudson, B. (1979) Aesthetic ideology and urban design, *Annals of the Association of American Geographers*, **69**, 339–61.

Rueschemeyer, D., Stephens, E.H. and Stephens, J.D. (1992) *Capitalist Development and Democracy*, Polity Press, Cambridge.

Rusten, G. (1997) The role of geographic concentration in promoting competitive advantage: the Norwegian furniture industry, *Norsk geogr. Tidsskr.*, **51**, 173–85.

S

Sabel, C. (1989) Flexible specialisation and the re-emergence of regional economies, in Hirst, P and Zeitlin, J. (eds) *Industrial Decline? Industrial Structure and Policy in Britain and her Competitors,* Berg, Oxford.

Sachs, W. (1992) *The Development Dictionary*, Zed, London.

Sachs, J.D. and Warner, A.M. (2001) Natural resources and economic development: the curse of natural resources, *European Economic Review*, **45**, 827–38.

Sack, R. (1986) *Human Territoriality: Its Theory and History*, Cambridge University Press, Cambridge.

Sack, R. (1992) *Place, Modernity and the Consumers World*, Johns Hopkins University Press, Baltimore, MD.

Saint-Laurent, C. (1996) The forest principles and the Inter-Governmental Panel on forests, in Dodds, F. (ed.) *The Way Forward: Beyond Agenda 21*, Earthscan, London, 65–82.

Salway, P. (1981) Roman Britain, in *Oxford History of England*, Oxford University Press, Oxford.

Sandbach, F. (1980) *Environment, Ideology and Policy*, Basil Blackwell, Oxford.

Sant, M. and Simons, P. (1993) The conceptual basis of counterurbanisation: critique and development, *Australian Geographical Studies*, **31**, 113–26.

Santos, M. (1974) Geography, Marxism and underdevelopment, *Antipode*, **6**, 1–9.

Sarre, P. and Blunden, J. (1995) *An Overcrowded World? Population, Resources and Environment*, Oxford University Press, Oxford.

Sassen, S. (1991) *The Global City: New York, London, Tokyo*, Princeton University Press, Princeton, NJ.

Sassen, S. (1994) *Cities in a World Economy*, Pine Forge Press, London.

Sassen, S. (2001) *The Global City: New York, London, Tokyo*, Princeton University Press, Princeton.

Schilder, A. and Boeve, B. (1996) An institutional

approach to local economic development, *Regional Development Dialogue*, **17**, 94–117.

Schneider, G. and Schubert, P. (1997) *Wer baut wo? Dies grossen Bauprojekte im Überblick*, Jaron Verlag, Berlin.

Schramm, G. (1993) Rural electrification in LDCs as a tool for economic development, *OPEC Review*, **17**, 501–17.

Schumpeter, J. (1939) *Business Cycles: A Theoretical, Historical and Statistical Analysis of the Capitalist Process*, McGraw-Hill, London.

Schuurman, F.J. (2001) Globalization and development studies: introducing the challenges, in Schuurman, F.J. (ed.) *Globalization and Development: Challenges for the 21st Century*, Sage, London, 3–16.

Scott, A.J. (1988) *New Industrial Space,* Pion Press, London.

Scott, A.J. and Brown, E.R. (eds) (1993) *South-Central Los Angeles: Anatomy of an Urban Crisis*, The Ralph and Goldy Lewis Center for Regional Policy Studies, Working Paper Series, 06. Available at: http://repositories.cdlib.org/lewis/wps/06

Scott, J. and Simpson, P. (eds) (1991) *Sacred Spaces and Profane Spaces: Essays in the Geographics of Judaism, Christianity and Islam*, Greenwood Press, New York.

Sells, M.A. (1998) *The Bridge Betrayed. Religion and Genocide in Bosnia*, University of California Press, Berkeley.

Sennett, R. (1990) *The Conscience of the Eye: Design and Social Life in Cities*, Faber and Faber, London.

Serow, W.J. (1991) Recent trends and future prospects for urban-rural migration in Europe, *Sociologia Ruralis*, **31**, 269–80.

Serrão, J.V. (1979) *História de Portugal*, vol. 1, Verbo, Lisbon.

Service, E.R. (1971) *Primitive Social Organization*, 2nd edition, Random House, New York.

Seton-Watson, H. (1977) *Nations and States: An Enquiry into the Origins of Nations and the Politics of Nationalism*, Westview Press, Boulder, CO.

Shakespeare, T. (1994) Cultural representations of disability: dustbins for disavowal, *Disability and Society*, **9**, 283–99.

Shanin, T. (ed.) (1971) *Peasants and Peasant Societies*, Penguin, Harmondsworth.

Sharp, J. (1999) Critical geopolitics, in Cloke, D., Crang, P. and Goodwin, M. (eds) *Introducing Human Geographies*, Arnold, London, 179–88.

Shaw, A. (1994) Urban growth and land-use conflicts: the case of New Bombay, *Bulletin of Concerned Asian Scholars*, **26**, 33–44.

Sherratt, A. (ed.) (1980) *The Cambridge Encyclopaedia of Archaeology*, Cambridge University Press, Cambridge.

Shields, R. (1991) *Places on the Margin*, London, Routledge.

Shen, J. (1995) Rural development and rural to urban migration in China 1978–1990, *Journal of Rural Studies*, **26**, 395–409.

Shiva, V. (1991) *The Violence of the Green Revolution: Third World Agriculture, Ecology and Politics*, Zed, London.

Shiva, V. (1998) *Biopiracy: The Plunder of Nature and Knowledge*, Green Books, Devon.

Shiva, V. (1999) We will drive Monsanto out of the country, *Asian Age*, 6 March, 9.

Shiva, V. (2002) *Water Wars: Privatization, Pollution, and Profit*, Pluto Books, London.

Shiva, V. and Moser, I. (1995) *Biopolitics: A Feminist and Ecological Reader on Biotechnology*, Zed, London.

Short, J.R. (1989) Yuppies, yuffies and the new urban order, *Transactions of the Institute of British Geographers*. NS, **14**, 173–88.

Short, J.R. (1996) *The Urban Order: An Introduction to Cities, Culture and Power*, Blackwell, Oxford.

Shucksmith, M. (1993) Farm household behaviour and the transition to post-productivism, *Journal of Agricultural Economics*, **44**, 466–78.

Shucksmith, M., Chapman, P., Clark, G. and Black, S. (1994) Social welfare in rural Europe, in Copus, A. and Marr, P. (eds) *Rural Realities, Trends and Choices*, Proceedings of the 35th EAAE Seminar, Scottish Agricultural College, Aberdeen, 199–219.

Shurmer-Smith, P. (ed.) (2002) *Doing Cultural Geography*, Sage, London

Shurmer-Smith, P. and Hannam, K. (1994) *Worlds of Desire, Realms of Power: A Cultural Geography*, Arnold, London.

Sibley, D. (1995) *Geographies of Exclusion: Society and Difference in the West*, Routledge, London.

Sibley, D. (1999) Creating geographies of difference, in D. Massey, J. Allen and P. Sarre (eds) *Human Geography Today*, Cambridge, Polity, 115–28.

Sidaway, J.D. (1994) Geopolitics, geography and 'terrorism' in the Middle East, *Environment and Planning D: Society and Space*, **12**, 357–72.

Sidaway, J.D. (1998) What is in a Gulf? From the 'arc of crisis' to the Gulf War, in Ó Tuathail, G. and Dalby, S. (eds) *Rethinking Geopolitics*, Routledge, London and New York, 224–39.

Sidaway, J.D. (1999) American power and the Portuguese Empire, in Slater, D. and Taylor, P.J. (eds) *The American Century*, Blackwells, Oxford, 195–209.

Sidaway, J.D. (2000) Iberian geopolitics, in Dodds, K. and Atkinson, D. (eds) *Geopolitical Traditions: A*

Century of Geopolitical Thought, Routledge, London, 118–49.

Sidaway, J.D. and Pryke, M. (2000) The free and the unfree: 'emerging markets', the Heritage Foundation and the 'index of economic freedom', in Bryson, J.R., Daniels, P.W., Henry, N.D. and Pollard, J.S. (eds) Knowledge, Space, Economy, Routledge, London.

Siddiquee, N.A. (1996) Health for all by 2000? An assessment of the delivery of health care services in rural Bangladesh, South Asia, **19**, 49–70.

Simmel, G. (1978) The Philosophy of Money, Routledge & Kegan Paul, London.

Simmel, G. (1991) Money in modern culture, Theory, Culture and Society, **8**, 17–31.

Simmons, I.G. (1996) Changing the Face of the Earth: Culture, Environment, History, 2nd edition, Blackwell, Oxford.

Simpson, E.S. (1994) The Developing World: An Introduction, 2nd edition, Longman, Harlow.

Sittirak, S. (1998) The Daughters of Development: Women and the Changing Environment, Zed, London.

Sjöberg, K. (1993) The Return of the Ainu: Cultural Mobilization and the Practice of Ethnicity in Japan, Harwood Academic Publishers, London.

Skelton, T. and Valentine, G. (eds) (1998) Cool Places: Geographies of Youth Cultures, Routledge, London.

Skidmore, T.E. and Smith, P.H. (1988) Modern Latin America, 2nd edition, Oxford University Press, New York.

Skocpol, T. (1992) Protecting Soldiers and Mothers: The Political Origins of the American Welfare State, Harvard University Press, Cambridge, MA.

Slater, D. (1992) On the borders of social theory: learning from other regions, Environment and Planning D: Society and Space, **10**, 307–27.

Slater, D. (1993) The geopolitical imagination and the enframing of development theory, Transactions of the Institute of British Geographers, NS, **18**, 419–37.

Slater, D. (1997) Consumer Culture and Modernity, Polity Press, Cambridge.

Slater, D. (2003) Beyond Euro-Americanism: democracy and post-colonialism, in Anderson, K., Domosh, M., Pike, S. and Thrift, N. (eds) Handbook of Cultural Geography, Sage, London, 420–32.

Smith, A. (1776) An Inquiry into the Nature and Causes of the Wealth of Nations, edited by Skinner, A. (1974) Penguin, London.

Smith, A. (1977) The Wealth of Nations, edited by Skinner, A., Penguin, Harmondsworth.

Smith A.D. (1986) The Ethnic Origin of Nations, Blackwell, Oxford.

Smith, D.M. (1992) Apartheid in South Africa, 3rd edition, Cambridge University Press, Cambridge.

Smith, D.M. (1995) Geography and Social Justice, Oxford, Blackwells.

Smith, M.D. (1996) The empire filters back, consumption, production, and the politics of Starbucks coffee, Urban Geography, **17**, 502–24.

Smith, M. (1999) Burma: Insurgency and the Politics of Ethnicity, Zed Books, London.

Smith, N. (1984) Isiah Bowman: political geography and geopolitics, Political Geography Quarterly, **3**, 69–76.

Smith, N. (1996) The New Urban Frontier: Gentrification and the Revanchist City, London, Routledge.

Smith, N. (2002) New globalism, new urbanism, gentrification as global urban strategy, Antipode, **34**, 434–57.

Smith, N. (2003) American Empire: Roosevelt's Geographer and the Prelude to Globalization, University of California Press, Berkeley.

Smith, R.A. (1852) On the air and rain of Manchester, Memoirs and Proceedings of the Manchester Literary and Philosophical Society, **2**, 207–17.

Smith, S. (1989) The Politics of Race and Residence, Cambridge, Polity.

Smith, S.J. (1999) Society-space, in Cloke, P., Crang, P. and Goodwin, M. (eds) Introducing Human Geographies, Arnold, London, 12–23.

Sogge, D. (2002) Give and Take: What's the Matter with Foreign Aid?, Zed Books, London.

Soja, E.J. (1989) Postmodern Geographies: The Reassertion of Space in Critical Social Theory, Verso, London.

Soja, E. (1996) Thirdspace: Journeys to Los Angeles and Other Real and Imagined Places, Blackwell, Oxford.

Soja, E. (1997) Six discourses on the postmetropolis, in Westwood, S. and Williams, J. (eds) Imagining Cities: Scripts, Signs, Memory, Routledge, London, 19–30.

Sondheimer, S. (ed.) (1991) Women and the Environment: A Reader, Zed Books, London.

Spain, D. (1992) Gendered Spaces, University of North Carolina Press, Chapel Hill.

Speer, A. (1995) Inside the Third Reich, Phoenix, London.

Spencer, D. (1995) Counterurbanisation: the local dimension, Geoforum, **26**, 153–73.

Spillius, A. (2003) Dialing Tone, Telegraph Magazine, 19 April, 38–43

Spivak, G.C. (1985) Three women's texts and a critique of imperialism, Critical Inquiry, **12**, 243–61.

Spivak, G.C. (1993) Outside in the Teaching Machine, Routledge, London.

Staehli, L., Kodras, J. and Flint, C. (eds) (1997) State Devolution in America, Sage, Beverly Hills, CA.

Stamp, L.D. and Beaver, S. (1933) The British Isles: A Geographic and Economic Survey, Longman, London.

Stamp, L.D. and Beaver, S. (1963) *The British Isles: A Geographic and Economic Survey*, 5th edition, Longman, London.

Stempel, J.D. (1981) *Inside the Iranian Revolution*, Indiana University Press, Bloomington.

Stenning, A. and Bradshaw, M. (1999) Globalization and transformation: the changing geography of the Post-Socialist world, in Bryson, J., Henry, N., Keeble, D. and Martin, R. (eds) *The Economic Geography Reader: Producing and Consuming Global Capitalism*, Wiley, Chichester, 88–96.

Stevens, P. (2003) *Resource Impact – Curse or Blessing? A Literature Survey*, Centre for Energy, Petroleum and Mineral Law and Policy, electronic journal, available at: www.cepmlp.org

Stienberg P. E. (2001) *The Social Construction of the Ocean*, Cambridge University Press, Cambridge.

Stille, A. (1997) *Excellent Cadavers*, Vintage, London.

Stöhr, W.B. and Taylor, D.R.F. (1981) *Development from Above or Below? The Dialectics of Regional Planning in Developing Countries*, John Wiley, Chichester.

Stoker, G. (1995) Regime theory and urban politics, in Judge, D., Stoker, G. and Wolman, H. (eds) *Theories of Urban Politics*, Sage, London.

Storey, D.J. (1997) *Understanding the Small Business Sector*, Thomson, London.

Storey, D. (2001) *Territory: The Claiming of Space*, Prentice Hall, Harlow.

Storper, M. (1993) Regional worlds of production: learning and innovation in the technology districts of France, Italy and the USA, *Regional Studies*, **27**, 433–55.

Storper, M. (1997) *The Regional World: Territorial Development in a Global Economy*, The Guilford Press, New York.

Strange, S. (1996) *Casino Capitalism*, Blackwell, Oxford.

Strange, S. (1999) *Mad Money: When Markets Outgrow Governments*, University of Michigan Press, Ann Arbor.

Stutz, F.P. and De Souza, A.R. (1998) *The World Economy: Resources, Location, Trade and Development*, 3rd edition, Prentice-Hall, Upper Saddle River, NJ.

Sugden, D. (1982) *Arctic and Antarctic*, Basil Blackwell, Oxford.

Sugrue, T. (1996) *The Origins of the Urban Crisis: Race and Inequality in Postwar Detroit*, Princeton University Press, Princeton, NJ.

Sukarno, I. (1995) Modern History Sourcebook: President Sukarno of Indonesia: Speech at the opening of the Bandung conference, 18 April 1995, online, available at www.fordham.edu/halsall/mod/1955sukarno-bandong.html [accessed 7 April 2003].

Sweetman, C. (ed.) (2000) *Gender in the Twenty-first Century*, Oxfam, Oxford.

Swyngedouw, E. (2000) Authoritarian governance, power, and the politics of rescaling, *Environment and Planning D: Society and Space*, **18**, 63–76.

Sylvester, C. (1996) Picturing the Cold War: an art graft/eye graft, *Alternatives*, **21**, 393–418.

T

Takahashi, L.M. and Dear, M.J. (1997) The changing dynamics of community opposition to human service facilities, *Journal of the American Planning Association*, **63**, 79–93.

Takeuchi, K., Katoh, K., Nan, Y. and Kou, Z. (1995) Vegetation change in desertified Kerqin sandy lands, *Inner Mongolia/Geographical Reports*, Tokyo Metropolitan University, **30**, 1–24.

Tarrant, J. (1980) The geography of food aid, *Transactions of the Institute of British Geographers*, NS, **5**, 125–40.

Taussig, M. (1997) *The Magic of the State*, Routledge, London.

Taylor, I. (1997) Running on empty, *The Guardian*, 14 May, 2–6.

Taylor, M. (1998) Combating the social exclusion of housing estates, *Housing Studies*, **13**, 819–32.

Taylor, P.J. (1989) *Political Geography*, Longman, London.

Taylor, P.J. (1994) From heartland to hegemony: changing the world in political geography, *Geoforum*, **25**, 403–17.

Taylor, P.J. and Flint, C. (2000) *Political Geography: World-Economy, Nation State and Locality*, Prentice Hall, Harlow.

Taylor, P. and Johnston, R. (1979) *The Geography of Elections*, Penguin, London.

Tesfahuney, M. (1999) Mobility, racism and geopolitics, *Political Geography*, **17**, 499–515.

Thiongo, N.W. (1986) *Decolonising the Mind: The Politics of Language in African Literatures*, James Currey, London.

Thomas, D.S.G. and Middleton, N. (1994) *Desertification: Exploding the Myth*, John Wiley, Chichester.

Thompson, E.P. (1963) *The Making of the English Working Class*, Penguin, London and New York.

Thompson, E.P. (1994) The moral economy of the English crowd in the eighteenth century, in Thompson, E.P. (ed.) *Customs in Common*, Penguin, Harmondsworth.

Thompson, G. (ed.) (1999) *Economic Dynamism in the Asia Pacific Region*, Routledge, London.

Thrift, N. (1996) A hyperactive world, in Johnston, R.J., Taylor, P.J. and Watts, M. (eds) *Geographies of Global Change*, Blackwell, Oxford.

Thrift, N. (1998) Virtual capitalism: the globalisation of reflexive business knowledge, in Carrier, J.G. and Miller, D. (eds) *Virtualism: A New Political Economy*, Berg, Oxford.

Thrift, N. (2001) It's the romance, not the finance, that makes the business worth pursuing': disclosing a new market culture, *Economy and Society*, **30**, 412–32.

Tickamyer, A. and Duncan, C. (1990) Poverty and opportunity structure in rural America, *Annual Review of Sociology*, **16**, 67–86.

Tickell, A. (1996) Making a melodrama out of a crisis: reinterpreting the collapse of Barings Bank, *Environment and Planning D: Society and Space*, **14**, 5–33.

Tickell, A. (2003) Cultures of money, in Anderson, K, Domosh, M, Pile, S and Thrift, N. (eds) *Handbook of Cultural Geography*, Sage, London, 116–30.

Tiffin, M, Mortimore, M, Gickuki, F. (1994) *More People, Less Erosion: Environmental Recovery in Kenya*, John Wiley, Chichester.

Tilly, C. (1984) Social movements and national politics, in Bright, C. and Harding, S. (eds) *Statemaking and Social Movements: Essays in History and Theory*, University of Michigan Press. Ann Arbor, MI.

Tilly, C. (1985) War-making and state-making as organized crime, in Evans, P., Rueschemeyer, R. and Skocpol, T. (eds) *Bringing the State Back In*, Princeton University Press, Princeton, NJ.

Tilly, C. (1986) *The Contentious French*, Harvard University Press, Cambridge, MA.

Time (1998) 26 October, 74.

Times Atlas of World History (1989) 3rd edition, Guild Publishing, London.

Toennies, F. (1887) *Community and Society* (1957 edn), Michigan State University Press, East Lansing, MI.

Tomlinson, J. (1991) *Cultural Imperialism: A Critical Introduction*, Pinter, London.

Tønnesson, S. and Antlöv, S. (1996) Asian in theories of nationalism and national identity, in Tønnesson, S. and Antlöv, H. (eds) *Asian Forms of the Nation*, Curzon, Richmond, 1–40.

Tordoff, W. (1997) *Government and Politics in Africa*, 3rd edition, Macmillan, London.

Touraine, A. (1981) *The Voice and the Eye: An Analysis of Social Movements*, Cambridge University Press, Cambridge.

Tranberg-Hansen, K. (1994) Dealing with used clothing: 'Salaula' and the construction of identity in Zambia's Third Republic, *Public Culture*, **6**, 503–23.

Tratado de Limites (1866) *Tratado de Limites entre Portugal e Espanha assinado em Lisboa aos 29 de Setembro de 1864*, Imprensa Nacional, Lisbon.

Truman, H. (1949) *Public Papers of the President, 20 January*, US Government Printing Offices, Washington, DC.

Tuan, Yi-Fu (2002) *Dear Colleague: Common and Uncommon Observations,* Minneapolis MI: University of Minnesota Press.

Tunstall, J. (1977) *The Media are American*, Constable, London.

Turner II, B.L., Clark, W.C., Kates, R.W., Richards, J.F., Matthews, J.T. and Meyer, W.B. (1990) *The Earth as Transformed by Human Action*, Cambridge University Press, Cambridge.

Tyler May, E. (1989) Cold War – Warm Hearth: politics and the family in postwar America, in Fraser, S. and Gerstle, G. (eds) *The Rise and Fall of the New Deal Order, 1930*, Princeton University Press, Princeton, NJ, 153–81.

Tyler, M. and Abbott, P. (1998) Chocs away: weight watching in the contemporary airline industry, *Sociology*, **32**, 433–50.

U

Understanding Cities (1999) DD304, Open University, Milton Keynes.

UN-HABITAT (2003) *The UN-HABITAT Strategic Vision*, The United Nations Human Settlements Programme, Nairobi.

UN Millennium Development Goals (2000) Available at: http://www.millenniumgoals.org/

UNESCO (1994) *Statistical Yearbook*, United Nations Educational, Scientific and Cultural Organization, Paris.

United Nations (UN) (2002) *International Migration Report 2002*, Department of Economic and Social Affairs, United Nations, New York.

United Nations (UN) (2003) *World Population Prospects: The 2002 Revision*, United Nations Population Division, New York.

United Nations Conference on Trade and Development (UNCTAD) (1998) *World Investment Report 1998*, UNCTAD, Geneva.

United Nations Conference on Trade and Development (UNCTAD) (1999) *World Investment Report 1999: Foreign Direct Investment and the Challenge of Development*, UNCTAD, Geneva.

United Nations Conference on Trade and Development (UNCTAD) (2001) *Estimated FDI Flows in 2001 and Impact of the Events in the United States*, UNCTAD, Geneva.

United Nations Conference on Trade and Development (UNCTAD) (2002a) *World Investment Report 2002: Transnational Corporations and Export Competitiveness*, UNCTAD, Geneva.

United Nations Conference on Trade and Development (UNCTAD) (2002b) *Ecommerce and Development Report 2002*, UNCTAD, Geneva.

United Nations Conference on Trade and Development (UNCTAD) (2003) *World Investment Report 2002: Transnational Corporations and Export Competititveness: An Overview*, UNCTAD, New York and Geneva, available at: www.unctad.org [accessed 1 August, 2003].

United Nations Children's Fund (UNICEF) (2000) *The Progress of Nations 2000*, online, available at: www.unicef.org [accessed 22 February 2002].

United Nations Children's Fund (UNICEF) (2003) *State of the World's Children Report*, online, available at: www.unicef.org/sowc03/ [accessed 9 June 2004].

United Nations Development Programme (UNDP) (1993) *Human Development Report 1993*, Oxford University Press, Oxford.

United Nations Development Programme (UNDP) (1997) *Human Development Report 1997*, Oxford University Press, Oxford.

United Nations Development Programme (UNDP) (1998) *Human Development Report 1998*, Oxford University Press, Oxford.

United Nations Development Programme (UNDP) (1999) *Human Development Report 1999*, Oxford University Press, Oxford.

United Nations Environment Programme (UNEP) (1997) *Global Environemntal Outlook 1*, UNEP, Nairobi, available at: www.grida.no/geol/

United Nations Environment Programme (UNEP) (1999) *Global Environmental Outlook 2000*, UNEP-Earthscan Publications, London.

United Nations Environment Programme (UNEP) (2000) *Global Environemntal Outlook 2000*, UNEP, Nairobi, available at: www.grida.no/geo2000/

United Nations Environment Programme (UNEP) (2003) *GEO-3: Global Environment Outlook 2003*, Earthscan, London.

Unwin, T. (ed.) (1994) *Atlas of World Development*, Wiley, Chichester.

Urry, J. (1990) *The Tourist Gaze: Leisure and Travel in Contemporary Societies*, Sage, London.

US Department of Commerce (1998) *The Emerging Digital Economy*, US Department of Commerce, Washington, DC, available at: www.commerce.gov/emerging.htm

V

Valentine, G. (1989) The geography of women's fear, *Area*, **21**, 385–90.

Valentine, G. (1995) Out and about: geographies of lesbian landscapes, *International Journal of Urban and Regional Research*, **19**, 96–111.

Valentine, G. (2001) *Social Geographies: Space and Society*, Prentice Hall, Harlow.

van de Kaa, D.J. (1987) Europe's second demographic transition, *Population Bulletin*, **42**, 1–57.

Vastoia, A. (1997) Perceived quality and certification: the case of organic fruit, in Schiefer, G. and Helbig, R. (eds) *Quality Management and Process Improvement for Competitive Advantage in Agriculture and Food*, Proceedings of the 49th Seminar of the European Association of Agricultural Economists, Bonn.

Veldman, J. (1984) Proposal for a theoretical basis for the human geography of rural areas, in Clark, G., Groenendijk, J. and Thissen, F. (eds) *The Changing Countryside*, Geo Books, Norwich, 17–26.

Verba, S., Schlozman, K.L. and Brady, H.E. (1995) *Voice and Equality: Civic Voluntarism in American Politics*, Harvard University Press, Cambridge, MA.

Vining, D. and Kontuly, T. (1978) Population dispersal from major metropolitan regions: an international comparison, *International Regional Science Review*, **3**, 49–73.

Vitebsky, P. (1996) The northern minorities, in Smith, G. (ed.) *The Nationalities Question in the Post-Soviet States*, Longman, London, 94–112.

Vitousek, P.M., Mooney, H.A., Lubchenco, J. and Melillo, J.M. (1997) Human domination of Earth's ecosystems, *Science*, **277**, 494–9.

W

Wacquant, L.J.D. (1993) Urban outcasts: stigma and division in the black American ghetto and the French urban periphery, *International Journal of Urban and Regional Research*, **17**, 366–83.

Wacquant, L.J.D. (1995) The ghetto, the state and the new capitalist economy, in Kasinitz, P. (ed.) *Metropolis. Centre and Symbol of our Times*, Macmillan, London, 418–49.

Wade, R. (2001) Winners and losers, *The Economist*, April, 79–82.

Wait, G., McGuirk, Dunn, K., and Hartig, K. (eds) (2000) *Introducing Human Geography: Globalization, Difference and Inequality*, Longman, French's Forest, NSW.

Walby, S. (1994) Is citizenship gendered?, *Sociology*, **28**, 379–95.

Walford, N. (2003) Productivism is allegedly dead, long live productivism: evidence of continued productivist attitudes and decision-making in south-east England, *Journal of Rural Studies*, **19**, 491–502.

Walker, A.S. (1999) Women's initiatives and activities worldwide, *IWTC Women's Globalnet #126*, International Women's Tribune Centre, 23 July, available at: http://womensnet.org.za/news.

Walker, R. (1996) Another round of globalization in San Francisco, *Urban Geography*, **17**, 60–94.

Wall, D (1999) *Earth First! and the Anti-roads Movement*, Routledge, London.

Wallace, D. and Wallace, R. (1998) *A Plague on Your Houses: How New York was Burned Down and how National Public Health Crumbled*, London, Methuen.

Wallace, D. and Wallace, R. (2000) Life and death in Upper Manhattan and the Bronx: toward an evolutionary perspective on catastrophic social change, *Environment and Planning A*, **32**, 1245–66.

Wallace, R. and Wallace, D. (1990) US Apartheid and the spread of AIDS to the suburbs: a multi-city analysis of the political economy of spatial epidemic threshold, *Social Science and Medicine,* **41**, 433–45.

Wallace, I. (1985) Towards a geography of agribusiness, *Progress in Human Geography*, **9**, 491–514.

Wallerstein, I. (1979) *The Capitalist World Economy*, Cambridge University Press, Cambridge.

Wallerstein, I. (1980) Imperialism and development, in Bergeson, A. (ed.) *Studies of the Modern World System*, Academic Press, New York.

Wallerstein, I. (1991) *Geopolitics and Geoculture: Essays on the Changing World System*, Cambridge University Press, Cambridge.

Wallerstein, I. (1994) Development: lodestar or illusion?, in Sklair, L. (ed.) *Capitalism and Development*, Routledge, London, 3–20.

Walmsley, D.J., Epps, W.R. and Duncan, C.J. (1995) *The New South Wales North Coast 1986–1991: Who Moved Where, Why and with What Effect?*, Australian Government Publishing Service, Canberra.

Wang, W.C., Yung, Y.L., Lacis, A.A., Mo, T. and Hansen, J.E. (1976) Greenhouse effect due to manmade perturbations of other gases, *Science*, **194**, 685–90

Ward, D. (1987) Population growth, migration, and urbanization, 1860–1920, in Mitchell R.D. and Groves, P.A. (eds) *North America: the Historical Geography of a Changing Continent*, Hutchinson, London.

Ward, D. (1989) *Poverty, Ethnicity and the American City, 1840–1925*, Cambridge University Press, Cambridge.

Warf, B. (1999) The hypermobility of capital and the collapse of the Keynesian state, in Martin, R.L. (ed.) *Money and the Space Economy*, Wiley, Chichester.

Watney, S. (1987) *Policing Desire: Pornography, AIDS and the Media*, Methuen, London.

Watnick, M. (1952–3) The appeal of Communism to the peoples of underdeveloped areas, *Economic Development and Cultural Change*, **1**, 22–36.

Watson, J.L. (1997a) Introduction: transnationalism, localization, and fast foods in East Asia, in Watson J.L. (ed.) *Golden Arches East: McDonalds in East Asia*, Stanford University Press, Stanford, CA, 1–38.

Watson, J.L. (1997b) McDonalds in Hong Kong: consumerism, dietary change, and the rise of a children's culture, in Watson, J.L. (ed.) *Golden Arches East: McDonalds in East Asia*, Stanford University Press, Stanford, CA, 77–109.

Watson, S. (1999) City politics, in Pile, S., Brook, C. and Mooney, G. (eds) *Unruly Cities? Order/Disorder*, Routledge, London, 201–45.

Watts, M. (1993a) Development I: power, knowledge, discursive practice, *Progress in Human Geography*, **17**, 257–72.

Watts, M. (1993b) The geography of post-colonial Africa: space, place and development in sub-Saharan Africa 1960–1993, *Singapore Journal of Tropical Geography*, **14**, 173–90.

Weber, A. (1929) *Alfred Weber's Theory of the Location of Industries*, Chicago University Press, Chicago.

Webster, D. (1988) *Looka Yonder! The Imaginary America of Populist Culture*, Comedia, London.

Webster, P. (2003) Last chance for Kyoto, *New Scientist,* 25 October, 42–5.

Weeks, J.R. (1988) The demography of Islamic nations, *Population Bulletin*, **43**, 1–54.

Werbner, P. (1997) Introduction: The dialectics of cultural hybridity, in Werbner, P. and Modood, T. (eds) *Debating Cultural Hybridity: Multi-Cultural Identities and the Politics of Anti-Racism*, Zed, London, 1–26.

Werbner, P. and Modood, T. (eds) (1997) *Debating Cultural Hybridity: Multi-Cultural Identities and the Politics of Anti-Racism*, Zed, London.

Whatmore, S. (1995) From farming to agribusiness: the global agro-food system, in Johnston, R., Taylor, P. and Watts, M. (eds) *Geographies of Global Change*, Blackwell, Oxford, 36–49.

Whatmore, S. and Thorne, L. (1997) Nourishing networks: alternative geographies of food, in Goodman, D. and Watts, M. (eds) *Globalising Food*, Routledge, London, 287–304.

Whittaker, D.J. (1997) *United Nations in the Contemporary World*, Routledge, London.

Whittlesey, D. (1936) Major agricultural regions of the Earth, *Annals of the Association of American Geographers*, **26**, 199–240.

Wiener, M.J. (1985) *English Culture and the Decline of the Industrial Spirit, 1850–1980*, Penguin, Harmondsworth.

Wilkins, P. (1983) *Women and Engineering in the Plymouth Area: Job Segregation and Training at Company Level*, Equal Opportunities Commission, London.

Williams, C.C. (1996) Local exchange and trading systems: a new source of work and credit for the poor and unemployed?, *Environment and Planning A*, **28**, 1395–415.

Williams, C.L. (1993) Introduction, in Williams, C.L. (ed.) *Doing 'Women's Work': Men in Nontraditional Occupations*, Sage, London, 1–9.

Williams, K. (1998) *Get me a Murder a Day! A History of Mass Communication in Britain*, Arnold, London.

Williams, P. and Hubbard, P. (2001) Who is disadvantaged? Retail change and social exclusion, *International Review of Retail, Distribution and Consumer Research*, **11**, 267–86.

Willis, P. (1990) *Uncommon Culture*, Open University Press, Milton Keynes.

Wilson, A. (1992) Technological utopias, world's fairs and theme parks, in Wilson, A. (ed.) *The Culture of Nature: North American Landscape from Disney to the Exxon Valdez*, Blackwell, Oxford, 157–90.

Wilson, G. (2001) From productivism to post-productivism ... and back again? Exploring the (un)changed natural and mental landscapes of European agriculture, *Transactions of the Institute of British Geographers*, **26**, 77–102.

Wilson, G. and Hart, K. (2001) Farmer participation in agri-environmental schemes: towards conservation-oriented thinking? *Sociologia Ruralis*, **41**, 254–74.

Wilson, J.S. (1858) The general and gradual desiccation of the Earth and atmosphere, Report of the Proceedings of the British Association for the Advancement of Science, 155–6.

Wilton, R. (1996) Diminished worlds? The geography of everyday life with HIV/AIDS, *Health and Place*, **2**, 69–83.

Winckler, V. (1987) Women and work in contemporary Wales, *Contemporary Wales: An Annual Review of Economic and Social Research*, **1**, 53–72.

Winchester, H. and White, P. (1988) The location of marginalised groups in the inner city, *Environment and Planning D: Society and Space*, **6**, 37–54.

Winchester, H., Kong, L. and Dunn, K. (2003) *Landscapes: Ways of Imagining the World*, London, Prentice Hall.

Winichakul, T. (1996) Maps and the formation of the geo-body of Siam, in Tønnesson, S. and Antlöv, H. (eds) *Asian Forms of the Nation*, Curzon Press, Richmond, 67–91.

Winter, M. (2003) Embeddedness, the new food economy and defensive localism. *Journal of Rural Studies*, **19**, 23–32.

Wirth, L. (1938) Urbanism as a way of life, *American Journal of Sociology*, **44**, 1–24, reprinted in Hart, P.K. and Reiss, A.J. (eds) (1957) *Cities and Society*, Free Press, Chicago.

Wolch, J. (1991) Urban homelessness, an agenda for research, *Urban Geography*, **12**, 99–104.

Wolf, E.R. (1966) *Peasants*, Prentice-Hall, Englewood Cliffs, NJ.

Wolfe, T. (1988) *Bonfire of the Vanities*, New York, Bantam.

Wolfe-Phillips, L. (1987) Why Third World – origins, definitions and usage, *Third World Quarterly*, **9**, 1311–19.

Wolters, O.W. (1982) *History, Culture, and Religion in Southeast Asian Perspectives*, Institute of Southeast Asian Studies, Singapore.

Women and Geography Study Group of the Institute of British Geographers (1984) *Geography and Gender*, Hutchinson, London.

Women and Geography Study Group (1997) *Feminist Geographies: Explorations in Diversity and Difference*, Longman, Harlow.

Wood, A., Steadman-Edwards, P. and Mang, J. (2000) *The Root Causes of Biodiversity Loss*, Earthscan, London.

Wood, A. and Welch, C. (1998) *Policing the Policemen: The Case for an Independent Evaluation Mechanism for the IMF*, Bretton Woods Project, Friends of the Earth US, London.

Wood, G. (1985) *Labelling in Development Policy: Essays in Honour of Bernard Schaffer*, Sage, London.

Woodward, K. (1997) Concepts of identity and difference, in Woodward, K. (ed) *Identity and Difference*, London, Sage, 8–59.

World Bank (1997) *The State in a Changing World*, World Development Report, World Bank, Washington, DC.

World Bank (1998a) *World Bank Atlas 1998*, World Bank, Washington, DC.

World Bank (1998b) *World Development Indicators 1998*, World Bank, Washington, DC.

World Bank (1999) *World Development Indicators*, World Bank, Washington, DC, also available as *World Development Indicators on CD-Rom*, World Bank, Washington, DC.

World Bank (2001a) *World Development Indicators*, World Bank, Washington, DC.

World Bank (2001b) *Attacking Poverty: World Development Report 2000/2001*, Oxford University Press and World Bank, Washington, DC.

World Bank (2002a) *World Development Indicators on CD-Rom*, World Bank, Washington, DC.

World Bank (2002b) *World Development Report 2002:*

Building Institutions for Markets, World Bank, Washington, DC.

World Bank (2003) *World Development Indicators on CD-ROM*, World Bank, Washington, DC.

World Bank Poverty Net. Online, available at: www.worldbank.org/poverty/index.htm

World Commission on Environment and Development (WCED) (1987) *Our Common Future*, Oxford University Press, Oxford.

World Economic Forum (2003) *Global Information Technology Report 2002–2003*, World Economic Forum, Geneva.

World Resources Institute (WRI) (2003) *World Resources 2002–2004*, WRI, Washington, DC.

World Trade Organization (WTO) (1997) *Annual Yearbook*, WTO, Geneva.

Worldwatch Institute (2003) *State of the World 2003*, Earthscan, London.

Wright, R. (1995) *The Color Curtain: A Report on the Bandung Conference*, Banner Books, New York.

Wrigley, N. and Lowe, M. (eds) (1996) *Retailing, Consumption and Capital: Towards the New Retail Geography*, Longman, Harlow.

Wrigley, N., Warm, D. and Margetts, B. (2003) Deprivation, diet and food-retail access: findings from the Leeds 'food deserts' study, *Environment and Planning A*, **35**, 151–88.

www-cfr.jims.cam.ac.uk/archive/PRESENTATIONS/ seminars/lent2001/wra.pdf [online, accessed 24 November 2003].

www.epaynew.com/statistics/bankstats.html [online, accessed 3 November 2003].

www.lafleurs.com/catalog/4-2003_lottery_almanac_ download_505976 [online, accessed 24 November 2003].

www.lotteryinsider.com/stats/agent.htm [online, accessed 24 November 2003].

Wuyts, M., Mackintosh, M. and Hewitt, T. (eds) (1992) *Development Policy and Public Action*, Oxford University Press, Oxford.

Y

Yan, Y. (1997) McDonalds in Beijing: the localization of Americana, in Watson, J.L. (ed.) *Golden Arches East: McDonalds in East Asia*, Stanford University Press, Stanford, CA, 39–76.

Yawnghwe, C.T. (1987) *The Shan of Burma: Memoirs of a Shan Exile*, Institute of Southeast Asian Studies, Singapore.

Young, I.M. (1989) Polity and group difference: a critique of the ideal of universal citizenship, *Ethics*, **9**, 250–74.

Young, J. (1990) *Post Environmentalism*, Belhaven Press, London.

Young, M. and Wilmott, P. (1957) *Family and Kinship in East London*, Routledge & Kegan Paul, London.

Yuval-Davis, N. (1997) *Gender and Nation*, Sage, London.

Yuval-Davis, N. (2003) Citizenship, territoriality and the gendered construction of difference, in Brenner, N., Jessop, B., Jones, M. and MacLeod, G. (eds) *State/Space: A Reader*, Blackwell, Oxford, 309–25.

Z

Zelinsky, W. (1971) The hypothesis of the mobility transition, *Geographical Review*, **61**, 219–49.

Zelizer, V. (1989) The social meaning of money: 'special monies', *American Journal of Sociology*, **95**, 342–77.

Zlotnik, H. (1998) International migration 1965–96: an overview, *Population and Development Review*, **24**, 429–68.

Zouza, L.B. (2002) Vital interests and budget deficits: US Foreign Aid after September 11, *Middle East Insight*, March–April, 25–9.

Zukin, S. (1995) *The Cultures of Cities*, Blackwell, Oxford.

INDEX